COMPUTATIONAL PROSPECTS OF INFINITY

Part II: Presented Talks

LECTURE NOTES SERIES
Institute for Mathematical Sciences, National University of Singapore

Series Editors: Louis H. Y. Chen and Ka Hin Leung
Institute for Mathematical Sciences
National University of Singapore

ISSN: 1793-0758

Published

Vol. 4   An Introduction to Stein's Method
edited by A. D. Barbour & Louis H. Y. Chen

Vol. 5   Stein's Method and Applications
edited by A. D. Barbour & Louis H. Y. Chen

Vol. 6   Computational Methods in Large Scale Simulation
edited by K.-Y. Lam & H.-P. Lee

Vol. 7   Markov Chain Monte Carlo: Innovations and Applications
edited by W. S. Kendall, F. Liang & J.-S. Wang

Vol. 8   Transition and Turbulence Control
edited by Mohamed Gad-el-Hak & Her Mann Tsai

Vol. 9   Dynamics in Models of Coarsening, Coagulation, Condensation and Quantization
edited by Weizhu Bao & Jian-Guo Liu

Vol. 10   Gabor and Wavelet Frames
edited by Say Song Goh, Amos Ron & Zuowei Shen

Vol. 11   Mathematics and Computation in Imaging Science and Information Processing
edited by Say Song Goh, Amos Ron & Zuowei Shen

Vol. 12   Harmonic Analysis, Group Representations, Automorphic Forms and Invariant Theory — In Honor of Roger E. Howe
edited by Jian-Shu Li, Eng-Chye Tan, Nolan Wallach & Chen-Bo Zhu

Vol. 13   Econometric Forecasting and High-Frequency Data Analysis
edited by Roberto S. Mariano & Yiu-Kuen Tse

Vol. 14   Computational Prospects of Infinity — Part I: Tutorials
edited by Chitat Chong, Qi Feng, Theodore A Slaman, W Hugh Woodin & Yue Yang

Vol. 15   Computational Prospects of Infinity — Part II: Presented Talks
edited by Chitat Chong, Qi Feng, Theodore A Slaman, W Hugh Woodin & Yue Yang

*For the complete list of titles in this series, please go to
http://www.worldscibooks.com/series/lnsimsnus_series.shtml*

Lecture Notes Series, Institute for Mathematical Sciences,
National University of Singapore

Vol.
15

COMPUTATIONAL PROSPECTS OF INFINITY

Part II: Presented Talks

Editors

Chitat Chong
National University of Singapore, Singapore

Qi Feng
Chinese Academy of Sciences, China & National University of Singapore, Singapore

Theodore A. Slaman, W. Hugh Woodin
University of California at Berkeley, USA

Yue Yang
National University of Singapore, Singapore

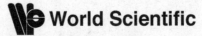

World Scientific

NEW JERSEY · LONDON · SINGAPORE · BEIJING · SHANGHAI · HONG KONG · TAIPEI · CHENNAI

Published by

World Scientific Publishing Co. Pte. Ltd.

5 Toh Tuck Link, Singapore 596224

USA office: 27 Warren Street, Suite 401-402, Hackensack, NJ 07601

UK office: 57 Shelton Street, Covent Garden, London WC2H 9HE

British Library Cataloguing-in-Publication Data
A catalogue record for this book is available from the British Library.

Lecture Notes Series, Institute for Mathematical Sciences,
National University of Singapore – Vol. 15
COMPUTATIONAL PROSPECTS OF INFINITY
Part II: Presented Talks

Copyright © 2008 by World Scientific Publishing Co. Pte. Ltd.

ISBN-13 978-981-279-654-7
ISBN-10 981-279-654-1

Printed in Singapore.

CONTENTS

Foreword vii

Preface ix

Generating Sets for the Recursively Enumerable Turing Degrees
 Klaus Ambos-Spies, Steffen Lempp and Theodore A. Slaman 1

Coding into $H(\omega_2)$, Together (or Not) with Forcing Axioms.
A Survey
 David Asperó 23

Nonstandard Methods in Ramsey's Theorem for Pairs
 Chi Tat Chong 47

Prompt Simplicity, Array Computability and Cupping
 Rod Downey, Noam Greenberg, Joseph S. Miller and
 Rebecca Weber 59

Lowness for Computable Machines
 Rod Downey, Noam Greenberg, Nenad Mihailović and
 André Nies 79

A Simpler Short Extenders Forcing — Gap 3
 Moti Gitik 87

Limit Computability and Constructive Measure
 Denis R. Hirschfeldt and Sebastiaan A. Terwijn 131

The Strength of Some Combinatorial Principles Related to
Ramsey's Theorem for Pairs
 Denis R. Hirschfeldt, Carl G. Jockusch, Jr.,
 Bjørn Kjos-Hanssen, Steffen Lempp and
 Theodore A. Slaman 143

Absoluteness for Universally Baire Sets and the Uncountable II
*Ilijas Farah, Richard Ketchersid, Paul Larson and
Menachem Magidor*   163

Monadic Definability of Ordinals
Itay Neeman   193

A Cuppable Non-Bounding Degree
Keng Meng Ng   207

Eliminating Concepts
André Nies   225

A Lower Cone in the wtt Degrees of Non-Integral
Effective Dimension
André Nies and Jan Reimann   249

A Minimal rK-Degree
Alexander Raichev and Frank Stephan   261

Diamonds on $\mathcal{P}_\kappa \lambda$
Masahiro Shioya   271

Rigidity and Biinterpretability in the Hyperdegrees
Richard A. Shore   299

Some Fundamental Issues Concerning Degrees of Unsolvability
Stephen G. Simpson   313

Weak Determinacy and Iterations of Inductive Definitions
MedYahya Ould MedSalem and Kazuyuki Tanaka   333

A tt Version of the Posner-Robinson Theorem
W. Hugh Woodin   355

Cupping Computably Enumerable Degrees in the
Ershov Hierarchy
Guohua Wu   393

FOREWORD

The Institute for Mathematical Sciences at the National University of Singapore was established on 1 July 2000. Its mission is to foster mathematical research, both fundamental and multidisciplinary, particularly research that links mathematics to other disciplines, to nurture the growth of mathematical expertise among research scientists, to train talent for research in the mathematical sciences, and to serve as a platform for research interaction between the scientific community in Singapore and the wider international community.

The Institute organizes thematic programs which last from one month to six months. The theme or themes of a program will generally be of a multidisciplinary nature, chosen from areas at the forefront of current research in the mathematical sciences and their applications.

Generally, for each program there will be tutorial lectures followed by workshops at research level. Notes on these lectures are usually made available to the participants for their immediate benefit during the program. The main objective of the Institute's Lecture Notes Series is to bring these lectures to a wider audience. Occasionally, the Series may also include the proceedings of workshops and expository lectures organized by the Institute.

The World Scientific Publishing Company has kindly agreed to publish the Lecture Notes Series. This Volume, "Computational Prospects of Infinity, Part II: Presented Talks", is the fifteenth of this Series. We hope that through the regular publication of these lecture notes the Institute will achieve, in part, its objective of promoting research in the mathematical sciences and their applications.

January 2008

Louis H.Y. Chen
Ka Hin Leung
Series Editors

PREFACE

The Workshop on *Computational Prospects of Infinity* was held at the Institute for Mathematical Sciences, National University of Singapore, from 20 June to 15 August 2005.

The focus in the first month of the Workshop was on set theory, with two tutorials running in parallel, by John Steel ("Derived Models Associated to Mice") and W. Hugh Woodin ("Suitable Extender Sequences"). There were also 21 talks given during this period. The second half of the Workshop was devoted to recursion theory, with two tutorials given respectively by Rod Downey ("Five Lectures on Algorithmic Randomness") and Ted Slaman ("Definability of the Turing Jump"). In addition, there were 42 talks delivered over four weeks.

This volume constitutes Part II of the *proceedings* of the Workshop. It contains refereed articles, both contributed and invited, based on talks presented at the Workshop. Written versions of the tutorials are collected in Part I of the *proceedings* which appears as a separate volume.

The workshop provided a platform for researchers in the logic community from many parts of the world to meet in Singapore, to discuss mathematics and to experience a city that is a meeting point of East and West. We thank the Department of Mathematics and the Institute for Mathematical Sciences at the National University of Singapore for their support and hospitality extended to all participants.

November 2007

Chi Tat Chong
National University of Singapore, Singapore

Qi Feng
Chinese Academy of Sciences, China and
National University of Singapore, Singapore

Theodore A. Slaman
University of California at Berkeley, USA

W. Hugh Woodin
University of California at Berkeley, USA

Yue Yang
National University of Singapore, Singapore
Editors

GENERATING SETS FOR THE RECURSIVELY ENUMERABLE TURING DEGREES

Klaus Ambos-Spies

Department of Mathematics and Computer Science
University of Heidelberg
D-69120 Heidelberg, Germany
E-mail: ambos@math.uni-heidelberg.de

Steffen Lempp[*]

Department of Mathematics
University of Wisconsin–Madison
Madison, WI 53706-1388, USA
E-mail: lempp@math.wisc.edu

Theodore A. Slaman[†]

Department of Mathematics
University of California–Berkeley
Berkeley, CA 94720-3840, USA
E-mail: slaman@math.berkeley.edu

We give an example of a subset of the recursively enumerable Turing degrees which generates the recursively enumerable degrees using meet and join but does not generate them using join alone.

1. Introduction

One of the recurrent themes in the area of the recursively enumerable (r.e.) degrees has been the study of the *meet operator*. While, trivially, the partial ordering of the r.e. degrees is an upper semi-lattice, i.e., the join

---

[*]Lempp was partially supported by NSF grant DMS-0140120 and a Mercator Guest Professorship of the Deutsche Forschungsgemeinschaft.

[†]Slaman was partially supported by the Alexander von Humboldt Foundation and by National Science Foundation Grant DMS-9988644.

operator is total, the meet of two incomparable r.e. degrees may or may not exist (Lachlan (1966), Yates (1966)). The asymmetry between joins and meets is further illustrated by the fact that, by Sacks' splitting theorem (Sacks (1963)), every nonzero r.e. degree is join-reducible, i.e., is the join of two lesser degrees, whereas there are both, meet-reducible (branching) and meet-irreducible (nonbranching), incomplete r.e. degrees (Lachlan (1966)).

The existence of meets and the failure of meets are densely distributed in the partial ordering of the r.e. degrees. So Fejer (1983) showed that the non-branching degrees are dense while Slaman (1991) showed that the branching degrees are dense. Similarly, every interval of the r.e. degrees contains an incomparable pair of degrees without meet (Ambos-Spies (1984)) and an incomparable pair of degrees with meet (Slaman (1991)). That, actually, the lack of meets is more common than the existence of meets has been demonstrated by Ambos-Spies (1984) and, independently, by Harrington (unpublished) who showed that, for any nonzero, incomplete r.e. degree **a**, there is an incomparable degree **b** such that the meet of **a** and **b** does not exist, but also that there is such a degree **a** such that, for any incomparable degree **b**, the meet of **a** and **b** does not exist. More evidence, that the failure of meets is more typical than their existence, was given by Jockusch (1985) who showed that, given r.e. degrees **a**, **b** and **c** such that **a** and **b** are incomparable and **c** is the meet of **a** and **b**, none of these degrees is e-generic.

Another way to look at the join and meet operators in the r.e. degrees is to study *generating sets*, i.e., sets of r.e. degrees which generate all the recursively enumerable degrees under (finitely many applications of) join and meet. The question now arises naturally whether both the join operation and the meet operation are needed here. As observed in Ambos-Spies (1985), the above results on nonbranching degrees easily imply that the join operation is indeed necessary, namely there is a subset of the re-cursively enumerable degrees which generates all recursively enumerable degrees using join and meet but not using meet alone. Ambos-Spies, how-ever, left open the question of whether the meet operation is necessary (see Ambos-Spies (1985), Problem 1). The above mentioned negative results on meets by Fejer (1983), Ambos-Spies (1984) and Jockusch (1985) may suggest a negative answer to this question. More evidence in this direction has been obtained by Ambos-Spies (1985) who showed that any generating set intersects any notrivial initial segment of the r.e. degrees and, more recently, by Ambos-Spies, Ding and Fejer (unpublished) who showed that any generating set generates the high r.e. degrees using join alone. Despite this negative evidence, in this paper, we answer Ambos-Spies' question affirmatively by the following

Theorem 1.1 *There exists a subset* $\mathbf{G}$ *of the recursively enumerable Turing degrees which generates the recursively enumerable Turing degrees using meet and join but does not generate them using join alone.*

Proof. Our theorem follows by our technical result, Theorem 2.1, below, using a nonconstructive definition of the set $\mathbf{G}$. Fix the recursively enumerable degree $\mathbf{a}$ from Theorem 2.1. Let $\{\mathbf{x}_n\}_{n\in\omega}$ be a (noneffective) enumeration of all recursively enumerable degrees $\leq \mathbf{a}$. We now define a (noneffective) sequence of recursively enumerable degrees $\mathbf{0} = \mathbf{y}_0 \leq \mathbf{y}_1 \leq \mathbf{y}_2 \leq \cdots < \mathbf{a}$ as follows: Set $\mathbf{y}_0 = \mathbf{0}$. Given $\mathbf{y}_n < \mathbf{a}$, check whether $\mathbf{y}_n \cup \mathbf{x}_n = \mathbf{a}$. If not, then set $\mathbf{y}_{n+1} = \mathbf{y}_n \cup \mathbf{x}_n$. Otherwise, let $\mathbf{b}$ be the recursively enumerable degree given by Theorem 2.1 using $\mathbf{x} = \mathbf{x}_n$ and $\mathbf{y} = \mathbf{y}_n$, and set $\mathbf{y}_{n+1} = \mathbf{y}_n \cup \mathbf{b}$. Finally, we define

$$\mathbf{G} = \{\mathbf{x} \mid \mathbf{x} \not\leq \mathbf{a} \text{ or } \exists n\, (\mathbf{x} \leq \mathbf{y}_n)\}.$$

By Theorem 2.1, the degree $\mathbf{a}$ is clearly not the join of any finite set of degrees in $\mathbf{G}$. On the other hand, fix any recursively enumerable degree $\mathbf{x}$ and assume $\mathbf{x} \notin \mathbf{G}$. Then $\mathbf{x} \leq \mathbf{a}$, and so $\mathbf{x} = \mathbf{x}_n$ for some $n \in \omega$. Since $\mathbf{x} \notin \mathbf{G}$, we have $\mathbf{x} \not\leq \mathbf{y}_{n+1}$ and so $\mathbf{x} \cup \mathbf{y} = \mathbf{a}$ for $\mathbf{y} = \mathbf{y}_n$. Fix $\mathbf{b}, \mathbf{c}, \mathbf{d}$, and $\mathbf{e}$ as in Theorem 2.1. Then $\mathbf{x} = \mathbf{b} \cup (\mathbf{d} \cap \mathbf{e})$ where all of $\mathbf{b}, \mathbf{d}$, and $\mathbf{e}$ are in $\mathbf{G}$ since $\mathbf{b} \leq \mathbf{y}_{n+1}$ and $\mathbf{d}, \mathbf{e} \not\leq \mathbf{a}$. $\square$

2. The technical theorem and some intuition for its proof

Starting with this section, we will prove the technical theorem needed to establish Theorem 1.1:

Theorem 2.1 *There is a nonrecursive, recursively enumerable set A such that for every pair of recursively enumerable sets X and Y, if X and Y are recursive in A and A is recursive in XY then one of the following conditions holds.*

1. A is recursive in Y.

2. There are recursively enumerable sets B, C, D, and E such that

    (a) X has the same Turing degree as BC,

(b) *D and E are not recursive in A and the degree of C is the infimum of the degrees of DC and EC, and*

(c) *A is not recursive in BY.*

2.1. *Requirements and simple strategies*

We disassemble the statement of Theorem 2.1 into requirements as follows. First, A must be nonrecursive and so we must satisfy all the requirements $\Theta \neq A$, where Θ is a recursive function.

Second, for each X, Y, $\Lambda_{a,x}$, $\Lambda_{a,y}$, and $\Lambda_{xy,a}$, we associate the principal equations $\Lambda_{a,x}(A) = X$, $\Lambda_{a,y}(A) = Y$, and $\Lambda_{xy,a}(XY) = A$. We can satisfy our requirement on X, Y, $\Lambda_{a,x}$, $\Lambda_{a,y}$, and $\Lambda_{xy,a}$ in any of several ways. If the principal equations are not valid then our requirement is satisfied.

Anticipating that the principal equations actually are valid, we enumerate the sets B, C, D, and E and recursive functionals $\Gamma_{x,b}$, $\Gamma_{x,c}$, and $\Gamma_{bc,x}$. We ensure that $\Gamma_{x,b}(X) = B$, $\Gamma_{x,c}(X) = C$, and $\Gamma_{bc,x}(BC) = X$. Now, our requirement is satisfied in one of two ways.

For every recursive functional Θ_{by}, if $\Theta_{by}(BY) = A$ then there is a $\Delta_{y,a}$, which we enumerate during our construction, such that $\Delta_{y,a}(Y) = A$. If there is a $\Delta_{y,a}$ such that $\Delta_{y,a}(Y) = A$, then again our requirement is satisfied.

Otherwise, we ensure that every instance of the following family of requirements is satisfied.

1. For all Θ_a, $\Theta_a(A) \neq D$ and $\Theta_a(A) \neq E$.

2. For all Ψ_{cd} and Ψ_{ce}, if $\Psi_{cd}(CD) = \Psi_{ce}(CE)$ then there is a Ξ_c such that $\Xi_c(C) = \Psi_{cd}(CD) = \Psi_{ce}(CE)$.

2.2. *Strategies*

2.2.1. *Making $\Theta \neq A$: $\sigma_0(\Theta)$*

We ensure that $\Theta \neq A$ by choosing a number n, keeping n out of A until seeing $\Theta(n) = 0$, and then enumerating n into A. This strategy σ_0 is one of the standard methods to satisfy requirements of this form.

2.2.2. *Measuring whether the equations hold: $\sigma_1(X, Y, \Lambda_{a,x}, \Lambda_{a,y}, \Lambda_{xy,a})$*

Now, we consider the more complicated requirements. Suppose that X, Y, $\Lambda_{a,x}$, $\Lambda_{a,y}$, and $\Lambda_{xy,a}$ are given.

Our strategy $\sigma_1(X, Y, \Lambda_{a,x}, \Lambda_{a,y}, \Lambda_{xy,a})$ approximates if the principal equations hold for X, Y, $\Lambda_{a,x}$, $\Lambda_{a,y}$, and $\Lambda_{xy,a}$. We will abbreviate by σ_1 the strategy $\sigma_1(X, Y, \Lambda_{a,x}, \Lambda_{a,y}, \Lambda_{xy,a})$ and use similar conventions throughout this section. Essentially, σ_1 measures expansionary stages in the approximation to these equalities. For technical reasons, explained below, σ_1 waits for something more than simple expansion. In the following, a_1 and a_2 are variables of the strategy which enumerates pairs (a_1, a_2) into a list of pairs of witnesses.

1. If a_1 is undefined and it is possible to do so, choose a value for a_1 that is larger than $\lambda_{a,x}(A, x)[s]$ for every x previously mentioned in the construction during a σ_1-expansionary stage. Let x_1 be the smallest number x such that $\lambda_{a,x}(A, x)[s]$ is greater than a_1. Suspend the enumeration of any functionals associated with B, C, D or E. (We may assume that we have not enumerated any computations from BC of X at arguments greater than or equal to x_1.)

   Wait until the first stage s such that $(\Lambda_{xy,a}(XY) \restriction a_1 + 1 = A \restriction a_1 + 1)[s]$, and $(\Lambda_{a,x}(A) = X)[s]$ and $(\Lambda_{a,y}(A) = Y)[s]$ on all numbers less than or equal to the maximum of $\lambda_{xy,a}(XY)[s] \restriction a_1 + 1$. At this stage, we let a_2 equal the supremum of $(\lambda_{a,x}(A) = X)[s]$ and $(\lambda_{a,y}(A) = Y)[s]$ on all numbers less than or equal to the maximum of $\lambda_{xy,a}(XY)[s] \restriction a_1 + 1$. We enumerate the pair (a_1, a_2) into our list and let the strategies of lower priority resume the enumeration of any functionals associated with B, C, D or E. (The (a_1, a_2) notation will be convenient below.) Go to Step 2.

2. At the next stage when σ_1 is active, we say that a_1 is undefined, and go to Step 1.

Consider the possibilities. The strategy σ_1 could reach a limit in Step 1. In this case, one of the principal equations fails and the requirement is satisfied.

If σ_1 does not reach a limit in Step 1 then it enumerates infinitely many stable pairs and has no other effect on the construction.

For the remainder of this section, we assume that σ_1 does not reach a finite limit and that all subsequent strategies act during the stages when σ_1 enumerates a new pair. We call such stages σ_1-expansionary.

2.2.3. Computations between B, C, and X: $\sigma_2(X, Y, \Lambda_{a,x}, \Lambda_{a,y}, \Lambda_{xy,a})$

Our strategy σ_2 builds functionals $\Gamma_{x,b}$ and $\Gamma_{x,c}$ and ensures that if the principal equations are valid then for each n there are infinitely many s such that $(\Gamma_{x,b}(X, n) = B(n))[s]$ and $(\Gamma_{x,c}(X, n) = C(n))[s]$. This, combined with our preserving A, B, and C, will be sufficient to conclude that B and C are recursive in X.

We ensure their correctness by imposing the constraint on all lower priority strategies τ that if $\Gamma_{x,b}(X, n)[s]$ or $\Gamma_{x,c}(X, n)[s]$ is defined while τ acts then τ cannot enumerate n into B or C, respectively, during that stage.

Similarly, we ensure that X is recursive in BC by enumerating a functional $\Gamma_{bc,x}$ and ensuring that if the principal equalities hold then for all n, $\Gamma_{bc,x}(BC, n) = X(n)$ during infinitely many stages of the construction.

We have complete freedom to define the uses of these functionals, but the construction does not require a subtle decision. During σ_1-expansionary stages, we enumerate new computations into $\Gamma_{bc,x}$. If n enters X during stage s and $\Gamma_{bc,x}(BC, n) = 0[s]$ then we must enumerate a number less than or equal to $\gamma_{bc,x}(BC, n)[s]$ into either B or C. We set the uses of these functions to be larger than any number previously used in the construction.

In the case of maintaining $\Gamma_{bc} = X$, we also have the freedom to decide which of B and C to change when recording a change in X. The choice made is irrelevant to σ_2. In our construction, we will leave the decision to the highest priority strategy for which it is relevant. See the discussion of the strategies of type σ_6.

2.2.4. Making C the infimum of CD and CE:
$$\sigma_3(X, Y, \Lambda_{a,x}, \Lambda_{a,y}, \Lambda_{xy,a}, \Psi_{cd}, \Psi_{ce})$$

We will use the branching strategies from Fejer (1982) and attempt to make the degree of C equal to the infimum of the degrees of CD and CE. Suppose that Ψ_{cd} and Ψ_{ce} are given and let σ_3 denote our branching strategy associated with this pair. Then, σ_3 enumerates a functional Ξ_c. Say that s is σ_3-expansionary if and only if the least number n such that $(\Psi_{cd}(CD, n) \neq \Psi_{ce}(CE, n))[s]$ is larger than at any earlier stage.

First, during stage s, if there is an n such that $\Xi_c(C, n)[s]$ is defined and a strategy of priority less than or equal to that of σ_3 enumerates numbers into C, D, or E so that neither $(\Psi_{cd}(CD, n) = \Xi_c(C, n))[s]$ nor $(\Psi_{ce}(CE, n) = \Xi_c(C, n))[s]$, then σ_3 must enumerate a number less than or equal to $\xi_c(C,n)[s]$ into C. (We will have to argue that this enumeration is compatible with C's being recursive relative to X.)

Second, if s is σ_3-expansionary then for the least n such that $\xi_c(C,n)$ is not defined, we choose a value for $\xi_c(C,n)[s]$ which is larger than any number previously mentioned in the construction and enumerate a computation into Ξ_c setting $\Xi_c(C[s],n) = \Psi_{cd}(CD[s],n)$ with use $\xi_c(C,n)[s]$.

If $\Psi_{cd}(CD) = \Psi_{ce}(CE)$ then there will be infinitely many σ_3-expansionary stages. Since we will be preserving computations from CD and CE, the converse will also be true. So, if $\Psi_{cd}(CD) \neq \Psi_{ce}(CE)$, then σ_3 will act finitely often. Otherwise, it produces a functional Ξ_c from C which is defined infinitely often to agree with the common value of $\Psi_{cd}(CD)$ and $\Psi_{ce}(CE)$. Again, since we are preserving the sets that we construct, this will be sufficient to ensure that $\Xi_c(C)$ is equal to this common value.

We will assume that there are infinitely many σ_3-expansionary stages and describe the appropriate strategies to follow. These strategies act only during σ_3-expansionary stages.

Instability in C and compatibility between σ_2 and σ_3. The strategies σ_3 introduce an instability to the initial segments of C. Namely, suppose that a strategy τ enumerates a number c into C. Then, c enters both CD and CE and could change the common value of $\Psi_{cd}(CD,c_1)$ and $\Psi_{ce}(CE,c_1)$. In response, τ enumerates $\xi_c(\check{C},c_1)$ into C, possibly changing C at a number less than c, and the effect can propagate. We call the set of numbers that enter C in this way the cascade initiated by c.

When combined with the strategy to ensure that C is recursive in X, the branching strategies make it difficult to enumerate any number at all into C. If $\Gamma_{x,c}(X,m)$ is defined then we cannot enumerate any c into C unless we can be sure that the instability in C will not propagate to the point of requiring that m enter C. We will use some of the ideas of Slaman (1991) to work within this constraint.

Definition 2.2 A number c is σ_3-*stable at stage* s if for all m, if $\xi_c(C,m)[s] < c$ then either $\psi_{cd}(CD,m) < c$ or $\psi_{ce}(CE,m) < c$.

We note that if c is σ_3-stable at stage s then any cascade initiated by a number greater than or equal to c during stage s does not include any number less than c. To prove this claim, consider the recursive propagation of a cascade initiated by a number greater than or equal to c, and let m be the first number less than c to appear in the cascade. Earlier in the propagation of the cascade, C would have to change below the minimum of $\psi_{cd}(CD, m)[s]$ and $\psi_{ce}(CE, m)[s]$. By the stability of c, this minimum is less than c and we have contradicted m's being first.

Thus, if c is stable and $\Gamma_{x,c}(X, c)[s]$ is not defined, then we can enumerate c into C and respect both σ_2 and σ_3. We will design all the strategies to follow so that they enumerate only stable numbers into C. Of course, there is no such constraint on B, since B is not constructed to be branching.

2.2.5. Making $\Theta_a(A) \neq D$ and $\Theta_a(A) \neq E$: $\sigma_4(X, Y, \Lambda_{a,x}, \Lambda_{a,y}, \Lambda_{xy,a}, \Theta_a)$ and $\sigma_5(X, Y, \Lambda_{a,x}, \Lambda_{a,y}, \Lambda_{xy,a}, \Theta_a)$

We use a variation, σ_4, on the basic Friedberg strategy to ensure that $\Theta_a(A) \neq D$. (The strategy σ_5 for E is similar.) We choose n larger than any number previously mentioned in the construction and constrain n from entering D. We wait for a stage s such that $(\Theta_a(A, n) = 0)[s]$. By our assumption, s will be σ_1 and σ_3-expansionary.

Then, we enumerate n into D and constrain any number less than s from entering any set under construction other than D. We note that these actions are consistent. Since we did not enumerate anything into A and A's computation of X exists on a longer interval than ever before, X cannot change at any number m such that $\Gamma_{bc,x}(BC, m)[s]$ is defined. So, σ_2 will not require any change in B or C. Since s is σ_3-expansionary, both $\Psi_{cd}(CD)[s]$ and $\Psi_{ce}(CE)[s]$ were defined (before we changed D), agreed on a longer interval than ever before, and agreed with $\Xi_c(C)[s]$ where $\Xi_c(C)[s]$ was defined. Since we did not change E, $\Psi_{ce}(CE)[s]$ is still equal to $\Xi_c(C)[s]$ where the latter is defined and σ_3 does not require any change in C.

2.2.6. If $\Theta_{by}(BY) = A$ then $\Delta_{y,a}(Y) = A$: $\sigma_6(X, Y, \Lambda_{a,x}, \Lambda_{a,y}, \Lambda_{xy,a}, \Theta_{by})$

Now we come to the crux of the proof of Theorem 2.1. Our strategy σ_6 must either diagonalize $\Theta_{by}(BY)$ against A, or it must determine that A is recursive in Y. Since Y is an arbitrary set below A, both cases are possible.

In the context of the construction, σ_6 can assume that every stage is σ_1- and σ_2-expansionary. Now, this implies that if we enumerate a number a into A during stage s, there is a later stage t during which A recomputes X and Y and either $X[s] \neq X[t]$ or $Y[s] \neq Y[t]$ and the least m at which the inequality occurs is less than $\lambda_{xy,a}(XY, a)[s]$. In other words, if we change A then one of X or Y must change in order to correct $\Lambda_{xy,a}(XY)$.

To establish $\Theta_{by}(BY) \neq A$, at least once we would have to change A without having to change B and without Y's having changed. This could happen, since the change in A could be recorded in X and we could record the change in X (for the sake of σ_3) in C. If, on the other hand, this diagonalization is not possible then we must conclude that A is recursive in Y. The conclusion is not unreasonable since every change in A results in a change in Y. But, if even one change in A results in a change in X, then we must be able to record that change in C.

We have reached the technical problem to be solved to prove the theorem. For every A-change allowed by σ_6, if it results in a change in X, then we must be able to record that change in X and in C. Now remember that we are only able to enumerate numbers into C which are σ_3-stable during their stage of enumeration. So we must ensure that changes in X can be recorded in C by the enumeration of such numbers. This is the purpose in our strategy σ_6.

Configurations. Consider a possible stage-s situation as depicted in Figure 1. In this picture, we illustrate a number a_1 which we intend to enumerate into A; x_1 is the least number x such that a_1 is less than $\lambda_{a,x}(A, x)[s]$ and hence the least number which might enter X when a_1 enters A; x_2 is equal to $\lambda_{xy,a}(XY, a_1)[s]$ and a_1's entering A would cause a change in XY below x_2; a_2 is the supremum of $\lambda_{a,x}(A, x_2)[s]$ and $\lambda_{a,y}(A, x_2)[s]$, so our preserving A on numbers less than a_2 will preserve the relationship between a_1, x_1, and x_2; c is $\gamma_{bc,x}(BC, x_1)[s]$ and so enumerating c into C would correct the computation of X from BC on every argument at which X might change; we intend that c be σ_3-stable at stage s and so, for any m, if $\xi_c(m)$ is less than c then one of $\psi_{cd}(CD, m)[s]$ or $\psi_{ce}(CE, m)[s]$ is also less than c; finally, x_3 is $\gamma_{x,c}(X, c)$, the use of X's computation of C at argument c. Note that we can preserve the relationships between these numbers by preserving A up to a_2, and B,C,D, and E up to c. (We can keep x_3 above x_2 by enumerating our functions so that the uses of new computations are at least as great as the uses of earlier computations at the same argument.)

Suppose that, in this situation, we were to enumerate a_1 into A and Y did not change below x_2 to record that fact (for example, if $Y \not\geq A$). Then X would change below x_2, allowing us to enumerate the σ_3-stable number c into C and thereby correct BC's computations for any change in X allowed by a_1's entering A.

We say that the situation depicted in Figure 1 is a σ_1-*configuration for* a_1. Anticipating that $\Theta_{by}(BY) = A$, we must ensure that for all but a

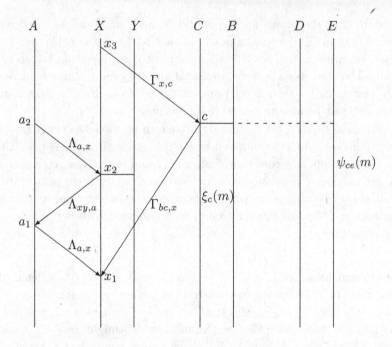

Figure 1: Configuration for a_1.

recursive set of numbers a, if a enters A then it does so in the role of a_1 with a configuration as above.

Generating configurations. Though configurations seem artificial at first, they are very common. In fact, a new configuration can be produced during every σ_1-expansionary stage.

Note that, by the constraint imposed by σ_3, at the beginning of every stage t, for every m, if $\Xi_c(C, m)[t]$ is defined then one of $\Psi_{cd}(CD, n)[t]$ and $\Psi_{ce}(CE, n)[t]$ is also defined with the same value.

Now consider a σ_1-expansionary stage s. Let c be the least strict upper bound on the range of $\xi_c(C)[s]$. By the observation above, c is σ_3-stable. Since s is σ_1-expansionary, σ_1 enumerated a pair (a_1, a_2) related as in Figure 1. Further, by the choice of a_1 and a_2, a_1 is greater than $\lambda_{a,x}(A, x)[s]$ for every x such that $\Gamma_{bc,x}(BC, x)$ has ever been defined. Consequently, there is no computation in $\Gamma_{bc,x}[s]$ which applies to the argument x_1. Then, we can use σ_2 to enumerate computations into $\Gamma_{x,b}$, $\Gamma_{x,c}$, and $\Gamma_{bc,x}$ so that $\gamma_{bc,x}(BC, x_1) > c$, and $\gamma_{x,c}(X, c) > \lambda_{xy,a}(X, a_1)$. In Figure 1, $\lambda_{xy,a}(X, a_1)$

would be x_2 and $\gamma_{x,c}(X,c)$ would be x_3. In short, at the beginning of each stage s, the whole of $\xi_c[s]$ is stable and at a σ_2-expansionary stage, we can use the pair (a_1, a_2) enumerated by σ_1 and enumerate new computations into our functionals to extend $\xi_c[s]$ to a configuration for a_1.

Restricting to configured numbers. We satisfy the requirement

$$\Theta_{by}(BY) = A \implies \Delta_{y,a}(Y) = A$$

as follows.

1. If a is undefined and we are not preserving an inequality between $\Theta_{by}(BY)$ and A, then choose a value for a such that a is larger than any value previously chosen and such that there is a configuration for a. We restrain a from entering A and preserve the configuration for a until we find a stage s such that $(\Theta_{by}(BY,a) = 0)[s]$ and that $\Lambda_{a,y}(A)[s]$ is equal to $Y[s]$ on all numbers less than or equal to $\theta_{by}(BY,a)[s]$. At stage s, we go to Step 2.

2. Let a_0 be the largest number that we have previously enumerated as *allowed to enter A with a certified configuration* (or 0 if there is no such number). For each number n between a_0 and a, if $n \notin A[s]$ then we restrain n from ever entering A. We enumerate a into the set of numbers still allowed to enter A and we say that the current configuration for a together with the current computation $(\Theta_{by}(BY,a) = 0)[s]$ is the certified configuration associated with a during stage s.

   For all strategies τ of lower priority, require that if τ enumerates a into A during stage t then the certified configuration associated with a during stage s must also exist during stage t. That is, the initial segments of the sets involved in the configuration for a and the computation from BY must not have changed. Further, if during the next σ_1-expansionary stage u it happens that $Y \restriction \theta_{by}(BY)[t]$ is equal to $Y \restriction \theta_{by}(BY)[u]$ (i.e., Y did not change) then the change in X is recorded in C and we preserve the inequality $\Theta_{by}(BY,a) \neq A(a)$ by preserving the appropriate initial segments of the sets under construction.

   We say that a is now undefined and go to Step 1.

Clearly, if for every Θ_{by} we can conclude that $\Theta_{by}(BY) \neq A$ then we have satisfied our requirement. Further, each of the strategies will act at most finitely often and cause little trouble to the rest of the construction.

Suppose this is not the case and consider the effects of the above strategy σ_6 when $\Theta_{by}(BY) = A$. Assume that the strategy is never injured (or be willing to accept finitely many exceptions). Then σ_6 enumerates an infinite increasing sequence of numbers a as still being allowed to enter A. Call this sequence the σ_6-stream of numbers. For each number n, if n is not an element of the σ_6-stream then n is an element of A if and only if n is enumerated into A before any number greater than n is enumerated into the σ_6-stream. Thus, the restriction of A to the numbers not in the σ_6-stream is recursive. Now, consider a number a which is enumerated into the σ_6-stream, say at stage s_a.

By the action of σ_6, for every number a in the σ_6-stream, if a enters A during a stage s greater than or equal to the one during which a was enumerated in the sequence, then the configuration existent when a was enumerated into the stream by σ_6 is still available during stage s. Since $\Theta_{by}(BY) = A$ and there are infinitely many σ_1-expansionary stages, it must be the case that during the interval from stage t to the next σ_1-expansionary stage after t, Y changed below $\lambda_{xy,a}(XY,a)[s]$. Thus, if a is an element of the σ_6-stream and is enumerated into the stream at stage s, then a enters A no later than the first stage u after stage s such that $Y[u] \upharpoonright \lambda_{xy,a}(XY,a)[s] = Y \upharpoonright \lambda_{xy,a}(XY,a)[s]$. It follows that A is recursive relative to Y.

2.2.7. *One sequence $(X, Y, \Lambda_{a,x}, \Lambda_{a,y}, \Lambda_{xy,a})$*

If we were to work only with one sequence $(X, Y, \Lambda_{a,x}, \Lambda_{a,y}, \Lambda_{xy,a})$ or equivalently have only one strategy of type σ_1, then our construction would be particularly simple. We would start with the strategies of type σ_1 and σ_2 and follow them with the strategies of type σ_3, σ_4, σ_5, and σ_6 (as well as nonrecursiveness strategies σ_0). In the simplest case, one of these strategies could have a finite outcome and we could conclude that our requirement is satisfied. In the simplest case, the σ_1-strategy could have a finite outcome and we could conclude that our requirement is satisfied. If not, then either each of the σ_6-strategies would have a finite outcome and we would have satisfied all the necessary requirements $\Theta_{by}(BY) \neq A$, or one of our strategies would have an infinite outcome and we would conclude that our requirement is satisfied by virtue of A's being recursive relative to Y.

In this last case, how can we conclude that A is not recursive? The σ_6-strategy that generates an infinite stream associates with these numbers an infinite stream of certified configurations. The strategy to ensure

$\Theta \neq A$ chooses a number a, preserves its configuration and preserves its certification by preserving B and preserving enough of A to ensure that Y cannot change on any relevant number. If at a later stage t it happens that $\Theta(a)[t] = 0$ then the diagonalization strategy enumerates a into A.

3. The global construction

In the previous section, we analyzed the combinations of the strategies associated with a single sequence $(X, Y, \Lambda_{a,x}, \Lambda_{a,y}, \Lambda_{xy,a})$. We now combine the strategies for all possible such sequences and thereby present a proof of Theorem 2.1.

3.1. *Interactions between σ-strategies*

In fact, there is very little interaction between the strategies associated with different sequences. For the most part, their constraints apply to different B's, C's, D's, and E's and so are mutually compatible. The only set that they have in common is A and the only constraints that they put upon A are the finite ones associated with successful diagonalization and the infinite one constraining the enumeration of new elements of A to the conditions of a σ_6-stream.

Consider a strategy τ constrained to work within an infinite σ_6-stream. The new constraint on τ is that at stage t, τ can enumerate element a into A if and only if a was enumerated into the σ_6-stream at a stage $s < t$ and the configuration associated with a during the stage s still exists during stage t.

If τ is associated with a σ_0-strategy ensuring $\Theta \neq A$ then when τ chooses its number with which to diagonalize, τ chooses that number a from the σ_6-stream. While τ is waiting for $\Theta(a) = 0$, τ preserves enough of A to preserve the σ_6-configuration associated with a. If $\Theta(a)$ is seen to be equal to 0 then τ can enumerate a into A consistently with the constraint of σ_6.

By inspection of the strategies, this is the only way by which numbers enter A and so we need not make many internal changes within our families of strategies.

3.2. *The tree of strategies*

We fix recursive enumerations $(\Theta^i : i \in \omega)$ of all recursive functionals relative to the empty set, $((X^i, Y^i, \Lambda^i_{a,x}, \Lambda^i_{a,y}, \Lambda^i_{xy,a}) : i \in \omega)$ of all sequences as described in σ_1, and, for each i, $(\Psi^{i,j} : j \in \omega)$, $(\Theta_a^{i,j} : j \in \omega)$, and $(\Theta_{by}^{i,j} : j \in \omega)$ of all recursive functionals with one set argument. Of course, the enumerations $(\Psi^{i,j} : j \in \omega)$, $(\Theta_a^{i,j} : j \in \omega)$, and $(\Theta_{by}^{i,j} : j \in \omega)$ need not depend on i, but the notation will be convenient below. Let $((i,j) : i, j \in \omega)$ be a recursive ordering of $\omega \times \omega$ of order type ω. We will assume that for all j, i is less than or equal to the position of (i,j) in this ordering.

We define a tree T of sequences of pairs of strategies and outcomes using recursion. We will also order the immediate extensions of each node from left to right. Ordering by first difference, we have a left to right ordering for all incompatible sequences in T. As usual, shorter nodes or nodes to the left will be assigned higher priority than those below or to the right. For $\eta \in T$, we will speak of the extensions of η as being below η in T. We start with the empty sequence as an element of T.

Suppose that $\eta = ((\tau_k, o_k) : k < \ell)$ is an element of T.

Definition 3.1 Suppose $k < \ell$ and τ_k is of the form $\sigma_1(X^i, Y^i, \Lambda_{a,x}^i, \Lambda_{a,y}^i, \Lambda_{xy,a}^i)$. Then

1. τ_k is *in effect* at η if and only if o_k is Π_2, and

2. τ_k is *unresolved* at η if and only if for all j,
$$(\sigma_6(X^i, Y^i, \Lambda_{a,x}^i, \Lambda_{a,y}^i, \Lambda_{xy,a}^i, \Theta_{by}^{i,j}), \Pi_2) \notin \eta.$$

Definition 3.2 Suppose $k < \ell$ and τ_k is of the form $\sigma_6(X, Y, \Lambda_{a,x}, \Lambda_{a,y}, \Lambda_{xy,a}, \Theta_{by})$. Then, τ_k is *in effect* at η if and only if o_k is Π_2.

Strategies below η in T are based on the assumption that there will be infinitely many expansionary stages for the σ_1- and σ_6-strategies in effect at η. If τ_k is unresolved at η then no strategy in η has determined that A is recursive relative to $\mathbb{Y}$.

Let i_{max} be the largest i such that there is a $k < \ell$ such that τ_k is equal to $\sigma_1(X^i, Y^i, \Lambda_{a,x}^i, \Lambda_{a,y}^i, \Lambda_{xy,a}^i)$. (If there is no such i, let $i_{max} = -1$.)

Case 1. If there is a pair (i^*, j^*) among the first i_{max} many such pairs such that

     A. $\sigma_1(X^{i^*}, Y^{i^*}, \Lambda_{a,x}^{i^*}, \Lambda_{a,y}^{i^*}, \Lambda_{xy,a}^{i^*})$ is in effect at η, unresolved at η, and

     B. one of $\sigma_2(X^{i^*}, Y^{i^*}, \Lambda_{a,x}^{i^*}, \Lambda_{a,y}^{i^*}, \Lambda_{xy,a}^{i^*})$,
        $\sigma_3(X^{i^*}, Y^{i^*}, \Lambda_{a,x}^{i^*}, \Lambda_{a,y}^{i^*}, \Lambda_{xy,a}^{i^*}, \Psi^{i^*,j^*})$,

$$\sigma_4(X^{i^*}, Y^{i^*}, \Lambda_{a,x}^{i^{'*}}, \Lambda_{a,y}^{i^*}, \Lambda_{xy,a}^{i^*}, \Theta_a^{i^*,j^*}),$$

$$\sigma_5(X^{i^*}, Y^{i^*}, \Lambda_{a,x}^{i^*}, \Lambda_{a,y}^{i^*}, \Lambda_{xy,a}^{i^*}, \Theta_a^{i^*,j^*}),$$

or $\sigma_6(X^{i^*}, Y^{i^*}, \Lambda_{a,x}^{i^*}, \Lambda_{a,y}^{i^*}, \Lambda_{xy,a}^{i^*}, \Theta_{by}^{i^*,j^*})$ does not appear in (the first coordinate of an element of) η, $\,\frown$

then let (i, j) be the least such (i^*, j^*). We determine the immediate successor of η in T by the first of the following conditions which applies.

1. If $\sigma_2(X^i, Y^i, \Lambda_{a,x}^i, \Lambda_{a,y}^i, \Lambda_{xy,a}^i)$ does not appear in η then

$$\eta^\frown(\sigma_2(X^i, Y^i, \Lambda_{a,x}^i, \Lambda_{a,y}^i, \Lambda_{xy,a}^i), \Pi_1) \cdot \in T.$$

2. If $\sigma_3(X^i, Y^i, \Lambda_{a,x}^i, \Lambda_{a,y}^i, \Lambda_{xy,a}^i, \Psi^{i,j})$ does not appear in η then

$$\eta^\frown(\sigma_3(X^i, Y^i, \Lambda_{a,x}^i, \Lambda_{a,y}^i, \Lambda_{xy,a}^i, \Psi^{i,j}), \Sigma_2) \in T,$$

$$\eta^\frown(\sigma_3(X^i, Y^i, \Lambda_{a,x}^i, \Lambda_{a,y}^i, \Lambda_{xy,a}^i, \Psi^{i,j}), \Pi_2) \in T,$$

and the Σ_2-extension of η is to the right of the Π_2-extension.

3. If $\sigma_4(X^i, Y^i, \Lambda_{a,x}^i, \Lambda_{a,y}^i, \Lambda_{xy,a}^i, \Theta_a^{i,j})$ does not appear in η then

$$\eta^\frown(\sigma_4(X^i, Y^i, \Lambda_{a,x}^i, \Lambda_{a,y}^i, \Lambda_{xy,a}^i, \Theta_a^{i,j}), \Sigma_1) \in T,$$

$$\eta^\frown(\sigma_4(X^i, Y^i, \Lambda_{a,x}^i, \Lambda_{a,y}^i, \Lambda_{xy,a}^i, \Theta_a^{i,j}), \Pi_1) \in T,$$

and the Π_1-extension of η is to the right of the Σ_1-extension.

4. If $\sigma_5(X^i, Y^i, \Lambda_{a,x}^i, \Lambda_{a,y}^i, \Lambda_{xy,a}^i, \Theta_a^{i,j})$ does not appear in η then

$$\eta^\frown(\sigma_5(X^i, Y^i, \Lambda_{a,x}^i, \Lambda_{a,y}^i, \Lambda_{xy,a}^i, \Theta_a^{i,j}), \Sigma_1) \in T,$$

$$\eta^\frown(\sigma_5(X^i, Y^i, \Lambda_{a,x}^i, \Lambda_{a,y}^i, \Lambda_{xy,a}^i, \Theta_a^{i,j}), \Pi_1) \in T,$$

and the Π_1-extension of η is to the right of the Σ_1-extension.

5. If $\sigma_6(X^i, Y^i, \Lambda_{a,x}^i, \Lambda_{a,y}^i, \Lambda_{xy,a}^i, \Theta_{by}^{i,j})$ does not appear in η then

$$\eta^\frown(\sigma_6(X^i, Y^i, \Lambda_{a,x}^i, \Lambda_{a,y}^i, \Lambda_{xy,a}^i, \Theta_{by}^{i,j}), \Sigma_2) \in T,$$

$$\eta^\frown(\sigma_6(X^i, Y^i, \Lambda_{a,x}^i, \Lambda_{a,y}^i, \Lambda_{xy,a}^i, \Theta_{by}^{i,j}), \Pi_2) \in T,$$

and the Σ_2-extension of η is to the right of the Π_2-extension.

Case 2. If there is no such pair (i, j) as above then we set $i = i_{max} + 1$ and determine the immediate successor of η in T as follows.

1. If $\sigma_0(\Theta^i)$ does not appear in η then

$$\eta^\frown(\sigma_0(\Theta^i), \Sigma_1) \in T,$$
$$\eta^\frown(\sigma_0(\Theta^i), \Pi_1) \in T,$$

and the Σ_1-extension of η is to the left of the Π_1-extension.

2. Otherwise,

$$\eta^\frown(\sigma_1(X, Y, \Lambda_{a,x}, \Lambda_{a,y}, \Lambda_{xy,a}), \Sigma_2) \in T,$$
$$\eta^\frown(\sigma_1(X, Y, \Lambda_{a,x}, \Lambda_{a,y}, \Lambda_{xy,a}), \Pi_2) \in T,$$

and the Σ_2-extension of η is to the right of the Π_2-extension.

3.2.1. η-configurations

Notice that if $\eta \in T$ then there is a unique strategy σ such that σ appears as the first component in the last element of the immediate successors of η.

Definition 3.3 Suppose that η is an element of T.

1. Let

$$\{\sigma_1(X^{i_j}, Y^{i_j}, \Lambda_{a,x}^{i_j}, \Lambda_{a,y}^{i_j}, \Lambda_{xy,a}^{i_j}) : j < \ell_1\}$$

be the sequence of σ_1-strategies σ in effect at η. Then an η-*configuration for a_1* is a finite initial segment A and the sets associated with these strategies such that for each $j < \ell_1$, there is a $\sigma_1(X^{i_j}, Y^{i_j}, \Lambda_{a,x}^{i_j}, \Lambda_{a,y}^{i_j}, \Lambda_{xy,a}^{i_j})$-configuration for a_1 within this initial segment.

2. We say that an η-configuration is *certified* if in addition to the above, for every σ_6-strategy τ_k which is in effect at η, the computation setting $\Theta_{by}(BY, a) = 0$ has not changed since the stage during which τ_k enumerated the configuration for a as certified.

For example, if η has only one σ_1-strategy $\sigma_1(X, Y, \Lambda_{a,x}, \Lambda_{a,y}, \Lambda_{xy,a})$ with a Π_2-outcome, then an η-configuration for a_1 is the same as a $\sigma_1(X, Y, \Lambda_{a,x}, \Lambda_{a,y}, \Lambda_{xy,a})$-configuration for a_1, as described in Figure 1. With n such strategies, an η-configuration is described by n copies of Figure 1, one for each strategy and involving the sets associated with that strategy, sharing a common value for a_1.

The i_jth component of an η-configuration for a_1 is the initial segment of A, B^{i_j}, C^{i_j}, D^{i_j}, and E^{i_j} which makes up the $\sigma_1(X^{i_j}, Y^{i_j}, \Lambda_{a,x}^{i_j}, \Lambda_{a,y}^{i_j}, \Lambda_{xy,a}^{i_j})$-configuration for a_1.

3.3. *The construction*

We organize our construction by stages s, where s is greater than or equal to 1. Each stage s is divided into at most s many substages t, where t is also greater than or equal to 1. We proceed as follows during stage s.

Let $\eta[s, 0]$ equal the empty sequence.

Given $\eta[s, t-1]$ with t less than or equal to s, let σ be the strategy which appears in the first component of the immediate successors of $\eta[s, t-1]$. We may assume that σ has been assigned a number a_1 and an $\eta[s, t-1]$-configuration for a_1 during an earlier stage and that no component of that configuration has changed since the stage during which it was assigned. (Otherwise, a strategy simply ends the stage since by its hypothesis, it will eventually be assigned such a number a_1 by a σ_2- or σ_5-strategy as described below.)

We follow the instructions of σ, which depend on its type as described below. At the end of its action, either σ ends stage s and we go to stage $s+1$ with $t = 0$ or σ determines a value for $\eta[s, t]$. In the second case, if t is less than s then we continue with substage $t+1$ of stage s.

We adapt the pure strategies described in the previous section to work within the full construction as follows.

3.3.1. *Adding an A-diagonalization strategy σ_0*

Suppose that σ is a diagonalization strategy $\sigma_0(\Theta^i)$ to ensure that $\Theta^i \neq A$.

If $\Theta^i(a_1)[s]$ is not equal to 0 then we restrain any number from entering any set involved in our η-configuration for a_1. We let $\eta[s, t]$ be $\eta[s, t-1]^\frown(\sigma, \Pi_1)$.

If $\Theta^i(a_1)[s]$ is equal to 0 and a_1 is not an element of $A[s]$, then we enumerate a_1 into A and end stage s.

If $\Theta^i(a_1)[s]$ is equal to 0 and we have already enumerated a_1 into A, then we let $\eta[s, t]$ be $\eta[s, t-1]^\frown(\sigma, \Sigma_1)$.

3.3.2. *Adding a $\sigma_1(X, Y, \Lambda_{a,x}, \Lambda_{a,y}, \Lambda_{xy,a})$*

Suppose that σ is $\sigma_1(X^i, Y^i, \Lambda_{a,x}^i, \Lambda_{a,y}^i, \Lambda_{xy,a}^i)$. We alter the pure σ_1-strategy (described in the previous section) in the following way.

First, we measure σ-expansions in terms of stages when $\eta[s, t-1]$ is active in the construction. Second, while waiting for a σ-expansionary stage, we preserve the $\eta[s, t-1]$-configuration for a_1.

If s is not σ-expansionary in the above sense then let $\eta[s, t]$ be $\eta[s, t-1]^\frown(\sigma, \Sigma_2)$.

Otherwise, we enumerate the pair (a_1, a_2) as in the pure σ_1-strategy, we let $\eta[s, t]$ be $\eta[s, t-1]^\frown(\sigma, \Pi_2)$, and we cancel all strategies on nodes to the right of $\eta[s, t]$.

3.3.3. Adding a $\sigma_2(X, Y, \Lambda_{a,x}, \Lambda_{a,y}, \Lambda_{xy,a})$

Suppose that σ is $\sigma_2(X^i, Y^i, \Lambda^i_{a,x}, \Lambda^i_{a,y}, \Lambda^i_{xy,a})$. By the definition of T and the previous paragraph, we may assume that the last strategy mentioned in $\eta[s, t-1]$ is of the form $\sigma_1(X^i, Y^i, \Lambda^i_{a,x}, \Lambda^i_{a,y}, \Lambda^i_{xy,a})$ and that s is expansionary for that strategy (i.e., the sequence $\eta[s, t-1]$ ends with the pair $(\sigma_1(X^i, Y^i, \Lambda^i_{a,x}, \Lambda^i_{a,y}, \Lambda^i_{xy,a}), \Pi_2)$).

We let $\eta[s, t]$ be $\eta[s, t-1]^\frown(\sigma, \Pi_1)$, and we alter the pure σ_2-strategy in the following way.

First, we may need to change BC to record a change in X. If during the previous stage s' during which σ_2 acted, some strategy associated with an extension of η enumerated a number a into A, then let μ be the node in T associated with that strategy. There are two cases to consider. In the first case, there is no strategy $\sigma_6(X^i, Y^i, \Lambda^i_{a,x}, \Lambda^i_{a,y}, \Lambda^i_{xy,a}, \Theta_{by})$ above μ with $B^i = B$ and $C^i = C$ which certified a. In this case, we record the change in X by changing B accordingly. Otherwise, if Y did not change below $\lambda^i_{xy,a}(a)$ between stage s' and the current stage, then X must have changed there. This allows us to record all changes in X by enumerating c into C, where c is the number depicted in Figure 1, and we do so without changing B.

Next, let (a_1, a_2) be the pair just enumerated by $\sigma_1(X^i, Y^i, \Lambda^i_{a,x}, \Lambda^i_{a,y}, \Lambda^i_{xy,a})$. We extend the definitions of the functionals $\Gamma_{x,b}$, $\Gamma_{x,c}$, and $\Gamma_{bc,x}$ so that we have a $\sigma_1(X^i, Y^i, \Lambda^i_{a,x}, \Lambda^i_{a,y}, \Lambda^i_{xy,a})$-configuration for a_1 and thus an $\eta[s, t]$-configuration for a_1. If there is a μ such that $\eta[s, t] \subseteq \mu$, μ was active during a previous stage, the set of strategies in effect at μ is equal to the set of strategies in effect at $\eta[s, t]$, and μ does not currently have an $\eta[s, t]$-configuration assigned to it, then we assign the configuration for a_1 to the leftmost and shortest such μ (that is, the one of highest priority). We cancel all strategies to the right of μ and end stage s.

3.3.4. *Adding a* $\sigma_3(X, Y, \Lambda_{a,x}, \Lambda_{a,y}, \Lambda_{xy,a}, \Psi_{cd}, \Psi_{ce})$

Our only alteration to the pure σ_3-strategy is to make it measure expansionary stages taking into account only those stages during which it is active. For $\sigma = \sigma_3(X, Y, \Lambda_{a,x}, \Lambda_{a,y}, \Lambda_{xy,a}, \Psi_{cd}, \Psi_{ce})$ we let $\eta[s, t]$ be $\eta[s, t-1]^\frown(\sigma, \Pi_2)$ if s is σ-expansionary and $\eta[s, t-1]^\frown(\sigma, \Sigma_2)$ otherwise.

3.3.5. *Adding a* $\sigma_4(X, Y, \Lambda_{a,x}, \Lambda_{a,y}, \Lambda_{xy,a}, \Theta_a)$ *or a* $\sigma_5(X, Y, \Lambda_{a,x}, \Lambda_{a,y}, \Lambda_{xy,a}, \Theta_a)$

We use the pure σ_4- and σ_5-strategies without change. For a σ_4-strategy σ, we let $\eta[s, t]$ be $\eta[s, t-1]^\frown(\sigma, \Sigma_1)$ if the diagonalization witness n has been enumerated into D and $\eta[s, t-1]^\frown(\sigma, \Sigma_2)$ otherwise. (For a σ_5-strategy σ, D is replaced by E.)

3.3.6. *Adding a* $\sigma_6(X, Y, \Lambda_{a,x}, \Lambda_{a,y}, \Lambda_{xy,a}, \Theta_{by})$

We do not alter the first phase of the pure σ_6-strategy. We start with a number a for which we have a certified $\eta[s, t-1]$-configuration. We restrain a from entering A and preserve its configuration. We wait for a stage s such that $(\Theta_{by}(BY, a) = 0)[s]$. If the current s is not such a stage then, for $\sigma = \sigma_6(X, Y, \Lambda_{a,x}, \Lambda_{a,y}, \Lambda_{xy,a}, \Theta_{by})$ we set

$$\eta[s, t] = \eta[s, t-1]^\frown(\sigma, \Sigma_2).$$

Otherwise, we set

$$\eta[s, t] = \eta[s, t-1]^\frown(\sigma, \Pi_2).$$

If there is an a such that we wait forever for an s such that $(\Theta_{by}(BY, a) = 0)[s]$, then $\Theta_{by} \neq A$ and the requirement is satisfied. Otherwise, according to the pure σ_6-strategy, we should restrict the enumeration of numbers into A to those which appear in the stream it generates.

We must describe how the numbers in that stream are distributed to the strategies associated with nodes extending $\eta[s, t] = \eta[s, t-1]^\frown(\sigma, \Pi_2)$. For this, we make the same alteration which we made for the σ_2-strategies.

We can assume that σ has been assigned a certified $\eta[s, t-1]$-configuration for a number a_1. When σ finds a computation setting $\Theta_{by}(BY, a_1) = 0$ and for which there is are $\Lambda_{a,y}(A)$ computations agreeing with Y below $\theta_{by}(BY, a_1)$, then we say that these computations certify the $\eta[s, t-1]$-configuration for a_1 with respect to σ. Thus, a_1 now has a certified $\eta[s, t]$-configuration.

If there is a μ such that $\eta[s,t] \subseteq \mu$, μ was active during a previous stage, the set of strategies in effect at μ is equal to the set of strategies in effect at η, and μ does not currently have an $\eta[s,t]$-configuration assigned to it, then we assign the configuration for a_1 to the leftmost and shortest such μ (that is, the one of highest priority). We cancel all strategies η' to the right of μ and end stage s. If there is no such μ then we cancel all strategies to the right of $\eta[s,t]$.

3.4. Analyzing the construction

Let η^∞ be the path through T such that

1. for infinitely stages s and infinitely many substages t, $\eta[s,t]$ is a subsequence of η^∞, and

2. for at most finitely many stages s and substages t, $\eta[s,t]$ is to the left of η^∞.

Following convention, we say that η^∞ is the true path of the construction.

Lemma 3.4 *The true path η^∞ is an infinite path in T.*

Proof. Suppose that η is a finite initial segment of η^∞. We will argue that there is a proper extension of η which is also contained in η^∞.

Note that T is a finite branching tree. Consequently, if there are infinitely many s during which η acts and does not end the stage, then there is a leftmost proper extension which acts infinitely often and the claim is proven.

There are four cases in which η peremptorily ends a stage during which it acts. Firstly η might end with a strategy which ends the stage since it is not assigned a number, which can happen at most finitely often in a row by the strategy's assumption on outcomes of strategies above it. Next, η might end with a strategy for making $\Theta \neq A$, in which case this strategy can end the stage at most once without being initialized. Otherwise, either η ends with a σ_2-strategy, or it ends with a σ_6-strategy, and, in both cases, η allocates an η-certified configuration to a strategy below it. But, if μ is eligible to be assigned a configuration by η then μ must have been active during an earlier stage of the construction. If η were to end all but finitely many of the stages during which it finds a new certified η-configuration, then there can only be finitely many such μ's. Eventually, every such μ will have a certified η-configuration assigned to it. But then the next time that

η finds a new certified η-configuration it will not end the stage and some proper extension of η will be active.

Lemma 3.5 *For each finite $\eta \subset \eta^{\infty}$, the following conditions hold.*

1. *We cancel η during the last stage s_{η} during which there is a t such that $\eta[s_{\eta}, t]$ is to the left of η.*

2. *We let η act infinitely often.*

3. *During every stage greater than or equal to s_{η}, we respect all of the constraints imposed by η during any earlier stage.*

Proof. Routine. □

Lemma 3.6 *Our construction satisfies all of the requirements of Section 2.1.*

Proof. This follows as in the analysis of the individual strategies in Section 2.2. □

Theorem 2.1 follows directly from Lemma 3.6.

References

Ambos-Spies, K. (1984). On pairs of recursively enumerable degrees. *Trans. Amer. Math. Soc. 283*(2), 507–531. MR MR737882 (85d:03083).

Ambos-Spies, K. (1985). Generators of the recursively enumerable degrees. In *Recursion theory week (Oberwolfach, 1984)*, Volume 1141 of *Lecture Notes in Math.*, pp. 1–28. Berlin: Springer. MR 87i:03081.

Fejer, P. A. (1982). Branching degrees above low degrees. *Trans. Amer. Math. Soc. 273*(1), 157–180. MR 84a:03044.

Fejer, P. A. (1983). The density of the nonbranching degrees. *Ann. Pure Appl. Logic 24*(2), 113–130. MR MR713296 (85g:03062).

Jockusch, Jr., C. G. (1985). Genericity for recursively enumerable sets. In *Recursion theory week (Oberwolfach, 1984)*, Volume 1141 of *Lecture Notes in Math.*, pp. 203–232. Berlin: Springer. MR MR820782 (87f:03117).

Lachlan, A. H. (1966). Lower bounds for pairs of recursively enumerable degrees. *Proc. London Math. Soc. (3) 16*, 537–569. MR MR0204282 (34 #4126).

Sacks, G. E. (1963). *Degrees of unsolvability*. Princeton, N.J.: Princeton University Press. MR MR0186554 (32 #4013).

Slaman, T. A. (1991). The density of infima in the recursively enumerable degrees. *Ann. Pure Appl. Logic 52*(1-2), 155–179. International Symposium on Mathematical Logic and its Applications (Nagoya, 1988). MR 92e:03061.

Yates, C. E. M. (1966). A minimal pair of recursively enumerable degrees. *J. Symbolic Logic 31*, 159–168. MR MR0205851 (34 #5677).

CODING INTO $H(\omega_2)$, TOGETHER (OR NOT) WITH FORCING AXIOMS. A SURVEY

David Asperó

Institució Catalana de Recerca i Estudis Avançats (ICREA)
and
Departament de Lògica, Història i Filosofia de la Ciència
,Universitat de Barcelona
Baldiri Reixac, s/n, 08028 Barcelona, Spain
E-mail: david.aspero@icrea.es
URL: http://www.icrea.es/pag.asp?id=David.Aspero

This paper is mainly a survey of recent results concerning the possibility of building forcing extensions in which there is a simple definition, over the structure $\langle H(\omega_2), \in \rangle$ and without parameters, of a prescribed member of $H(\omega_2)$ or of a well-order of $H(\omega_2)$. Some of these results are in conjunction with strong forcing axioms like PFA^{++} or MM, some are not. I also observe (Corollary 4:4) that the existence of certain objects of size $\aleph_1$ follows outright from the existence of large cardinals. This observation is motivated by an (unsuccessful) attempt to extend a PFA^{++} result to one mentioning MM^{++}.

1. Main starting questions and some pieces of notation

The work presented here deals mostly with the problem of finding optimal definitions of well-orders of the reals and other objects. More precisely, it addresses the following two questions.[1]

Question 1: Suppose A is a subset of ω_1. Suppose we are given the task of going over to a *nice* set-forcing extension[2] in which A admits a *simple* definition $\Phi(x)$, without parameters, over the structure $\langle H(\omega_2), \in \rangle$ or over

---

[1] As quoted from [As2].

[2] So, if *nice* is to be interpreted as preserving stationary subsets of ω_1 in the ground model and A is a stationary and co-stationary subset of ω_1, then A will remain stationary and co-stationary in the extension.

some natural (definable) extension of this structure, like $\langle H(\omega_2), \in, NS_{\omega_1}\rangle$. What is the lowest degree of logical complexity that can be attributed to a definition $\Phi(x)$ for which we can perform the above task?

Question 2: What is the lowest degree of logical complexity of formulas for which there is a formula $\Phi(x, y)$ (again without parameters) with that complexity and with the property that we can go over to a set–forcing extension in which the set of real numbers admits a well–order defined by $\Phi(x, y)$ (again over the structure $\langle H(\omega_2), \in \rangle$ or over some natural extension of it)?

For some background on these problems the reader is referred to [As2]. Logical complexity – for formulas of a language extending the language of set theory – will be measured in this paper by the familiar Levy hierarchy $\bigcup_{n<\omega}\{\Sigma_n, \Pi_n\}$. Recall that a formula is Σ_0 (equivalently, Π_0) if all of its quantifiers are restricted[3] and that, for $n > 0$, a formula is Σ_n (respectively, Π_n) if it is of the form $(\exists x)\varphi$ for a Π_{n-1} formula φ (respectively, if it is of the form $(\forall x)\varphi$ for a Σ_{n-1} formula φ). Note that, in any model M of ZF without the Power Set Axiom, if P is a definable class in M and $\varphi(x_0, \ldots x_k)$ is a formula in the language of the structure $\langle M, \in, P\rangle$, then there is some formula $\psi(x_0, \ldots x_k) \in \bigcup_{n<\omega}\{\Sigma_n, \Pi_n\}$ (in the same language) such that, in $\langle M, \in, P\rangle$, $\varphi(x_0, \ldots x_k)$ is logically equivalent to $\psi(x_0, \ldots x_k)$.[4] In other words, the Levy hierarchy provides a classification, up to logical equivalence, of all formulas over structures $\langle M, \in, P\rangle$ as above. Also, note that, for every $n < \omega$, every formula in $\Sigma_n \cup \Pi_n$ is logically equivalent to a formula in $\Sigma_{n+1} \cap \Pi_{n+1}$.

Throughout this paper, $H(\omega_2)$ denotes the set of all sets whose transitive closure has size at most $\aleph_1$, and NS_{ω_1} denotes the nonstationary ideal on ω_1. Given a regular cardinal κ, $cf(\kappa)$ is the class of all ordinals of cofinality κ. $\mathcal{L}$ will denote the first order language of the structure $\langle H(\omega_2), \in, NS_{\omega_1}\rangle$. Given a set X of ordinals, $ot(X)$ will denote the order type of X. Recall that a partial order $\mathcal{P}$ is proper if for every regular cardinal $\theta > |TC(\mathcal{P})|$, every countable $N \prec H(\theta)$ containing $\mathcal{P}$ and every $p \in \mathcal{P} \cap N$ there is some $q \in \mathcal{P}$ extending p such that q is $(N, \mathcal{P})$–generic, i.e. such that $q \Vdash_{\mathcal{P}} \tau \in \check{N}$ whenever $\tau \in N$ is a $\mathcal{P}$–name for an ordinal. Also, $\mathcal{P}$ is semiproper in case for every θ, N and p as above there is a condition q extending p such that

---

[3]In other words, if all its quantifiers occur in a subformula of the form $(\forall x)(x \in y \to \varphi)$ or $(\exists x)(x \in y \land \varphi)$.

[4]That is, $\langle M, \in, P\rangle \models (\forall x_0, \ldots x_k)\,(\varphi(x_0, \ldots x_k) \leftrightarrow \psi(x_0, \ldots x_k))$.

q is $(N, \mathcal{P})$–semigeneric, i.e. such that $q \Vdash_{\mathcal{P}} \tau \in \check{N}$ for every name $\tau \in N$ for an ordinal in ω_1^V. Every proper partial order is semiproper and, if $\mathcal{P}$ is semiproper, then every stationary subset of ω_1 remains stationary after forcing with $\mathcal{P}$.

$L(\mathbb{R})$ is the $\subseteq$–minimal transitive inner model of ZF containing all reals and all ordinals. Some arguments in Section 4, and in the proofs of Theorems 2.8 and 2.9 in Section 2, involve $\mathbb{P}_{max}$ forcing. $\mathbb{P}_{max}$ is a poset belonging to $L(\mathbb{R})$ and definable in $L(\mathbb{R})$ (without parameters). If $x^{\dagger}$ exists for every real x,[5] then $\mathbb{P}_{max}$ is a homogeneous forcing and is σ–closed (in V and in $L(\mathbb{R})$). In particular, forcing with it over $L(\mathbb{R})$ does not add new reals. The standard reference for $\mathbb{P}_{max}$ forcing is [W].

It will be convenient to fix a notion of incompatibility, for pairs of formulas, which is absolute with respect to sufficiently arbitrary models of set theory. We will say that two $\mathcal{L}$–formulas $\Phi_0(x)$ and $\Phi_1(x)$ are *ZFC–provably incompatible* if ZFC proves that for every uncountable regular cardinal κ and every $x \in H(\kappa^+)$, $\langle H(\kappa^+), \in, NS_{\kappa} \rangle \models \neg(\Phi_0(x) \wedge \Phi_1(x))$. Also, for an $\mathcal{L}$–formula in two free variables $\Phi(x, y)$, we will say that $\Phi(x, y)$ is *ZFC–provably antisymmetric* if ZFC proves that for every uncountable regular cardinal κ and every $x, y \in H(\kappa^+)$, $x \neq y$, $\langle H(\kappa^+), \in, NS_{\kappa} \rangle \models \neg(\Phi(x, y) \wedge \Phi(y, x))$.[6]

Acknowledgment. Many of the results in this paper were presented in two talks that I gave in Singapore in July 2005. These talks were part of the IMS workshop "Computational Prospects of Infinity". I thank the members of the Organizing Committee for inviting me.

2. Results not mentioning forcing axioms

The following theorems are proved in [As4].

Theorem 2.1 *([As4]) There are Σ_3 $\mathcal{L}$–formulas $\Phi_0(x)$ and $\Phi_1(x)$ and Π_3 $\mathcal{L}$–formulas $\Psi_0(x)$ and $\Psi_1(x)$ with the following two properties.*

(1) $(\Phi_0(x), \Phi_1(x))$ and $(\Psi_0(x), \Psi_1(x))$ are two pairs of ZFC–provably incompatible formulas over the structure $\langle H(\omega_2), \in, NS_{\omega_1} \rangle$.

---

[5]Which follows from all large cardinal assumptions used in the arguments alluded to here.

[6]The notions of provably incompatible pairs of formulas over $\langle H(\omega_2), \in, NS_{\omega_1} \rangle$ and of provably incompatible formulas in the language of set theory (over $\langle H(\omega_2), \in \rangle$) are of course defined in the natural way. And the same goes for the corresponding notions of provably antisymmetric formulas.

(2) Given any $A \subseteq \omega_1$ there is a proper poset forcing that

    (a) A is defined, over $\langle H(\omega_2), \in, NS_{\omega_1} \rangle$, by $\Phi_0(x)$ and by $\Psi_0(x)$, and

    (b) $\omega_1 \backslash A$ is defined, over $\langle H(\omega_2), \in, NS_{\omega_1} \rangle$, by $\Phi_1(x)$ and by $\Psi_1(x)$.

Theorem 2.2 *([As4]) There is a Σ_3 $\mathcal{L}$–formula $\Phi(x, y)$ and a Π_3 $\mathcal{L}$–formula $\Psi(x, y)$ with the following two properties.*

(1) $\Phi(x, y)$ and $\Psi(x, y)$ are ZFC–provably antisymmetric formulas over the structure $\langle H(\omega_2), \in, NS_{\omega_1} \rangle$.

(2) If there is an inaccessible cardinal, then there is a proper poset $\mathcal{P}$ forcing the existence of a well–order $\leq$ of $H(\omega_2)$ of order type ω_2 such that $\leq$ is defined, over $\langle H(\omega_2), \in, NS_{\omega_1} \rangle$, both by $\Phi(x, y)$ and by $\Psi(x, y)$.

In fact, Theorem 2.1 can be easily derived[7] from the following result.

Theorem 2.3 *([As4]) There is a Σ_2 $\mathcal{L}$–formula $\Phi(x)$ such that for every uncountable regular cardinal κ and every $A \subseteq \kappa$ there is a poset $\mathcal{P}$ with the following properties.*

(1) $\mathcal{P}$ is κ–distributive,[8] proper, and preserves κ. Also, if $2^\mu = \mu^+$ whenever μ is an infinite cardinal with $\mu^+ < \kappa$, then $\mathcal{P}$ preserves all stationary subsets of κ. Finally, if $2^{<\kappa} = \kappa$ and $2^\kappa = \kappa^+$, then $\mathcal{P}$ has the κ^+–chain condition.

(2) $\mathcal{P}$ forces

$$A = \{\xi < \kappa \,:\, \langle H(\kappa^+), \in, NS_\kappa \rangle \models \Phi(\xi)\}$$

Similarly, Theorem 2.2 is a consequence of the following result, also proved in [As4], by taking $\kappa = \omega_1$.

Theorem 2.4 *([As4]) There is a Σ_3 $\mathcal{L}$–formula $\Phi(x, y)$ and a Π_3 $\mathcal{L}$–formula $\Psi(x, y)$ satisfying (1) and (2) below.*

---

[7]By taking $\kappa = \omega_1$ and by taking $\Phi_0(x)$ and $\Psi_0(x)$ to be $\Phi(x)$ and $\Phi_1(x)$ and $\Psi_1(x)$ to be $\neg\Phi(x)$.

[8]Recall that a forcing notion is κ–distributive if and only if it does not add new sequences of ordinals of length less than κ.

(1) $\Phi(x,y)$ *and* $\Psi(x,y)$ *are ZFC–provably antisymmetric formulas.*

(2) Given any uncountable regular cardinal κ there is a poset $\mathcal{P}$ with the following properties.

    (a) $\mathcal{P}$ preserves ω_1 and, if $\kappa > \omega_1$, then it satisfies (1) from Theorem 2.3.

    (b) $\mathcal{P}$ forces that

$$\{(x,y) \in H(\kappa^+) \times H(\kappa^+) : \langle H(\kappa^+), \in, NS_\kappa \rangle \models \Phi(x,y)\}$$

    is equal to

$$\{(x,y) \in H(\kappa^+) \times H(\kappa^+) : \langle H(\kappa^+), \in, NS_\kappa \rangle \models \Psi(x,y)\}$$

    and is a well–order of $H(\kappa^+)$ of order type κ^+.

Note that none of Theorems 2.3 and 2.4 can hold for $\kappa = \omega$: By [Mar-St], Projective Determinacy holds if there are infinitely many Woodin cardinals. In particular, under this large cardinal assumption there can be no well–order of the reals definable over $\langle H(\omega_1), \in \rangle$ (even allowing parameters). And it can be seen that if $\delta < \kappa$ are such that δ is a limit of infinitely many Woodin cardinals and κ is a measurable cardinal, then given any poset $\mathcal{P}$ of size less than δ, any $\mathcal{P}$–generic filter G over V and any real $r \in V[G] \backslash V$, r is not definable over $\langle H(\omega_1)^{V[G]}, \in \rangle$ by any formula with a real number in V as parameter.[9]

Given a regular cardinal $\kappa \geq \omega_1$, the proofs of Theorems 2.3 and 2.4 involve the manipulation, by forcing, of certain weak club–guessing properties for club–sequences defined on stationary subsets of κ, in such a way that the Σ_2 theory of $\langle H(\kappa^+), \in, NS_\kappa \rangle$ with ordinals in κ as parameters codes any prescribed subset of κ.

Given an ordinal γ, $Lim(\gamma)$ denotes the set of nonzero limit ordinals in γ. A *club–sequence* will be a sequence of the form $\bar{\alpha} = \langle \alpha_\delta : \delta \in Lim(\gamma) \rangle$ – for some ordinal γ – such that each α_δ is a subset of δ. The set S of $\delta \in Lim(\gamma)$ such that α_δ is a club of δ is called the *domain of $\bar{\alpha}$*. It will also be denoted by $dom(\bar{\alpha})$. We may say that $\bar{\alpha}$ *is defined on S*. If $\bar{\alpha}$ is a club–sequence and γ is such that $sup(dom(\bar{\alpha})) = \gamma$, then we may say that $\bar{\alpha}$ *is a club–sequence on γ*. If τ is an ordinal such that the order type of α_δ

---

[9]This follows from a result of Woodin to the effect that the theory of $L(\mathbb{R})$ with real numbers as parameters cannot be changed by forcing with $\mathcal{P}$ whenever $\mathcal{P}$ is a poset with $|\mathcal{P}| < \delta$ and $\delta < \kappa$ are as above (see [L], Theorem 3.1.12 for a proof).

is τ for each $\delta \in dom(\overline{\alpha})$, then we say that the *height of* $\overline{\alpha}$ is τ. $ht(\overline{\alpha})$ will denote the height of $\overline{\alpha}$ (if it exists). As in [A-Sh] (for ladder systems, that is, for club–sequences of height ω), if $\overline{\alpha}$ is a club–sequence and $\delta \in dom(\overline{\alpha})$, then α_δ will denote $\overline{\alpha}(\delta)$ (and similarly with other Greek letters). We will say that $\overline{\alpha}$ is a *coherent club–sequence* if there is a club–sequence $\overline{\beta}$ with $dom(\overline{\alpha}) \subseteq dom(\overline{\beta})$ and $\overline{\beta} \restriction dom(\overline{\alpha}) = \overline{\alpha} \restriction dom(\overline{\alpha})$,[10] and such that $\overline{\beta}$ is coherent in the usual sense, that is, such that for every $\delta \in dom(\overline{\beta})$ and every limit point γ of β_δ, $\gamma \in dom(\overline{\beta})$ and $\beta_\delta \cap \gamma = \beta_\gamma$.

The concepts in this paragraph are defined in [A-Sh] for ladder systems.[11] Let $\overline{\alpha}$ be a club–sequence on an ordinal γ of uncountable cofinality. We say that $\overline{\alpha}$ is *guessing* in case for every club $C \subseteq \gamma$ there is some $\delta \in C \cap dom(\overline{\alpha})$ such that $\alpha_\delta \backslash C$ is bounded in δ. Furthermore, we say that $\overline{\alpha}$ is *strongly guessing* if for every club $C \subseteq \gamma$ there is a club $D \subseteq \gamma$ such that $\alpha_\delta \backslash C$ is bounded in δ for every $\delta \in D \cap dom(\overline{\alpha})$.[12] $\overline{\alpha}$ is *avoidable* if there is a club $C \subseteq \gamma$ such that $\alpha_\delta \cap C$ is bounded in δ for each $\delta \in dom(\overline{\alpha}) \cap C$. Given two club–sequences $\overline{\alpha}$ and $\overline{\beta}$ on the same ordinal γ, γ of uncountable cofinality, $\overline{\beta}$ is *disjoint from* $\overline{\alpha}$ if $\beta_\delta \cap \alpha_\delta = \emptyset$ for every $\delta \in dom(\overline{\alpha}) \cap dom(\overline{\beta})$. Given a strongly guessing club–sequence $\overline{\alpha}$ on an ordinal γ and a set $X \subseteq \gamma$ including $dom(\overline{\alpha})$, $\overline{\alpha}$ is said to be *maximal for X* in case every ladder system defined on X and disjoint from $\overline{\alpha}$ is avoidable.

The proofs of Theorems 2.3 and 2.4 make use of a certain weak club–guessing property for club–sequences,[13] which is best defined after introducing the following strong version of intersection of two sets of ordinals: Given two sets of ordinals, X and Y, $X \cap^* Y$ is defined as the set of $\delta \in X \cap Y$ such that δ is not a limit point of X. Now, given a club–sequence $\overline{\alpha}$ on an ordinal γ of uncountable cofinality, we will say that $\overline{\alpha}$ *is type-guessing* in case for every club $C \subseteq \gamma$ there is some $\delta \in C \cap dom(\overline{\alpha})$ with $ot(\alpha_\delta \cap^* C)$ as high as possible, that is, with $ot(\alpha_\delta \cap^* C) = ot(\alpha_\delta)$. We will say that $\overline{\alpha}$ *is strongly type-guessing* in case for every club $C \subseteq \gamma$ there is a club $D \subseteq \gamma$ such that $ot(\alpha_\delta \cap^* C) = ot(\alpha_\delta)$ for every $\delta \in D \cap dom(\overline{\alpha})$.

Also, for a set X of ordinals and an ordinal δ, the *Cantor–Bendixson rank of δ with respect to X*, $rnk_X(\delta)$, is defined by specifying that $rnk_X(\delta) = 0$ if and only if δ is not a limit point of X, that $rnk_X(\delta) \geq 1$

---

[10]Where, given a club–sequence $\overline{\alpha}$ and a set X, the restriction of $\overline{\alpha}$ to X, to be denoted by $\overline{\alpha} \restriction X$, is that club–sequence which is equal to $\overline{\alpha}$ on X and is $\emptyset$ elsewhere.

[11]They will occur in Definition 2.3.

[12]Note that a strongly guessing club–sequence is guessing if and only if its domain is stationary.

[13]Defined in [As4].

if and only if δ is a limit point of X and, for each ordinal $\eta \geq 1$, that $rnk_X(\delta) > \eta$ if and only if δ is a limit ordinal and there is a sequence $(\delta_\xi)_{\xi < ot(\delta)}$ converging to δ such that $rnk_X(\delta_\xi) \geq \eta$ for every ξ.[14] An ordinal δ will be said to be *perfect* if $rnk_\delta(\delta) = \delta$.[15] Note that $rnk_\delta(\delta) \leq \delta$ for every ordinal δ and that, given any uncountable regular cardinal κ, the set of perfect ordinals below κ is a club.

Definition 2.1 *([As4]) Given an uncountable regular cardinal κ, $\mathcal{A} = \{\overline{\alpha}^\nu : \nu < \lambda\}$ (for $1 \leq \lambda \leq \kappa$) is an almost specifiable set of club–sequences on κ (asscs on κ, for short) if and only if*

(a) *there is a one–to–one sequence $\langle \tau_\nu : \nu < \lambda \rangle$ of perfect ordinals below κ such that, for each ν, τ_ν has countable cofinality and $\overline{\alpha}^\nu$ is a coherent club–sequence on κ of height τ_ν,*

(b) *$\langle dom(\overline{\alpha}^\nu) : \nu < \lambda \rangle$ is a sequence of pairwise disjoint stationary sets,*

(c) *each $\overline{\alpha}^\nu$ is strongly type–guessing, and*

(d) *given any coherent club–sequence $\overline{\beta}$ on κ with stationary domain, if $\overline{\beta}$ has height υ of countable cofinality and $\upsilon \neq ht(\overline{\alpha}^\nu)$ for every $\nu < \lambda$, then $\overline{\beta}$ is not strongly type–guessing.*

Lemma 2.5 gives a justification for the use of the phrase 'almost specifiable' in Definition 2.1. It implies that $\{ht(\overline{\alpha}^\nu) : \nu < \lambda_0\} = \{ht(\overline{\beta}^\nu) : \nu < \lambda_1\}$ holds whenever $\{\overline{\alpha}^\nu : \nu < \lambda_0\}$ and $\{\overline{\beta}^\nu : \nu < \lambda_1\}$ are two asscs's on the same κ.

Lemma 2.5 *([As4]) Suppose $\mathcal{A} = \{\overline{\alpha}^\nu : \nu < \lambda\}$ is an almost specifiable set of club–sequences on an uncountable regular cardinal κ. Then, $\{ht(\overline{\alpha}^\nu) : \nu < \lambda\}$ is equal to the set of perfect ordinals $\tau < \kappa$ of countable cofinality and such that there is a strongly type–guessing coherent club–sequence on κ of height τ and with stationary domain.*

The above lemma follows immediately from the definition of asscs. Theorem 2.3 is an immediate consequence of Lemma 2.5 and of the following result.

---

[14]Thus, for example, $rnk_X(\delta) = 1$ if and only if δ is a limit point of ordinals in X but not a limit point of limit points of X.

[15]Thus, with this definition, the first perfect ordinal is 0, the second is $\epsilon_0 = sup\{\omega, \omega^\omega, \omega^{(\omega^\omega)}, \omega^{(\omega^{(\omega^\omega)})}, \ldots\}$, etc.

Theorem 2.6 *([As4]) Let $\kappa \geq \omega_1$ be a regular cardinal, let A be a subset of κ and let $(\xi_\eta)_{\eta<\kappa}$ be the strictly increasing enumeration of all perfect ordinals ξ less than κ with $cf(\xi) = \omega$. Then there is a poset $\mathcal{P}$ satisfying the niceness properties in (1) from Theorem 2.3 and forcing that there is an asscs $\{\overline{\alpha}^\nu : \nu < ot(A)\}$ on κ such that $A = \{\eta < \kappa : (\exists \nu < ot(A))(ht(\overline{\alpha}^\nu) = \xi_\eta)\}$.*

Theorem 2.3 follows then since, for any given regular $\kappa \geq \omega_1$ and any $A \subseteq \kappa$, the poset given by the above theorem forces that A is the set of $\eta < \kappa$ with the property that there is a coherent strongly type-guessing club–sequence with stationary domain included in κ and of height ξ_η (where $(\xi_\eta)_{\eta<\kappa}$ is as in the above theorem).[16] The strategy for building the poset $\mathcal{P}$ in Theorem 2.6 is quite as one would expect: $\mathcal{P}$ is the limit of a forcing iteration of length κ^+ built with supports of size less than κ. We may assume that $2^{<\kappa} = \kappa$ and $2^\kappa = \kappa^+$ hold in the ground model. In the first step of the iteration one adds, by initial segments, a set $\mathcal{A}$ of coherent club–sequences with the relevant heights. Then one kills, along the iteration, all possible obstacles to $\mathcal{A}$ being an asscs. In fact, it suffices to force, for all clubs $C \subseteq \kappa$ arising during the iteration, with the natural poset for shooting a club $D \subseteq \kappa$ such that $ot(\alpha_\delta \cap^* C) = ot(\alpha_\delta)$ for all $\delta \in D \cap dom(\overline{\alpha})$ (for all $\overline{\alpha} \in \mathcal{A}$). This is enough to ensure that (d) from Definition 2.1 holds for $\mathcal{A}$. The bulk of the proof is of course the verification that this poset $\mathcal{P}$ has the desired properties. Also, several of the technicalities involves in the coding — for example the restriction to perfect ordinals of countable cofinality or the consideration of the operation $\cap^*$ — are there precisely to make (d) hold.

The proof of Theorem 2.4 involves a certain principle which provides a simple way of encoding members of $H(\kappa^+)$ (for some κ) by ordinals in κ^+, quite in the spirit of the principle ψ_{AC} (for $H(\omega_2)$) from [W].

Definition 2.2 *([As4]) Let κ be an uncountable regular cardinal, let F be a function from κ into $\mathcal{P}(\kappa)$, and let $\mathcal{S} = \langle S_i : i < \kappa \rangle$ be a sequence of pairwise disjoint stationary subsets of κ. $\overline{\Psi}_{AC}^{F,\mathcal{S}}$ is then the statement that there is an enumeration $\langle B_\zeta : \kappa \leq \zeta < \kappa^+ \rangle$ of all subsets of κ such that, for each ζ, there is a club $E \subseteq [\zeta]^{<\kappa}$ with the property that for every $X \in E$ and every $i < \kappa$,*

---

[16] Noting that this property can indeed be expressed by a Σ_2 $\mathcal{L}$–formula over the structure $\langle H(\kappa^+), \in, NS_\kappa \rangle$.

(a) $ot(X) \in F(X \cap \kappa)$ *if* $X \cap \kappa \in S_i$ *and* $i \in B_\zeta$, *and*

(b) $ot(X) \notin F(X \cap \kappa)$ *if* $X \cap \kappa \in S_i$ *and* $i \notin B_\zeta$.

It is easy to see that there is a Σ_1 formula $\Theta(x, y, z)$ with the property that if κ, F and S are such that

(1) κ is an uncountable regular cardinal, $F : \kappa \longrightarrow \mathcal{P}(\kappa)$ is a function and S is a κ–sequence of pairwise disjoint stationary subsets of κ, and

(2) $\overline{\Psi}_{AC}^{F,S}$ holds,

then $\leq_{F,S} := \{ \langle x, y \rangle \in H(\kappa^+) \times H(\kappa^+) : \langle H(\kappa^+), \in \rangle \models \Theta(x, y, \langle F, S \rangle) \}$ is a well–order of $H(\kappa^+)$ of order type κ^+. Moreover, this formula can be taken so that ZFC proves that there are no κ, F and S as in (1) for which there are any distinct x, $y \in H(\kappa^+)$ with $H(\kappa^+) \models \Theta(x, y, \langle F, S \rangle) \wedge \Theta(y, x, \langle F, S \rangle)$.

The following result is proved in [As4].

Theorem 2.7 *Let* $\kappa \geq \omega_1$ *is a regular cardinal, and suppose* $2^{<\kappa} = \kappa$ *and* $2^\kappa = \kappa^+$ *hold. Then there is a* κ–distributive poset $\mathcal{P}$ *adding a function* $F : \kappa \longrightarrow [\kappa]^{<\kappa}$ *and a* κ–sequence S *of pairwise disjoint stationary subsets of* κ *such that* $\overline{\Psi}_{AC}^{F,S}$ *holds. Furthermore,* $\mathcal{P}$ *has the* κ^+–chain condition *and, if* $2^\mu = \mu^+$ *whenever* μ *is an infinite cardinal with* $\mu^+ < \kappa$, *then* $\mathcal{P}$ *preserves all stationary subsets of* κ.

When $\kappa > \omega_1$, Theorem 2.4 is proved by combining the forcing construction for proving the above result with the one for proving Theorem 2.6 with respect to a subset of κ coding the parameter (F, S) for which we force $\overline{\Psi}_{AC}^{F,S}$: We may assume that $2^{<\kappa} = \kappa$ and $2^\kappa = \kappa^+$ hold in the ground model. We start by adding, by forcing with initial segments, a function $F : \kappa \longrightarrow [\kappa]^{<\kappa}$ and a κ–sequence S of mutually disjoint stationary subsets of κ. Then we build a forcing iteration of length κ^+ with supports of size less than κ in which we simultaneously perform the tasks of

(1) forcing $\overline{\Psi}_{AC}^{F,S}$ (as in the proof of Theorem 2.7), and

(2) adding a suitable asscs on κ (as in the proof of Theorem 2.6) coding a fixed subset of κ that encodes (in some Σ_1 way) the pair (F, S).

The desired forcing is then the limit of this iteration. Now, the $\mathcal{L}$–formula $\Phi(x, y)$ witnessing Theorem 2.4 can be taken to express the following property $P_0(x, y)$: "There is a maximal set T of perfect ordinals

$\tau < \kappa$ of countable cofinality with the property that there is a coherent strongly type-guessing club–sequence with stationary domain and with height τ such that T encodes a pair (F, S) with F a function from κ into $\mathcal{P}(\kappa)$ and S a κ–sequence of mutually disjoint stationary subsets of κ, and $x \leq_{F,S} y$[17]".

$P_0(x, y)$ can also be expressed, in a slightly more convoluted way, by saying that there is a set T such that

(a) for every $\tau \in T$, τ is a perfect ordinal in κ of countable cofinality and there is a coherent club–sequence $\overline{\alpha}$ of height τ, with $dom(\overline{\alpha}) \subseteq \kappa$ and $dom(\overline{\alpha}) \notin NS_\kappa$, and such that for every club $C \subseteq \kappa$, $\{\delta \in dom(\overline{\alpha}) : ot(\alpha_\delta \cap^* C) \neq \tau\} \in NS_\kappa$,

(b) for every $\tau \in \kappa \cap cf(\omega)$, $\tau \in T$ or else for every coherent club–sequence $\overline{\alpha}$ of height τ, either $dom(\overline{\alpha})$ is not a stationary subset of κ or there is a club $C \subseteq \omega_1$ such that $\{\delta \in dom(\overline{\alpha}) : ot(\alpha_\delta \cap^* C) \neq \tau\} \notin NS_\kappa$, and such that

(c) T encodes a pair (F, S) with F a function from κ into $\mathcal{P}(\kappa)$ and S a κ–sequence of mutually disjoint stationary subsets of κ, and $x \leq_{F,S} y$.

Clearly, (a) and (b) are, respectively, a Σ_2 property of T over the structure $\langle H(\kappa^+), \in, NS_\kappa \rangle$ and a Π_2 property of T, also over $\langle H(\kappa^+), \in, NS_\kappa \rangle$. And, since $\leq_{F,S}$ is as described right after Definition 2.2, (c) is a Σ_1 property, over $\langle H(\kappa^+), \in \rangle$, about T, x and y. Thus, $P_0(x, y)$, being expressible over $\langle H(\kappa^+), \in, NS_\kappa \rangle$ as $(\exists T)[\Phi_a(T) \wedge \Phi_b(T) \wedge \Phi_c(T, x, y)]$, with $\Phi_a(u)$ a Σ_2 $\mathcal{L}$–formula, $\Phi_b(u)$ a Π_2 $\mathcal{L}$–formula, and $\Phi(u, v, w)$ a Σ_1 formula in the language of set theory, can be written as a Σ_3 $\mathcal{L}$–formula over $\langle H(\kappa^+), \in, NS_\kappa \rangle$.

As to $\Psi(x, y)$, it can be taken to express the property – let us call it $P_1(x, y)$ – that every maximal set T of ordinals $\tau < \kappa$ with the property stated in the description of $\Phi(x, y)$ encodes a pair (F, S) as before, and $x \leq_{F,S} y$. Let us call $Q(T)$ the property of T expressed by the conjunction of (a) and (b) in the above description. $P_1(x, y)$ can be written as

$$(\forall T)[Q(T) \rightarrow (T \text{ encodes a pair } (F, S) \text{ and } x \leq_{F,S} y)].$$

$Q(T)$ is a Σ_3 property of T over $\langle H(\kappa^+), \in, NS_\kappa \rangle$, and the expression to the right of the implication sign is a Σ_1 property of T, x and y. It follows then that $P_1(x, y)$ can be written as a Π_3 formula over $\langle H(\kappa^+), \in, NS_\kappa \rangle$.

---

[17]Where $\leq_{F,S}$ is as in the paragraph right after Definition 2.2.

The idea behind the proof when $\kappa = \omega_1$ is the same. The proof in this case combines the forcing iteration for Theorem 2.6 with a certain iteration of proper posets, due to Moore [Mo] (see the next section), which codes subsets of ω_1 by ordinals using some fixed parameter p. Now we force this coding while at the same time making p definable. In the extension, the inaccessible cardinal becomes ω_2.

One final word on the proofs of Theorems 2.6 and 2.4: They do not depend on any general forcing iteration lemmas. Instead, they rely on direct constructions depending quite closely on the actual definition of the iteration. By a forcing iteration lemma here I mean a statement typically asserting that if $\langle \mathcal{P}_\xi : \xi \leq \lambda \rangle$ is *any* forcing iteration built with some fixed kind of supports, based on a sequence $\langle \dot{\mathcal{Q}}_\xi : \xi < \lambda \rangle$ of names for posets,[18] and each $\dot{\mathcal{Q}}_\xi$ is forced, by $\mathcal{P}_\xi$, to have a certain property P, then $\mathcal{P}_\lambda$ also has property P. It is well–known[19] that fairly general iteration lemmas can be obtained for countable support iterations (or for some reasonable variation of this type of iterations). On the other hand, there are serious obstacles to proving similar general lemmas for iterations built using uncountable supports, which are precisely the kind of iterations one is typically faced with when forcing some statement about subsets of κ, for some $\kappa > \omega_2$, while at the same time preserving ω_1 and ω_2.

It turns out that Theorems 2.1 and 2.2 are optimal as stated from the point of view of the Levy hierarchy. More specifically, by appealing mainly to results of Woodin one can prove that, in the presence of sufficiently strong large cardinals,[20] 3 cannot be replaced by 2 in the statement of either Theorem 2.1 or Theorem 2.2. In fact, one cannot prove a version of either Theorem 2.1 or Theorem 2.2 in which Σ_3 (equivalently, Π_3) is replaced by Π_2. Theorems 2.8 and 2.9 present more precise formulations of this claim.

Theorem 2.8 *([As2]) Given any stationary and co-stationary $A \subseteq \omega_1$, if AD, the Axiom of Determinacy, holds in $L(\mathbb{R})$ $(AD^{L(\mathbb{R})})$ and if there is a Woodin cardinal below a measurable cardinal, then there is no pair $\Phi_0(x)$, $\Phi_1(x)$ of necessarily incompatible Π_2 formulas over $\langle H(\omega_2), \in, NS_{\omega_1}, r \rangle_{r \in \mathbb{R}}$ such that A is defined, over $\langle H(\omega_2), \in, NS_{\omega_1}, r \rangle_{r \in \mathbb{R}}$, by $\Phi_0(x)$ and $\omega_1 \backslash A$ is defined, over $\langle H(\omega_2), \in, NS_{\omega_1}, r \rangle_{r \in \mathbb{R}}$, by $\Phi_1(x)$.*

---

[18] That is, for all ξ with $\xi + 1 \leq \lambda$, $\mathcal{P}_{\xi+1}$ is the set of $\xi + 1$–sequences p such that $p \upharpoonright \xi \in \mathcal{P}_\xi$ and such that $p \upharpoonright \xi \Vdash_\xi p(\xi) \in \dot{\mathcal{Q}}_\xi$.

[19] See for example [Sh].

[20] For example a proper class of Woodin cardinals.

Theorem 2.9 *([As2]) Assume $AD^{L(\mathbb{R})}$ and suppose there is a Woodin cardinal with a measurable cardinal above. Then there is no necessarily antisymmetric Π_2 formula $\Phi(x,y)$ over the structure $\langle H(\omega_2), \in, NS_{\omega_1}, r \rangle_{r \in \mathbb{R}}$ such that $\Phi(x,y)$ defines, over $\langle H(\omega_2), \in, NS_{\omega_1}, r \rangle_{r \in \mathbb{R}}$, a well–order of $\mathbb{R}$.*

In Theorem 2.8 above, two formulas $\Phi_0(x)$ and $\Phi_1(x)$ in the language of the structure $\langle H(\omega_2), \in, NS_{\omega_1}, r \rangle$ – where r is a real number – are said to be *necessarily incompatible* if, for every generic extension M of $L(\mathbb{R})$ satisfying ZFC,

$$\langle H(\omega_2), \in, NS_{\omega_1}, r \rangle^M \models \neg(\Phi_0(x) \wedge \Phi_1(x))$$

for every $x \in H(\omega_2)^M$. Also, in Theorem 2.9, a formula $\Phi(x,y)$ in the language of the structure $\langle H(\omega_2), \in, NS_{\omega_1}, r \rangle$ – where, again, r is a real number – is *necessarily antisymmetric* in case for every generic extension M of $L(\mathbb{R})$ satisfying ZFC,

$$\langle H(\omega_2), \in, NS_{\omega_1}, r \rangle^M \models (\neg \exists x, y)(x \neq y \wedge \Phi(x,y) \wedge \Phi(y,x))$$

Theorems 2.8 and 2.9 are easily proved using the theory of $\mathbb{P}_{max}$ forcing by arguments very much contained in the proof of Observation 4.3 in Section 4.[21] From Theorem 2.8 it follows that if there is a proper class of Woodin cardinals and A is a stationary and co-stationary subset of ω_1, then there is no pair $(\Phi_0(x), \Phi_1(x))$ of ZFC–provably incompatible Π_2 $\mathcal{L}$–formulas for which there is a poset $\mathcal{P}$ such that $\mathcal{P}$ preserves the stationarity of both A and $\omega_1 \backslash A$ and such that $\mathcal{P}$ forces that A and $\omega_1 \backslash A$ are defined over $\langle H(\omega_2), \in, NS_{\omega_1} \rangle$ by, respectively, $\Phi_0(x)$ and $\Phi_1(x)$.[22] Likewise, Theorem 2.9 implies that, under the same large cardinal assumption, there is no poset forcing the existence of a well–order of $H(\omega_2)$ definable over the structure $\langle H(\omega_2), \in, NS_{\omega_1} \rangle$ by a ZFC–provably antisymmetric Π_2 $\mathcal{L}$–formula.

One may ask whether it is possible to drop the predicate NS_{ω_1} in the statement of either Theorem 2.1 or 2.2. Concerning this question, there is a version of the above theorems with $\langle H(\omega_2), \in \rangle$ replacing the more expressive $\langle H(\omega_2), \in, NS_{\omega_1} \rangle$. The coding techniques employed in the proof of these theorems are quite different from the ones used in the proofs of Theorems 2.6 and 2.4. These results use $ZFC +$ "There is an inaccessible limit of

---

[21]Using the fact that, under $AD^{L(\mathbb{R})}$, $\mathbb{P}_{max}$ is homogeneous and σ–closed and using, respectively, the fact that AD prohibits the existence of stationary and co-stationary subsets of ω_1, and the fact that AD prohibits the existence of well–orders of $\mathbb{R}$.

[22]Since, by a result of Woodin, $AD^{L(\mathbb{R})}$ follows from the existence of infinitely many Woodin cardinals with a measurable cardinal above them.

measurable cardinals" as base theory, rather than just ZFC (+ there is an inaccessible cardinal).

Recall that, if $\alpha < \omega_2$ is an ordinal and $\pi : \omega_1 \longrightarrow \alpha$ is a surjection, the function $g : \omega_1 \longrightarrow \omega_1$ defined by letting $g(\nu) = ot(\pi``\nu)$ for each ν is called a *canonical function for* α. This name is justified by the fact that any two functions thus obtained (for the same α) coincide on a club of ω_1. By a canonical function will be meant a canonical function for some ordinal below ω_2. Given $S \subseteq \omega_1$ and two functions $f, g : S \longrightarrow \omega_1$, we will say that g *dominates* f *on* S *mod. a club* (equivalently, f *is dominated by* g *on* S *mod. a club*) if there is a club $C \subseteq \omega_1$ such that $f(\nu) < g(\nu)$ for every $\nu \in S \cap C$.

The following notion is defined in [As5].

Definition 2.3 *([As5]) Given an ordinal* λ, $1 \le \lambda \le \omega_1$, $\langle S, \langle S_i : i < \lambda\rangle, f, \overline{\alpha}\rangle$ *is a simple decoding object if*

(a) $\{S_i : i < \lambda\} \cup \{S, \omega_1 \backslash (S \cup \bigcup_{i<\lambda} S_i)\}$ *is a collection of pairwise disjoint stationary subsets of* ω_1,

(b) *every function from* S *into* ω_1 *is dominated on* S *mod. a club by some canonical function,*

(c) f *is a function from* $\omega_1 \backslash (S \cup \bigcup_{i<\lambda} S_i)$ *into* ω_1 *dominating every canonical function on* $\omega_1 \backslash (S \cup \bigcup_{i<\lambda} S_i)$ *mod. a club,*

(d) $\overline{\alpha}$ *is a ladder system defined on* $\bigcup_{i<\lambda} S_i$ *which is strongly guessing and maximal for* ω_1, *and* $\alpha_\delta \cap \bigcup_{i<\lambda} S_i = \emptyset$ *for every* $\delta \in \bigcup_{i<\lambda} S_i$, *and*

(e) *there is a sequence* $\langle r_i : i < \lambda\rangle$ *such that for every* $i < \lambda$ *and every* $\delta \in S_i$, r_i *is the set of* $k < \omega$ *for which there are infinitely many* $n < \omega$ *such that* $\{\alpha_\delta(n), \alpha_\delta(n+k+1)\} \subseteq S$ *and* $\{\alpha_\delta(n+j) : 1 \le j \le k\} \cap S = \emptyset$.

If $\langle S, \langle S_i : i < \lambda\rangle, f, \overline{\alpha}\rangle$ *is a simple decoding object, then we let* $code(S, \langle S_i : i < \lambda\rangle, f, \overline{\alpha}) = \{r_i : i < \lambda\}$, *where* $\langle r_i : i < \lambda\rangle$ *witnesses (e)* *for* $\langle S, \langle S_i : i < \lambda\rangle, f, \overline{\alpha}\rangle$. *We will say that* $\langle S, \langle S_i : i < \lambda\rangle, f, \overline{\alpha}\rangle$ *encodes* $\{r_i : i < \lambda\}$.

Lemma 2.10 shows that simple decoding objects are unique in a quite strong sense.

Lemma 2.10 *([As5]) Suppose* $\langle S^0, \langle S^0_i : i < \lambda_0 \rangle, f_0, \overline{\alpha}^0 \rangle$ *and* $\langle S^1, \langle S^1_i : i < \lambda_1 \rangle, f_1, \overline{\alpha}^1 \rangle$ *are simple decoding objects. Then there are clubs* $D \subseteq C$ *of* ω_1 *such that*

(i) $S^0 \cap C = S^1 \cap C$ *and* $\bigcup_{i < \lambda_0} S^0_i \cap C = \bigcup_{i < \lambda_1} S^1_i \cap C$,

(ii) *for every* $\delta \in D$, *both* $\alpha^0_\delta \Delta \alpha^1_\delta$ *and* $\alpha^0_\delta \backslash C$ *are finite.*

In particular, $code(S^0, \langle S^0_i : i < \lambda_0 \rangle, f_0, \overline{\alpha}^0) = code(S^1, \langle S^1_i : i < \lambda_1 \rangle, f_1, \overline{\alpha}^1)$.

The following theorem is proved in [As5].

Theorem 2.11 *([As5']) Suppose* κ *is an inaccessible limit of measurable cardinals. Let* λ *be an ordinal,* $1 \leq \lambda \leq \omega_1$, *and let* $\langle r_i : i < \lambda \rangle$ *be a sequence of sets of integers. There is a semiproper poset* $\mathcal{P} \subseteq V_\kappa$ *forcing that there is a simple decoding object* $\langle S, \langle S_i : i < \lambda \rangle, f, \overline{\alpha} \rangle$ *such that* code $(S, \langle S_i : i < \lambda \rangle, f, \overline{\alpha}) = \{r_i : i < \lambda\}$.

The first of the following two results is a corollary of Theorem 2.11, and the second follows from combining the forcing construction for proving Theorem 2.11 with, for example, the one for proving Theorem 2.7.[23]

Theorem 2.12 *([As5]) There are* Σ_3 *formulas* $\Phi_0(x)$ *and* $\Phi_1(x)$ *in the language of set theory such that*

(1) $(\Phi_0(x), \Phi_1(x))$ *are ZFC–provably incompatible formulas over the structure* $\langle H(\omega_2), \in \rangle$, *and*

(2) *given any* $A \subseteq \omega_1$, *if there is an inaccessible limit* κ *of measurable cardinals, then there is a semiproper poset* $\mathcal{P} \subseteq V_\kappa$ *forcing that* A *and* $\omega_1 \backslash A$ *are defined over* $\langle H(\omega_2), \in \rangle$ *by, respectively,* $\Phi_0(x)$ *and* $\Phi_1(x)$.

Theorem 2.13 *([As5]) There is a* Σ_3 *formula* $\Phi(x, y)$ *in the language of set theory such that*

(1) $\Phi(x, y)$ *is a ZFC–provably antisymmetric formula over* $\langle H(\omega_2), \in \rangle$, *and*

(2) *if there is an inaccessible limit* κ *of measurable cardinals, then there is a semiproper poset* $\mathcal{P} \subseteq V_\kappa$ *forcing that* $\Phi(x, y)$ *defines, over* $\langle H(\omega_2), \in \rangle$, *a well–order of* $H(\omega_2)$ *of order type* ω_2.

---

[23]Very much as in the proof of Theorem 2.4.

For Theorem 2.12, each formula $\Phi_\epsilon(x)$ (for $\epsilon \in \{0,1\}$) will be a Σ_3 formulas expressing the property $P_\epsilon(x)$, where $P_0(x)$ and $P_1(x)$ are defined by, respectively, "x is a countable ordinal and there is a real r encoding x (in some Σ_1 way), together with a simple decoding object encoding a set of reals to which r belongs" and "x is a countable ordinal and there is a simple decoding object encoding a set of reals X such that $r \notin X$ whenever r is a real encoding x (in the same way as before)".

It is easy to see that both properties can be expressed by Σ_3 formulas over $\langle H(\omega_2), \in \rangle$: For $P_0(x)$, since "$x \in \omega_1$" and "r encodes x" are Σ_1 properties of the relevant objects, the verification will be finished once we see that

$$(\exists S, (S_i)_{i<\lambda}, f, \overline{\alpha})\, Q(S, (S_i)_{i<\lambda}, f, \overline{\alpha}, r)$$

can be written as a Σ_3 sentence over $\langle H(\omega_2), \in \rangle$ with r as parameter, where $Q(S, (S_i)_{i<\lambda}, f, \overline{\alpha}, r)$ expresses that $(S, (S_i)_{i<\lambda}, f, \overline{\alpha})$ is a simple decoding object and that $r \in code(S, (S_i)_{i<\lambda}, f, \overline{\alpha})$. But $Q(S, (S_i)_{i<\lambda}, f, \overline{\alpha}, r)$ can be expressed by saying that (a)–(d) from Definition 2.3 hold for the relevant parameters, and that $r \in code(S, (S_i)_{i<\lambda}, f, \overline{\alpha})$. Since ω_1 is Σ_2 definable over $\langle H(\omega_2), \in \rangle$, (a) from Definition 2.3 is a Σ_2 property of S and $(S_i)_{i<\lambda}$ (over $\langle H(\omega_2), \in \rangle$). (b) from Definition 2.3 is clearly a Π_2 property of S, and (c) is also a Π_2 property (of f, S and $(S_i)_{i<\lambda}$). (d) is expressed by saying that $\overline{\alpha}$ is a ladder system with $dom(\overline{\alpha}) = \bigcup_{i<\lambda} S_i$, that for every club $C \subseteq \omega_1$ there is a club $D \subseteq \omega_1$ such that $\alpha_\delta \backslash C$ is bounded in δ for every $\delta \in D \cap dom(\overline{\alpha})$, and that for every ladder system $\overline{\beta}$ disjoint from $\overline{\alpha}$ there is a club $C \subseteq \omega_1$ such that $\beta_\delta \cap C$ is finite for every $\delta \in C \cap dom(\overline{\beta})$; hence, it can be written as a Π_2 formula over $\langle H(\omega_2), \in \rangle$. Finally, "$r \in code(S, (S_i)_{i<\lambda}, f, \overline{\alpha})$" can be expressed by saying that there is some $i < \lambda$ such that $r = \{k < \omega : (\exists^\infty n < \omega)(\{\alpha_\delta(n), \alpha_\delta(n+k+1)\} \subseteq S \wedge \{\alpha_\delta(n+j) : 1 \leq j \leq k\} \cap S = \emptyset)\}$ for every $\delta \in S_i$, and so it is a Σ_0 property of r, S and $(S_i)_{i<\lambda}$. It follows that $P_0(x)$ can be written as a Σ_3 formula over $\langle H(\omega_2), \in \rangle$. The verification for $P_1(x)$ is along the same lines.

The formula $\Phi(x,y)$ witnessing Theorem 2.13 can be taken to say that there is a simple decoding object encoding a set of reals X such that X encodes (in some simple standard way) a pair $(F, \mathcal{S})$ as in Definition 2.2 (for $\kappa = \omega_2$) and $x \leq_{F,\mathcal{S}} y$, for the relation $\leq_{F,\mathcal{S}}$ described after Definition 2.2. It is easy to see, by an analysis as before, that there is indeed a Σ_3 formula expressing the above property of x and y over $\langle H(\omega_2), \in \rangle$.

Question 2.1 *Is it possible to prove versions of either Theorem 2.12 or 2.13 with Π_3 replacing Σ_3?*

3. Results mentioning strong forcing axioms

The forcing constructions presented in the previous section are flexible enough to accommodate posets with the countable chain condition. Thus, all models built there can be taken to be models of MA_{ω_1}. However, it is not possible to modify those constructions so as to produce models of, for example, $BPFA$.[24] In fact, the techniques presented there for coding a fixed subset of ω_1 are incompatible with $BPFA$. The reason for this is that $BPFA$ implies that every club–sequence is avoidable.[25]

PFA^{++} is the following strong form of PFA:

Given any proper poset $\mathcal{P}$, any sequence $\langle D_i : i < \omega_1 \rangle$ of dense subsets of ω_1 and any sequence $\langle \tau_i : i < \omega_1 \rangle$ of $\mathcal{P}$–names for stationary subsets of ω_1 there is a filter $G \subseteq \mathcal{P}$ such that, for each $i < \omega_1$, $G \cap D_i \neq \emptyset$ and $\{\nu < \omega_1 : (\exists p \in G)(p \Vdash_{\mathcal{P}} \nu \in \tau_i)\}$ is a stationary subset of ω_1.

The first main result in this section is the following.

Theorem 3.1 *([As3]) Suppose κ is a supercompact cardinal and A is a subset of ω_1. Then there is a semiproper poset $\mathcal{P} \subseteq V_\kappa$ such that*

(1) $\mathcal{P}$ forces PFA^{++},

(2) $\mathcal{P}$ forces that A is definable, over the structure $\langle H(\omega_2), \in, NS_{\omega_1} \rangle$, by a Σ_5 formula without parameters, and

(3) $\mathcal{P}$ forces the existence of a well–order of $H(\omega_2)$ definable, over the structure $\langle H(\omega_2), \in, NS_{\omega_1} \rangle$, by a Σ_5 formula without parameters.

This time the proofs involve the manipulation of certain guessing properties of functions $F : S \longrightarrow \mathcal{P}(\omega_1)$[26] with respect to canonical functions.

---

[24] $BPFA$ is the assertion that $\langle H(\omega_2), \in \rangle$ is a Σ_1–elementary substructure of the structure $\langle H(\omega_2), \in \rangle$ as computed in any forcing extension via a proper poset. $BPFA$ is a trivial consequence of the Proper Forcing axiom (PFA).

[25] One can easily verify that the standard poset for introducing, by initial conditions, a club avoiding a fixed club–sequence defined on a subset of ω_1 and whose height exists is always proper.

[26] Where $S \subseteq \omega_1$ and where, for every $\nu \in S$, $ot(F(\nu))$ is in some prescribed interval of countable ordinals.

Definition 3.1 *([As3]) Let S be a stationary subset of ω_1. Given $I \subseteq \omega_1$, S has guessing density I if for every stationary $S^* \subseteq S$,*

(a) *there is a function $F : S^* \longrightarrow \mathcal{P}(\omega_1)$ such that $ot(F(\nu)) \in I$ for all $\nu \in S^*$ and such that $\{\nu \in S^* : g(\nu) \in F(\nu)\}$ is stationary for every $\alpha < \omega_2$ and every canonical function g for α, and*

(b) *given any function $F' : S^* \longrightarrow \mathcal{P}(\omega_1)$, if $ot(F'(\nu)) < min(I)$ for all $\nu \in S^*$, then there is an ordinal $\alpha < \omega_2$ such that $\{\nu \in S^* : g(\nu) \in F'(\nu)\}$ is nonstationary for every canonical function g for α.*

Note that every stationary $S \subseteq \omega_1$ has density ω_1 and that there is no such thing as the unique guessing density of S: if S has guessing density I_0 and $I_1 \subseteq \omega_1$ is such that $min(I_1) \leq min(I_0)$ and $sup(I_0) \leq sup(I_1)$, then S also has guessing density I_1. Also, it is easy to see that for every stationary $S \subseteq \omega_1$, the assumption that $\Diamond(S^*)$[27] holds for every stationary $S^* \subseteq S$ implies that S has guessing density $\{1\}$. Finally, BMM[28] implies that no stationary subset of ω_1 has guessing density bounded in ω_1.

The following theorem is proved in [As3].

Theorem 3.2 *([As3]) Suppose κ is an inaccessible cardinal which is a limit of measurable cardinals. Let $\langle S_i : i < \omega_1 \rangle$ be a sequence of pairwise disjoint stationary subsets of ω_1 and let $\langle \alpha_i : i < \omega_1 \rangle$ be a sequence of nonzero countable ordinals.*

Then there is a semiproper poset $\mathcal{P} \subseteq V_\kappa$ forcing, for every $i < \omega_1$, that S_i has guessing density the interval $[\alpha_i, \omega^{\alpha_i \cdot \omega})$[29] if $\alpha_i > 1$ and that S_i has guessing density $\{1\}$ if $\alpha_i = 1$.

The proof of Theorem 3.2 involves an analysis of iterations of models of set theory relative to sequences of (possibly different) measurable cardinals. It uses a generalization of a theorem of Kunen ([Ku]) saying that for every ordinal ϵ there are only finitely many measurable cardinals γ for which there is a normal measure U on γ such that ϵ is not a fixed point by the elementary embedding, of the universe, derived from U.

---

[27] Namely the statement that there is a sequence $\langle X_\alpha : \alpha \in S^* \rangle$ with $X_\alpha \subseteq \alpha$ for all α and $\{\alpha \in S^* : X \cap \alpha = X_\alpha\}$ stationary for each $X \subseteq \omega_1$.

[28] Namely the statement that the structure $\langle H(\omega_2), \in \rangle$ is a Σ_1–elementary substructure of $\langle H(\omega_2), \in \rangle^{V^{\mathcal{P}}}$ for every partial order $\mathcal{P}$ preserving stationary subsets of ω_1.

[29] α_i^ω denotes ordinal exponentiation.

The forcing iteration for proving Theorem 3.2 consists of semiproper posets. In fact, the posets $\mathcal{Q}$ used there are quite close to being proper, in the sense that, although it may not be true that, for an arbitrary structure N containing $\mathcal{Q}$ and an arbitrary $p \in \mathcal{Q} \cap N$, there is an $(N, \mathcal{Q})$–generic condition extending p, it is nevertheless true that for every N and p as above there is a name $\tilde{N}$ for a structure including N and there is a condition extending p which is $(\tilde{N}, \mathcal{Q})$–generic (in a natural way) and, moreover, there is sufficient control in V on what the order type of the interpretation of $\tilde{N} \cap \overline{\kappa}$ (for the relevant $\overline{\kappa}$) is going to be. This ensures that things work as desired.[30] This type of argument also shows that our forcing iteration is robust enough to accommodate arbitrary proper posets (of size less than κ), even on a suitable club with complement unbounded in κ.

By a result of Moore ([Mo]), $BPFA$ implies the existence, given any ladder system $\overline{e}$ on ω_1 and any sequence $(U_i)_{i<\omega_1}$ of pairwise disjoint stationary subsets of ω_1, of a well–order of $H(\omega_2)$ definable over $\langle H(\omega_2), \in \rangle$ by a Σ_2 formula with $p = (\overline{e}, (U_i)_{i<\omega_1})$ as parameter. The proof of Theorem 3.1 follows from the above considerations: Suppose we start from a supercompact cardinal instead of just an inaccessible limit of measurable cardinals. Suppose $\langle S_i : i < \omega_1 \rangle$ and $\langle \alpha_i : i < \omega_1 \rangle$ (in the statement of Theorem 3.2) are such that $S_i \cap (i + 1) = \emptyset$ for all $i > 0$ and such that $A^* = \{\alpha_i : i < \omega_1\}$ is a sparse enough set of countable ordinals coding in some simple (say, Σ_1) way, both a prescribed $A \subseteq \omega_1$ and a parameter $(\overline{e}, (U_i)_{i<\omega_1})$ as in Moore's theorem. We build a forcing iteration in which we perform the usual Baumgartner construction for PFA^{++} on a suitable club C of κ, and in which we force as in Theorem 3.2 on the complement of C. In the end we obtain a model in which PFA^{++} holds and in which A^* is defined as the set of $\alpha < \omega_1$ for which there is a stationary subset of ω_1 with guessing density equal to the interval $[\alpha, \omega^{\alpha \cdot \omega})$.[31] It follows easily that in this model A^* – and therefore also A – is definable over $\langle H(\omega_2), \in \rangle$ by a Σ_5 formula without parameters. Since A^* also codes a parameter as in Moore's theorem, it follows by that theorem that in the resulting model there is a well–order of $H(\omega_2)$ definable over $\langle H(\omega_2), \in \rangle$ also by a Σ_5 formula without parameters. This proves Theorem 3.1.

---

[30]It is worth mentioning that, unlike the forcing construction for proving Theorem 2.11 – which is a revised countable support iteration –, the construction for Theorem 3.2 works with a countable support iteration.

[31]Note that, if ξ and ξ' are distinct countable ordinals and a stationary subset of ω_1 has guessing density $[\omega^{\omega^{1+\xi}}, \omega^{\omega^{1+\xi+1}})$, then it cannot have guessing density $[\omega^{\omega^{1+\xi'}}, \omega^{\omega^{1+\xi'+1}})$.

The second main result in this section is due to P. Larson and involves a strong form of Martin's Maximum (MM). Recall that MM is the following provably maximal forcing axiom for collections of $\aleph_1$–many dense:[32]

Suppose $\mathcal{P}$ is a poset such that forcing with $\mathcal{P}$ preserves the stationarity of all stationary subsets of ω_1 and suppose $\langle D_i : i < \omega_1 \rangle$ is a sequence of dense subsets of $\mathcal{P}$. Then there is a filter $G \subseteq \mathcal{P}$ such that $G \cap D_i \neq \emptyset$ for every $i < \omega_1$.

MM is a maximal forcing axiom in the sense that, on the one hand, if $\mathcal{P}$ is a poset forcing that some stationary subset of ω_1 from the ground model is no longer stationary, then one can easily find a collection $\langle D_i : i < \omega_1 \rangle$ of dense subsets of ω_1 such that $D_i \cap G = \emptyset$ for some i whenever $G \subseteq \mathcal{P}$ is a filter; whereas, on the other hand, MM can be forced over any model with a supercompact cardinal. In an older version of this paper I was asking whether the hypothesis that there is a supercompact cardinal (or some other reasonable large cardinal assumption) implies that there is a partial order forcing Martin's Maximum, together with the existence of a well–order of $H(\omega_2)$ definable, over $\langle H(\omega_2), \in \rangle$, by a formula without parameters (or even by a formula with a real number as parameter). Regarding this question, P. Larson has recently proved the following result ([L2]).

Theorem 3.3 *(Larson) Suppose κ is a supercompact limit of supercompact cardinals. Then there is semiproper poset $\mathcal{P} \subseteq V_\kappa$ such that*

(1) *$\mathcal{P}$ forces $MM^{+\omega}$,[33] and*

(2) *$\mathcal{P}$ forces that there is well–order of $H(\omega_2)$ definable, over $\langle H(\omega_2), \in \rangle$, by a formula without parameters.*

4. Open questions and some consequences of large cardinal axioms

In the model of Theorem 3.3, MM^{++} fails necessarily. As far as I know, the following questions remain open.

---

[32] Defined and proved consistent in [F-M-Sh].

[33] $MM^{+\omega}$ is the strengthening of MM saying that for every poset $\mathcal{Q}$ preserving stationary subsets of ω_1, every sequence $\langle D_i : i < \omega_1 \rangle$ of dense subsets of $\mathcal{Q}$ and every sequence $\langle \tau_n : n < \omega \rangle$ of $\mathcal{Q}$–names for stationary subsets of ω_1 there is a filter $G \subseteq \mathcal{Q}$ such that $G \cap D_i \neq \emptyset$ for every $i < \omega_1$ and such that $\{\alpha < \omega_1 : (\exists p \in G)(p \Vdash_\mathcal{Q} \alpha \in \tau_n)\}$ is stationary for every $n < \omega$. MM^{++} is the strengthening of $MM^{+\omega}$ incorporating sequences of length ω_1 – instead just of length ω – of names for stationary subsets of ω_1.

Questions 4.1 *Assume some reasonable large cardinal hypothesis. Is it possible to force in such a way that MM^{++} holds in the extension, together with the existence of a well–order of $H(\omega_2)$ definable, over $\langle H(\omega_2), \in \rangle$, by a formula without parameters (or even by a formula with a real number as parameter)?*

Does MM^{++} imply that there is a well–order of $H(\omega_2)$ definable, over $\langle H(\omega_2), \in \rangle$, by a formula with at most a real number as parameter?

Let $\mathcal{D}$ denote the class of all subsets of $H(\omega_2)$ which are definable, over $\langle H(\omega_2), \in \rangle$, by a formula with at most a real number as parameter. I will finish this paper with a remark motivated by a failed attempt (so far) to construct a model of MM^{++} in which there is a well–order of $H(\omega_2)$ belonging to $\mathcal{D}$ (thus providing an answer to the first question above).

One first observation – due to Larson – I want to mention is that, in the presence of MM, the existence of a subset $A \subseteq \omega_2 \cap cf(\omega)$ such that A belongs to $\mathcal{D}$ and such that both A and $(\omega_2 \cap cf(\omega)) \backslash A$ are stationary suffices to prove the existence of a stationary and co-stationary subset of ω_1 belonging to $\mathcal{D}$. In fact,

Fact 4.1 *(Larson) Suppose NS_{ω_1} is saturated and suppose $(\mathcal{P}(\omega_1))^{\sharp}$ exists. Suppose as well that for every stationary subset A of $\omega_2 \cap cf(\omega)$ there is some $\delta < \omega_2$ of cofinality ω_1 such that both A and $\omega_2 \backslash A$ reflect at δ. If there is some $A \subseteq \omega_2 \cap cf(\omega)$ such that $A \in \mathcal{D}$ and such that both A and $(\omega_2 \cap cf(\omega)) \backslash A$ are stationary subsets of ω_2, then there is a stationary and co-stationary $S \subseteq \omega_1$ such that $S \in \mathcal{D}$.*

Proof: Let r_0 be a real such that A is definable over $\langle H(\omega_2), \in \rangle$ by a formula with only r_0 as parameter. Let $\delta < \omega_2$ be an ordinal of uncountable cofinality such that both $A \cap \delta$ and $\delta \backslash A$ are stationary subsets of δ. From the saturation of NS_{ω_1} and the existence of $(\mathcal{P}(\omega_1))^{\sharp}$ it follows, by [W], Theorem 3.17, that the second uniform indiscernible[34] (u_2) is ω_2. In particular, we may fix a real r such that $|\delta|^{L[r]} = \omega_1^V$ and such that δ is definable, in $L[r]$, from ω_1^V. Notice that $cf(\delta)^{L[r]}$ is then exactly ω_1^V. Let C be the $<_{L[r]}$–least club of δ of order type ω_1^V and let $(\alpha_\nu)_{\nu < \omega_1^V}$ be its strictly increasing enumeration. Then, since $A \cap \delta$ and $\delta \backslash A$ are both stationary, $S := \{ \nu < \omega_1 : \alpha_\nu \in A \}$ is a stationary and co-stationary subset of ω_1 which is definable, over $\langle H(\omega_2), \in \rangle$, by a formula with r_0, δ

---

[34] An ordinal is a uniform indiscernible if it is a Silver indiscernible for $L[r]$ for every real r.

and r as parameters, and therefore also by a formula with r_0 and r as parameters. $\square$

The hypotheses of Fact 4.1 follow from MM: The saturation of NS_{ω_1} follows from [F-M-Sh], the existence of the sharp of every set follows in fact from BMM by a result of Schindler ([S]), and the simultaneous reflection of pairs of stationary subsets of $\omega_2 \cap cf(\omega)$ follows from [F-M-Sh].

A second observation is that, again in the presence of MM, the existence of a stationary and co-stationary subset belonging to $\mathcal{D}$ suffices to prove that in fact every subset of ω_1 belongs to $\mathcal{D}$:

Fact 4.2 *([As1], Observation 1.1) Suppose that BMM holds and NS_{ω_1} is saturated. Then, given any stationary and co-stationary $S \subseteq \omega_1$ and any $A \subseteq \omega_1$ there is a real r such that $A \in L[r, S]$. In particular, if $S \in \mathcal{D}$, then A is also in $\mathcal{D}$.*

Finally, MM implies the existence, given a sequence $\langle S_\alpha : \alpha < \omega_1 \rangle$ of pairwise disjoint stationary subsets of ω_1, of a well–order of $H(\omega_2)$ which is definable over $\langle H(\omega_2), \in \rangle$ by a formula with $\langle S_\alpha : \alpha < \omega_1 \rangle$ as parameter (for example by [W], Theorem 5.14 and Lemma 5.13).

A consequence of these observations is that, under MM, the existence of a well–order of $H(\omega_2)$ belonging to $\mathcal{D}$ follows from – and in fact is equivalent to – the existence of a stationary $A \subseteq \omega_2 \cap cf(\omega)$ belonging to $\mathcal{D}$ and such that $(\omega_2 \cap cf(\omega)) \backslash A$ is also stationary. Now, suppose we are in a context in which MM holds and we want to argue that there is a well–order of $H(\omega_2)$ in $\mathcal{D}$. One possibility would be to prove that there is some formula $\Phi(x)$, perhaps with a real parameter, such that $A_{\Phi(x)}$ and $(\omega_2 \cap cf(\omega)) \backslash A_{\Phi(x)}$ are both stationary for $A_{\Phi(x)} := \{\alpha \in \omega_2 \cap cf(\omega) : \langle H(\omega_2), \in \rangle \models \Phi(\alpha)\}$.

This strategy asks for the stationarity of sets of ordinals of countable cofinality. On the other hand, Namba forcing (Nm) preserves stationary subsets of ω_1 and forces $cf(\omega_2^V) = \omega$. It certainly forces the following property about $x = \omega_2^V$, for any collection Δ of axioms of set theory of bounded Levy complexity:[35] "$cf(x) = \omega$, there is a transitive set $M \models \Delta$ such that $\omega_1^M = \omega_1$ and such that x is ω_2^M, and there is a Nm^M–generic filter over M". By general arguments involving forcing axioms it easily follows then that, by MM, there are stationarily many $\alpha < \omega_2$ of countable cofinality such that this statement holds with $x = \alpha$. Thus, it seemed to be a good idea to choose as $\Phi(x)$ a formula expressing something like the above property. In other words, it seemed a good idea to try to argue that the

---

[35]Or even for $\Delta = ZFC$ if, for example, some rank–initial segment of the universe satisfies ZFC.

complement, relative to $\omega_2 \cap cf(\omega)$, of the set defined by the above property
– or some other related property – must be stationary. For example, one
could have expected to start with a model with a supercompact cardinal
κ, perform the usual MM^{++}–forcing construction by a forcing with the
κ–chain condition, as in [F-M-Sh], and argue that, in the extension, the
above property fails for (say) stationarily many $\alpha < \omega_2 = \kappa$ of countable
cofinality in the ground model. However, the observations I am about to
present show that this hope is sterile.

It is well–known that certain properties for definable sets of reals follow
from the existence of large cardinals. The first observation I want to make
extends this type of results to the level of $H(\omega_2)$. In other words, it shows
that the existence, in the universe, of certain objects of size $\aleph_1$ *follows*
outright from large cardinal axioms.

Observation 4.3 *Suppose there are cardinals $\delta < \kappa$ such that δ is a limit of
infinitely many Woodin cardinals and κ is measurable. Let $\Phi(x)$ be either a
Σ_1 formula or a Π_1 formula (in the language $\mathcal{L}^*$ for the structure $\langle H(\omega_2),
\in, NS_{\omega_1}, r \rangle_{r \in \mathbb{R}}$), let $\lambda \in \{\omega, \omega_1\}$ and let $\Phi^*(x)$ be a Σ_1 $\mathcal{L}^*$–formula or a
Π_1 $\mathcal{L}^*$–formula expressing $\Phi(x) \wedge cf(x) = \lambda$. Suppose there is a poset
$\mathcal{P} \in V_\delta$ forcing that there are stationarily many $\alpha < u_2$ such that $\langle H(\omega_2), \in,
NS_{\omega_1}, r \rangle_{r \in \mathbb{R}} \models \Phi^*(\alpha)$. Then, in V, there is a club $C \subseteq u_2$ in $L(\mathbb{R})$ such
that $\langle H(\omega_2), \in, NS_{\omega_1}, r \rangle_{r \in \mathbb{R}} \models \Phi^*(\alpha)$ holds, in V, for every $\alpha \in C$.*

Proof: Let $\Phi(x)$, λ and $\mathcal{P}$ provide a counterexample. Suppose $\Phi^*(x)$
is a Σ_1 formula (the argument when $\Phi^*(x)$ is Π_1 is the same). Let A be,
in $V^{\mathcal{P}}$, the set of $\alpha < u_2$ such that $\langle H(\omega_2), \in, NS_{\omega_1}, r \rangle_{r \in \mathbb{R}} \models \Phi^*(\alpha)$. In
$V^{\mathcal{P}}$, there are infinitely many Woodin cardinals with a measurable above
them. In particular, $AD^{L(\mathbb{R})}$ holds and there is a Woodin cardinal with a
measurable above. Hence, by the proof of [W], Theorem 4.65, in $V^{\mathcal{P}}$, $\mathbb{P}_{max}$
forces $\langle H(\omega_2), \in, NS_{\omega_1}, r \rangle_{r \in \mathbb{R}} \models \Phi^*(\alpha)$ over $L(\mathbb{R})^{V^{\mathcal{P}}}$ whenever $\alpha \in A$.[36]
In particular, by the definability of the forcing relation over ZF–models,
$A \in L(\mathbb{R})^{V^{\mathcal{P}}}$. If $u_2 \backslash A$ were stationary in $V^{\mathcal{P}}$, then A and $(\omega_2 \cap cf(\lambda)) \backslash A$
would be stationary subsets of ω_2 $(= u_2)$ in $L(\mathbb{R})^{V^{\mathcal{P}}}$.[37] But this contradicts
a result, of Martin and Paris, saying that $\{C \cap cf(\lambda) : C$ a club of $\omega_2\}$ generates an ultrafilter of $\omega_2 \cap cf(\lambda)$ under AD (see [K], p. 395). Hence, in
$L(\mathbb{R})^{V^{\mathcal{P}}}$ it holds that there is a club $C \subseteq u_2$ of ordinals of cofinality λ such

---

[36]Since every $\alpha < u_2$ is Σ_1 definable from ω_1 together with a real.

[37]As $L(\mathbb{R})^{V^{\mathcal{P}}}$ computes u_2 and cofinalities below u_2 correctly and as $AD \models u_2 = \omega_2$.

that $\mathbb{P}_{max}$ forces $\langle H(\omega_2), \in, NS_{\omega_1}, r \rangle_{r \in \mathbb{R}} \models \Phi(\alpha)$ for every $\alpha \in C$ of cofinality λ. By another result of Woodin referred to already in Section 2, forcing with $\mathcal{P}$ does not change the theory of $L(\mathbb{R})$ with real numbers as parameters. Hence, in $L(\mathbb{R})^V$ there is also a club $C \subseteq u_2$ with the above property. Take any $\alpha \in C$ of cofinality λ and suppose, towards a final contradiction, that $\langle H(\omega_2), \in, NS_{\omega_1}, r \rangle_{r \in \mathbb{R}} \models \neg\Phi(\alpha)$ holds in V. Again by the proof of [W], Theorem 4.65, this time applied in V, $\langle H(\omega_2), \in, NS_{\omega_1}, r \rangle_{r \in \mathbb{R}} \models \neg\Phi(\alpha)$ also holds in $L(\mathbb{R})^{\mathbb{P}_{max}}$. Contradiction. $\square$

The following consequence of Observation 4.3 is relevant to the possible scenario presented in this section for building a model of MM^{++} with a well–order of $H(\omega_2)$ belonging to $\mathcal{D}$.

Corollary 4.4 *Suppose there is a supercompact cardinal. Then there is a club $C \subseteq u_2$, $C \in L(\mathbb{R})$, with the property, in V, that for every $\lambda \in \{\omega, \omega_1\}$ and every formula $\varphi(x)$, if ZFC proves that $\varphi(x)$ defines a poset – call it $\mathcal{P}_{\varphi(x)}$ – preserving stationary subsets of ω_1 and forcing $cf(\omega_2^V) = \lambda$, then for every $\alpha \in C$ of cofinality λ there is a transitive model M of ZFC computing stationary subsets of ω_1 correctly and with $\omega_2^M = \alpha$ and there is a $(\mathcal{P}_{\varphi(x)})^M$–generic filter over M.*

Proof: Fix $\lambda \in \{\omega, \omega_1\}$ and a formula $\varphi(x)$ a above. Since $cf(u_2) \geq \omega_1$, it suffices to show that there is a club $C \subseteq u_2$ such that every $\alpha \in C \cap cf(\lambda)$ has the following property (in V):

$P(\alpha)$: There is a transitive model M of ZFC computing stationary subsets of ω_1 correctly and such that $\omega_2^M = \alpha$ and there is a $(\mathcal{P}_{\varphi(x)})^M$– generic filter over M.

By the construction in [F-M-Sh] there is an iteration $\langle \mathcal{P}_\alpha : \alpha \leq \kappa \rangle$, $\mathcal{P}_\kappa \subseteq V_\kappa$, such that each $\mathcal{P}_\kappa / \dot{G}_\alpha$ is semiproper in $V^{\mathcal{P}_\alpha}$, and forcing both $u_2 = \omega_2$ and that there are stationarily many $\alpha < \omega_2 = \overline{\kappa}$ such that $\alpha = \omega_2^{V^{\mathcal{P}_\alpha}}$ and such that there is a $(\mathcal{P}_{\varphi(x)})^{V^{\mathcal{P}_\alpha}}$–generic filter over $V^{\mathcal{P}_\alpha}$. Hence, by reflection, we may fix $\overline{\kappa} < \kappa$ and a semiproper poset $\mathcal{P} \subseteq V_{\overline{\kappa}}$ forcing $u_2 = \omega_2$ and, since κ is inaccessible in $V^{\mathcal{P}_{\overline{\kappa}}}$, forcing that there are stationarily many $\alpha < \omega_2 = \overline{\kappa}$ such that $P(\alpha)$. Now we can apply Observation 4.3 since $P(\alpha)$ can be written as a Σ_1 sentence, over $\langle H(\omega_2), \in, NS_{\omega_1} \rangle$, with α as parameter. $\square$

Remember the approach pointed out before for producing, in the presence of MM and using Namba forcing, a stationary $A \subseteq \omega_2 \cap cf(\omega)$ in $\mathcal{D}$ with $(\omega_2 \cap cof(\omega)) \backslash A$ also stationary. Corollary 4.4 shows that this approach cannot work (just take $\varphi(x)$ to be a definition of Namba forcing).

References

[A-Sh] U. Abraham, S. Shelah, *Coding with ladders a well ordering of the reals.* J. of Symbolic Logic, 67 (2002), 579–597.

[As1] D. Asperó, *The nonexistence of robust codes for subsets of ω_1.* Fundamenta Mathematicae, vol. 186 (2005), no. 3, 215–231.

[As2] D. Asperó, *Coding by club–sequences.* Ann. Pure and Applied Logic, 142 (2006), 98–114.

[As3] D. Asperó, *Guessing and non-guessing of canonical functions.* Ann. Pure and Applied Logic, 146 (2007), 150–179.

[As4] D. Asperó, *Coding into $H(\kappa^+)$ without parameters.* Submitted.

[As5] D. Asperó, *Forcing a simply definable well–order of $H(\omega_2)$.* Preprint.

[K] A. Kanamori, *The Higher Infinite: Large Cardinals in Set Theory from their Beginnings.* Springer-Verlag, Berlin (1994).

[Ku] K. Kunen, *A model for the negation of the Axiom of Choice*, in Cambridge Summer School in Mathematical Logic (A. Mathias and H. Rogers Jr., eds.), Lecture Notes in Math., vol. 337, Springer, Berlin (1973), 489–494.

[F-M-Sh] M. Foreman, M. Magidor, S. Shelah, *Martin's Maximum, saturated ideals and non-regular ultrafilters, I.* Ann. of Mathematics, 127 (1988), 1–47.

[L1] P. Larson, *The Stationary Tower. Notes on a Course by W. Hugh Woodin*, AMS University Lecture Series, 32. Providence, RI (2004).

[L2] P. Larson, *Martin's Maximum and definability in $H(\aleph_2)$.* Preprint.

[Mar-St] D. Martin, J. Steel, *A proof of projective determinacy*, J. American Math. Soc., 2 (1989), 71–125.

[Mo] J. Moore, *Set mapping reflection.* J. Math. Logic, 5 (2005), 87–98.

[S] R. Schindler, *Semi-proper forcing, remarkable cardinals, and Bounded Martin's Maximum*, Math. Logic Q., 50 (2004), 527–532.

[Sh] S. Shelah, *Proper and Improper Forcing*, 2nd. ed., Perspectives in Mathematical Logic, Springer, Berlin (1998).

[W] H. Woodin, *The axiom of Determinacy, Forcing Axioms, and the Non-stationary ideal.* De Gruyter Series in Logic and its Applications, Number 1. Berlin, New York (1999).

NONSTANDARD METHODS IN
RAMSEY'S THEOREM FOR PAIRS

C. T. Chong

Department of Mathematics
National University of Singapore
Singapore 117542
E-mail: chongct@math.nus.edu.sg

We discuss the use of nonstandard methods in the study of Ramsey type problems, and illustrate this with an example concerning the existence of definable solutions in models of $B\Sigma_2^0$ for the combinatorial principles of Ramsey's Theorem for pairs and cohesiveness.

1991 *Mathematics Subject Classification.* 03D20, 03F30, 03H15

1. Introduction

Ramsey's Theorem (RT_k^n) states that every partition of the n-element subsets of $\mathbb{N}$ into k classes has an infinite homogeneous set. The proof-theoretic strength of RT_k^n is an area of active research by recursion theorists in recent years.

Taking the system of second order arithmetic RCA_0 as the base theory, an earlier result of Jockusch [8] implies that for $n > 2$, the arithmetic comprehension axiom is a consequence of RT_k^n. This was shown not to hold under RT_2^2 by Seetapun and Slaman [11], so that over RCA_0, RT_2^2 is strictly weaker than RT_k^n for $n > 2$.

In a systematic study of RT_2^2, Cholak, Jockusch and Slaman [1] introduced two combinatorial principles related to RT_2^2, defined over a model

---

The author wishes to thank the Department of Mathematics, University of California, Berkeley, for its hospitality during his visit from October to December 2005, where part of the research was done. This paper is a revised version of a talk given in August 2005 at the Program in Computational Prospects of Infinity held at the Institute for Mathematical Sciences, National University of Singapore.

$\mathcal{M} = \langle M, \mathbb{X}, +, \cdot, 0, 1 \rangle$ of RCA$_0$, where $\mathbb{X}$ is the collection of second order objects of $\mathcal{M}$:

(1) The principle of stable Ramsey's Theorem for pairs (SRT$_2^2$): A partition f of $[M]^2$ into two classes (sometimes also called "two colors") is said to be *stable* if $\lim_s f(x, s)$ exists for all $x \in M$. SRT$_2^2$ states that every stable partition of $[M]^2$ into two classes has an infinite homogeneous set in $\mathcal{M}$.

(2) The principle of cohesiveness (COH): Let Y be an M-infinite set in $\mathcal{M}$ and let $Y_s = \{t | (s, t) \in Y\}$. $\langle Y_s \rangle_{s \in M}$ is called a Y-array. Then a set C is Y-cohesive if for all s, either $C \cap Y_s$ is $\mathcal{M}$-finite or $C \cap \bar{Y}_s$ is $\mathcal{M}$-finite. COH states that every Y-array, where $Y \in \mathcal{M}$, has a Y-cohesive set $C \in \mathcal{M}$.

It is known ([1]) that over the base theory RCA$_0$, Ramsey's Theorem for pairs is equivalent to SRT$_2^2$ + COH. In view of the definition of stable 2-coloring, it is natural to conjecture that SRT$_2^2$ is strictly weaker than RT$_2^2$. Much effort has been expended on establishing this conjecture. First of all, it is not difficult to see that, given a stable partition $f \in \mathcal{M}$ of $[M]^2$ into two colors, there is an f-homogeneous set Δ_2^0 in the parameter defining f. Jockusch [8] exhibited a recursive partition f_J of $[\mathbb{N}]^2$ into two classes that has no Δ_2 homogeneous solution (thus the partition is necessarily nonstable). A first attempt to proving the relative strength of SRT$_2^2$ and RT$_2^2$ is to show that every stable recursive partition of $[\mathbb{N}]^2$ into two colors has a low homogeneous solution (i.e. whose jump is $\emptyset'$). An iteration of the construction will then produce a model of RCA$_0$ + SRT$_2^2$ that does not include a homogeneous solution to f_J according to Jockusch's result. Downey, Hirschfeldt, Lempp and Solomon [5] showed that this approach failed by demonstrating the existence of a Δ_2 set A of integers with no infinite low subset in A or $\bar{A}$. Thus to produce a model $\mathcal{M} = \langle M, \mathbb{X}, +, \cdot, 0, 1 \rangle$ of RCA$_0$ + SRT$_2^2$ without RT$_2^2$ where $M = \mathbb{N}$ (i.e. an ω-model) requires a more sophisticated approach. In [6], Hirschfeldt, Jockusch, Kjos-Hanssen, Lempp and Slaman show that while a low homogeneous set is not always guaranteed for a stable 2-coloring of pairs, it is nevertheless true that an incomplete Δ_2 homogeneous solution exists. The authors also propose ways of attacking the SRT$_2^2$ problem by considering different possible ω-models that could be constructed to avoid RT$_2^2$.

In this paper we discuss the use of nonstandard models in investigations of Ramsey type problems. From the proof-theoretic point of view, there is no reason to restrict oneself to ω-models in studying subsystems of second order arithmetic. Since there exist nonstandard models of RCA_0 that exhibit simpler structures in the Turing degrees, Ramsey type problems may be analyzed from a different perspective and this could shed light on the problems themselves. For example, there exist models of $RCA_0 + B\Sigma_2^0$ in which every incomplete Turing degree is low. This means that the counter-example given in [5] does not apply to these models. It will be useful to study the problem of SRT_2^2 in such a situation. This is discussed further later in the paper.

Our main result gives an illustration of the link between Ramsey type problems and nonstandard models. Namely, we are concerned with the question of existence of solutions for RT_2^2 or COH that is recursive in the double Turing jump of the set parameter defining the 2-coloring or array. Classical results of Jockusch and others show that this is true for ω-models. In general, however, the answer is determined by the underlying first order inductive strength of the model being considered (Corollaries 3.1, 3.2 and 3.3).

We assume that the reader is familiar with recursion theory in models of fragments of Peano arithmetic as well as subsystems of second order arithmetic (cf. [3] and [13] for details). Here we will only present a summary of the facts that are relevant to the discussion below.

2. $B\Sigma_2^0$ models

Let $I\Sigma_n^0$ denote the induction scheme for Σ_n^0 formulas (with number and set parameters), where $n \geq 0$. All models $\mathcal{M}$ considered in this paper satisfy at least $I\Sigma_1^0$ ($I\Sigma_n^0$ is denoted Σ_n^0-IND in [13]). A bounded set in $\mathcal{M}$ is $\mathcal{M}$-finite if it is coded in $\mathcal{M}$. Otherwise it is called $\mathcal{M}$-infinite. An unbounded set in $\mathcal{M}$ is necessarily $\mathcal{M}$-infinite, although the converse is not always true (for example $\mathbb{N}$ is not $\mathcal{M}$-finite in any nonstandard model $\mathcal{M}$).

Let $B\Sigma_n^0$ denote the scheme which states that every Σ_n^0 definable function maps an $\mathcal{M}$-finite set onto an $\mathcal{M}$-finite set. The result of Kirby and Paris [9] for first order formulas may be generalized to show that for all $n \geq 0$, $I\Sigma_{n+1}^0$ is strictly stronger than $B\Sigma_{n+1}^0$, which is in turn strictly stronger than $I\Sigma_n^0$. Indeed, the results stated without proof in this section were originally shown for first order fragments of Peano arithmetic. Their generalization to subsystems of second order arithmetic is immediate.

Proposition 2.1. *If $\mathcal{M} \models I\Sigma_n^0$, then every bounded $\Sigma_n^0(\mathcal{M})$ set is $\mathcal{M}$-finite.*

If $\mathcal{M} \models B\Sigma_n^0$, then a cut $I \subset M$ is a set that is closed downwards as well as under the successor function. I is a Σ_n^0 cut if it is Σ_n^0 definable over $\mathcal{M}$. Proposition 2.1 implies that $\mathcal{M} \models I\Sigma_n^0$ if and only if there is no proper (i.e. bounded) Σ_n^0 cut. We consider only proper Σ_n^0 cuts in this paper.

If $\mathcal{M} \models B\Sigma_n^0$ but not $I\Sigma_n^0$, we call it a $B\Sigma_n^0$ model. In this case, there is a $\Sigma_n^0(\mathcal{M})$ function mapping a Σ_n^0 cut cofinally into M.

We identify a number in M with the set of its predecessors.

Definition 2.1. *Let $\mathcal{M}$ be a model of* RCA$_0$. *A set $X \subset M$ is regular if $X \restriction a$ is $\mathcal{M}$-finite for each $a \in M$.*

Proposition 2.2. *Let $\mathcal{M}$ be a $B\Sigma_2^0$ model of* RCA$_0$. *Then*

  (i) *Every $\Delta_2^0(\mathcal{M})$ set is regular;*
  (ii) *X is $\Delta_2^0(\mathcal{M})$ if and only if X is recursive in Y', where Y is the set parameter occurring in the definition of X.*
  (iii) *Every $X \in \mathbb{X}$ is regular.*

Proposition 2.3. *Let $\mathcal{M}$ be a model of* RCA$_0$ *and $X \in \mathbb{X}$. Then $\mathcal{M} \models B\Sigma_2^0(X)$ if and only if $\mathcal{M} \models B\Sigma_1^0(X')$.*

Definition 2.2. *Let $\mathcal{M} \models$* RCA$_0$ *and $X \subset M$. Then X is hyperregular if the image of every bounded set under a function that is weakly recursive in X (i.e. $\Delta_1(X)$) is bounded.*

Given a model $\mathcal{M}$ of RCA$_0$, and $A \subset M$, let $\mathcal{M}[A]$ denote the structure generated from A over $\mathcal{M}$ by closing under functions recursive in $A \oplus X$, where $X \in \mathbb{X}$.

Proposition 2.4. (Mytilinaios and Slaman [10]) *Let $\mathcal{M}$ be a $B\Sigma_2^0$ model of* RCA$_0$ *and $A \subset M$. The following are equivalent:*

  (i) *A is hyperregular;*
  (ii) *$\mathcal{M}[A] \models I\Sigma_1^0$.*

The notion of coding in $B\Sigma_2^0$ models of RCA_0 is central to the analysis of Ramsey type problems in this paper. Although the original definition was given for first order structures, it carries over to second order structures in an obvious way.

Definition 2.3. *Let A be a subset of M, where $\mathcal{M} \models \mathrm{RCA}_0$. A set $X \subseteq A$ is* coded *on A if there is an $\mathcal{M}$-finite set $\hat{X}$ such that $\hat{X} \cap A = X$.*

Definition 2.4. *Let A be a subset of $\mathcal{M}$. We say that a set X is Δ_n^0 on A if both $A \cap X$ and $A \cap \overline{X}$ are $\Sigma_n^0(\mathcal{M})$.*

The following result is a generalization of Mytilinaios and Slaman [10] to models of RCA_0. It will be used repeatedly in the sequel:

Proposition 2.5. *There is a $B\Sigma_n^0$ model $\mathcal{M}_0$ of RCA_0 in which ω is a Σ_2^0 cut and every subset is coded on ω.*

3. Ramsey's Theorem in $B\Sigma_2^0$ models of RCA_0

We first prove a general theorem for the model $\mathcal{M}_0$ as given in the above proposition. Let I denote ω. Also fix $g : I \to M_0$ to be an increasing Σ_2^0 cofinal function, with set parameter $Y \in \mathbb{X}$.

Theorem 3.1. *If $C \leq_T Y''$ is in $\mathbb{X}$, then $C' \leq_T Y'$.*

We first prove a lemma.

Lemma 3.1. *If $C \leq_T Y''$ is in $\mathbb{X}$, then $(C \oplus Y)' \leq_T I \oplus Y'$.*

Proof. Let $\Phi^{Y''} = C$. We identify a set with its characteristic function. By Proposition 2.2 (iii), C is regular. Hence for each $i \in I$, there is a neighborhood condition $\langle P_i, N_i \rangle$ of Y'' such that $\langle C \restriction g(i), P_i, N_i \rangle \in \Phi$.

Claim 1. $(C \oplus Y)'$ is regular.

Proof of Claim 1. Since C and Y are in $\mathbb{X}$, $B\Sigma_2^0(C \oplus Y)$ holds in $\mathcal{M}_0$. Fix $i \in I$. For $x \in g(i) \cap (C \oplus Y)'$, let $h(x)$ be the least y such that x is in $(C \oplus Y)'_y$, the set enumerated into $(C \oplus Y)'$ by stage y using $C \oplus Y$ as an oracle (recall that $(C \oplus Y)'$ is $\Sigma_1(C \oplus Y)$), and let it be 0 otherwise. Then by $B\Sigma_2^0(C \oplus Y)$, the image of $g(i)$ under h is $\mathcal{M}_0$-finite and has a largest element denoted y_0. Then for $x < g(i)$, $x \in (C \oplus Y)'$ if and only if $x \in (C \oplus Y)'_{y_0}$.

Claim 2. If $D \subset Y''$ is $\mathcal{M}_0$-finite, then there is a j_D such that $D \subset Y''_{g(j_D)}$, where $Y''_{g(j_D)}$ is the set of elements enumerated into Y'' by stage $g(j_D)$ using Y' as oracle.

Proof of Claim 2. By Proposition 2.3, $B\Sigma_1^0(Y')$ holds. The map taking each $x \in D$ to the least y such that $x \in Y''_y$ is Y'-recursive, hence $\mathcal{M}_0$-finite and bounded below a y_0. Let j_D be the least $j \in I$ such that $g(j) > y_0$.

Claim 3. If $X \leq_T Y''$ is regular, then $X \leq_T I \oplus Y'$.

Proof of Claim 3. Suppose $\Psi^{Y''} = X$. Then for each $i \in I$, there exists $\langle P_i, N_i \rangle$ such that $P_i \subset Y''$ and $N_i \subset \bar{Y}''$ with $\langle X \restriction g(i), P_i, N_i \rangle \in \Psi$. By Claim 2, there is a $j \in I$ such that $P_i \subset Y''_{g(j)}$ and $N_i \subset \bar{Y}''_{g(j)}$. Hence the set

$$Z = \{(i,j,j') \in I \times I \times I | \forall \langle P, N \rangle [P \subset Y''_{g(j)} \ \& \ N \subset \bar{Y}''_{g(j)} \ \& \\ \Psi^{\langle P,N \rangle} \restriction g(i) \text{ total} \to (N \cap Y''_{g(j')} \neq \emptyset)]\} \tag{1}$$

satisfies the property that for each i, the set $\{(j,j')|(i,j,j') \in Z\}$ is bounded in $I \times I$. Furthermore, Z is $\Delta_2^0(Y)$ on $I \times I \times I$ and therefore coded on $I \times I \times I$ by an $\mathcal{M}_0$-finite set $\hat{Z}$ according to Proposition 2.5. We may assume that for each (i,j), j' is the least such that $(i,j,j') \in \hat{Z}$. Let j_i be the greatest j for $(i,j,j') \in Z$. Then for $j_i + 1$, $(j_i + 1)' \in \bar{I}$. In particular, there is a $\langle P, N \rangle \subset Y''_{g(j_i+1)} \times \bar{Y}''_{g(j_i+1)}$ such that $N \subset \bar{Y}''$ and $\Psi^{\langle P,N \rangle} \restriction g(i)$ is total, and so equal to $X \restriction g(i)$. Furthermore, since $X \restriction g(i)$ is $\mathcal{M}_0$-finite, all the computations in $\Psi_{g(j_i+1)}$ that disagree with $X \restriction g(i)$ will be identified at stage $g(j^*)$ for some $j^* > j_i$. The least such j^*, denoted j_i^*, is where there is only one output for $\Psi_{g(j^*)}^{\langle P,N \rangle} \restriction g(i)$ for any $\langle P, N \rangle \in Y''_{g(j_i+1)}, \bar{Y}''_{g(j_i+1)}$.

Now we may compute C from $I \oplus Y'$ as follows: To decide $C \restriction y$, use Y' to choose an i such that $g(i) > y$. Use I to find the least j such that $(i,j,j') \in \hat{Z}$ and $j' \in \bar{I}$. This is $j_i + 1$. Then $C \restriction g(i)$ is the set D such that there is an $N \subset \bar{Y}''_{g(j_i^*)}$ with $\langle D, Y''_{g(j_i+1)}, N \rangle \in \Psi_{g(j_i+1)} \cap \Psi_{g(j_i^*)}$.

By Claim 1 and Claim 3, $C \oplus Y \leq_T I \oplus Y'$. Since $B\Sigma_2^0(C \oplus Y)$ holds, $I\Sigma_1^0(C \oplus Y)$ is true and therefore by Proposition 2.4, $C \oplus Y$ is hyperregular. This means that for all i, there is a least j, denoted $\hat{j}_i$, such that $(C \oplus Y)' \restriction$

$g(i) = (C \oplus Y)'_{g(j)} \restriction g(i)$. Now the set

$$V = \{(i,j) | (C \oplus Y)'_{g(j)} \restriction g(i) \neq (C \oplus Y)'_{g(j+1)} \restriction g(i)\}$$

is $\Delta^0_2(C \oplus Y)$ on $I \times I$ and therefore coded on $I \times I$ by an $\mathcal{M}_0$-finite set $\hat{V}$ according to Proposition 2.5. Furthermore, $\hat{j}_i$ is the least $j \in I$ such that $(i,j) \notin V$. Then using $\hat{V}$ as a parameter, we may compute $(C \oplus Y)'$ from $I \oplus Y'$ as follows: Given i, use I to obtain $\hat{j}_i$ and then use Y' to compute $(C \oplus Y)'_{g(\hat{j}_i)}$. Then $x \in (C \oplus Y)' \restriction g(i)$ if and only if $x \in (C \oplus Y)'_{g(\hat{j}_i)}$. This completes the proof of the lemma. $\qquad\Box$

Proof of Theorem 3.1. First note that just like Claim 1 of Lemma 3.1, one can show that Y' is regular. By Lemma 3.1 let $\Psi^{I \oplus Y'} = (C \oplus Y)'$. A neighborhood condition $\langle P, N \rangle$ of I may be identified with a pair (c,d) where $c \in I$ and $d \in \bar{I}$ (let c be the maximum of P and d be the minimum of N). We make the following claim.

Claim. For each $i \in I$ there exist $(j,j',j'') \in I \times I \times I$ such that

(i) *Existence.* There is a (c,d) with $c < j < j' \leq d$ and $\Psi^{(c,d) \oplus Y' \restriction g(j'')} \restriction g(i) = \Psi^{I \oplus Y'} \restriction g(i)$ and

(ii) *Consistency.* There is no $x < g(i)$ and (c',d') with $c' \leq j < j' \leq d'$ such that $\Psi^{(c',d') \oplus Y' \restriction g(j'')}(x) \downarrow \neq \Psi^{I \oplus Y' \restriction g(j'')}(x)$.

Proof of Claim. Otherwise, there is an $i \in I$ such that for all $(j,j',j'') \in I \times I \times I$, either (i) or (ii) is false. More precisely, there is an $i \in I$ such that for all $(j,j',j'') \in I \times I \times I$, either

(iii) For any (c,d) with $c \leq j < j' \leq d$, if $\Psi^{(c,d) \oplus Y' \restriction g(j'')} \restriction g(i) \downarrow$, then there is an $x < g(i)$ such that $\Psi^{(c,d) \oplus Y' \restriction g(j'')}(x) \neq \Psi^{I \oplus Y' \restriction g(j'')}(x)$, or

(iv) There is an $x < g(i)$ and a (c,d) with $c \leq j < j' \leq d$ such that $\Psi^{(c,d) \oplus Y' \restriction g(j'')}(x) \downarrow \neq \Psi^{I \oplus Y' \restriction g(j'')}(x)$ (in this case d has to be in I).

Since $(C \oplus Y)' \leq_T I \oplus Y'$, there exist (j_0, j'_0, j''_0) in $I \times I \times I$ and $c_0 \leq j_0 < j'_0 < d_0 \in \bar{I}$ such that $\Psi^{(c_0,d_0) \oplus Y' \restriction g(j''_0)} \restriction g(i) = \Psi^{I \oplus Y'} \restriction g(i)$. Hence for the triple (j_0, j'_0, j''_0), (iii) is false and (iv) must hold. Now if the set $J = \{j | \text{(iv) holds for } (j_0, j, j''_0) \text{ with } c \leq j_0 < j \leq d\}$ is bounded in I, say by j^*, then it implies that for any (c,d) such that $c \leq j_0 < j^* \leq d$, if $\Psi^{(c,d) \oplus Y' \restriction g(j''_0)}(x) \downarrow$ then it is equal to $\Psi^{I \oplus Y' \restriction g(j''_0)}(x)$. Since $c_0 < j^* < d_0 \in \bar{I}$, we see that (j_0, j^*, j''_0) satisfies (i) and (ii), contradicting the assumption that i is a counterexample to the claim.

Hence J is unbounded in I. But then this implies that for $d < d_0$, $d \in \bar{I}$ if and only if for all d' where $d \le d' \le d_0$, there is no $c \le j_0$ and $x < g(i)$ such that $\Psi^{(c,d') \oplus Y' \restriction g(j_0'')}(x) \downarrow \ne \Psi^{(c_0,d_0) \oplus Y' \restriction g(j_0'')}(x)$. This gives $\bar{I}$ a $\Pi_1^0(Y' \restriction g(j_0''))$ definition. Since $Y' \restriction g(j_0'')$ and $(C \oplus Y)' \restriction g(i)$ are both $\mathcal{M}_0$-finite, $\bar{I}$ is $\mathcal{M}_0$-finite, which is a contradiction, proving the Claim.

By the Claim, the set of quadruples (i, j, j', j'') satisfying (i) and (ii) is $\Pi_2^0(C \oplus Y)$ on $I \times I \times I \times I$, hence coded on $I \times I \times I \times I$ by an $\mathcal{M}_0$-finite set $\hat{h}$ which we may consider to be a function taking i to the least triple (j, j', j''). Then Y' computes $(C \oplus Y)'$ as follows: Given y, find an $i \in I$ using Y' such that $g(i) > y$. Now use $\langle \hat{h}_1(i), \hat{h}_2(i) \rangle \oplus Y' \restriction g(\hat{h}_3(i))$ to enumerate $(C \oplus Y)' \restriction g(i)$ via Ψ, where $\hat{h}(i) = \langle \hat{h}_1(i), \hat{h}_2(i), \hat{h}_3(i) \rangle$. This computation has to be correct by the choice of the function $\hat{h}$.

Since $(C \oplus Y)' \le_T Y'$, it follows that $C' \le_T Y'$. $\square$

We now apply Theorem 3.1 to study Ramsey's Theorem for pairs. The proof of the following result follows essentially Jockusch [8]. The only point to note is that the construction works under the assumption of $B\Sigma_2^0$.

Proposition 3.1. *Let $\mathcal{M}$ be a $B\Sigma_2^0$ model of* RCA$_0$*. Let $Y \in \mathbb{X}$. Then there is a Y-recursive partition f_J of $[M]^2$ into two colors with no homogeneous solution recursive in Y'.*

Definition 3.1. *Let $\varphi(X, Y)$ be a second order formula with X and Y as free set variables. A model $\mathcal{M}$ of* RCA$_0$ *is a double-jump model of $\forall Y \exists X \varphi(X, Y)$ if for all $Y \in \mathcal{M}$, there is an $X \in \mathcal{M}$ such that $X \le_T Y''$ and $\mathcal{M} \models \varphi(X, Y)$.*

Results of Stephan and Jockusch [14] and Cholak, Jockusch and Slaman [1] show that every ω-model of RCA$_0$ is an ω-submodel of one that is a double-jump model of RT$_2^2$, if $\forall Y \exists X \varphi(X, Y)$ is interpreted as asserting the existence of a homogeneous set X for a Y-recursive 2-coloring of pairs. The same conclusion also holds for COH. It turns out that the strength of first order induction plays a crucial role for this to be true.

Corollary 3.1. *$\mathcal{M}_0$ is not a double-jump model of* RT$_2^2$.

Proof. Suppose otherwise. Let $Y \in \mathcal{M}_0$ be chosen such that the $\Sigma_2^0(\mathcal{M}_0)$ cut I is defined with parameter Y. Let f_J be the Y-recursive partition given in Proposition 3.1. Then by assumption there is a homogeneous solution

$H_{f_J} \leq_T Y''$ for f_J. By Theorem 3.1, $H_J <_T Y'$. However, this contradicts Proposition 3.1. $\qquad\square$

The proof of the next result is essentially in Theorem 12.4 of [1].

Proposition 3.2. *Let $\mathcal{M}$ be a $B\Sigma_2^0$ model of* RCA_0. *Let $g : I \to M$ be Σ_2^0, increasing and cofinal with parameter Y. Then there is a $Z \leq_T Y$ with a Z-array that has no Z-cohesive set C satisfying $C' \leq_T Y'$.*

Proof. Let e be such that $\Phi_e^{Y'}$ is a $\{0, 1\}$-valued partial function that has no Y'-computable extension to a total function. Define $Z = \{(s, t)|\Phi_t^{Y_t'}(s) \downarrow= 0\}$. Let $f(s, t) = 0$ if $(s, t) \in Z$, and equal to 1 otherwise. Let $Z_s = \{t|(s, t) \in Z\}$. Then both f and Z are recursive in Y.

Suppose that C is cohesive for the array $\langle Z_s \rangle_{s \in M_0}$ and $C' \leq_T Y'$. Then $\hat{f}(s) = \lim_{t \in C} f(s, t)$ exists for each s and is computable in C', hence in Y' by assumption. Furthermore $\hat{f}$ is a total function extending $\Phi_e^{Y'}$, contradicting the choice of e. $\qquad\square$

Let $\forall Y \exists X \varphi(X, Y)$ be COH: For every $Y \in \mathcal{M}$ and $Z \leq_T Y$, every Z-array has a Z-cohesive set $X \in \mathcal{M}$. Theorem 3.1 leads to the following corollary.

Corollary 3.2. *$\mathcal{M}_0$ is not a double-jump model of* COH.

It follows from Corollary 3.3 that starting with a $B\Sigma_2^0$ model $\mathcal{M}_0$ of RCA_0, it is not possible to add second order objects to $\mathcal{M}_0$ within double jump to obtain a model of $RCA_0 + COH$ while preserving $B\Sigma_2^0$. This is expressed as follows. Let DB-COH denote the statement that every Y-array has a Y-cohesive set recursive in Y''.

Corollary 3.3. $RCA_0 + COH$ *does not prove* DB-COH.

Chong, Slaman and Yang have recently proved a Π_1^1-conservation theorem for $RCA_0 + COH + B\Sigma_2^0$ over $RCA_0 + B\Sigma_2^0$. A key step in the proof is to expand, for a given $\mathcal{M}$ and Y-array where $Y \in \mathcal{M}$, a Y-cohesive set in the generic extension that preserves $B\Sigma_2^0$. The Y-cohesive set thus obtained is not recursive in Y'' (in fact highly non-effective). Corollary 3.3 hints at the need for a non-Y''-effective approach.

Let $\forall Y \exists X \varphi(X, Y)$ be SRT_2^2. Theorem 3.1, when applied to SRT_2^2, says that if $\mathcal{M}_0$ is a model of SRT_2^2 then it necessarily contains only incomplete homogeneous sets (incomplete relative to the parameter defining the 2-coloring). Furthermore, $\mathcal{M}_0$ satisfies the following useful result whose origin (in the form of α-recursion theory) goes back to Shore [12] (See also Mytilinaios and Slaman [10] and Chong and Yang [3]):

Proposition 3.3. *Let $\mathcal{M}$ be a $B\Sigma_2^0$ model of RCA_0. Let I be a $\Sigma_2^0(\mathcal{M})$ cut with parameter $Y \in \mathcal{M}$. Then every $X <_T Y'$ in $\mathcal{M}$ is Y-low, i.e. $X' \leq_T Y'$.*

Proof. Let $\mathcal{M}$ and Y be as given, and let I be a $\Sigma_2^0(Y)$ cut. Let $X <_T Y'$ be in $\mathcal{M}_0$. Then $B\Sigma_2^0(X)$ holds in $\mathcal{M}$ and so X' is regular. Since $X' \leq_T Y''$, Lemma 3.1 implies that X' is recursive in $I \oplus Y'$. The argument in the proof of Theorem 3.1 (after Claim 3) now implies that $X' \leq_T Y'$. $\square$

It is not difficult to see that every stable 2-coloring of pairs has a homogeneous solution Δ_2^0 in the parameter defining the coloring. In a $B\Sigma_2^0$ setting, the crucial question is whether every stable 2-coloring of pairs has a homogeneous solution that is low relative to the parameter that defines the coloring. If the answer is yes, then one may generate a $B\Sigma_2^0$ model of $\mathrm{RCA}_0 + \mathrm{SRT}_2^2$ as follows: Begin with a countable $B\Sigma_2^0$ model $\mathcal{M}_0^*$ (with no second order objects). Let this be the ground model. At stage $n+1$, construct an incomplete homogeneous set for a stable 2-coloring defined over $\mathcal{M}_n^*$. One can arrange the stable 2-colorings in such a way that $\mathcal{M} = \bigcup_n \mathcal{M}_n^*$ is a model of SRT_2^2. Now $B\Sigma_2^0$ is preserved in every $\mathcal{M}_n^*$ since only low sets are added, guaranteeing that $\mathcal{M} \models B\Sigma_2^0$. Then Proposition 3.1 ensures that $\mathcal{M}$ is not a model of RT_2^2.

We end this paper with two questions.

Question 1. Is there a double-jump $B\Sigma_2^0$ model for SRT_2^2?

Question 2. Is there a $B\Sigma_2^0$ model of $\mathrm{RCA}_0 + \mathrm{RT}_2^2$ or COH obtained from a ground model by working within the nth jump, where $n > 2$?

References

[1] Peter Cholak, Carl Jockusch and Theodore A. Slaman. On the strength of Ramsey's Theorem. *J. Symbolic Logic*, 66: 1–55, 2001.

[2] C. T. Chong and K. J. Mourad. The degree of a Σ_n cut. *Ann. Pure Appl. Logic*, 48(3):227–235, 1990.

[3] C. T. Chong and Yue.Yang. Computability theory in arithmetic: Provability, structure and techniques. In M. Lerman, P. Cholak, S. Lempp and R. A. Shore, editors, *Computability Theory and Its Applications: Current Trends and Open Problems*, volume 257 of *Contemporary Mathematics*, pages 73–82, Providence RI, 2000. AMS.

[4] C. T. Chong and Yue Yang. The jump of a Σ_n cut. *J. London Math. Soc.*, 2007.

[5] Rod Downey, Denis Hirschfeldt, Steffen Lempp and Reed Solomon. A Δ_2^0 Set with no infinite low subset in either it or its complement. *J. Symbolic Logic*, 66: 1371–1381, 2001.

[6] Denis Hirschfeldt, Carl G. Jockusch, Bjorn Kjos-Hanssen, Steffen Lempp and Theodore A. Slaman. The strength of some combinatoral principles related to Ramsey's Theorem for pairs. In *Proc. Recursion Theory Wokshop, Singapore 2005*. World Scientific, to appear

[7] Jeff Hirst. *Combinatorics in Subsystems of Second Order Arithmetic*. PhD thesis, Pennsylvania State University, 1987.

[8] Carl G. Jockusch. Ramsey's Theorem and recursion theory. *J. Symbolic Logic*, 37: 268-280, 1972.

[9] L. A. Kirby and J. B. Paris. Σ_n-collection schemas in arithmetic. In *Logic Colloquium '77*, pages 199–209, Amsterdam, 1978. North–Holland Publishing Co.

[10] Michael E. Mytilinaios and Theodore A. Slaman. Σ_2-collection and the infinite injury priority method. *J. Symbolic Logic*, 53(1): 212–221, 1988.

[11] David Seetapun and Theodore A. Slaman. On the strength of Ramsey's Theorem. *Notre Dame J. Logic*, 36(4): 570–583, 1995.

[12] Richard A. Shore. On the jump of α-recursively enumerable sets. *Trans. Amer. Math. Soc.*, 217: 351-0363, 1976.

[13] Stephen G.Simpson. *Subsystems of Second Order Arithmetric*. Heidelberg, 1999. Springer Verlag.

[14] Frank Stephan and Carl Jockusch Jr. A cohesive set that is not high. *Math. Logic Quarterly*, 39: 515–530, 1993.

PROMPT SIMPLICITY,
ARRAY COMPUTABILITY AND CUPPING

Rod Downey

School of Mathematics
Statistics and Computer Science, Victoria University
P.O. Box 600, Wellington, New Zealand
E-mail: Rod.Downey@vuw.ac.nz

Noam Greenberg

School of Mathematics
Statistics and Computer Science, Victoria University
P.O. Box 600, Wellington, New Zealand
E-mail: greenberg@mcs.vuw.ac.nz

Joseph S. Miller

Department of Mathematics, University of Connecticut
196 Auditorium Road, Unit 3009, Storrs, CT 06269-3009, USA
E-mail: joseph.miller@math.uconn.edu

Rebecca Weber

Department of Mathematics, Dartmouth College
6188 Bradley Hall Hanover, NH 03755-3551, USA
E-mail: rweber@math.dartmouth.edu

We show that the class of c.e. degrees that can be joined to $0'$ by an array computable c.e. degree properly contains the class of promptly simple degrees.

The first, second and fourth authors' research was supported by the Marsden Fund of New Zealand, via postdoctoral fellowships. The second and third authors were supported by a grant from the Singapore Institute for Mathematical Sciences. Most of this research was carried out whilst the first, second and third authors were supported by the Singapore IMS during the *Computational Prospects at Infinity* programme, in 2005.

1. Introduction

The main class examined in this paper is the class of *array computable* degrees introduced by Downey, Jockusch and Stob in [11, 12]. We recall from [11] that a c.e. set A is array noncomputable iff for all $g \leqslant_{wtt} 0'$ there is a function $f \leqslant_T A$ that is not dominated by g; that is, for infinitely many x we have $f(x) > g(x)$. Whilst the original definition was in terms of "very strong arrays", the given characterization highlights the fact that being array noncomputable is akin to being non-low$_2$, where A is non-low$_2$ using the same definition, but replacing $\leqslant_{wtt}$ by $\leqslant_T$. Indeed in [12], the authors showed that the array noncomputable degrees share many of the properties of the non-low$_2$ degrees with respect to cupping, lattice embeddings and the like.

The importance of the notion of array noncomputability has been highlighted by recent work on randomness and domination/tracing properties in computability theory. (For examples, see Cholak, Coles, Downey and Herrmann [3], Downey, Hirschfeldt, Nies, Terwijn [10], Downey and Hirschfeldt [8], Kummer [15], Schaeffer [21], Stephan and Wu [18], Terwijn and Zambella [24].) For instance, Kummer [15] shows that the c.e. degrees containing containing c.e. sets of high Kolmogorov complexity are exactly the array noncomputable degrees. Nies [19] shows that the K-trivial degrees are all array computable. Ng, Stephan and Wu [18] prove the interesting result that a c.e. degree is array computable if and only if the degree consists of only reals in the field generated by the left c.e. reals. Ismukhametov [14] proves the remarkable result that the array computable c.e. degrees are the only c.e. degrees that have strong minimal covers in the Turing degrees, and hence the array computable degrees are definable in the Turing degrees if the computably enumerable degrees are definable.

In this paper we plan to add to our understanding of the lowness concept of array computability.

Two of the most influential concepts in the computability theory of the computably enumerable sets are the concepts of *lowness* and *prompt simplicity*.

The former concerns the intrinsic information content of a set; to say that a set is low for some class would indicate that it has little information content and resembles a computable set relative to the class in question. For instance, the classical notion of lowness, that $A' \equiv_T 0'$, says that the jump operator cannot distinguish between a low set and the empty set as oracles using Turing reducibility. There is a huge literature on low (c.e.) sets and how they resemble computable sets. For example, Soare [22] proved that if a c.e. set A is low then its lattice of c.e. supersets is effectively isomorphic to the lattice of all c.e. sets, and Robinson [20] proved that Sacks's splitting theorem can be carried out above any low c.e. degree, whereas Lachlan [16] demonstrated that this fact is not true for a general incomplete c.e. degree.

Prompt simplicity was introduced by Maass [17] in connection with automorphisms of the lattice of c.e. sets. Recall that a co-infinite c.e. set A is called promptly simple iff there is a computable function p, and an enumeration of $A = \cup_s A_s$, such that for all $e < \omega$, if the c.e. set W_e (the eth c.e. set in the canonical indexing) is infinite, then

$$\exists^\infty x, s \ [x \in W_{e,\,\text{at}\,s} \ \& \ x \in A_{p(s)}].$$

Thus, prompt simplicity is a *dynamic* property which expresses *how fast* elements can enter a set. It turns out that this concept, and variations, was the key to the solution of many longstanding questions about the lattice of c.e. sets as discussed in Harrington and Soare [13].

Roughly speaking, a promptly simple set resembles the halting problem in its dynamic properties. In a beautiful paper, Ambos-Spies, Jockusch, Shore and Soare [1] showed that lowness and promptness are intimately related. In particular, they showed that whilst promptly simple sets might not be Turing complete, they did indeed resemble complete sets in that they were *low cuppable*.

Theorem 1.1 ([1]). *If A is promptly simple then there is a low set B such that*

$$A \oplus B \equiv_T 0'.$$

In fact, the main result of [1] says a lot more. The promptly simple degrees coincide with the low cuppable ones; and the promptly simple degrees and the cappable degrees (i.e. those $\mathbf{a}$ for which there is a $\mathbf{b} \neq \mathbf{0}$ with $\mathbf{a} \cap \mathbf{b} = \mathbf{0}$) form an algebraic decomposition of the c.e. degrees into a strong filter and an ideal.

In this paper we will re-examine Theorem 1.1 with respect to array computability. Recent work on concepts like K-triviality (e.g., Nies [19]) and almost deep degrees (Cholak, Grozsek, Slaman [4]) as well as older work of Bickford and Mills [2], shows that the low degrees have a deep and poorly understood structure.

We prove the following (everything here is c.e.).

Theorem 1.2. *Every promptly simple degree is cuppable to $0'$ by an array computable degree.*

The proof of this result uses a new technique involving certain priority re-arrangements which is of independent technical interest. This result might lead the reader to believe that AC cupping could be used to characterize the promptly simple degrees, but this hope fails.

Theorem 1.3. *There is a degree that is cuppable by array computable degrees but which is not promptly simple.*

One question we have not been able to answer is Nies' question of whether every promptly simple degree is *superlow* cuppable (recall that A is *superlow* if $A' \equiv_{tt} 0'$). A positive answer would supersede our result because all superlow c.e. degrees are array computable (Schaeffer [21]).[1] We remark that in some sense an affirmative answer would be the best one could hope for, in terms of the concepts mentioned. A degree **a** is almost deep if for all low **b**, $\mathbf{a} \cup \mathbf{b}$ is also low, and hence, since there are low promptly simple degrees, they certainly cannot all be cupped by an almost deep degree; moreover, the K-trivial degrees are bounded above by an incomplete (in fact, a low$_2$) degree, and hence again (using the fact that no upper cone of the c.e. degrees can avoid the promptly simple degrees, [1]) there are promptly simple degrees that cannot be cupped to $\mathbf{0}'$ by a K-trivial degree.

A related question is whether the low c.e. degrees and the super-low c.e. degrees are elementarily equivalent. In a later paper, Downey, Greenberg and Weber [7] show that this is not the case by showing that no array non-computable degree can bound a 1-3-1, whereas there are low embeddings of 1-3-1. We also remark that the present paper led the authors to examine other permitting notions relating to array computability and domination. It turns out that another related notion (of being totally ω-c.e.) corresponds exactly to embeddability of certain upper semilattices in the c.e. degrees. These results and further generalizations can be found in Downey, Greenberg and Weber [7] and Downey and Greenberg [6]. These notions of permitting arise as a modification of the following characterization of array computability:

Definition 1.4. Let $f \colon \omega \to \omega$. A Δ_2^0 function g is f-c.e. if there is an effective approximation $g(x, s)$ of g (that is, $g(\cdot) = \lim_s g(\cdot, s)$) such that for all x,

$$|\{s : g(x, s) \neq g(x, s + 1)\}| \leq f(x).$$

---

[1]An intermediate question that we do not know the answer to is whether every promptly simple degree is cuppable by a *low* array computable degree.

Lemma 1.5 ([11]). *Let $f\colon \omega \to \omega$ be strictly increasing and computable. Then a c.e. set A is array computable iff every $g \leqslant_T A$ is f-c.e.*

In the sequel we use either the identity function or $x \mapsto x + 1$ for f.

1.1. Notation

Notation is standard and follows Soare [23, XIV, s.4].

2. PS $\subseteq$ AC cuppable

In this section we prove Theorem 1.2.

We are given a computably enumerable set A that permits promptly. In response, we enumerate a set B that will be array computable and join A up to $0'$.

We define moving markers $\gamma(m)$. At any given time, a marker may be defined or undefined; when we *erase* a marker it means that we make it undefined.

Let $\langle \Psi_e \rangle$ be an effective enumeration of all Turing functionals. To ensure that B is array computable, the requirement Q_e will, in the case that $\Psi_e(B) = g$ is total, construct an id-c.e. approximation for g. The strategy for Q_e will not be to impose restraint (since we cannot restrain the global join requirement) but rather to remove potentially dangerous $\Psi_e(B)$ computations by "disengaging" γ-markers from computations.

The construction is done on a tree of strategies — the strategy for Q_e measures whether or not $\Psi_{e'}(B)$ is total, for all $e' < e$. The eth level is devoted to Q_e, and each node has two outcomes, the infinite and the finite (the infinite is stronger). As is customary, if α is a node working for Q_e then we let $\Psi_\alpha = \Psi_e$.

By the recursion theorem (and the slowdown lemma), let p be a computable function that witnesses that A permits promptly with respect to an array of c.e. sets which are enumerated during the construction (an enumeration occurs when a number is tested for prompt permission; in a flashback-like narrative, we will specify the sets we need during the verifications).

General instructions about erasure are: if $\gamma(m)$ is enumerated into B, then it is immediately erased; if $\gamma(m)$ is erased, then all $\gamma(n)$ for $n > m$ are also immediately erased.

For any Q_e-node α and any input k we determine, at stage s, whether to accept a $\Psi_\alpha(B, k)$ computation or not. A computation will not be accepted if there are too many γ-markers below its use. For each k we will determine a number n (which decreases with s, until α is initialised) such that $\Psi_\alpha(B,k)$ is believed only if for $m \geqslant n$, there are no $\gamma(m)$-markers below the use $\psi_\alpha(k)$. The way this is done during the construction is to define, at every stage, a (non-decreasing) sequence $\langle k_\alpha(m) \rangle$ such that $k_\alpha(m)$ is the least input k that does not want to disengage m. So a computation $\Psi_\alpha(B,k) \downarrow$ is α-confirmed if for all m, $\gamma(m) \downarrow < \psi_\alpha(k)$ implies $k \geqslant k_\alpha(m)$.

2.1. Construction

At stage 0, we begin by defining, for every node α of level e on the tree, $k_\alpha(m) = 0$ for $m \leqslant e$ and $k_\alpha(m) = m - e$ otherwise.

At stage s, if m enters $0'$, then we put $\gamma(m)[s]$ into B (and $\gamma(m)$ remains undefined for ever).

Next, we construct the path of nodes accessible at stage s. We also describe which nodes are *expansionary* at stage s.

Suppose that a node α is accessible at s; let t be the last α-expansionary stage (0 if there is no such stage). Then s is α-expansionary if $\operatorname{dom} \Psi_\alpha(B)[s] > t$.

If s is not α-expansionary then the finite outcome of α is accessible at s; α does not act at s.

The computation $\Psi_\alpha(B)[s]$ is *unconfirmed due to m* if for some $k < \operatorname{dom} \Psi_\alpha(B)$ we have both $k < k_\alpha(m)$ and $\gamma(m) < \psi_\alpha(k)$ (in particular, $\gamma(m)$ is defined). Let m be the smallest number such that $\Psi_\alpha(B)$ is unconfirmed due to m. (What to do if $\Psi_\alpha(B)$ is confirmed? Not likely, but then $\alpha^\frown\infty$ is accessible and α does nothing).

First, α asks for prompt permission from A: it looks for a change in $A \upharpoonright \gamma(m)$ between stages s and $p(s)$. If permission is granted, then α erases $\gamma(m)$, and the infinite outcome of α is accessible.

If permission is not granted, then α enumerates $\gamma(m)$ into B (and erases it). In this case, we end the stage — there are no more accessible nodes. Also, we update k_β for all nodes β such that $\beta^\frown\infty \subseteq \alpha$. We redefine $k_\beta(m) = s$ (and to keep things in order, redefine $k_\beta(m+i) = s+i$ for all $i < \omega$). [This is because $\Psi_\beta(B)$ computations which may have been confirmed were nonetheless injured by $\gamma(m)$ entering B. To avoid a repeated injury (by the same m), the injured computations need to get less tolerant; from now, they too wish to disengage m.]

At the end of the stage, all nodes β which *lie to the right* of the path of accessible nodes are *initialised*. This means that we redefine $k_\beta(m) = 0$ for all $m < s$.

Finally, for every $m < s$ which is not yet in $0'$ and such that $\gamma(m)$ is not defined, we redefine $\gamma(m)$ with large value (keeping the sequence $\langle \gamma(m) \rangle$ increasing).

2.2. Verifications

Define the *true path* to be the leftmost path of nodes that are accessible infinitely often.

Lemma 2.1. *The true path is infinite; each node on the true path is eventually never initialised.*

Proof. Let α be a node on the true path. If there are finitely many α-expansionary stages then the finite outcome of α is on the true path and is eventually never initialised.

Suppose that there are infinitely many α-expansionary stages. If at infinitely many of those stages we find $\Psi_\alpha(B)$ to be confirmed then $\alpha^\frown\infty$ is on the true path. Otherwise, we enumerate an auxiliary set $U = U_\alpha$; if at an expansionary stage s, α asks for permission from A to erase $\gamma(m)$, then we enumerate $\gamma(m)$ into U at s. Then U is infinite. By the properties of p, there are infinitely many stages at which A gives α permission to erase the marker; at each such stage, $\alpha^\frown\infty$ is accessible; so $\alpha^\frown\infty$ is on the true path. $\square$

Claim 2.2. *Let α be a node and let $m < \omega$. Suppose that at some stage t, $k_\alpha(m) = 0$. Then at no stage $s \geqslant t$ does α erase $\gamma(m)$.*

Proof. Denote by $k_\alpha(m)\,[s]$ the value of $k_\alpha(m)$ at the beginning of stage s.

By induction on $s < \omega$, we show that for all α such that $k_\alpha(m) = 0\,[s]$,

  (1) For all β extending α, $k_\beta(m) = 0\,[s]$.
  (2) α does not erase $\gamma(m)$ at s.
  (3) $k_\alpha(m) = 0\,[s+1]$.

First note that (1) holds at stage $s = 0$ by the initial definition of k_α and k_β. (2) and (3) hold at 0 because at stage 0 nobody does anything.

Suppose that (1)-(3) hold for stage $s - 1$. Suppose that $k_\alpha(m) = 0\,[s]$. Let β be any node extending α. If $k_\alpha(m) = 0\,[s-1]$ then by (1)$(\alpha, s-1)$, $k_\beta(m) = 0\,[s-1]$; and then by (3)$(\beta, s-1)$, $k_\beta(m) = 0\,[s]$. Otherwise, $k_\alpha(m)$ was set to be 0 during stage $s - 1$, which means that α was initialised at stage $s - 1$. Then β was also initialised at stage $s - 1$ and so $k_\beta(m) = 0\,[s]$. This establishes (1)(s).

Now at s, α only wants to disengage from m if for some $k < k_\alpha(m)\,[s]$ we have $\gamma(m) < \psi_\alpha(k)$. Since there is no such k, α does not erase m at s and so $(2)(s)$ holds.

$k_\alpha(m)$ is increased at stage s only if some node β extending α erases m at s. By $(1)(s)$ and $(2)(s)$, no node β extending α erases m at s. It follows that $k_\alpha(m)\,[s+1] = k_\alpha(m)\,[s] = 0$.   □

Claim 2.3. *For every m, $\gamma(m)$ is erased only finitely often.*

Proof. We show this by induction on m. Assume the claim holds up to $m - 1$.

Let $\alpha_0 \subset \alpha_1 \subset \cdots \subset \alpha_{m-1}$ be the first m nodes on the true path.
Let $s^* = s^*(m)$ be a stage as follows:

  (1) α_{m-1} is never initialised after s^*;
  (2) α_{m-1} is accessible at s^* (so $s^* > m$ and every node β which lies to the right of α_{m-1} is initialised at s^*);
  (3) No $\gamma(n)$ for $n < m$ is ever erased after s^*.

The assumptions on s^*, together with the initial definitions of k_β and with Claim 2.2, show that after s^*, only nodes among $\alpha_0, \ldots, \alpha_{m-1}$ may erase $\gamma(m)$. By *reverse* induction on $i < m$, we show that after some stage, α_i does not erase $\gamma(m)$.

Suppose that after some stage $s_i > s^*$, no node extending α_i erases $\gamma(m)$. Of course s_{m-1} exists.

At s_i, $k_{\alpha_i}(m)$ reaches a fixed value.

Suppose for contradiction that α_i erases $\gamma(m)$ infinitely often. At any stage s at which α_i erases $\gamma(m)$, we enumerate $\gamma(m)\,[s]$ into an auxiliary set $U = U_{\alpha_i, m}$. By our assumptions on p, there is some stage $s > s_i$ at which dom $\Psi_{\alpha_i}(B) > k_{\alpha_i}(m)$ and at which α_i receives permission from A to erase $\gamma(m)$ without enumerating it into B (there are infinitely many stages at which permission is granted).

The computation $\Psi_{\alpha_i}(B) \upharpoonright k_{\alpha_i}(m)$ which we have at stage s is not injured at s and in fact will never be injured and always be confirmed. Thus after stage s, α_i never erases $\gamma(m)$.   □

Corollary 2.4. $0' \leqslant_T A \oplus B$.

id-*c.e. approximations.* Suppose that $\Psi_e(B)$ is total. Let α be on the eth level of the true path (so $\Psi_\alpha = \Psi_e$). We know that there are infinitely many α-expansionary stages and hence $\alpha^\frown\infty$ is on the true path.

Let r^* be the last stage at which α is initialised. Thus we have $k_\alpha(m) = 0$ from stage r^* onwards exactly for $m \leqslant e$ or $m \leqslant r^*$ (and we may assume that $r^* > e$). For $m \geqslant r^*$ we have $k_\alpha(m) \geqslant m - e$ from stage r^* onwards.

Let s^* be a stage after which the marker $\gamma(r^*)$ is never erased.

Definition 2.5. We *believe* a computation $\Psi_\alpha(k)[s]$ if $s > s^*$, $\alpha^\frown\infty$ is accessible at s, and if for all β such that $\beta^\frown\infty \subseteq \alpha$ and for all m such that $k \geqslant k_\alpha(m)[s]$ we have $\mathrm{dom}\,\Psi_\beta(B) > k_\beta(m)[s]$.

We can effectively recognise believable computations.

Claim 2.6. *The correct $\Psi_\alpha(k)$ computation is eventually believable at every stage at which $\alpha^\frown\infty$ is accessible.*

Proof. For all m such that the final value of $k_\alpha(m)$ is at most k and each β as above, $k_\beta(m)$ eventually stabilizes, and $\mathrm{dom}\,\Psi_\beta(B)$ goes to infinity on the stages at which $\alpha^\frown\infty$ is accessible. As there are only finitely many such m ($m < r^*$ or $m \leqslant e + k$), there is a stage at which all such $k_\beta(m)$ have stabilized and $\mathrm{dom}\,\Psi_\beta(B)$ has surpassed them. □

Claim 2.7. *Suppose that $\Psi_\alpha(k)$ is believable at s. Then it is confirmed at s (perhaps after α erased a marker).*

Proof. Because $\alpha^\frown\infty$ is accessible at s. □

Claim 2.8. *Suppose that $\Psi_\alpha(k)$ is believable at stage s, and that the computation is injured at stage $t \geqslant s$, by some $\gamma(m)[t]$. Then either m enters $0'$ at t, or $\gamma(m)$ is enumerated into B by some node extending $\alpha^\frown\infty$ (so $k < k_\alpha(m)$ from t onwards).*

This shows that different believable $\Psi_\alpha(k)$ computations are injured by $\gamma(m)$, for any given m, at most once: no number goes into $0'$ more than once; and if $k < k_\alpha(m)$ from t onwards then no computation $\Psi_\alpha(k)[s']$ with $\gamma(m) < \psi_\alpha(k)[s']$ is believable. Also, there are at most k such m's: suppose that $\Psi_\alpha(k)[s]$ is believable and that $\gamma(m) < \psi_\alpha(k)[s]$ injures the computation later. Then $k \geqslant k_\alpha(m)[s] \geqslant m - e$ (as $m < r^*$ is impossible). Overall we see that the believable computations form an id-c.e. approximation for $\Psi_e(B)$.

Proof of Claim 2.8. Assume m does not enter $0'$ at t. We show that the node that enumerates $\gamma(m)$ into B at stage t must extend $\alpha^\frown\infty$.

We have $m < \gamma(m)[t] < \psi_\alpha(k)[s] < s$. Every β which lies to the right of $\alpha^\frown\infty$ is initialised at stage s and so we set $k_\beta(m) = 0$ at s. By claim 2.2, β never erases $\gamma(m)$ after s.

Of course, α is not initialised after s and so $\gamma(m)$ cannot be erased at t by some node β such that α extends the finite outcome of β. It remains to see that no β such that $\beta^\frown \infty \subseteq \alpha$ erases $\gamma(m)$ at t.

If it does, then there is some l such that $\Psi_\beta(l) \downarrow [t]$ and $\gamma(m) < \psi_\beta(l)[t]$ but $l < k_\beta(m)[t]$. Now no-one erased $\gamma(m)$ between stages s and t so $\gamma(m)[s] = \gamma(m)[t]$ and $k_\beta(m)[s] = k_\beta(m)[t]$. Thus $l < k_\beta(m)[s]$. The computation $\Psi_\alpha(k)[s]$ is believable so it is confirmed: $k \geqslant k_\alpha(m)[s]$. By the definition of believability we have dom $\Psi_\beta(B)[s] > k_\beta(m)[s] > l$.

At s, the computation $\Psi_\beta(l)$ is confirmed. This means that we must have $\gamma(m)[s] \geqslant \psi_\beta(l)[s]$, for $k_\beta(m)[s] > l$. No markers $\gamma(n)$ for $n \leqslant m$ were erased between s and t; so B didn't change between s and t on numbers below $\psi_\beta(l)[s]$. Thus $\psi_\beta(l)[s] = \psi_\beta(l)[t]$ and altogether we get a contradiction to $\gamma(m)[t] < \psi_\beta(l)[t]$. $\square$

3. AC cuppable $\neq$ PS

In this section we prove Theorem 1.3.

To do this, we construct a c.e. set A; we construct a c.e. set B such that **a** and **b** form a minimal pair (thus ensuring that **a** is not promptly simple — see Ambos-Spies, Jockusch, Shore and Soare [1]); and we construct a c.e. set C that is array computable, and a Turing functional Γ such that $\Gamma(A \oplus C) = 0'$.

We need to meet the following requirements:

$$P_e : \quad \overline{B} \neq W_e$$
$$N_e : \quad \Phi_e(A) = \Phi_e(B) = g \Rightarrow g \equiv_T 0$$
$$Q_i : \quad \Psi_i(C) \text{ total} \Rightarrow \Psi_i(C) \text{ is } (\text{id} + 1)\text{-c.e.}$$

Note that the fact that B is not computable implies that neither A nor C are complete or computable.

3.1. The strategy

We first discuss the basic strategy for meeting each requirement. The first two are familiar: P_e is met by following the Friedberg-Muchnik strategy of picking a follower x and holding onto it until it enters W_e, in which case it is enumerated into B as well. N_e is met by following the Lachlan strategy of monitoring the length of agreement, and allowing only one side of the computations $\Phi_e(A)$ and $\Phi_e(B)$ be injured at a time; our construction will be done on a tree of strategies, and so the restraint will be implicit in the machinery of the tree.

To meet a Q_i requirement, we construct an $(\mathrm{id} + 1)$-c.e. approximation for $\Psi_i(C)$ by preserving each computation $\Psi_i(C, x)$. To do this, we prevent most numbers from entering C.

As mentioned, we make use of a tree of strategies. The tree is used to arrange the restraints imposed by N_e and Q_i requirements, which are of infinitary type.

We order the requirements effectively and let each level of the tree be devoted to a single requirement. Each Q- or N-node on the tree has two children representing two possible outcomes. The outcomes are f (finite) and ∞ (infinite), the latter guessing totality of $\Psi_i(C)$ or that $\Phi_e(A) = \Phi_e(B)$ are total. The P-nodes do not impose any restraint and so have a single outcome.

The priority ordering on the tree is the lexicographic ordering generated by the ordering $\infty < f$. We say that a node α lies to the left of node β if α is stronger than β but they are not $\subset$-comparable.

The driving force behind the construction is numbers entering $0'$. At stage s we have an increasing sequence of markers (Γ-uses) $\gamma(0, s), \ldots, \gamma(s, s)$. When at some stage s, a number m enters $0'$, we must put $\gamma(m) = \gamma(m, s)$ into either A or C. Some nodes will have a preference between A and C: for example, Q-nodes may want to prevent $\gamma(m)$ from entering C, and so will favour A. Another possible scenario is that an N-node, currently during an expansionary stage, allows us to put number into either A or B but not both. A longer P-node might wish to enumerate a follower into B; it would thus prefer $\gamma(m)$ to go into C rather than A. The third possibility is that an N-node impose finite restraint on A while at a non-expansionary stage; again it will have an opinion as to where $\gamma(m)$ should go.

As expected, the final decision lies with the strongest node that has any preference.

The first scenario described (a Q-node α wants to protect a computation and so puts a marker into A) needs more elaboration. As described so far, it is possible that α acts infinitely many times, each time injuring some N-node $\beta \supseteq \alpha^\frown \infty$. To avoid this, α needs to take preventative action. Thus, instead of simply removing markers when they pose problems, whenever α sees a new computation (on x, say) that it wants to preserve, it puts all markers that are potentially dangerous in the future (that is, all markers between x and its $\Psi_i(C)$-use) into A straight away. Injury for β is avoided because at this stage (compared with the stage at which we really would have needed to put markers into A), the markers are still large.

We note a further delicacy. As in the proof of Theorem 1.2, it is not enough to verify that the markers are "confirmed"; we need to take care of inputs between markers by actively confirming larger numbers. For this, a *test point* $d(\alpha)$ is appointed for a Q-node α; it is for convergence of $\Psi_i(C)$ on that test point that we wait, and this point may be larger than the relevant marker. Naïvely applying this strategy may result in a single marker being driven to infinity. We need to find a compromise; thus the test point $d(\alpha)$ may at some stages be *mobile*, where we keep lifting its value to the next marker (to make sure that markers are not raised infinitely often); but when it really needs to act, it becomes *stationary*.

The rules for test points are as follows. A mobile test point $d(\alpha)$ must always have the value of some marker. Thus if the marker is enumerated into A or C, we raise the value of the test point to be the next marker (which will be the least one chosen at the end of the stage). However, if a marker *smaller* than the test point is enumerated, then the test point becomes stationary and will not point at a marker; indeed it will not move until α acts to confirm it.

3.2. Construction

At stage s, we define the path of nodes accessible at s; for each node α, we define how it acts. At stage s, a new number m is enumerated into $0'$. The stage always ends with some accessible node enumerating $\gamma(m)$ into either A or C. The usual conventions for working with markers apply; if $\gamma(n)$ is enumerated into a set then all of $\gamma(k)$ for $k \in (n, s)$ are also enumerated into the same set and are redefined with large values at the end of the stage.

For any node β, let $r(\beta)[s]$ be the last stage before s at which β was initialised.

The first node accessible at any stage is the root. Suppose that α is accessible at s; the requirement to which α is assigned determines its actions.

α *is assigned to* P_e. If α does not have a follower (this is the first time we visit α since it was initialised), we appoint a fresh (large) follower for α.

Let x be α's follower. If $x \in B$ or if $x \notin W_e$ then α does nothing; its single child is accessible next.

Otherwise, α enumerates x into B and $\gamma(m)$ into C. (The stage is now ended).

α *is assigned to* N_e. If $\Phi_e(A)(x)\!\downarrow = \Phi_e(B)(x)\!\downarrow$, then let

$$u_e(x) = \max\{\psi_e(A)(x), \psi_e(B)(x)\}.$$

Input x is *believable* at s if there is no Q-node β such that $\beta^\frown\infty \subseteq \alpha$ and $d(\beta)[s]$ is smaller than $u_e(x)[s]$.

The modified *length of agreement*, $\ell(e)[s]$, is the maximal y such that on every $x < y$ there is a believable computation on x at stage s.

Let $r = r(\alpha^\frown f)[s]$. If $\ell(e, s) > \ell(e, r)$ then we say that s is *expansionary* for α and let ∞ be α's outcome.

Suppose that s is not expansionary. If $\gamma(m) \leqslant r$ then α enumerates $\gamma(m)$ into C and halts the stage. Otherwise, the outcome is finite, and $\alpha^\frown f$ is accessible.

α *is assigned to* Q_i. If s is the first stage since $r(\alpha)[s]$ at which α is accessible, then $d(\alpha)[s]$ is not yet defined; we define it with value equal to the least marker greater than $r(\alpha)[s]$, and set it to a mobile state.

Let t be the last stage at which α was accessible. There are two cases.

(1) $t > r(\alpha)[s]$ and at t, α acted to confirm $d(\alpha)[t]$ (case 2b(i) below). In this case we let $\alpha^\frown\infty$ be accessible. [We believe totality as the "length of confirmation" has increased. We are not worried about any of the markers entering C because we already protected computations by removing dangerous markers. We now let weaker nodes act.]

(2) Otherwise. There are two cases.

    (a) $\Psi_i(C)(d(\alpha)[s]) \uparrow$. In this case there is nothing we can do on behalf of the requirement.

        (i) If $\gamma(m) > r(\alpha^\frown f)[s]$ then we let $\alpha^\frown f$ be accessible.

        (ii) If $\gamma(m) \leqslant r(\alpha^\frown f)[s]$ then α imposes finite restraint by enumerating $\gamma(m)$ into C.

    (b) $\Psi_i(C)(d(\alpha)[s]) \downarrow$. In this case we want to make progress.

        (i) If $\gamma(m) > d(\alpha)[s]$ then we *confirm* $d(\alpha)$: we enumerate all markers greater than $d(\alpha)[s]$ into A and halt the stage. Note that $d(\alpha)[s]$ itself (if it is a marker), is not enumerated. As described later, we now pick new large markers; we raise the value of $d(\alpha)$ and redefine $d(\alpha)[s + 1]$ to be the least marker among the new ones just chosen; we set it to be mobile.
[At the next stage at which α is accessible, we follow case 1.]

        (ii) However, if $\gamma(m) \leqslant d(\alpha)[s]$ then there is no point in confirming $d(\alpha)[s]$ as smaller numbers will have to enter a set at this stage. We let $\alpha^\frown\infty$ be accessible and do nothing else.

At the end of the stage we initialise nodes that lie to the right of the accessible nodes (and all nodes of length $> s$), and reassign a large value (greater than s) to any marker $\gamma(k)$ ($k \leqslant s$) which is not defined ($\gamma(k)$ was just enumerated into a set or $k = s$). [Note we do not initialise *extensions* of the final accessible node.] Let $\gamma(k)$ be the least marker enumerated at stage s. If $\gamma(k) = d(\alpha)[s]$ for a Q-node α, define $d(\alpha)[s+1]$ to point to a new marker. On the other hand, if $\gamma(k) < d(\alpha)[s]$, then $d(\alpha)$ will no longer point to a marker value and hence has become *stationary* until α acts to confirm it.

3.3. Verification

Claim 3.1. *Suppose that α is accessible at stage s, and that m enters $0'$ at s. Then $r(\alpha) \leqslant \gamma(m)$ $[s]$.*

Proof. This is by induction on α. For the root, we know that $r(\langle\rangle)[s] = 0$ for all s. Suppose that α is accessible at s. If α is a P-node then $r(\alpha) = r(\beta)$ $[s]$ where β is α's only child. If α is a Q- or N-node, then $r(\alpha) = r(\alpha^\frown \infty)$ $[s]$. If $\gamma(m) < r(\alpha^\frown f)$ $[s]$ then $\alpha^\frown f$ cannot be accessible at s: instead of going to $\alpha^\frown f$, α enumerates $\gamma(m)$ into C and halts the stage. $\qquad\square$

Corollary 3.2. *If a node β is initialised at some stage r, then after r, β never enumerates a number smaller than r into any set.*

Proof. There are three kinds of enumerations: enumeration of $\gamma(m)$ (where m enters $0'$); enumeration of a follower into B (by a P-node); and enumeration of markers into A by a Q-node that is confirming its test point.

By claim 3.1, if at stage $s > r$, m enters $0'$ and β enumerates $\gamma(m)$, then (as β is accessible) $\gamma(m) \geqslant r(\beta)[s] \geqslant r$.

If β is a P-node, then after r it picks new followers, all greater than r.

And if β is a Q-node that is confirming $d(\beta)$, then by definition, $d(\beta) \geqslant r(\beta)$ $[s]$ (it is initially assigned to a larger value and can then only increase), and only markers greater than $d(\beta)$ are enumerated. $\qquad\square$

Lemma 3.3. *For every m, the marker $\gamma(m)$ is eventually fixed.*

Proof. Suppose that after some stage s_0, the markers $\gamma(0), \ldots, \gamma(m)$ do not change. Also suppose that after s_0, $m + 1$ does not enter $0'$, so the only way $\gamma(m+1)$ can enter a set after stage s_0 is when some Q-node confirms a smaller $d(\alpha)[s]$ (necessarily $d(\alpha)[s] \geqslant \gamma(m)$), in which case $\gamma(m+1)$ will be enumerated into A.

Let α be such a node. After confirming at stage s, a new value is set for $\gamma(m+1)$, which is also the new value of $d(\alpha)$. At s, $d(\alpha)$ is set to be mobile; and it can never be reverted to be stationary (only $\gamma(m)$ enumeration can do that). So after s, we always have $d(\alpha) \geqslant \gamma(m+1)$ (even if they increase together). Therefore, after stage s_0, there is at most one stage s at which α's action causes $\gamma(m+1)$ to be enumerated into A.

Consider a Q-node of length $\geqslant s_0, \gamma(m)$. Any test point it ever appoints is at least $\gamma(m+1)$ and as in the previous paragraph, if it does appoint it to be $\gamma(m+1)$ it remains mobile until it is lifted to be greater than $\gamma(m+1)$. So such a node never enumerates any $\gamma(m+1)$ into A. This shows that after s_0, only finitely many nodes enumerate a $\gamma(m+1)$ into A.

[An alternative construction could have been: do not allow nodes of length k to enumerate $\gamma(k)$ (i.e., always set $d(\alpha) \geqslant \gamma(k)$). Then you would need this current lemma to show finite injury on the true path.] □

Corollary 3.4. $0' \leqslant_T A \oplus C$.

Proof. Whenever m goes into $0'$, $\gamma(m)$ enters either A or C, and is redefined with a large value. □

Lemma 3.5. *Suppose that α is a node on the tree that is accessible infinitely often and that is eventually never initialised. Then there is an immediate successor of α on the tree that is accessible infinitely often and is eventually never initialised.*

Since the root is always accessible and is never initialised, inductively applying the lemma shows the existence of an infinite true path.

Proof. Let $r_0 = r(\alpha)[\omega]$ be the last stage at which α is initialised. Of course, everything depends on the requirement to which α is assigned.

α *is assigned to P_e.* After stage r_0, α is assigned a follower x; that follower is never cancelled. After r_0, α acts at most once; after that, whenever α is accessible, so is its child. Also, the child is never initialised without α also being initialised.

α *is assigned to N_e.* Suppose that there is a last α-expansionary stage $r \geqslant r_0$. After r, $\alpha {}^\frown f$ is never initialised. After r, α only enumerates a marker $\gamma(m)$ into C if $\gamma(m) \leqslant r$, which is to say, finitely often. At other stages, whenever α is accessible, so is $\alpha {}^\frown f$.

If there are infinitely many expansionary stages, then $\alpha {}^\frown \infty$ is accessible infinitely often; it is not initialised after r_0.

α *is assigned to* Q_i. If there are infinitely many stages at which α confirms its test point, then $\alpha\frown\infty$ is accessible infinitely often.

Suppose that there is a last stage $r_1 > r_0$ at which α confirms $d(\alpha)$. Suppose that at the end of r_1, we set $d(\alpha) = \gamma(k)$. Let $r_2 > r_1$ be a stage after which the value of $\gamma(k)$ is fixed. Then after r_2, the value of $d(\alpha)$ is also fixed (it may or may not have become stationary before r_2). After r_2, case (2b(ii)) only holds if $\gamma(m) \leqslant d(\alpha)$, so finitely many times. So there is a stage $r_3 > r_2$ after which $\alpha\frown\infty$ is never accessible (so $\alpha\frown f$ is not initialised after stage r_3). Later, case (2a(ii)) applies only if $\gamma(m) \leqslant r_3$; again, finitely many times. So after some stage $r_4 > r_3$, whenever α is accessible, so is $\alpha\frown f$. $\qquad\square$

Corollary 3.6. *Every P-requirement is met, so B is not computable.*

Proof. Standard. $\qquad\square$

Lemma 3.7. *Every Q_i requirement is met. Hence C is array computable.*

Proof. Let α on the true path work for Q_i. Suppose that $\Psi_i(C)$ is total. There are infinitely many stages at which α confirms its test point. For otherwise, as argued in the proof above, there would be a final fixed value for $d(\alpha)$, and a stage after which case (2b(ii)) doesn't hold; but this contradicts $\Psi_i(C)$ converging permanently on that final value.

Definition 3.8. We say that a number $x < \omega$ is *confirmed* at stage s if $\Psi_i(C)(x)\downarrow [s]$ with use smaller than the least marker greater than x.

The point, of course, is that a confirmed computation can only be injured by a number not greater than the input.

Suppose that α is last initialised at stage r_0. At the next stage r_1 at which α is accessible, α sets $d(\alpha) = \gamma(k_0)$ for some k_0. Let $r_2 \geqslant r_1$ be the least stage after which $\gamma(k_0)$ is always fixed. At the end of r_2 we have $x_0 := d(\alpha)[r_2 + 1] \leqslant \gamma(k_0)[r_2 + 1]$. We in fact must have equality, because at a later stage α confirms its test point.

We prove three claims that suggest a way to approximate $\Psi_i(C)$.

Claim 3.9. *Let $s > r_2$ be a stage at which α acts (to confirm its test point). Suppose that $x \in [x_0, d(\alpha)[s]]$ is confirmed at the end of s. Then it is confirmed at any later stage at which α acts.*

Proof. Suppose that at the end of stage s, $\gamma(m-1) < x \leqslant \gamma(m)$. We have $\gamma(m) \leqslant d(\alpha)$ $[s+1]$. By assumption, $\psi_i(C)(x) < \gamma(m)$ $[s+1]$. Let $t > s$ be the next stage at which α acts. If $\gamma(m-1)$ is not enumerated between s and t, then the computation is preserved and so it is of course still confirmed. Otherwise, let $\gamma(n)$ be the smallest marker enumerated between t and s, say at stage u. Then at the end of u we fix $d(\alpha)$ to be stationary and redefine a greater $\gamma(n)$. At t, $d(\alpha)$ is confirmed; since there is no marker in $(x, d(\alpha)[t]]$, x is also confirmed. $\qquad\square$

Claim 3.10. *Let $s > r_2$ be a stage at which α acts and let $n \in [x_0, d(\alpha)[s]]$ be a marker. Then n is confirmed at the end of s.*

Proof. If $d(\alpha)[s] = n$, then n is confirmed at stage s. Otherwise, let $t < s$ be the least stage at the end of which $d(\alpha) > n$. Then $d(\alpha)[t] = n$ (n is already a marker at t) and α acts at t and confirms n. $\qquad\square$

Claim 3.11. *Let $s > r_2$ be a stage at which α acts. Let $x \in [x_0, d(\alpha)[s]]$. Suppose that $\Psi_i(C)(x)[s]$ is injured at a later stage u by some $y > x$. Then at the end of the next stage t at which α acts, x is confirmed.*

Proof. Let $\gamma(m-1)[s] < x \leqslant \gamma(m)$ $[s]$. We must have $\gamma(m) \leqslant d(\alpha)$ $[s]$; otherwise, $\gamma(m)[s]$ is redefined at s to be larger than $\psi_i(C)(x)$ and then x is confirmed. Then $\gamma(m)$ is confirmed at the end of s and so $y = \gamma(m)[s]$. Also at the end of s we have $d(\alpha) > y$. It follows that at the end of u we have $\gamma(m-1) < x < d(\alpha) < \gamma(m)$ (since $d(\alpha)$ is stationary). We thus see that at t, the confirmation of $d(\alpha)$ is also a confirmation of x. $\qquad\square$

The following is an approximation for $g = \Psi_i(C)$: at a stage $s > r_2$ at which α confirms $d(\alpha)$, guess $g(x) = \Psi_i(C)(x)$ for all $x \leqslant d(\alpha)[s]$. Then the claims ensure that a guessed computation can be injured by some $y > x$ at most once after s. It follows that this approximation is $\mathrm{id} + 1$-c.e. $\qquad\square$

Lemma 3.12. *Every N_e requirement is met. As $B >_T 0$, it follows that A and B form a minimal pair.*

Proof. Let α on the true path work for N_e. Suppose that $\Phi_e(A) = \Phi_e(B) = g$ are total and equal. We first argue that $\ell(e)[s] \to \infty$ (and so $\alpha^\frown\infty$ is on the true path). This is because for all x, eventually there is a permanent $\Phi_e(A)(x) = \Phi_e(B)(x)$ with total use u; for Q-nodes β such that $\beta^\frown\infty \subseteq \alpha$ we know that $d(\beta)[s] \to \infty$; and there are are finitely many such β (admissibility rocks).

Assume that after stage r_1, α is never initialised and no P-node $\beta \subset \alpha$ enumerates a follower into B.

The familiar argument of Lachlan's that shows that g is computable follows through, provided that we can show that:

(1) If $s > r_1$ is an α-expansionary stage, then at s perhaps numbers enter A or numbers enter B but not both.

(2) If $s > r_1$ is an α-expansionary stage and $x < \ell(e)[s]$, then numbers below $u_e(x)[s]$ do not enter either A or B until the next α-expansionary stage.

The first is immediate; it holds at every stage of the construction. We verify the second point. Now suppose that $s > r_1$ is α-expansionary and let $x < \ell(e)[s]$.

- Nodes that lie to the right of $\alpha^\frown \infty$ are initialised at stage s; they never enumerate anything smaller than s (which is in turn greater than $u_e(x)[s]$) into any set.
- Nodes that lie to the left of α are not accessible after r_1.
- P-nodes $\beta \subset \alpha$ do not enumerate numbers into B after stage r_1.
- Nodes extending $\alpha^\frown \infty$ are not accessible between s and the next α-expansionary stage.

Thus the only possible culprits are Q-nodes $\beta \subset \alpha$ that confirm $d(\beta)$. If $\beta^\frown f \subseteq \alpha$ then β does not confirm $d(\beta)$ after r_1; if it does, then at the next stage at which it is accessible, α will be initialised. However, if β confirms $d(\beta)$ at $t > s$, then it only enumerates numbers greater than $d(\beta)[t] \geqslant d(\beta)[s]$. But x is believable at s, which means $u_e(x)[s] \leqslant d(\beta)[s]$.

$\square$

References

[1] K. Ambos-Spies, C. Jockusch Jr., R.A. Shore, and R.I. Soare, *An algebraic decomposition of recursively enumerable degrees and the coincidence of several degree classes with the promptly simple degrees*, Trans. Amer. Math. Soc., Vol. 281 (1984), 109-128.

[2] M. Bickford and C. Mills, *Lowness properties of r.e. sets*, typewritten unpublished manuscript.

[3] P. Cholak, R. Coles, R. Downey, and E. Herrmann, *Automorphisms of the lattice of Π_0^1 classes: Perfect thin classes and anr degrees*, Trans. Amer. Math. Soc. Vol. 353 (2001), 4899-4924.

[4] P. Cholak, M. Groszek, and T. Slaman, *An almost deep degree*, J. Symbolic Logic, Vol. 66(2) (2001), 881-901.

[5] P. Cholak, R. Downey, and M. Stob, *Automorphisms of the lattice of recursively enumerable sets: promptly simple sets*, Trans. American Math. Society, 332 (1992), 555-570.

[6] R. Downey and N. Greenberg, *Domination and definiability II: The ω^ω case*, in preparation.

[7] R. Downey, N. Greenberg and R. Weber, *Domination and definability I: The ω case*, in preparation.

[8] R. Downey and D. Hirschfeldt, *Algorithmic Randomness and Complexity*, Springer-Verlag, to appear.

[9] R. Downey, D. Hirschfeldt, A. Nies, and F. Stephan, *Trivial reals*, extended abstract in *Computability and Complexity in Analysis* Malaga, (Electronic Notes in Theoretical Computer Science, and proceedings, edited by Brattka, Schröder, Weihrauch, FernUniversität, 294-6/2002, 37-55),July, 2002. Final version appears in *Proceedings of the 7th and 8th Asian Logic Conferences*, (R. Downey, Ding Decheng, Tung Shi Ping, Qiu Yu Hui, Mariko Yasuugi, and Guohua Wu (eds)), World Scientific, Singapore (2003) 103-131.

[10] R. Downey, D. Hirschfeldt, A. Nies, and S. Terwijn, *Calibrating randomness*, to appear, Bulletin Symbolic Logic.

[11] R. Downey, C. Jockusch, and M. Stob, *Array nonrecursive sets and multiple permitting arguments*, in *Recursion Theory Week* (Ambos-Spies, Muller, Sacks, eds.) Lecture Notes in Mathematics 1432, Springer-Verlag, Heidelberg, 1990, 141-174.

[12] R. Downey, C. Jockusch, and M. Stob, *Array nonrecursive degrees and genericity*, in *Computability, Enumerability, Unsolvability* (Cooper, Slaman, Wainer, eds.), London Mathematical Society Lecture Notes Series 224, Cambridge University Press (1996), 93-105.

[13] L. Harrington and R. Soare, *Post's Program and incomplete recursively enumerable sets*, Proc. Natl. Acad. of Sci. USA 88 (1991), 10242-10246.

[14] S. Ishmukhametov, *Weak recursive degrees and a problem of Spector,* in *Recursion Theory and Complexity*, (ed. M. Arslanov and S. Lempp), de Gruyter, (Berlin, 1999), 81-88.

[15] M. Kummer, *Kolmogorov complexity and instance complexity of recursively enumerable sets*, SIAM Journal of Computing, Vol. 25 (1996), 1123-1143.

[16] A. Lachlan, *A recursively enumerable degree which will not split over all lesser ones*, Ann. Math. Logic, Vol. 9, (1975), 307-365.

[17] W. Maass, *Characterization of the recursively enumerable sets with supersets effectively isomorphic to all recursively enumerable sets,* Trans. Amer. Math. Soc., Vol. 279 (1983), 311-336.

[18] Ng Keng Meng, F. Stephan and G. Wu, The degrees of weakly computable reals, in preparation.

[19] A. Nies, *Lowness properties and randomness*, Advances in Mathematics Vol. 197 (2005), 274-305.

[20] R.W. Robinson, *Jump restricted interpolation in the recursively enumerable degrees*, Annals of Math., Vol 93 (1971), 586-596.

[21] B. Schaeffer, *Dynamic notions of genericity and array noncomputability*, Ann. Pure Appl. Logic, Vol. 95(1-3) (1998), 37-69.

[22] R.I. Soare, *Automorphisms of the lattice of recursively enumerable sets Part II: Low sets*, Annals of of Math. Logic, Vol. 22 (1982), 69-107.

[23] R.I. Soare, *Recursively enumerable sets and degrees* (Springer, Berlin, 1987).

[24] S. Terwijn and D. Zambella, *Algorithmic randomness and lowness*, Journal of Symbolic Logic, vol. 66 (2001), 1199-1205.

LOWNESS FOR COMPUTABLE MACHINES

Rod Downey

School of Mathematics, Statistics and Computer Science, Victoria University
P.O. Box 600, Wellington, New Zealand
E-mail: Rod.Downey@vuw.ac.nz

Noam Greenberg

School of Mathematics, Statistics and Computer Science, Victoria University
P.O. Box 600, Wellington, New Zealand
E-mail: greenberg@mcs.vuw.ac.nz

Nenad Mihailović

Mathematical Institute, University of Heidelberg
Im Neuenheimer Feld 294, 69120 Heidelberg, Germany
E-mail: mihailovic@math.uni-heidelberg.de

André Nies

Department of Computer Science, University of Auckland
Private Bag 92019, Auckland, New Zealand
E-mail: andre@cs.auckland.ac.nz

Two lowness notions in the setting of Schnorr randomness have been studied (lowness for Schnorr randomness and tests, by Terwijn and Zambella [19], and by Kjos-Hanssen, Stephan, and Nies [7]; and Schnorr triviality, by Downey, Griffiths and LaForte [3, 4] and Franklin [6]). We introduce lowness for computable machines, which by results of Downey and Griffiths [3] is an analog of lowness for K. We show that the reals

---

The first, second and fourth authors are partially supported by the New Zealand Marsden Fund for basic research. This work was carried out while Mihailović was visiting Victoria University and was also partially supported by the Marsden Fund and by a "Doktorandenstipendium" from the DAAD (German Academic Exchange Service).

that are low for computable machines are exactly the computably traceable ones, and so this notion coincides with that of lowness for Schnorr
randomness and for Schnorr tests.

1. Introduction

A central set of results in the theory of algorithmic randomness were established by Nies and his co-authors. They prove the coincidence of a number of natural "anti-randomness" classes associated with prefix-free Kolmogorov complexity. Recall that A is called *low for K* if for all x, $K^A(x) \geq$
$K(x) - O(1)$,[a] A is called *K-trivial* if for all n, $K(A \restriction n) \leq K(n) + O(1)$, and
A is called *low for Martin-Löf randomness* if the collection of reals Martin-
Löf random relative to A is the same as the collection of Martin-Löf random
reals. We have the following.

Theorem 1.1. (Nies, Hirschfeldt, [12, 13]) *For every real A, the following
are equivalent.*

  (i) *A is low for K.*
 (ii) *A is K-trivial.*
(iii) *A is low for Martin-Löf randomness.*

The situation for other notions of randomness is less clear. In this paper
we look at the situation for Schnorr randomness. Recall that a real A is
said to be *Schnorr random* iff for all Schnorr tests $\{U_n : n \in \mathbb{N}\}$, $A \notin \cap_n U_n$,
where a Schnorr test is a Martin Löf test such that $\mu(U_n) = 2^{-n}$ for all n.
(Of course 2^{-n} is a convenience. As Schnorr [16] observed, any uniformly
computable sequence of reals with effective limit 0 would do.)

The reader might note that there are two possible lowness notions associated with Schnorr randomness. A real A is *low for Schnorr randomness*
if no Schnorr random real becomes non-Schnorr-random relative to A. But
since there is no universal Schnorr test, we can also define the stronger
(and more technical) notion of lowness for *tests*; a real A is *low for Schnorr
tests* if for every A-Schnorr test $\{U_n^A : n \in \mathbb{N}\}$, there is a Schnorr test
$\{V_n : n \in \mathbb{N}\}$ such that $\cap_n U_n^A \subseteq \cap_n V_n$.

---

[a]In this paper K will denote prefix-free Kolmogorov complexity and we will refer to
members $A = a_0 a_1 \ldots$ of Cantor space as *reals*, with $A \restriction n$ being the first n bits of A.
We assume that the reader is familiar with the theory of algorithmic randomness. For
details we refer to the monographs of Li and Vitányi [10], of Downey and Hirschfeldt [5],
and of Nies [14].

Terwijn and Zambella [19] proved that there are reals that are low for Schnorr tests. In fact, they classified the collection of reals which are low for Schnorr tests.

For any n, we let D_n denote the nth canonical finite set.

Definition 1.2. (Terwijn and Zambella [19]) We say that a real A is *computably traceable* if there is a computable function $h(x)$ such that for all functions $g \leq_T A$, there is a computable collection of canonical finite sets $D_{r(x)}$ with $|D_{r(x)}| \leq h(x)$ and such that $g(x) \in D_{r(x)}$.

We remark that (as noticed by Terwijn and Zambella) if A is computably traceable then for the witnessing function h we can choose any computable, non-decreasing and unbounded function.

Terwijn and Zambella proved the following attractive result.

Theorem 1.3. (Terwijn and Zambella [19]) *A is low for Schnorr tests iff A is computably traceable.*

We remark that while all K-trivials are Δ_2^0 by a result of Chaitin [1], the computably traceable reals are all hyperimmune-free, and there are $2^{\aleph_0}$ many of them.

Subsequently, Kjos-Hanssen, Stephan, and Nies [7] proved that A is low for Schnorr randomness iff A is low for Schnorr tests.

The reader might wonder about analogs of the other results for K. The other members of the coincidence involve K-triviality and lowness for K. What about the Schnorr situation? We want some analog of the characterization of Martin-Löf randomness in terms of prefix-free complexity. (R is Martin-Löf random iff for all n, $K(R \restriction n) \geq n - O(1)$.) Such a characterization was discovered by Downey and Griffiths [3]. They define a prefix-free Turing machine M to be *computable* if the domain of M has computable measure, that is, $\sum_{\{\sigma : M(\sigma)\downarrow\}} 2^{-|\sigma|}$ is a computable real. They then establish the following:

Theorem 1.4. (Downey and Griffiths [3]) *R is Schnorr random iff for all computable machines M, for all n, $K_M(R \restriction n) \geq n - O(1)$.*[b]

---

[b]Note that since the range of M need not be all of $2^{<\omega}$, we need to let $K_M(x) = \infty$ for all strings x not in the range of M.

The quantification over machines is necessary because (as in the situation for Schnorr tests), there is no universal computable machine. With this result we are in a position to define a real A to be *Schnorr trivial* if for every computable machine N there is a computable machine M such that for all n, $K_M(A \upharpoonright n) \leq K_N(n) + O(1)$. This notion was initially explored by Downey and Griffiths [3] and Downey, Griffiths and LaForte [4], who showed that this class does not coincide with the reals that are low for Schnorr randomness. For instance, there are Turing complete Schnorr trivial reals. Johanna Franklin [6] established the following.

Theorem 1.5. (Franklin [6])

  (i) *There is a perfect set of Schnorr trivials.*
 (ii) *Every degree above $\mathbf{0}'$ contains a Schnorr trivial.*
(iii) *Every real that is low for Schnorr randomness is also Schnorr trivial.* [c]

Thus the relationship between lowness for Schnorr randomness and Schnorr triviality is quite different from the analogous situation for Martin-Löf randomness.

The last piece of the puzzle is the analog for lowness for K. Armed with the machine characterization of Schnorr randomness, we give the following definition.

Definition 1.6. A real A is *low for computable machines* iff for all A-computable machines M there is a computable machine N such that for all x,

$$K_M^A(x) \geq K_N(x) - O(1).$$

The reader might be concerned about whether for an A-computable machine M^A as in the definition above, M^B is B-computable for other oracles B. However, given a such a machine, we can obtain another oracle

---

[c]Interestingly, Franklin also showed that the reals that are low for Schnorr randomness are not closed under join. The referee points out that a proof from Lerman [9] can be used to establish Franklin's result. To wit, the minimal degrees generate the Turing degrees under meet and join, and the referee points out that the proof (in [9]) also shows that such degrees can be chosen computably traceable, in the same way that the standard construction of a minimal degree is automatically computably traceable.

machine $\widetilde{M}$ such that $M^A = \widetilde{M}^A$, and such that $\widetilde{M}^B$ is prefix-free and B-computable for every oracle B.[d]

A relativized version of the Kraft-Chaitin Theorem (Lemma 2.1) can be used to show that Theorem 1.4 relativizes. Namely, we have that R is A-Schnorr random iff for all A-computable machines M, for all n, $K_M^A(R \restriction n) \geq n - O(1)$. Therefore, every real A that is low for computable machines is low for Schnorr randomness, and by the results quoted above it follows further that A is low for Schnorr tests and thus is computably traceable. In this paper we show that unlike the situation for triviality, the coincidence of the reals low for Martin-Löf randomness and the low for K ones carries over to the Schnorr case:

Theorem 1.7. *A real A is low for computable machines iff A is computably traceable.*

We remark that part (iii) of Theorem 1.5 above is a consequence of Theorem 1.7, since every real A that is low for computable machines is Schnorr trivial. For let N be a computable machine. Let L be an A-computable machine such that for all n, $K_L^A(A \restriction n) = K_N(n)$ (for all x, if $N(x) = n$ then let $L(x) = A \restriction n$.) Then there is some computable machine M such that for all x, $K_M(x) \leq K_L^A(x) + O(1)$; M is as required to witness that A is trivial.

2. The Proof

We note that if we enumerate a Kraft-Chaitin set with a computable sum then the machine produced is computable:

Lemma 2.1. (Kraft-Chaitin) *Let $\langle d_0, \tau_0 \rangle, \langle d_1, \tau_1 \rangle, \dots$ be a computable list of pairs consisting of a natural number and a string. Suppose that $\sum_{i < \omega} 2^{-d_i}$ is a computable real (in particular, is finite). Then there is a computable machine N such that for all i, $K_N(\tau_i) \leq d_i + O(1)$.*

---

[d]Indeed, define the machine $\widetilde{M}$ as follows. First, we may assume that for every oracle B, M^B is prefix-free. Now let F be a computable functional such $F(A)$ is total and the measure of the set $\{x \leq F(A, n) : M^A(x) \text{ is defined after } F(A, n) \text{ steps}\}$ approximates $\mu(M^A)$ to within 2^{-n}. Define $\widetilde{M}^B$ inductively: at stage n, first wait for $F(B, n)$ to halt (in the meantime, no new $\widetilde{M}^B$-computations are recognised.) Next, allow M^B to run for $F(B, n)$ many steps and accept new computations as $\widetilde{M}^B$-computations; if at a later stage we see that $\mu(M^B) > \mu(M^B)[F(B, n)] + 2^{-n}$ then we stop accepting new $\widetilde{M}^B$-computations altogether. Then move to stage $n + 1$. Note that the construction is uniform in M, F but not in M alone.

(See Downey [2] for a proof of the Kraft-Chaitin theorem; the fact that we get a computable machine is immediate from the proof.)

To prove Theorem 1.7 we need to show that every computably traceable set A is low for computable machines. So let A be a computably traceable set and let M be an oracle machine such that M^A is A-computable. The idea (somewhat following Terwijn and Zambella) is to "break up" the machine M^A into small and finite pieces which we trace. We view M^A as a function from strings to strings. We will partition M^A into finite pieces $g, f_0, f_1, f_2, \ldots$ where for $n < \omega$, the measure of the domain of f_n is smaller than some small rational ε_n. We then trace the sequence $\langle f_n \rangle$; so for every n, we get $h(n)$ many candidates for f_n, each with domain with measure smaller than ε_n. If we keep $\sum_n h(n)\varepsilon_n$ finite, the union of all of the candidates can be translated into a Kraft-Chaitin set that produces the machine we want.

Let h be the computable function given by Definition 1.2 (again we remark that we can pick any reasonable function; it does not matter for this proof). Fix a computable, decreasing sequence of positive rationals $\varepsilon_0, \varepsilon_1, \ldots$ such that $\sum_{n<\omega} h(n)\varepsilon_n$ is finite; moreover, we want the convergence to be quick, say for every $m < \omega$,

$$\sum_{n \geq m} h(n)\varepsilon_n < 2^{-m}.$$

Let $\langle (\sigma_i, \tau_i) \rangle_{i<\omega}$ be an A-computable enumeration of M^A. We let M_s^A, the machine M^A at stage s, be $\{(\sigma_i, \tau_i) : i < s\}$, and similarly let $M_{\geq s}^A = M^A \setminus M_s^A = \{(\sigma_i, \tau_i) : i \geq s\}$, and for $s < t$, $M_{[s,t)}^A = M_t^A \setminus M_s^A$.

Let t_n be the least stage t such that $\mu(\mathrm{dom}\, M_{\geq t}^A) < \varepsilon_n$. We let $g = M_{t_0}^A$; for $n < \omega$, we let $f_n = M_{[t_n, t_{n+1})}^A$. The point is that the sequence $\langle t_n \rangle$, and so the sequence $\langle f_n \rangle$, are A-computable, as $\mu(\mathrm{dom}\, M_{\geq t}^A) = \mu(\mathrm{dom}\, M^A) - \mu(\mathrm{dom}\, M_t^A)$; the first number is A-computable by assumption, and the latter a rational, computable from the sequence $\langle (\sigma_i, \tau_i) \rangle$ and so from A. For all $n < \omega$, $\mu(\mathrm{dom}\, f_n) < \varepsilon_n$.

Each f_n is a finite function (and so has a natural number code.) We can thus computably trace the sequence $\langle f_n \rangle$; there is a computable sequence of finite sets $\langle X_n \rangle_{n<\omega}$ (i.e. $X_n = D_{r(n)}$ where r is computable) such that for each n, $|X_n| \leq h(n)$, and for each n, (the code for) $f_n \in X_n$. By weeding out elements, we may assume that for each $n < \omega$, every element of X_n is a code for a finite function f from strings to strings whose domain is prefix-free and has measure at most ε_n.

Enumerate a Kraft-Chaitin set L as follows. Let $\langle d, \tau \rangle \in L$ if there is some σ such that $|\sigma| = d$, and one of the following holds:

- $(\sigma, \tau) \in g$;
- For some n and for some $f \in X_n$, $(\sigma, \tau) \in f$.

The set L is computably enumerable. Further, the total of the requests $s = \sum_{(d,\tau) \in L} 2^{-d}$ is a finite, computable real, as we know that for any m,

$$\sum \{2^{-|\sigma|} : (\exists n \geq m)(\exists f \in X_n)[\sigma \in \operatorname{dom} f]\} \leq \sum_{n \geq m} h(n)\varepsilon_n \leq 2^{-m}.$$

From the "computable" Kraft-Chaitin theorem we get a computable machine N such that for some constant c, if $(d, \tau) \in L$, then $K_N(\tau) \leq d + c$. On the other hand, we know that if τ is in the range of M^A then $(K_M^A(\tau), \tau) \in L$ because $f_n \in X_n$ for all n. Thus N is as required.

References

1. Chaitin, G., *Information-theoretical characterizations of recursive infinite strings,* Theoretical Computer Science, Vol. 2 (1976), 45–48.
2. Downey, R., *Five lectures on algorithmic randomness,* this volume.
3. Downey, R. and E. Griffiths, *On Schnorr randomness,* Journal of Symbolic Logic, Vol. 69, No. 2 (2004), 533–554.
4. Downey, R., E. Griffiths, and G. LaForte, *On Schnorr and computable randomness, martingales, and machines,* Mathematical Logic Quarterly, Vol. 50, No. 6 (2004), 613–627.
5. Downey, R. and D. Hirschfeldt, *Algorithmic Randomness and Complexity,* Springer-Verlag Monographs in Computer Science, to appear (preliminary version www.mcs.vuw.ac.nz/~downey).
6. Franklin, J., Ph.D. Dissertation, University of California at Berkeley, in preparation.
7. Kjos-Hanssen, B., F. Stephan, and A. Nies. *Lowness for the class of Schnorr random reals,* SIAM Journal on Computing, Vol. 35, No. 3 (2006), 647–657.
8. Kučera, A. and S. Terwijn, *Lowness for the class of random sets,* Journal of Symbolic Logic, Vol. 64 (1999), 1396–1402.
9. Lerman, M., *Degrees of Unsolvability,* Springer-Verlag Berlin Heidelberg, 1983.
10. Li, M. and P.M.B. Vitányi, *An Introduction to Kolmogorov Complexity and Its Applications,* Springer-Verlag, New York, second edition, 1997.
11. Martin-Löf, P., *The definition of random sequences,* Information and Control, Vol. 9 (1966), 602–619.
12. Nies, A., *Reals which compute little,* to appear, *Proceedings of CL 2002.*
13. Nies, A., *Lowness properties and randomness,* Advances in Mathematics 197, Vol. 1 (2005), 274–305.
14. Nies, A., *Computability and Randomness,* monograph in preparation.

15. Schnorr, C. P., *A unified approach to the definition of a random sequence,* Mathematical Systems Theory, Vol. 5 (1971), 246–258.

16. Schnorr, C. P., *Zufälligkeit und Wahrscheinlichkeit,* Lecture Notes in Mathematics, Vol. 218, 1971, Springer-Verlag, New York.

17. Terwijn, S. A., *Computability and Measure,* Ph.D. Thesis, University of Amsterdam, 1998.

18. Terwijn, S. A., *Complexity and Randomness,* Notes for a course given at the University of Auckland, March 2003. Published as research report CDMTCS-212, University of Auckland.

19. Terwijn, S. A. and D. Zambella, *Computational randomness and lowness,* Journal of Symbolic Logic, Vol. 66, No. 3 (2001), 1199–1205.

A SIMPLER SHORT EXTENDERS FORCING — GAP 3

Moti Gitik

School of Mathematical Sciences
Raymond and Beverly Sackler Faculty of Exact Science
Tel Aviv University
Ramat Aviv 69978, Israel
E-mail: gitik@post.tau.ac.il

Assume GCH. Let $\kappa = \bigcup_{n<\omega} \kappa_n$, $\langle \kappa_n \mid n < \omega \rangle$ increasing, each κ_n is κ_n^{+n+3} – strong witnessed by $(\kappa_n, \kappa_n^{+n+3})$ – extender E_n. We would like to force $2^\kappa = \kappa^{+3}$ preserving all the cardinals and without adding new bounded subsets to κ. It was done first in [2, Sec.3]. Here we present a different method of doing this. The advantage of the present construction is that the preparation forcing is split completely from the main one. This makes the presentation much simpler and likely to allow a possibility of extensions to arbitrary gaps preserving large cardinals (which was not the case in [2]).

1. The Preparation Forcing

Miyamoto and Sharon pointed out that the forcing below recalls a simplified morass. Indeed it implies Velleman's simplified morass with linear limits [3] in a generic extension. We do not know if the objects are equivalent and think that it is not the case due to the intersection properties below. Also it is unclear if such structure exists in L and bigger inner models.

Definition 1.1 The set $\mathcal{P}'$ consists of elements of the form

$$\langle \langle A^{0\kappa^+}, A^{1\kappa^+}, C^{\kappa^+} \rangle, A^{1\kappa^{++}} \rangle$$

so that the following hold:

(1) $A^{1\kappa^{++}}$ is a closed subset of κ^{+3} of cardinality at most κ^{++}.
(2) $A^{0\kappa^+} \prec H(\kappa^{+3})$ of cardinality κ^+

(3) $A^{1\kappa^+}$ is a set of elementary submodels of $A^{0\kappa^+}$ each has cardinality κ^+ and includes κ^+ so that

- (a) $A^{0\kappa^+} \in A^{1\kappa^+}$
- (b) each element of $A^{1\kappa^+} \backslash \{A^{0\kappa^+}\}$ belongs to $A^{0\kappa^+}$
- (c) (well foundedness of inclusion) if $B, C \in A^{1\kappa^+}$ and $B \subsetneq C$, then $B \in C$.

In particular $\langle A^{1\kappa^+}, \subset \rangle$ is well-founded.

Define a function otp_{κ^+} on $A^{1\kappa^+}$ as follows:

$$otp_{\kappa^+}(A) = \sup\{otp(\langle C, \subset \rangle) \mid C \subseteq \mathcal{P}(A) \cap A^{1\kappa^+}$$
$$\text{is a chain under the inclusion}\}.$$

Note that for such A we have by (3c) that $\mathcal{P}(A) \cap A^{1\kappa^+} = A \cap A^{1\kappa^+} \cup \{A\}$.

(4) $C^{\kappa^+} : A^{1\kappa^+} \longrightarrow \mathcal{P}(A^{1\kappa^+})$ is so that

- (a) for each $A \in A^{1\kappa^+}$ we require that $C^{\kappa^+}(A)$ is a closed chain (under inclusion) of elements of $\mathcal{P}(A) \cap A^{1\kappa^+}$ of the length $otp_{\kappa^+}(A)$ and there is no chain in $\mathcal{P}(A) \cap A^{1\kappa^+}$ that properly includes $C^{\kappa^+}(A)$.

  In particular this means that there are chains of the maximal length (i.e. $otp_{\kappa^+}(A)$ which was defined as supremum is really a maximum) and $C^{\kappa^+}(A)$ is one of them. Also note that A is always the largest element of $C^{\kappa^+}(A)$. So $otp_{\kappa^+}(A)$ is always a successor ordinal.

- (b) (Coherence)
  if $B \in C^{\kappa^+}(A)$ then $C^{\kappa^+}(B)$ is the initial segment of $C^{\kappa^+}(A)$ starting with B

- (c) if $otp_{\kappa^+}(A) - 1$ is a limit ordinal (in such cases we shall refer to A as a limit model and otherwise like to a successor one) then each element of $A \cap A^{1\kappa^+} \backslash \{A\}$ is included (and hence also belongs) to one of the members of $C^{\kappa^+}(A)$.

(5) for all $A \in A^{1\kappa^+}$ if $\alpha \in A$ then $A^{1\kappa^{++}} \cap \alpha \in A$

(6) for all $A \in A^{1\kappa^+}$ if $\delta \in A^{1\kappa^{++}}$ and $\delta < \sup A$ then $\min(A \backslash \delta) \in A^{1\kappa^{++}}$

(7) if $A, B \in A^{1\kappa^+}$ then $otp(A \cap \kappa^{+3}) = otp(B \cap \kappa^{+3})$ iff $otp_{\kappa^+}(A) = otp_{\kappa^+}(B)$

Further we shall confuse A's with $A \cap \kappa^{+3}$.

(8) (isomorphism condition)

Let $A, B \in A^{1\kappa^+}$ and $otp(A) = otp(B)$ then the structures

$$\langle A, \in, \subseteq, \kappa, C^{\kappa^+}(A), A^{1\kappa^+} \cap A \,,\, C^{\kappa^+} \upharpoonright (A^{1\kappa^+} \cap A), \, A^{1\kappa^{++}} \cap A, f_A \rangle$$

and

$$\langle B, \in, \subseteq, \kappa, C^{\kappa^+}(B), A^{1\kappa^+} \cap B, \, C^{\kappa^+} \upharpoonright (A^{1\kappa^+} \cap B), \, A^{1\kappa^{++}} \cap B, \, f_B \rangle$$

are isomorphic over $A \cap B$, i.e. the isomorphism π_{AB} between them is the identity on $A \cap B$, where $f_A : \kappa^+ \longleftrightarrow A$, $f_B : \kappa^+ \longleftrightarrow B$ are some fixed in advance bijections.

Note that in particular we will have that $A \cap \kappa^{++} = B \cap \kappa^{++}$. Also, together with the next condition, we will have the opposite implication as well, i.e. $A \cap \kappa^{++} = B \cap \kappa^{++}$ implies $otp(A) = otp(B)$.

(9) (first intersection condition) if $A, B \in A^{1\kappa^+}$ $A \neq B$ and $otp(A) = otp(B)$ then there is $\alpha \in A \cap A^{1\kappa^{++}}$ s.t. $A \cap B = A \cap \alpha$

In particular this condition imply that $A \cap B = A \cap \sup(A \cap B)$ and $\alpha = \min(A \backslash A \cap B)$.

(10) (second intersection condition) if $A, B \in A^{1\kappa^+}$, $otp(A) \geq otp(B)$ and $B \not\subseteq A$, then there is $B' \in (A \cup \{A\}) \cap A^{1\kappa^+}$ s.t.

  (a) $otp(B') = otp(B)$

    Note that $otp(A) = otp(B)$ implies then that B' is A itself.

  (b) $A \cap B = B' \cap B$.

  (c) either

    (i) $\min(B' \backslash \sup(B \cap B') + 1) > \sup B$

    or

    (ii) $\min(B \backslash \sup(B \cap B') + 1) > \sup B'$ and then also $\min(B \backslash \sup(B \cap B') + 1) > \sup A$.

  (d) if $\alpha \in A \cap \sup B$, then $\alpha < \min(B \backslash \sup(B \cap B') + 1)$.

The meaning of (c) is that without the common part B and B' are basically one above another. If B is above B' then it is also above A. The condition (d) claims that A (that does not include B) cannot have ordinals in the interval $(\min(B \backslash \sup(B \cap B') + 1), \sup B)$. Note that by (9) above applied to B and B' there will be $\alpha \in B \cap A^{1\kappa^{++}}$ such that $B \cap B' = B \cap \alpha$. Then this $\alpha \geq \sup(B \cap B')$ and hence it is not in A by (d). Now, if $\sup(A) > \sup(B)$ then $\min(A \backslash \alpha) \in A^{1\kappa^{++}}$, by the condition (6) above.

(11) (immediate predecessors condition)

Let $B \in A^{1\kappa^+}$ be a successor model. Then

- (a) B has at most two immediate predecessors in $A^{1\kappa^+}$ (under the inclusion relation). They are required to have the same *otp*. In addition, if $Z \in B \cap A^{1\kappa^+}$ then either $Z = B_i$ or $Z \in B_i$ for $i = 0, 1$, where B_0, B_1 are the immediate predecessors of B in $A^{1\kappa^+}$.
- (b) if B' is an immediate predecessor of B and it is limit, then B' is the unique immediate predecessor of B.

So its impossible to split over a limit model. This technical condition will be useful further in 3.5.

(12) (closure of models)

Let $B \in A^{1\kappa^+}$ be a successor model. Then ${}^\kappa B \subseteq B$.

(13) If α is a successor element of $A^{1\kappa^{++}}$, then it has cofinality κ^{++}.

(14) $\max A^{1\kappa^{++}} \geq \sup(A^{0\kappa^+} \cap \kappa^{+3})$. $\square$

Now let $p = \langle \langle A^{0\kappa^+}, A^{1\kappa^+}, C^{\kappa^+} \rangle, A^{1\kappa^{++}} \rangle \in \mathcal{P}'$ and $B \in A^{1\kappa^+}$. Define $swt(p, B)$ (here swt stands for switch) to be

$$\langle \langle A^{0\kappa^+}, A^{1\kappa^+}, D^{\kappa^+} \rangle, A^{1\kappa^{++}} \rangle ,$$

where D^{κ^+} is obtained from C^{κ^+} as follows:

$D^{\kappa^+} = C^{\kappa^+}$ unless B has exactly two immediate predecessors in $A^{1\kappa^+}$. If $B_0 \neq B_1$ are such predecessors of B and, say $B_0 \in C^{\kappa^+}(B)$, then we set $D^{\kappa^+}(B) = C^{\kappa^+}(B_1)^\frown B$. Extend D^{κ^+} on the rest in the obvious fashion just replacing $C^{\kappa^+}(B_0)$ by $C^{\kappa^+}(B_1)$ for models including B and then moving over isomorphic models.

Intuitively, we switched here from B_0 to B_1.

Note that $swt(swt(p, B), B) = p$.

Define $q = swt(p, B_1, \ldots, B_n)$ by applying the operation swt n-times: $p_{i+1} = swt(p_i, B_i)$, for each $1 \leq i \leq n$, where $p_1 = p$ and $q = p_{n+1}$.

The following simple observation will be useful further (3.7).

Lemma 1.2 Let $p = \langle \langle A^{0\kappa^+}(p), A^{1\kappa^+}(p), C^{\kappa^+}(p) \rangle, A^{1\kappa^{++}}(p) \rangle \in \mathcal{P}'$ and $B \in A^{1\kappa^+}(p)$. Then there are $B_1, B_2, \ldots, B_n \in A^{1\kappa^+}$ such that $B \in C^{\kappa^+}(q)(A^{0\kappa^+}(p))$, where

$$q = \langle \langle A^{0\kappa^+}(p), A^{1\kappa^+}(p), C^{\kappa^+}(q) \rangle, A^{1\kappa^{++}}(p) \rangle = swt(p, B_1, B_2, \ldots, B_n).$$

Proof. If $B \in C^{\kappa^+}(p)(A^{0\kappa^+}(p))$, then let $q = p$. Otherwise, pick B_1 to be the smallest element of $C^{\kappa^+}(p)(A^{0\kappa^+})$ including B. Let B_{11} be the immediate predecessor of B_1 not in $C^{\kappa^+}(p)(B_1)$. If $B \in C^{\kappa^+}(p)(B_{11})$, then set $q = swt(p, B_1)$. Otherwise pick B_2 to be the the smallest element of $C^{\kappa^+}(p)(B_{11})$ including B. Note that $B_2 \in B_1$. So we go down on ranks. Hence after finitely many steps a model B_n with $B \in C^{\kappa^+}(p)(B_n)$ will be reached. Then $q = swt(p, B_1, B_2, ..., B_n)$ will be as desired. $\square$

Definition 1.3 Let $r, q \in \mathcal{P}'$. Then $r \geq q$ (r is stronger than q) iff there is $p = swt(r, B_1, \ldots, B_n)$ for some $B_1, \ldots, B_n$ appearing in r so that the following hold, where

$$p = \langle \langle A^{0\kappa^+}, A^{1\kappa^+}, C^{\kappa^+} \rangle, A^{1\kappa^{++}} \rangle$$

$$q = \langle \langle B^{0\kappa^+}, B^{1\kappa^+}, D^{\kappa^+} \rangle, B^{1\kappa^{++}} \rangle$$

(1) $\qquad A^{1\kappa^{++}} \cap (\max B^{1\kappa^{++}} + 1) = B^{1\kappa^{++}}$

(2) $\qquad A^{1\kappa^+} \supseteq B^{1\kappa^+}$

(3) $\qquad C^{\kappa^+} \restriction B^{1\kappa^+} = D^{\kappa^+}$

(4) $\qquad B^{0\kappa^+} \in C^{\kappa^+}(A^{0\kappa^+})$

(5) for each $A \in A^{1\kappa^+}$, $A \cap B^{0\kappa^+} \in B^{1\kappa^+}$ or there are $B \in B^{1\kappa^+}$ and $\alpha \in B^{1\kappa^{++}}$ such that $A \cap B^{0\kappa^+} = B \cap \alpha$.

Remarks (1) Note that if $t = swt(p, B_0, \ldots, B_n)$ is defined, then $t \geq p$ and $p = swt(swt(p, B_0, \ldots, B_n), B_n, B_{n-1}, \ldots, B_0) = swt(t, B_n, \ldots, B_0) \geq t$. Hence the switching produces equivalent conditions.

(2) We need to allow $swt(p, B)$ for the Δ-system argument. Since in this argument two conditions are combined into one and so C^0 should pick one of them only. Also it is needed for proving a strategical closure of the forcing.

(3) The use of finite sequences $B_0, \ldots, B_n$ is needed in order to insure transitivity of the order $\leq$ on $\mathcal{P}'$.

Let $p = \langle \langle A^{0\kappa^+}, A^{1\kappa^+}, C^{\kappa^+} \rangle, A^{1\kappa^{++}} \rangle \in \mathcal{P}'$. Set $p \backslash \kappa^{++} = A^{1\kappa^{++}}$. Define $\mathcal{P}'_{\geq \kappa^{++}}$ to be the set of all $p \backslash \kappa^{++}$ for $p \in \mathcal{P}'$.

The next lemma is obvious.

Lemma 1.4 $\langle \mathcal{P}'_{\geq \kappa^{++}}, \leq \rangle$ *is* κ^{+3}-*closed.*

Set $p \restriction \kappa^{++} = \langle\langle A^{0\kappa^+}, A^{1\kappa^+}, C^{\kappa^+}\rangle$ where $p = \langle\langle A^{0\kappa^+}, A^{1\kappa^+}, C^{\kappa^+}\rangle,$
$A^{1\kappa^{++}}\rangle \in \mathcal{P}'.$

Let $G(\mathcal{P}'_{\geq\kappa^{++}})$ be a generic subset of $\mathcal{P}'_{\geq\kappa^{++}}$. Define $\mathcal{P}'_{<\kappa^{++}}$ to be the set of all $p \restriction \kappa^{++}$ for $p \in \mathcal{P}'$ with $p\backslash\kappa^{++} \in \bar{G}(\mathcal{P}'_{\geq\kappa^{++}}).$

Let $p \in \mathcal{P}'$ and $q \in \mathcal{P}'_{\geq\kappa^{++}}$. Then $q\frown p$ denotes the set obtained from p by adding q to the last component of p, i.e. to A^{11}.

The following lemma is trivial.

Lemma 1.5 *Let $p \in \mathcal{P}'$, $q \in \mathcal{P}'_{\geq\kappa^{++}}$ and $q \geq \mathcal{P}'_{\geq\kappa^{++}} p\backslash\kappa^{++}$. Then $q\frown p \in \mathcal{P}'$ and $q\frown p \geq p.$*

It follows now that $\mathcal{P}'$ projects to $\mathcal{P}'_{<\kappa^{++}}.$
Let us turn to the chain condition.

Lemma 1.6 *The forcing $\mathcal{P}'_{<\kappa^{++}}$ satisfies κ^{+3}-c.c. in $V^{\mathcal{P}'_{\geq\kappa^{++}}}.$*

Proof. Suppose otherwise. Let us assume that

$$\emptyset\Vdash_{\mathcal{P}'_{\geq\kappa^{++}}} (\langle p_\alpha = \langle \underset{\sim}{A^{0\kappa^+}_\alpha}, \underset{\sim}{A^{1\kappa^+}_\alpha}, \underset{\sim}{C^{\kappa^+}_\alpha}\rangle \mid \alpha < \kappa^{+3}\rangle$$
$$\text{is an antichain in } \underset{\sim}{\mathcal{P}'}_{<\kappa^{++}}).$$

Without loss of generality we can assume that each $A^{0\kappa^+}_\alpha$ is forced to be a successor model, otherwise just extend conditions by adding one additional model on the top. Define by induction, using 1.3, an increasing sequence $\langle q_\alpha \mid \alpha < \kappa^{+3}\rangle$ of elements of $\mathcal{P}'_{\geq\kappa^{++}}$ and a sequence $\langle p_\alpha \mid \alpha < \kappa^{+3}\rangle$, $p_\alpha = \langle A^{0\kappa^+}_\alpha, A^{1\kappa^+}_\alpha, C^{\kappa^+}_\alpha\rangle$ so that for every $\alpha < \kappa^{+3}$

$$q_\alpha\Vdash_{\mathcal{P}'_{\geq\kappa^3}} \langle \underset{\sim}{A^{0\kappa^+}_\alpha}, \underset{\sim}{A^{1\kappa^+}_\alpha}, \underset{\sim}{C^{\kappa^+}_\alpha}\rangle = \check{p}_\alpha .$$

For a limit $\alpha < \kappa^{+3}$ let

$$\bar{q}_\alpha = \bigcup_{\beta<\alpha} q_\beta \cup \{\sup\bigcup_{\beta<\alpha} q_\beta\}$$

and q_α be its extension deciding $\underset{\sim}{p}^\alpha$. Also assume that $\max q_\alpha \geq \sup(A^{00}_\alpha \cap \kappa^{+3}).$

We form a Δ-system. By shrinking if necessary assume that for some stationary $S \subseteq \kappa^{+3}$ and $\delta < \kappa^{+3}$ we have the following for every $\alpha < \beta$ in S:

(a) $A^{0\kappa^+}_\alpha \cap \alpha = A^{0\kappa^+}_\beta \cap \beta \subseteq \delta$

(b) $A_\alpha^{0\kappa^+} \setminus \alpha \neq \emptyset$

(c) $\sup A_\alpha^{0\kappa^+} < \beta$

(d) $\sup \bar{q}_\alpha = \alpha + 1$

(e)

$$\langle A_\alpha^{0\kappa^+}, \in, \leq, \subseteq, \kappa, C_\alpha^{\kappa^+}, f_{A_\alpha^{0\kappa^+}}, A_\alpha^{1\kappa^+}, q_\alpha \cap A_\alpha^{0\kappa^+} \rangle$$

$$\langle A_\beta^{0\kappa^+}, \in, \leq, \subseteq, \kappa, C_\beta^{\kappa^+}, f_{A_\beta^{0\kappa^+}}, A_\beta^{1\kappa^+}, q_\beta \cap A_\beta^{0\kappa^+} \rangle .$$

are isomorphic over δ, i.e. by isomorphism fixing every ordinal below δ, where

$$f_{A_\alpha^{0\kappa^+}} : \kappa^+ \longleftrightarrow A_\alpha^{0\kappa^+}$$

and

$$f_{A_\beta^{0\kappa^+}} : \kappa^+ \longleftrightarrow A_\beta^{0\kappa^+}$$

are the fixed enumerations. Denote the isomorphism by $\pi_{\alpha\beta}$.

We claim that for $\alpha < \beta$ in S we can extend q_β to a condition forcing compatibility of p_α and p_β. Proceed as follows. Pick A to be an elementary submodel of cardinality κ^+ so that

(i) $A_\alpha^{1\kappa^+}, A_\beta^{1\kappa^+} \in A$

(ii) $C_\alpha^{\kappa^+}, C_\beta^{\kappa^+} \in A$

(iii) $q_\beta \in A$.

Extend q_β to $q = q_\beta \cup \sup(A \cap \kappa^{+3})$. Set $p = \langle A, A_\alpha^{1\kappa^+} \cup A_\beta^{1\kappa^+} \cup \{A\},$ $C_\alpha^{\kappa^+} \cup C_\beta^{\kappa^+} \cup \langle A, C_\beta^{\kappa^+}(A_\beta^{0\kappa^+})^\frown A \rangle \rangle$. Let us check that $\langle p, q \rangle \in \mathcal{P}'$. The conditions (1)-(6),(10),(11) of 1.1 hold trivially, (6) holds by (d) above. Let us check (8) and (9).

Suppose $X, Y \in A_\alpha^{1\kappa^+} \cup A_\beta^{1\kappa^+} \cup \{A\}$, $otp(X) = otp(Y)$ and $X \neq Y$. Then $X, Y \in A_\alpha^{1\kappa^+} \cup A_\beta^{1\kappa^+}$. If both X and Y belong to $A_\alpha^{1\kappa^+}$ or $A_\beta^{1\kappa^+}$ then we are done since $\langle p_\alpha, q_\beta \rangle$ and $\langle p_\beta, q_\beta \rangle$ satisfy 1.1(8). So, suppose $X \in A_\alpha^{1\kappa^+}$ and $Y \in A_\beta^{1\kappa^+}$. Let $X' \in A_\alpha^{1\kappa+}$ be the one corresponding to Y under (d), i.e. $\pi_{\alpha\beta}(X') = Y$. Then by 1.1(8) for $\langle p_\alpha, q_\alpha \rangle$ we will have $\xi \in X \cap q_\alpha$ such that $X \cap X' = X \cap \xi$. By (a) and $\pi_{\alpha\beta}(X') = Y$ we have $X' \cap Y = X' \cap \alpha = Y \cap \beta$. Then

$$X \cap Y = X \cap Y \cap \beta = X \cap X' \cap \alpha = X \cap \xi \cap \alpha = X \cap \min(\xi, \alpha).$$

If $\xi \leq \alpha$, then we are done since $\xi \in X$. If $\xi > \alpha$, then $\alpha \in \bar{q}_\alpha \cap \sup X$. Hence by 1.1(11), $\min(X \backslash \alpha)$ is in q_α and, clearly then

$$X \cap \alpha = X \cap \min(X \backslash \alpha) \, .$$

Let us check 1.1(9). Thus suppose that $X, Y \in A_\alpha^{1\kappa^+} \cup A_\beta^{1\kappa^+} \cup \{A\}$ and $otp(X) > otp(Y)$. Again we need only to consider the case when X and Y belong to different $A^{1\kappa^+}$'s. Assume for example that $X \in A_\alpha^{1\kappa^+}$ and $Y \in A_\beta^{1\kappa^+}$. As above, we pick $X' \in A_\alpha^{10}$ such that $\pi_{\alpha\beta}(X') = Y$. Now, by 1.1(9) for $\langle p_\alpha, q_\alpha \rangle$, find $X'' \in X \cap A_\alpha^{1\kappa^+}$ such that $otp(X'') = otp(X')$ and $X \cap X' = X'' \cap X'$. Then $X \cap Y = X \cap Y \cap \beta = X \cap X' \cap \alpha = X'' \cap X' \cap \alpha = X'' \cap Y \cap \beta = X'' \cap Y$, since $\sup(X'') < \sup(X) \leq \sup(A_\alpha^{0\kappa^+})$ which is below β by (c) above.

The condition 9(c) holds since the models are part of the Δ-system.

Clearly, $\langle p, q \rangle \geq \langle p_\beta, q_\beta \rangle$. $\langle p, q \rangle \geq \langle p_\alpha, q_\alpha \rangle$ follows using switching of $A_\beta^{0\kappa^+}$ to $A_\alpha^{0\kappa^+}$. $\qquad\square$

Lemma 1.7 $\mathcal{P}'$ *is κ^{++}-strategically closed.*

Proof. We define a winning strategy for the player playing at even stages. Thus suppose $\langle p_j \mid j < i \rangle$, $p_j = \langle \langle A_j^{0\kappa^+}, A_j^{1\kappa^+}, C_j^{\kappa^+} \rangle, A_j^{1\kappa^{++}} \rangle$ is a play according to this strategy upto an even stage $i < \kappa^{++}$. Set first

$$B_i^{0\kappa^+} = \bigcup_{j<i} A_j^{0\kappa^+}, B_i^{1\kappa^+} = \bigcup_{j<i} A_j^{1\kappa^+} \cup \{B_i^{0\kappa^+}\},$$

$$D_i^{\kappa^+} = \bigcup_{j<i} C_j^{\kappa^+} \cup \{\langle B_i^{0\kappa^+}, \{B_i^{0\kappa^+}\} \cup \{C_j^{\kappa^+}(A_j^{0\kappa^+}) \mid j \text{ is even}\}\rangle\}$$

and

$$B_i^{1\kappa^{++}} = \bigcup_{j<i} B_j^{1\kappa^{++}} \cup \{\sup \bigcup_{j<i} B_j^{1\kappa^{++}}\}.$$

Then pick $A_i^{0\kappa^+}$ to be a model of cardinality κ^+ such that

(a) $^\kappa A_i^{0\kappa^+} \subseteq A_i^{0\kappa^+}$

(b) $B_i^{0\kappa^+}, B_i^{1\kappa^+}, D_i^{\kappa^+}, B_i^{1\kappa^{++}} \in A_i^{0\kappa^+}$.

Set $A_i^{1\kappa^+} = B_i^{1\kappa^+} \cup \{A_i^{0\kappa^+}\}$, $C_i^{\kappa^+} = D_i^{\kappa^+} \cup \{\langle A_i^{0\kappa^+}, D_i^{\kappa^+}(B_i^{0\kappa^+}) \cup \{A_i^{0\kappa^+}\}\rangle\}$ and $A_i^{1\kappa^{++}} = B_i^{1\kappa^{++}} \cup \{\sup(A_i^{0\kappa^+} \cap \kappa^{+3}\}$. As an inductive assumption we assume that at each even stage $j < i$, p_j was defined in the same fashion. Then $p_i = \langle \langle A_i^{0\kappa^+}, A_i^{1\kappa^+}, C_i^{\kappa^+} \rangle, A_i^{1\kappa^{++}} \rangle$ will be a condition in $\mathcal{P}'$ stronger than each p_j for $j < i$. The switching may be required here once moving from an odd stage to its immediate successor even stage. $\qquad\square$

2. Types of Models

The basic approach here is as in [1] but instead of dealing with types of ordinals we shall consider elementary submodels of $H(\chi^{+k})$ for some χ big enough, $k \le \omega$ and types of such models.

Fix $n < \omega$. Fix using GCH an enumeration $\langle a_\alpha \mid \alpha < \kappa_n \rangle$ of $[\kappa_n]^{<\kappa_n}$ so that for every successor cardinal $\delta < \kappa_n$ the initial segment $\langle a_\alpha \mid \alpha < \delta \rangle$ enumerates $[\delta]^{<\delta}$ and every element of $[\delta]^{<\delta}$ appears stationarily many times in each cofinality $< \delta$ in the enumeration. Let $j_n(\langle a_\alpha \mid \alpha < \kappa_n \rangle) = \langle a_\alpha \mid \alpha < j_n(\kappa_n) \rangle$ where j_n is the canonical embedding of the $(\kappa_n, \kappa_n^{+n+3})$-extender E_n. Then $\langle a_\alpha \mid \alpha < \kappa_n^{+n+3} \rangle$ will enumerate $[\kappa_n^{+n+3}]^{<\kappa_n^{+n+3}}$ and we fix this enumeration. For each $k \le \omega$ consider a structure

$$\mathfrak{A}_{n,k} = \langle H(\chi^{+k}), \in, \subseteq, \le, E_n, \kappa_n, \chi, \langle a_\alpha \mid \alpha < \kappa_n^{+n+3} \rangle, 0, 1, \ldots, \alpha, \ldots \mid \alpha < \kappa_n^{+k} \rangle$$

in the appropriate language $\mathcal{L}_{n,k}$ with a large enough regular cardinal χ.

Remark It is possible to use κ_n^{++} here (as well as in [1]) instead of κ_n^{+k}. The point is that there are only κ_n^{++} many ultrafilters over κ_n and we would like that equivalent conditions use the same ultrafilter. The only parameter that that need to vary is k in $H(\chi^{+k})$.

Let $\mathcal{L}'_{n,k}$ be the expansion of $\mathcal{L}_{n,k}$ by adding a new constant c'. For $a \in H(\chi^{+k})$ of cardinality less than κ_n^{+n+3} let $\mathfrak{A}_{n,k,a}$ be the expansion of $\mathfrak{A}_{n,k}$ obtained by interpreting c' as a.

Let $a, b \in H(\chi^{+k})$ be two sets of cardinality less than κ_n^{+n+3}. Denote by $tp_{n,k}(b)$ the $\mathcal{L}_{n,k}$-type realized by b in $\mathfrak{A}_{n,k}$. Let $tp_{n,k}(a,b)$ be a the $\mathcal{L}'_{n,k}$-type realized by b in $\mathfrak{A}_{n,k,a}$. Note that coding a, b by ordinals we can transform this to the ordinal types of [1].

Lemma 2.1　(a) $|\{tp_{n,k}(b) \mid b \in H(\chi^{+k})\}| = \kappa_n^{+k+1}$.

(b) $|\{tp_{n,\kappa}(a,b) \mid a, b \in H(\chi^{+k})\}| = \kappa_n^{+k+1}$.

Proof. (a) The cardinality of the language $\mathcal{L}_{n,k}$ is κ_n^{+k} so the number of formulas is κ_n^{+k}. Now the number of types is $2^{\kappa_n^{+k}} = \kappa_n^{+k+1}$.

(b) The same argument. $\qquad\square$

This lemma implies immediately the following:

Lemma 2.2 Let $A \prec \mathfrak{A}_{n,k+1}$ and $|A| \ge \kappa_n^{+k+1}$. Then the following holds:

(a) for every $a, b \in H(\chi^{+k})$ there are $c, d \in A \cap H(\chi^{+k})$ with $tp_{n,k}(a,b) = tp_{n,k}(c,d)$

(b) for every $a \in A$ and $b \in H(\chi^{+k})$ there is $d \in A \cap H(\chi^{+k})$ so that $tp_{n,k}(a \cap H(\chi^{+k}),\ b) = tp_{n,k}(a \cap H(\chi^{+k}),\ d)$.

Proof. (a) Note that $tp_{n,k}(a,b) \in A$, by 2.1 and since $|A| \geq \kappa_n^{+k+1}$, so $A \supseteq \kappa_n^{+k+1}$. Now, $H(\chi^{+k+1}) \vDash (\exists x, y \in H(\chi^{+k}))\forall\varphi(v,u) \in tp_{n,k}(a,b)$ $(H(\chi^{+k}) \vDash \varphi(x,y)))$. But $A \prec H(\chi^{+k+1})$. So

$$A \vDash (\exists x, y \in H(\chi^{+k})\forall\varphi(x,y) \in tp_{n,k}(a,b)(H(\chi^{+k}) \vDash \varphi(x,y))) \ .$$

Pick $c, d \in A$ satisfying this formula. Then $c, d \in H(\chi^{+k})$ and $tp_{n,k}(c,d) = tp_{n,k}(a,b)$.

(b) Similar.   □

The next lemma will be crucial further for the chain condition arguments.

Lemma 2.3 *Suppose that $A \prec \mathfrak{A}_{n,k+1}, |A| \geq \kappa_n^{+k+1}$, $B \prec \mathfrak{A}_{n,k}$, and $C \in \mathcal{P}(B) \cap A \cap H(\chi^{+k})$. Then there is D so that*

(a) $D \in A$.

(b) $C \subseteq D$.

(c) $D \prec A \cap H(\chi^{+k}) \prec H(\chi^{+k})$.

(d) $tp_{n,k}(C, B) = tp_{n,k}(C, D)$.

Proof. As in 2.2., the following formula is true in $H(\chi^{+k+1})$:

$$\exists x \subseteq H(\chi^{+k})((x \prec H(\chi^{+k})) \wedge (x \supseteq C)$$
$$\wedge\ (\forall\varphi(y,z) \in tp_{n,k}(C, B)H(\chi^{+k}) \vDash \varphi(C, x))).$$

Then the same holds in A. Let D witness this. Hence $D \in A$, $D \supseteq C$, $D \prec A \cap H(\chi^{+k}) \prec H(\chi^{+k})$ and $tp_{n,k}(C, B) = tp_{n,k}(C, D)$.   □

Further we shall add models $B \cap H(\chi^{+k})$ with $B \prec H(\chi^{+k+1})$ or models realizing the same $tp_{n,k}(a, -)$ as those of elementary submodel of $H(\chi^{+k+1})$ intersected with $H(\chi^{+k})$ for any a inside. We will require that for every $k < \omega$, each condition p has an equivalent condition q with every model in it being an elementary submodel of $H(\chi^{+k})$.

3. The Main Forcing

Let $G(\mathcal{P}')$ be a generic subset of $\mathcal{P}'$.

Fix $n < \omega$. Following [2, Sec.3]. We define first Q_{n0}.

Definition 3.1 Let Q_{n0} be the set of the triples $\langle a, A, f \rangle$ so that:

(1) f is partial function from κ^{+3} to κ_n of cardinality at most κ

(2) a is a partial function of cardinality less than κ_n so that

(a) $\mathrm{dom}(a)$ consists of models and ordinals appearing in elements of $G(\mathcal{P}')$, i.e. if $X \in \mathrm{dom}(a)$, then for some $\langle \langle A^{0\kappa^+}, A^{1\kappa^+}, C^{\kappa^+} \rangle$, $A^{1\kappa^{++}} \rangle \in G(\mathcal{P}')$ we have $X = A^{0\kappa^+}$ or $X \in A^{1\kappa^{++}}$.

This means in particular that ordinals in $\mathrm{dom}(a)$ are taken from $A^{1\kappa^{++}}$ only

(b) for each $X \in \mathrm{dom}(a)$ there is $k \leq \omega$ so that $a(X) \subseteq H(\chi^{+k})$.

Also the following holds

  (i) $|X| = \kappa^+$ implies $|a(X)| = \kappa_n^{+n+1}$
  (ii) $|X| = \kappa^{++}$ implies $|a(X)| = \kappa_n^{+n+2}$ and $a(X) \cap \kappa_n^{+n+3} \in ORD$

Note that in (ii) X is an ordinal but $a(X)$ is not. Actually our main interest is in $a(X) \cap \kappa_n^{+n+3}$ which is required to be an ordinal.

Further passing from Q_{n0} to $\mathcal{P}$ we will require that for every $k < \omega$ for all but finitely many n's the n-th image of X will be an elementary submodel of $H(\chi^{+k})$. But in general just subsets are allowed here.

(c) if $A, B \in \mathrm{dom}(a)$, $A \in B$ (or $A \subseteq B$) and k is minimal such that $a(A) \subseteq H(\chi^{+k})$ or $a(B) \subseteq H(\chi^{+k})$, then $a(A) \cap H(\chi^{+k}) \in a(B) \cap H(\chi^{+k})$ (or $a(A) \cap H(\chi^{+k}) \subseteq a(B) \cap H(\chi^{+k})$).

The intuitive meaning is that a is supposed to preserve membership and inclusion. But we cannot literally require this since $a(A)$ and $a(B)$ may be substructures of different structures. So we first go down to the smallest of this structures and then put the requirement on the intersections.

(d) $\mathrm{dom}(a)$ has a maximal model (under inclusion) it is a member of $A^{1\kappa^+}$ for some condition in $G(\mathcal{P}')$. Its image $a(\max a)$ intersected with κ_n^{+n+3} is above all the rest of $\mathrm{rng}(a)$ restricted to κ_n^{+n+3} in the ordering of the extender E_n (via some reasonable coding by ordinals).

Recall that the extender E_n acts on κ_n^{+n+3} and our main interest is in Prikry sequences it will produce. So, parts of $\mathrm{rng}(a)$ restricted to κ_n^{+n+3} will play the central role.

(e) if $A, B \in \mathrm{dom}(a)$ and $otp(A) = otp(B)$ then

$$\langle a(A) \cap H(\chi^{+k}), \in \rangle \simeq \langle a(B) \cap H(\chi^{+k}), \in \rangle$$

where k is minimal such that $a(A) \subseteq H(\chi^{+k})$ or $a(B) \subseteq H(\chi^{+k})$.

Note that it is possible to have for example $a(A) \prec H(\chi^{+6})$ and $a(B) \prec H(\chi^{+18})$. Then we take $k = 6$.

Let π be the isomorphism between

$$\langle a(A) \cap H(\chi^{+k}), \in \rangle, \langle a(B) \cap H(\chi^{+k}), \in \rangle$$

and $\pi_{A,B}$ be the isomorphism between A and B given by 1.1(8). Require that for each $Z \in A \cap \mathrm{dom}(a)$ we have $\pi_{A,B}(Z) \in B \cap \mathrm{dom}(a)$ and

$$\pi(a(Z) \cap H(\chi^{+k})) = a(\pi_{A,B}(Z)) \cap H(\chi^{+k}).$$

(f) if $A, B \in \mathrm{dom}(a)$, $A \nsubseteq B$ and $B \nsubseteq A$ then models and ordinals witnessing 1.1(9,10) are in $\mathrm{dom}(a)$

(g) if $\alpha \in \mathrm{dom}(a)$ (i.e. a member of some $A^{1\kappa^{++}}$ for $A^{1\kappa^{++}}$ in $G(\mathcal{P}')$) then for each $A \in \mathrm{dom}(a)$ with $\alpha \in A$ the smallest model B of $C^{\kappa^+}(A)$ such that $\alpha \in B$ belongs to $\mathrm{dom}(a)$ as well as all its immediate predecessors. Note that by 1.1(11) there are at most two such models.

(h) if $A, B \in \mathrm{dom}(a)$ and $B \in A$, then the walk via $C^{\kappa^+}(A)$ from A to B is in $\mathrm{dom}(a)$, i.e. the least model $A_0 \in C^{\kappa^+}(A)$ such that $A_0 = B$ or $B \in A_0$ is in $\mathrm{dom}(a)$. If $B \in A_0$, then the immediate predecessor A_1 of A_0 with $B \in A_1$ or $B = A_1$ is in $\mathrm{dom}(a)$. Now, the least model $A_2 \in C^{\kappa^+}(A_1)$ such that $A_2 = B$ or $B \in A_2$ is in $\mathrm{dom}(a)$. If $B \in A_2$ then the immediate predecessor A_3 of A_2 with $B \in A_2$ or $B = A_2$ is in $\mathrm{dom}(a)$ and so on.

Note that only finitely many models are involved in such a walk.

(i) if $A \in \mathrm{dom}(a)$ then $C^{\kappa^+}(A) \cap \mathrm{dom}(a)$ is a closed chain. Let $\langle A_i | i < j \rangle$ be its increasing continuous enumeration. For each $l < j$ consider the final segment $\langle A_i | l \leq i < j \rangle$ and its image $\langle a(A_i) | l \leq i < j \rangle$. Find the minimal k so that

$$a(A_i) \subseteq H(\chi^{+k}) \text{ for each } i, l \leq i < j.$$

Then the sequence

$$\langle a(A_i) \cap H(\chi^{+k}) | l \leq i < j \rangle$$

is increasing and continuous.

Note that k here may depend on l, i.e. on the final segment.

(j) if $A, B \in \text{dom}(a)$, $A \not\subseteq B$, $B \not\subseteq A$ and $\sup A < \sup B$, then $\min(B \backslash \sup A) \in \text{dom}(a)$.

Note in this case necessarily $\min(B \backslash \sup A) \in A^{1\kappa^{++}}$. Thus, if $otp(A) = otp(B)$ then by 1.1(9) there is $\alpha \in B \cap A^{1\kappa^{++}}$ s.t. $A \cap B = B \cap \alpha$. By 1.1(10), this α will be as desired.

If $otp(A) > otp(B)$, then 1.1(10(c(ii))) applies and together with 1.1(9), we will have $\alpha = \min(B \backslash \sup(B \cap B') + 1)$ as desired.

If $otp(A) < otp(B)$, then find $A' \in B \cap A^{1\kappa^+}$ witnessing 1.1(10) for B and A. The assumption $\sup A < \sup B$ implies then that (i) of 1.1(10(c)) should hold. Pick $\tau \in A \cap A^{1\kappa^{++}}$ such that $A \cap \tau = A \cap A' = A \cap B$. Then $\tau \in A^{1\kappa^{++}} \backslash B$. By 1.1(6), $\min(B \backslash \tau) \in A^{1\kappa^{++}}$. Hence, $\alpha = \min(B \backslash \tau)$ will be equal to $\min(B \backslash \sup A)$, by 1.1(10(c(ii))) and we are done.

(k) if $A, \alpha \in \text{dom}(a)$ and $\sup A > \alpha$, then $\min(A \backslash \alpha) \in \text{dom}(a)$.

(l) if $A, B \in \text{dom}(a)$ and B is an immediate predecessor of A, then the other immediate predecessor of A is in $\text{dom}(a)$ as well.

(m) if $\langle \alpha_i | i < j \rangle$ is an increasing sequence of ordinals in $\text{dom}(a)$, then $\cup \{\alpha_i | i < j\} \in \text{dom}(a)$.

(n) if $A \in \text{dom}(a)$ is a limit model and $cof(otp_{\kappa^+}(A) - 1) < \kappa_n$ (i.e. the cofinality of the sequence $C^{\kappa^+}(A) \backslash \{A\}$ under the inclusion relation is less than κ_n) then a closed cofinal subsequence of $C^{\kappa^+}(A) \backslash \{A\}$ is in $\text{dom}(a)$. The images of its members under a form a closed cofinal in $a(A)$ sequence.

(o) if $\alpha \in \text{dom}(a)$ is a limit member of $A^{1\kappa^{++}}$ of cofinality less than κ_n, then a closed cofinal in α sequence from $A^{1\kappa^{++}}$ is in $\text{dom}(a)$ as well. The images of its members under a form a closed cofinal in $a(\alpha)$ sequence.

(3) $\{\alpha < \kappa^{+3} \mid \alpha \in \text{dom}(a)\} \cap \text{dom}(f) = \emptyset$.

(4) $A \in E_{n,a(\max(a))}$.

(5) for every ordinals α, β, γ which are elements of $\text{rng}(a)$ or the ordinals coding models of cardinality κ_n^{+n+1} in $\text{rng}(a)$ we have

$$\alpha \geq_{E_n} \beta \geq_{E_n} \gamma \quad \text{implies}$$
$$\pi_{\alpha\gamma}^{E_n}(\rho) = \pi_{\beta\gamma}^{E_n}(\pi_{\alpha\beta}^{E_n}(\rho))$$

for every $\rho \in \pi^{\,\text{``}}_{\max \text{rng}(a),\alpha}(A)$.

We define now Q_{n1} and $\langle Q_n, \leq_n, \leq_n^* \rangle$ as in [2, Sec. 2].

Definition 3.2 The set $\mathcal{P}$ consists of all sequences $p = \langle p_n \mid n < \omega \rangle$ so that

(1) for every $n < \omega$ $p_n \in Q_n$

(2) there is $\ell(p) < \omega$ such that

    (i) for every $n < \ell(p)$ $p_n \in Q_{n1}$,

    (ii) for every $n \geq \ell(p)$ we have $p_n = \langle a_n, A_n, f_n \rangle \in Q_{n0}$

    (iii) for every $n, m \geq \ell(p)$ $\max(\text{dom}(a_n)) = \max(\text{dom}(a_m))$ is a model of cardinality κ^+

(3) for every $n \geq m \geq \ell(p)$ $\text{dom}(a_m) \subseteq \text{dom}(a_n)$

(4) for every n, $\ell(p) \leq n < \omega$, and $X \in \text{dom}(a_n)$ we have that for each $k < \omega$ the set $\{m < \omega \mid \neg(a_m(X) \cap H(\chi^{+k}) \prec H(\chi^{+k}))\}$ is finite.

Lemma 3.3 *Suppose* $p = \langle p_k \mid k < \omega \rangle \in \mathcal{P}$, $p_k = \langle a_k, A_k, f_k \rangle$ *for* $k \geq \ell(p)$, X *is model or an ordinal appearing in an element of* $G(\mathcal{P}')$ *and* $X \notin \bigcup_{\ell(p) \leq k < \omega} \text{dom}(a_k) \cup \text{dom}(f_k)$. *Suppose that*

(a) if X is model then it is a successor model or if it is a limit one then $cof(otp_{\kappa^+}(X) - 1) > \kappa$.

(b) if X is an ordinal then it is a successor member of some $A^{1\kappa^{++}} \in G(\mathcal{P}')$ or it is a limit of cofinality above κ.

    Then there is a direct extension $q = \langle q_k \mid k < \omega \rangle$, $q_k = \langle b_k, B_k, g_k \rangle$ *for* $k \geq \ell(q)$, *of p so that starting with some $n \geq \ell(q)$ we have $X \in \text{dom}(b_k)$ for each $k \geq n$.*

Remark We would like to avoid at this stage adding limit models (or ordinals) of small cofinality since by 3.1(2(n,o)) this will require additional adding of sequences of models (or ordinals).

Proof. Note first that it is easy to add to p any A appearing in condition of $G(\mathcal{P}')$ of cardinality κ^+ so that the maximal model of p belongs to $C^{\kappa^+}(A)$. Just at each level $n \geq \ell(p)$ pick an elementary submodel of $H(\chi^{+\omega})$ of cardinality κ_n^{+n+1} including $\mathrm{rng}(a_n)$ as an element. Map A to such a model.

Suppose now we like to add to p some X which does not include the maximal model of p. Denote it by $\max(p)$. Without loss of generality we can assume that $X \in \max(p)$. Just otherwise using genericity of $G(\mathcal{P}')$ find A as above with $X \in A$. Pick a model $A \in \cup\{\mathrm{dom}(a_n) | n < \omega\}$ with $X \in A$ of the smallest possible otp_{κ^+} and among such models one of the least possible *rank* (the usual one). Suppose for simplicity that $A \in \mathrm{dom}(a_n)$ for each $n < \omega$.

Let $\langle A_i \mid i < otp_{\kappa^+}((A)) \rangle$ be increasing continuous enumeration of $C^{\kappa^+}(A)$.

Case A $X = A_i$ for some $i < otp_{\kappa^+}(A)$.

Let A_{i^*} be the first model of $C^{\kappa^+}(A)$ in $\bigcup_{n \geq \ell(p)} \mathrm{dom}(a_n)$ with $X \in A_{i^*}$.

Let $n < \omega$ be big enough so that $a_n(A_{i^*}) \prec H(\chi^{+n+3})$. Denote by $A_{i^{**}}$ the largest member of $C^{\kappa^+}(A_{i^*}) \setminus \{A_{i^*}\}$ inside $\mathrm{dom}(a_n)$, if $C^{\kappa^+}(A_{i^*}) \cap A_{i^*} \cap \mathrm{dom}(a_n) \neq \emptyset$. Notice that this set may vary once we change n.

Suppose that $A_{i^{**}}$ exists otherwise we may view it as empty and run the same argument.

Now for each $Y \in \mathrm{dom}(a_n)$ with $Y \not\supseteq A_{i^*}$ and $otpY \leq otpA_{i^*}$, we will have that $Y \cap A_{i^*} = Y \cap A_{i^{**}}$, since either

(i) $Y \in A_{i^*}$ and then $Y \in A_{i^{**}} \cup \{A_{i^{**}}\}$ (just use the walk from A_{i^*} to Y. It is supposed to be in the domain of a_n by 3.1(2(h))

or

(ii) $Y \notin A_{i^*}$ and then by 1.1(8) there is $Y' \in A_{i^*}$ with $Y \cap A_{i^*} = Y' \cap Y$. But by 3.1(2(f)), some such Y' must be in $\mathrm{dom}(a_n)$. By the choice of $A_{i^{**}}$ then $Y' \in A_{i^{**}} \cup \{A_{i^{**}}\}$. So, $Y \cap A_{i^*} = Y \cap A_{i^{**}}$.

If $Y \in \mathrm{dom}(a_n)$ with $Y \not\supseteq A_{i^*}$ and $otpY > otpA_{i^*}$, then by 3.1(2(f)), there are $Y' \in Y \cap A^{1\kappa^+} \cap \mathrm{dom}(a_n)$ so that $Y \cap A_{i^*} = Y' \cap A_{i^*}$. Again, by 3.1(2(f)), there is $\alpha \in A_{i^*} \cap A^{1\kappa^{++}} \cap \mathrm{dom}(a_n)$ such that $Y' \cap A_{i^*} = \alpha \cap A_{i^*}$. Then, by 3.1(2(g)), we have $\alpha \in A_{i^{**}}$, unless $A_{i^{**}}$ is the immediate predecessor of A_{i^*}. But then X must be equal to $A_{i^{**}}$.

Similar, if $\alpha \in \mathrm{dom}(a_n)$ and $\alpha < \sup A_{i^*}$ then $\min(A_{i^*} \setminus \alpha) \in A_{i^{**}}$.

Now pick X^* to be an element of $a_n(A_{i^*})$ such that $X^* \subseteq a_n(A_{i^*})$, $\kappa_n > X^* \subseteq X^*$, $X^* \prec H(\chi^{+n+2})$ and $a_n(A_{i^{**}}) \cap H(\chi^{+n+2}) \in X^*$. By the above, all the relevant information (intersections with models, ordinals etc.)

is already inside $a_n(A_{i^{**}})$. Hence we can extend a_n by adding to it the pair $\langle X, X^* \rangle$. Map X^* via all isomorphisms between $a_n(A_{i^*})$ and $a_n(B)$'s for each $B \in \mathrm{dom}(a_n)$ with $otp(B) = otp(A_{i^*})$.

Note that here is the place where we may drop to subsets of $H(\chi)$ and not elementary submodels of it. Just moving $a_n(A_{i^*})$ by isomorphisms may decrease degree of elementarity by one.

Let b_n be the result. Note that if X is an immediate predecessor of a model having another immediate predecessor X' or X is the immediate successor (in $C^{\kappa^+}(X)$) of a model in $\mathrm{dom}(a_n)$ and some model X' is another immediate predecessor of X, then adding of X may requires by 3.1(2(l)) adding of X' also. Let us delay the adding of such X' to the next case. Instead we deal with (or allow) a_n's which satisfy all the conditions of 3.1(2) but (l).

Claim A1 b_n satisfies all the conditions of 3.1(2) but (l). Moreover, if a_n satisfies all the conditions of 3.1(2) and X is not an immediate predecessor of a model having another immediate predecessor then also b_n satisfies all the conditions of 3.1(2).

Proof. Let X' be an image of X which was added to $\mathrm{dom}(a_n)$ and X'' be another model or ordinal in $\mathrm{dom}(b_n)$, which may be in $\mathrm{dom}(a_n)$ (and it is the case if it is an ordinal) or may be another image of X which was added to $\mathrm{dom}(a_n)$. We need to show that ordinals and models witnessing the intersection conditions for $X' \cap X''$ are in the domain. Split into a few cases.

Case A1.1 $X' = X$ and $X'' \in A_{i^*}$
     Then by (i) above $X'' \in A_{i^{**}} \cup A_{i^{**}}$. Hence, $X'' \in X = A_{i-1}$ and we are done.

Case A1.2 $X' = X$ and $X'' \notin A_{i^*}$
     Let us split this case into two.

Subcase A1.2.1 $X'' \in \mathrm{dom}(a_n)$
     If X'' is an ordinal, then either $X'' > \sup A_{i^*}$, then it is above $\sup X$ as well and we are done. Or $X'' < \sup A_{i^*}$ and then $\alpha = \min A_{i^*} \backslash X''$ is in $\mathrm{dom}(a_n)$. By the choice of $A_{i^{**}}$, then $\alpha \in A_{i^{**}}$. But, clearly, then $\alpha \in X$ and $\alpha = \min X \backslash X''$. So we are done.
     Assume now that X'' is a model.
     Let us point out that the walk from X'' to X is already in $\mathrm{dom}(a_n)$. We claim that the walk must terminate with A_{i^*}. Suppose otherwise. Thus let

Y be the first model of the walk which does not contain A. Compare Y and A_{i*}. By 1.1(9,10) and 3.1(2(f)), there is $\alpha \in A_{i*} \cap \mathrm{dom}(a_n)$ such that

$$Y \cap A_{i*} = A_{i*} \cap \alpha.$$

By the choice of A_{i**}, we must have $\alpha \in A_{i**}$. But on the other hand, $X \subseteq Y \cap A_{i*} = A_{i*} \cap \alpha$. In particular, $X \subseteq \alpha$. Which is clearly a contradiction, since $X \supseteq A_{i**}$.

If $X' \supseteq A_{i*}$ then the intersection properties are clear. Suppose that it is not the case. Compare X' with A_{i*}. By 1.1(9,10) and 3.1(2(f)), there is $\alpha \in A_{i*} \cap \mathrm{dom}(a_n)$ such that

$$X' \cap A_{i*} = A_{i*} \cap \alpha.$$

By the choice of A_{i**}, we must have $\alpha \in A_{i**}$. But

$$X' \cap X = X' \cap X \cap A_{i*} = X \cap A_{i*} \cap \alpha = X \cap \alpha$$

and, clearly $\alpha \in X$. In order to deal with the intersection on the other side compare $otpX'$ and $otpA_{i*}$. If $otpX' < otpA_{i*}$, then, by 3.1(2(f)) there is $Y \in A_{i*} \cap \mathrm{dom}(a_n)$ such that $otpY = otpX'$ and

$$X' \cap A_{i*} = X' \cap Y.$$

By the choice of A_{i**}, we must have $Y \in A_{i**} \cup \{A_{i**}\}$. But then $Y \subseteq X$. Hence

$$X' \cap X = X' \cap A_{i*} \cap X = X' \cap Y \cap X = X' \cap Y.$$

So we are done since both X' and Y are old.

If $otpX' \geq otpA_{i*}$, then, by 3.1(2(f)) there is $Y \in X' \cup \{X'\} \cap \mathrm{dom}(a_n)$ such that $otpY = otpA_{i*}$ and

$$X' \cap A_{i*} = Y \cap A_{i*}.$$

By 1.1(9,10), there are $\alpha \in A_{i*} \cap \mathrm{dom}(a_n)$ such that

$$Y \cap A_{i*} = A_{i*} \cap \alpha$$

and $\beta \in Y \cap \mathrm{dom}(a_n)$ such that

$$Y \cap A_{i*} = Y \cap \beta.$$

In this situation $Z = \pi_{A_{i*}Y}[X]$ will be added as an isomorphic image of X. Also, $\pi_{A_{i*}Y}[\alpha] = \beta$. Then

$$X' \cap X = X' \cap X \cap A_{i*} = Y \cap A_{i*} \cap X = Y \cap \beta \cap X = Z \cap \beta.$$

But as above, we must have $\alpha \in X$. Hence $\beta \in Z$ and we are done.

Subcase A1.2.2 $X'' \notin \operatorname{dom}(a_n)$

Then X'' is an isomorphic image of X. There is $B \in \operatorname{dom}(a_n)$ isomorphic to A_{i*} such that $\pi_{A_{i*}B}[X] = X''$. We need to take care only of the intersections properties of X and X'' one with another. By 1.1(9,10), there are $\alpha \in A_{i*} \cap \operatorname{dom}(a_n)$ such that

$$B \cap A_{i*} = A_{i*} \cap \alpha$$

and $\beta \in B \cap \operatorname{dom}(a_n)$ such that

$$Y \cap A_{i*} = Y \cap \beta.$$

Also, $\pi_{A_{i*}B}[\alpha] = \beta$. Then, as above, $\alpha \in X$ and so $\beta \in X''$. We have the following:

$$X'' \cap X = X'' \cap B \cap X \cap A_{i*} = X \cap \alpha = X'' \cap \beta.$$

Case A1.3 $X' \neq X$.

Then X' is an isomorphic image of X. There is $B \in \operatorname{dom}(a_n)$ isomorphic to A_{i*} such that $\pi_{A_{i*}B}[X] = X'$. The arguments of the previous cases work here completely the same after we replace A_{i*} with B and A_{i**} with $\pi_{A_{i*}B}[A_{i**}]$.
$\square$ of the claim.

Case B $X \notin C^{\kappa^{++}}(A)$.

By the previous case it is possible to add each of A_i's, for $i < otp_{\kappa^+}(A)$. Let us prove by induction on i that it is possible to add $X \in A_i$. Thus, if $i = 0$, then add first A_0. The only possibility for $X \in A_0$ is to be an ordinal. Also, A_0 has no predecessors. Let $X = \alpha$.

Subcase B1 Each $\beta \in A_0 \cap (\cup\{\operatorname{dom}(a_n)|n < \omega\})$ (if any) is less than α.

Assume that n is big enough such that $a_n(A_0) \prec H(\chi^{+k})$ for some $k \gg 2$. Pick now some $M \in a_n(A_0)$ such that

(1) $|M| = \kappa_n^{+n+2}$

(2) $M \prec H(\chi^{+k-1})$

(3) $M \supseteq \kappa_n^{+n+2}$

(4) $\operatorname{cof}(M \cap \kappa_n^{+n+3}) = \kappa_n^{+n+2}$

(5) for each $N \in \operatorname{rng}(a_n) \cap a_n(A_0)$ we have $N \cap H(\chi^{+k-1}) \in M$.

In particular, $M \cap \kappa_n^{+n+3}$ (the main part of M) is above each $N \cap \kappa_n^{+n+3}$, for every $N \in \operatorname{rng}(a_n) \cap a_n(A_0)$.

Define the image of α to be M. Move this setting to all the elements in $\mathrm{dom}(a_n)$ isomorphic to A_0 (if any). Denote the result by b_n.

Claim B1.1 b_n satisfies all the conditions of 3.1(2) but (l). Moreover, if a_n satisfies all the conditions of 3.1(2) then also b_n satisfies all the conditions of 3.1(2).

Proof. Let us check first 3.1(2(k)). Let Y be a model in $\mathrm{dom}(a_n)$ and α' be an image of α under isomorphism which was added to $\mathrm{dom}(a_n)$. We need to deal with the case when $\alpha' < \sup Y$ and show that $\min Y \backslash \alpha'$ is in $\mathrm{dom}(a_n)$. Thus let A_0' be a model in $\mathrm{dom}(a_n)$ isomorphic to A_0 such that $\alpha' = \pi_{A_0 A_0'}(\alpha)$. Compare Y with A_0'.

Case B1.1.1 $otp(Y) = otp(A_0')$.

We split into two subcases according to 1.1(10(c))(i) or (ii).

Subcase B1.1.1.1 $\min(A_0' \backslash \sup(A_0' \cap Y) + 1) > \sup Y$.

Then $\beta' = \min(A_0' \backslash \sup(A_0' \cap Y) + 1)$ is in $\mathrm{dom}(a_n)$ by 3.1(2(f)). Recall that α was above all the ordinals of $A_0 \cap \mathrm{dom}(a_n)$, hence α' will be such in $A_0' \cap \mathrm{dom}(a_n)$, by 3.1(2(e)). In particular, $\alpha' > \beta'$. But then $\alpha' > \sup Y$, which contradicts our assumption on α' and Y.

Subcase B1.1.1.2 $\min(Y \backslash \sup(A_0' \cap Y) + 1) > \sup A_0'$.

Again, $\beta' = \min(A_0' \backslash \sup(A_0' \cap Y) + 1)$ is in $\mathrm{dom}(a_n)$ by 3.1(2(f)). Also, $\alpha' > \beta'$. But $\alpha' < \sup A_0'$. Hence, $\alpha' < \min Y \backslash \alpha' = \min(Y \backslash \sup(A_0' \cap Y) + 1)$. But $\min(Y \backslash \sup(A_0' \cap Y) + 1) \in \mathrm{dom}(a_n)$, by 3.1(2(k)). So we are done.

Case B1.1.2 $otp(Y) < otp(A_0')$.

Then B' as in 1.1(10) must exist. But this is impossible since $otp_{\kappa^+}(A_0') = otp_{\kappa^+}(A_0) = 1$.

Case B1.1.3 $otp(Y) > otp(A_0')$.

Then we have a set B' as in 1.1(10) for Y and A_0 inside $\mathrm{dom}(a_n)$. Again we split into two cases according to (i) and (ii) of 1.1(10(c)).

Subcase B1.1.3.1 $\min(B' \backslash \sup(A_0' \cap B') + 1) > \sup A_0'$.

As before, α' should be above $\beta' = \min(A_0' \backslash \sup(A_0' \cap Y) + 1)$. By 1.1(10(d)), Y has no elements inside the interval

$$(\min(A_0' \backslash \sup(A_0' \cap B') + 1), \sup A_0).$$

We will need now the following useful claim:

Subclaim B1.1.3.1.1 There is $Z \in \mathrm{dom}(a_n)$ such that $Z \supseteq A_0'$ and $otp(Z) = otp(Y)$.

Proof. Consider the walks from $\max(a_n)$ to A_0' and to Y. Let B_0 be the first point where the walks split. Then B_0 must be a successor point with two immediate predecessors B_{00} and B_{01}. By 3.1(2(h)), then all these models B_0, B_{00}, B_{01} are in $\text{dom}(a_n)$. Assume without loss of generality that $A_0' \subseteq B_{00}$ and $Y \subseteq B_{01}$. If $Y = B_{01}$, then B_{00} will be as desired. Suppose that $Y \subset B_{01}$. Then just copy it to the B_{00} side by taking $Y_1 = \pi_{B_{01}B_{00}}[Y]$. Then $Y_1 \in \text{dom}(a_n)$, by 3.1(2(e)). If $Y_1 \supseteq A_0'$, then we are done. Otherwise consider the walks from B_{00} to A_0' and Y_1. After finitely many steps a model as desired will be reached.

$\square$ of the subclaim.

Compare now Y and Z. There is $\xi \in Y \cap \text{dom}(a_n)$ such that $Y \cap Z = Y \cap \xi$. Actually, $\xi = \min(Y \setminus \sup(Y \cap Z))$, by 1.1(9). Remember that $\alpha' \in Z$ and $\alpha' < \sup Y$. Then, by 1.1(10), $\xi = \min(Y \setminus \sup(Y \cap Z)) > \sup Z > \alpha'$. But now clearly, $\xi = \min(Y \setminus \alpha')$ and we are done.

Subcase B1.1.3.2 $\min(A_0' \setminus \sup(A_0' \cap B') + 1) > \sup B'$.

Then, by 1.1(10(c(ii))), we have also

$$\min(A_0' \setminus \sup(A_0' \cap B') + 1) > \sup Y.$$

Which implies that $\alpha' > \sup Y$ and contradicts our assumption.

This completes the check of 3.1(2(k)).

Let us turn to 3.1(2(h)). Suppose that Y is a model in $\text{dom}(a_n)$ and α' is an image of α added by the isomorphism between A_0 and some $A \in \text{dom}(a_n)$. Assume that $\alpha' \in Y$. We would like to show that the walk from Y to α' is already in $\text{dom}(a_n)$. We claim that the walk must terminate with A. Suppose otherwise. Thus let Z be the first model of the walk which does not contain A. Compare Z and A. By 1.1(9,10) and 3.1(2(f)), there is $\mu \in A \cap \text{dom}(a_n)$ such that

$$Z \cap A = A \cap \mu.$$

Then $\alpha' < \mu$, but recall that each $\beta \in A_0 \cap (\cup\{\text{dom}(a_n) | n < \omega\})$ (if any) is less than α. So, by isomorphism between A_0 and A, each $\beta \in A \cap (\cup\{\text{dom}(a_n) | n < \omega\})$ (if any) is less than α'. In particular, $\mu < \alpha'$. Contradiction.

Let us check now 3.1(2(m)). Thus let $\langle \alpha_i | i < j \rangle$ be a strictly increasing sequence of isomorphic images of α. For each $i < j$ there is a model $Y_i \in \text{dom}(a_n)$ isomorphic to A_0 such that $\alpha_i = \pi_{A_0 Y_i}(\alpha) = \alpha_i$. Note if $i, k < j$ are different then $\alpha_k \notin Y_i$. Just, by 1.1(8) the isomorphisms between models are identity on common parts of the models. Now, we pick for each $i < j$ the least ordinal $\tau_i \in Y_{i+1} \setminus Y_i$. There is such, since $\alpha_{i+1} \in Y_{i+1} \setminus Y_i$, $\alpha_{i+1} > \alpha_i \in Y_i$ and so, by 1.1(10(c)) we must have then $\min(Y_{i+1} \setminus \sup(Y_{i+1} \cap Y_i) + 1) > \sup Y_i$. Also, we have $\alpha_i < \sup Y_i < \tau_i \le \alpha_{i+1}$. By 3.1(2(f)), $\tau_i \in \mathrm{dom}(a_n)$ for each $i < j$. Hence,

$$\bigcup_{i<j} \alpha_i = \bigcup_{i<j} \tau_i \in \mathrm{dom}(a_n).$$

The rest of the conditions hold trivially.
$\square$ of the claim.

Subcase B2 $\kappa^{+3} \cap A_0 \cap (\cup\{\mathrm{dom}(a_n)|n < \omega\}) \setminus \alpha + 1$ is not empty.

Let $\delta = \min(\kappa^{+3} \cap A_0 \cap (\cup\{\mathrm{dom}(a_n)|n < \omega\}) \setminus \alpha + 1$. Pick n^* large enough so that for each $m \ge n^*$ we have $\delta, A_0 \in \mathrm{dom}(a_m)$ and $a_m(\delta) \prec H(\chi^{+k})$ and $a_n(A_0) \prec H(\chi^{+\ell})$ for $k, \ell \gg 2$. Fix $n \ge n^*$. Assume for simplicity that $a_n(\delta) \prec a_n(A_0) \prec H(\chi^{+k})$. Otherwise we just cut one of $a_n(\delta), a_n(A_0)$, i.e. we choose k to be the minimal so that $a_n(\delta) \subseteq H(\chi^{+k})$ or $a_n(A_0) \subseteq H(\chi^{+k})$ and intersect the one that not contained with $H(\chi^{+k})$.

Now we proceed as in Case B1 with $a_n(\delta)$ replacing $a_n(A_0)$. Thus pick some $M \in a_n(\delta) \cap a_n(A_0)$ such that

(1) $|M| = \kappa_n^{+n+2}$

(2) $M \prec H(\chi^{+k-1})$

(3) $M \supseteq \kappa_n^{+n+2}$

(4) $\mathrm{cof}(M \cap \kappa_n^{+n+3}) = \kappa_n^{+n+2}$

(5) for each $N \in \mathrm{rng}(a_n) \cap a_n(\delta)$ we have $N \cap H(\chi^{+k-1}) \in M$.

In particular, $M \cap \kappa_n^{+n+3}$ (the main part of M) is above each $N \cap \kappa_n^{+n+3}$, for every $N \in \mathrm{rng}(a_n) \cap a_n(\delta)$.

Define the image of α to be M. Move this setting to all the elements in $\mathrm{dom}(a_n)$ isomorphic to A_0 (if any). Denote the result by b_n.

Claim B2.1 b_n satisfies all the conditions of 3.1(2) but (1). Moreover, if a_n satisfies all the conditions of 3.1(2) then also b_n satisfies all the conditions of 3.1(2).

Proof. Let us check first 3.1(2(k)). Let Y be a model in $\mathrm{dom}(a_n)$ and α' be an image of α under isomorphism which was added to $\mathrm{dom}(a_n)$. We need to deal with the case when $\alpha' < \sup Y$ and show that $\min Y \setminus \alpha'$ is in $\mathrm{dom}(a_n)$. Thus let A_0' be a model in $\mathrm{dom}(a_n)$ isomorphic to A_0 such that $\alpha' = \pi_{A_0 A_0'}(\alpha)$. Denote $\pi_{A_0 A_0'}(\delta)$ by δ'. Compare Y with A_0'.

Case B2.1.1 $otp(Y) = otp(A_0')$.

We split into two subcases according to 1.1(10(c))(i) or (ii).

Subcase B2.1.1.1 $\min(A_0'\setminus \sup(A_0' \cap Y) + 1) > \sup Y$.

Then $\beta' = \min(A_0'\setminus \sup(A_0' \cap Y) + 1)$ is in $\mathrm{dom}(a_n)$ by 3.1(2(f)). Recall that α was above all the ordinals of $\delta \cap \mathrm{dom}(a_n)$, hence α' will be such in $\delta' \cap \mathrm{dom}(a_n)$, by 3.1(2(e)). In particular, if $\beta' < \delta'$ then $\alpha' > \sup Y$, which contradicts our assumption on α' and Y. Hence, $\beta' \geq \delta'$. By 1.1(9,10), we have $A_0' \cap Y = A_0' \cap \beta'$. So, $\alpha' \in Y$ and we are done.

Subcase B2.1.1.2 $\min(Y\setminus \sup(A_0' \cap Y) + 1) > \sup A_0'$.

Again, $\beta' = \min(A_0'\setminus \sup(A_0' \cap Y) + 1)$ is in $\mathrm{dom}(a_n)$ by 3.1(2(f)). By 1.1(9,10), we have $A_0' \cap Y = A_0' \cap \beta'$. Hence, if $\delta' \leq \beta'$ then $\alpha' \in Y$ and we are done. Suppose that $\delta' > \beta'$. Then also $\alpha' > \beta'$. It follows that $\alpha' < \min Y\setminus\alpha' = \min(Y\setminus \sup(A_0'\cap Y)+1)$. But $\min(Y\setminus \sup(A_0'\cap Y)+1) \in \mathrm{dom}(a_n)$, by 3.1(2(f)). So we are done.

Case B2.1.2 $otp(Y) < otp(A_0')$.

Then B' as in 1.1(10) must exists. But this is impossible since $otp_{\kappa^+}A_0' = otp_{\kappa^+}A_0 = 1$.

Case B2.1.3 $otp(Y) > otp(A_0')$.

Then we have a set B' as in 1.1(10) for Y and A_0' inside $\mathrm{dom}(a_n)$. Again we split into two cases according to (i) and (ii) of 1.1(10(c)).

Subcase B2.1.3.1 $\min(B'\setminus \sup(A_0' \cap B') + 1) > \sup A_0'$.

As before, α' should be above $\beta' = \min(A_0'\setminus \sup(A_0' \cap Y) + 1)$ unless it is already in Y. By 1.1(10(d)), Y has no elements inside the interval

$$(\min(A_0'\setminus \sup(A_0' \cap B') + 1), \sup A_0').$$

Let Z be as in Subclaim B1.1.3.1.1. Compare now Y and Z. There is $\xi \in Y\cap\mathrm{dom}(a_n)$ such that $Y\cap Z = Y\cap\xi$. Actually, $\xi = \min(Y\setminus \sup(Y\cap Z))$, by 1.1(9). Remember that $\alpha' \in Z$ and $\alpha' < \sup Y$. Then, by 1.1(10), $\xi = \min(Y\setminus \sup(Y \cap Z)) > \sup Z > \alpha'$. But now clearly, $\xi = \min(Y\setminus\alpha')$ and we are done.

Subcase B2.1.3.2 $\min(A_0'\setminus \sup(A_0' \cap B') + 1) > \sup B'$.

Then, by 1.1(10(c(ii))), we have also

$$\min(A_0'\setminus \sup(A_0' \cap B') + 1) > \sup Y.$$

We assumed that $\alpha' < \sup Y$, so $\alpha' \in A_0' \cap B'$. In particular, then $\alpha' \in Y$ and we are done.

This completes the checking of 3.1(2(k)).

Let us turn to 3.1(2(h)). Suppose that Y is a model in $\mathrm{dom}(a_n)$ and α' is an image of α added by the isomorphism between A_0 and some $A \in \mathrm{dom}(a_n)$. Assume that $\alpha' \in Y$. We would like to show that the walk from Y to α' is already in $\mathrm{dom}(a_n)$.

If the walk to α' must terminate with A then we are done. Suppose otherwise. Thus let Z be the first model of the walk which does not contain A. Compare Z and A. By 1.1(9,10) and 3.1(2(f)), there is $\mu \in A \cap \mathrm{dom}(a_n)$ such that

$$Z \cap A = A \cap \mu.$$

Also there is $S \in Z \cup \{Z\} \cap \mathrm{dom}(a_n)$ such that $otpS = otpA$ and $Z \cap A = S \cap A$ (remember that A is isomorphic to A_0 which is minimal). If S is the final model of the walk from Y to α' then we are done. Suppose otherwise. Let $T \neq S$ be such model.

Note that then necessarily $otpT = otpA$, since A is minimal the only other possibility is $otpT > otpA$. But if this happens then there will be $T' \in T$ isomorphic to A and with α' inside by 1.1(10). Which is impossible, since then T' must be one of the immediate predecessors of T or a member of them by 1.1(11). So it is possible to continue the walk contradicting to the choice of T as the final model.

We claim that $T \in \mathrm{dom}(a_n)$. Let us argue as follows. Pick Y_0 to be the last member of the common part of the walks from Y to S and to T. Then it should be a successor model. Let Y_{00}, Y_{01} be its immediate predecessors with $Y_{00} \in C^{\kappa^+}(Y_0)$. Then Y_{01} should include S, since otherwise the walk to a common point α' of both models S, T must go into direction of S, which is not the case. Now, Y_0 must be in $\mathrm{dom}(a_n)$. It follows from the definition of the walk and 3.1(2(h)) applied to $Y, S \in \mathrm{dom}(a_n)$. But then both Y_{00}, Y_{01} are in $\mathrm{dom}(a_n)$ by 3.1(2(h)) and (l). Hence also $S_0 = \pi_{Y_{00}, Y_{01}}[S]$ is in $\mathrm{dom}(a_n)$, by 3.1(2(e)). We have still $\alpha' \in S_0$. If $S_0 = T$, then we are done. Otherwise, pick Y_1 to be the last member of the common part of the walks from Y to S_0 and to T. It should be a successor model below Y_{00}. Let Y_{10}, Y_{11} be its immediate predecessors with $Y_{10} \in C^{\kappa^+}(Y_1)$. Then Y_{11} should include S_0, since otherwise the walk to a common point α' of both models S_0, T must go into direction of S_0, which is not the case. Now, Y_1 must be in $\mathrm{dom}(a_n)$. It follows from the definition of the walk and 3.1(2(h)) applied to $Y, S_0 \in \mathrm{dom}(a_n)$. But then both Y_{10}, Y_{11} are in $\mathrm{dom}(a_n)$ by 3.1(2(h)) and (l). Hence also $S_1 = \pi_{Y_{10}, Y_{11}}[S_0]$ is in $\mathrm{dom}(a_n)$, by 3.1(2(e)). We have still $\alpha' \in S_1$. If $S_1 = T$, then we are done. Otherwise, pick Y_2 to be the last member of the common part of the walks from Y to

S_1 and to T. It should be a successor model below Y_{10}. Continue as above and define $S_2 \in \mathrm{dom}(a_n)$. If $S_2 \neq T$, then we can continue to go down and to define Y_3 etc. After finally many stages T will be reached.

This completes the checking of 3.1(2(h)).

Let us check now 3.1(2(m)). Thus let $\langle \alpha_i | i < j \rangle$ be a strictly increasing sequence of isomorphic images of α. For each $i < j$ there is a model $Y_i \in \mathrm{dom}(a_n)$ isomorphic to A_0 such that $\alpha_i = \pi_{A_0 Y_i}(\alpha) = \alpha_i$. Note if $i, k < j$ are different then $\alpha_k \notin Y_i$. Just, by 1.1(8) the isomorphisms between models are identity on common parts of the models. Now, we pick for each $i < j$ the least ordinal $\tau_i \in Y_{i+1} \backslash Y_i$. There is such, since $\alpha_{i+1} \in Y_{i+1} \backslash Y_i$, $\alpha_{i+1} > \alpha_i \in Y_i$ and so, by 1.1(10(c)) we must have then $\min(Y_{i+1} \backslash \sup(Y_{i+1} \cap Y_i) + 1) > \sup Y_i$. Also, we have $\alpha_i < \sup Y_i < \tau_i \leq \alpha_{i+1}$. By 3.1(2(f)), $\tau_i \in \mathrm{dom}(a_n)$ for each $i < j$. Hence,

$$\bigcup_{i<j} \alpha_i = \bigcup_{i<j} \tau_i \in \mathrm{dom}(a_n).$$

The rest of the conditions hold trivialy.
□ of the claim.

Suppose now that $i > 0$ and for each $j < i$ it is possible to add elements of A_j. Let us show that it is possible to add elements of A_i. If i is limit, then this is clear since then $A_i = \cup \{A_j | j < i\}$. So assume that i is a successor ordinal and let $X \in A_i \backslash \cup \{A_j | j < i\}$. Suppose first that X is not an ordinal. Note that by 3.1(2(g)), in order to add an ordinal we need anyway to add models first.

We would like to run now a new induction on the length of a walk from A to X. Let us give a precise definition.

Suppose that $K, L \in A^{1\kappa^+}$ and $L \in K$, where as usual $A^{1\kappa^+}$ is taken from $G(\mathcal{P}')$. Set $wl(K, L) = 0$, if $L \in C^{\kappa^+}(K)$. Let $wl(K, L) = 1$, if there is $M \in C^{\kappa^+}(K)$ such that L is the immediate predecessor of M which is not in $C^{\kappa^+}(M)$. In general, set $wl(K, L) = 2n + 2$, if there is M such that $wl(K, M) = 2n + 1$ and $L \in C^{\kappa^+}(M)$; set $wl(K, L) = 2n + 1$, if there is M such that $wl(K, M) = 2n$ and L is the immediate predecessor of M which is not in $C^{\kappa^+}(M)$.

So the induction will be now on wl - the walk length from models in the domain to one that we like to add. Then the zero stage and all even stages are just as Case A. Stage one and all odd stages basically deal with the situation of adding X to A_i ones X is the immediate predecessor of A_i which is not in C^{κ^+}.

So let us concentrate on this case. Then A_i must have two immediate predecessors A_{i-1} and X. Again by Case A, we can assume that also $A_{i-1} \in \text{dom}(a_n)$. Note that it implies, in particular, that a_n fails to satisfy 3.1(2(l)). The problem will be fixed below by adding X.

Let A_{i**} be as in Case A but with A_{i-1} replacing A_{i*} there.

By 1.1(4(d)), $otpA_{i-1} = otpX$. Also there are $\alpha_1 \in A_{i-1} \cap A^{1\kappa^{++}}$, $\alpha_2 \in X \cap A^{1\kappa^{++}}$ such that $A_{i-1} \cap \alpha_1 = A_{i-1} \cap X = X \cap \alpha_2$, for some $A^{1\kappa^{++}}$ in $G(\mathcal{P}')$. We first add α_1 to $\text{dom}(a_n)$. Note that $\alpha_1 \in A_{i-1} \cap A^{1\kappa^{++}}$. So, the induction can be used to add it to the domain. Assume that already $\alpha_1 \in \text{dom}(a_n)$ and $a_n(\alpha_1) = M$ so that $M \in a_n(A_{i-1})$ and it is an elementary submodel of cardinality κ_n^{+n+2} of $H(\chi^{+k})$ for some $k \gg 2$. Find $X^* \in M$ such that $|X^*| = \kappa_n^{+n+1}$, $X^* \prec H(\chi^{+k-1})$ and X^* realizes over $a_n(A_{i-1}) \cap M$ the same $k-1$-type as $a_n(A_{i-1})$ does. It exists by elementarity, since we replace k by $k-1$.

If $\sup X < \alpha_1$, then we take X^* to be the image of X. Extend the condition by coping from $a_n(A_{i-1})$ to X^* all the elements of $a_n(A_{i-1})$ and $M \cap H(\chi^{k-1})$. If $\sup X > \alpha_1$, then $\beta_1 = \min(X \backslash A_{i-1} \cap X) > \sup A_{i-1}$. Work then inside $a_n(A_i)$ find a model M' above $\sup a_n(A_{i-1})$ of the same type as M and find X^{**} in it above $\sup a_n(A_{i-1})$ as well resembling X^* above. Add it to be the image of X.

Now copy everything from $A_{i-1} \cap \text{dom}(a_n)$ to X and move this setting to all the elements of $\text{dom}(a_n)$ isomorphic to A_i (if any). Denote the result by b_n.

The above is the heart of the argument. The basic idea goes back to [1]

Claim B3 b_n satisfies all the conditions of 3.1(2) but (l). Moreover, if a_n satisfies all the conditions of 3.1(2) but (l) only for A_i and the models isomorphic to it, then also b_n satisfies all the conditions of 3.1(2).

Proof. Let us check first 3.1(2(f)).

Suppose first that X', X'' are images of X and its elements (obtained by moving from $A_{i-1} \cap \text{dom}(a_n)$ to X) which were added to $\text{dom}(a_n)$.

We need to show witnessing the intersection conditions for $X' \cap X''$ are in the domain. Split the proof into a few cases.

Case B3.1 $X' = X$ and $X'' \in A_i$.

Assume that $X'' \neq X'$, otherwise every thing is just trivial. If $X'' \in X$ then we are done again. So we can assume that $X'' = A_{i-1}$ or $X'' \in A_{i-1} \cap \text{dom}(a_n)$. If $X'' = A_{i-1}$, then α_1 (from the definition of X) as well as its image- $\pi_{A_{i-1}X}(\alpha_1) = \alpha_X$ are in the domain of q and we are done. Suppose that $X'' \in A_{i-1} \cap \mathrm{dom}(a_n)$.

Subcase B3.1.1 X'' is not an ordinal.

Then

$$X \cap X'' = X \cap A_{i-1} \cap X'' = A_{i-1} \cap \alpha_1 \cap X'' = \alpha' \cap X''.$$

Hence only α' which is already in $\mathrm{dom}(a_n)$, is needed for the intersection of X'' and X. The opposite way- let $Z = \pi_{A_{i-1}X}[X'']$. Then

$$X \cap X'' = X \cap A_{i-1} \cap X'' = Z \cap A_{i-1} = Z \cap \alpha_X.$$

Again, Z and α_X were added so we are done.

Subcase B3.1.2 X'' is an ordinal.

Let us denote X'' by ξ. Suppose that $\xi < \sup X \cap \kappa^{+3}$. If $\xi < \alpha_1$, then ξ is in the common part of A_{i-1} and X. Otherwise, we will have $\alpha_X = \min(X \backslash \xi)$, by 1.1(10).

Case B3.2 $X' \in X$ and $X'' \in A_i$.

We allow the possibility that one of them is an ordinal. In this case we care only about the intersection on the model side.

Let $Z = \pi_{XA_{i-1}}[X']$. Then Z is in the domain of a_n. Also,

$$X' \cap X'' = X' \cap Z \cap X'' = \alpha_1 \cap Z \cap X''.$$

This takes care of the intersection of X' and X'' from the side of X'. Let us deal with the opposite side, i.e. X'. Take $Y = \pi_{XA_{i-1}}[X'']$. Then

$$X' \cap X'' = X' \cap Y \cap X'' = \alpha_X \cap Y \cap X'.$$

But both Y and α_X were added. So we are done.

Case B3.3 $X' \in X \cup \{X\}$ and $X'' \notin A_i$.

Then there is some $A'' \in \mathrm{dom}(a_n)$ isomorphic to A_i with $X'' \in A''$ being the image of an element of $X \cup \{X\}$ that was added under the isomorphism $\pi_{A_i A''}$. Let $Z = \pi_{A''A_i}[X'']$ and $Z' = \pi_{A_i A''}[X']$.

We deal first with the intersection of X' with X'' on the side of X'. Thus

$$X' \cap X'' = X' \cap A_i \cap A'' \cap X'' = X' \cap A_i \cap \alpha_{A_i A''} \cap Z,$$

where $\alpha_{A_i A''} = \min\{\delta | \delta \in A_i \backslash A''\}$. Then we can use Case B3.1 or B3.2, since all the components of the last intersection are in A_i.

Let us turn to the opposite side. Again,

$$X' \cap X'' = X' \cap A_i \cap A'' \cap X'' = X'' \cap A'' \cap \alpha_{A'' A_i} \cap Z',$$

where $\alpha_{A'' A_i} = \min\{\delta | \delta \in A'' \backslash A_i\}$.

Now we can move all the members of the last equality to A_i using $\pi_{A'' A_i}$ take care of the intersections using Case B3.1 or 3.2 and finally move back the result by $\pi_{A_i A''}$.

Case B3.4 $X' \not\in A_i$ and $X'' \not\in A_i$.

Then there are some A', $A'' \in \text{dom}(a_n)$ isomorphic to A_i with $X' \in A'$, $X'' \in A''$ and both X', X'' being images of elements of $X \cup \{X\}$ that were added under the isomorphisms $\pi_{A_i A'}$ and $\pi_{A_i A''}$. Let $Z = \pi_{A' A''}[X'']$ and $Z' = \pi_{A_i A''}[X']$.

We deal with the intersection of X' with X'' on the side of X'. The opposite side is similar. Thus

$$X' \cap X'' = X' \cap A' \cap A'' \cap X'' = X' \cap A' \cap \alpha_{A' A''} \cap Z,$$

where $\alpha_{A' A''} = \min\{\delta | \delta \in A' \backslash A''\}$. Note that such an ordinal exists by 3.1(f,k). Now the members of the last equality are all in A'. Move them to A_i using $\pi_{A'' A_i}$ take care of the intersections using Case B3.1 or 3.2 and finally move back the result by $\pi_{A_i A''}$.

Suppose now that only X' is an image of X or of its element and X'' is old, i.e. in $\text{dom}(a_n)$.

Then there is some $A' \in \text{dom}(a_n)$ isomorphic to A_i with $X' \in A'$ being the image of an element of $X \cup \{X\}$ that was added under the isomorphism $\pi_{A_i A'}$. Split into two cases according to $otpX''$.

Case B3.5 $otpX'' \leq otpA'$ or X'' is an ordinal.

Then we need only to deal with the intersection of X' and X'' on the side of X'.

Let $\alpha_{A' X''} = \min\{\delta | \delta \in A' \backslash X''\}$. Note that such ordinal is in $\text{dom}(a_n)$ by 3.1(f,k). Then

$$X' \cap X'' = X' \cap A' \cap X'' = X' \cap A' \cap \alpha_{A' X''}.$$

Now all the components of the last equality are in A', so we can deduce the conclusion as in Case B3.4.

Case B3.6 $otpX'' > otpA'$.

Find first $A'' \in X'' \cap \text{dom}(a_n)$ of the order type $otp\,A'$ such that $A' \cap X'' = A' \cap A''$. It exists by 3.1(f). Also let $\alpha_{A' A''} = \min\{\delta | \delta \in A' \backslash A''\}$ and $\alpha_{A'' A'} = \min\{\delta | \delta \in A'' \backslash A'\}$.

Deal first with the intersection of X' and X'' on the side of X'.

$$X' \cap X'' = X' \cap A' \cap X'' = X' \cap A' \cap A'' = X' \cap A' \cap \alpha_{A'A''}.$$

Now all the components of the last equality are in A', so we can deduce the conclusion as in Case B3.3.

Now we deal with the intersection of X' and X'' on the side of X'.

$$X' \cap X'' = X' \cap A' \cap X'' = X' \cap A' \cap A'' = \pi_{A'A''}[X'] \cap A' \cap \alpha_{A''A'}.$$

Now all the components of the last equality are in A'', so we can deduce the conclusion as in Case B3.3.

This completes checking of 3.1(2(f)).

Let us turn to 3.1(2(h)).

Suppose that Y, U appear in q and $U \in Y$. We claim that the walk from Y to U is also in q. We may assume that at least one of the elements Y, U is new (i.e. not in p). Split into a few cases.

Case B3.7 $Y \notin \mathrm{dom}(a_n)$.

Then there is $A \in \mathrm{dom}(a_n)$ isomorphic to A_i such that $Y \in A$ is the isomorphic image of a model X or of its element obtained by moving from $A_{i-1} \cap \mathrm{dom}(a_n)$ to X. We have $U \in Y$, so $U \in A$.

Without loss of generality we can assume that A is A_i, otherwise just move Y, U to A_i via the isomorphism run the the argument inside A_i and then move the result back to A.

Now, we apply the isomorphism $\pi_{XA_{i-1}}$ to Y, U. Let Y^*, U^* be the images. Then they are in $A_{i-1} \cup \{A_{i-1}\} \cap \mathrm{dom}(a_n)$. Hence the walk from Y^* to U^* is in $\mathrm{dom}(a_n)$. Its image under $\pi_{A_{i-1}X}$ will be then the walk from Y to U and we are done.

Case B3.8 $Y \in \mathrm{dom}(a_n)$.

There is $A \in \mathrm{dom}(a_n)$ isomorphic to A_i such that $U \in A$ is the isomorphic image of a model X or of its element obtained by moving from $A_{i-1} \cap \mathrm{dom}(a_n)$ to X.

Subcase B3.8.1 U is an image of X.

If the walk from Y to U terminates at A then we are done. Otherwise there must be a model Z on this walk inside $\mathrm{dom}(a_n)$ which does not contain A. We can take for example, $Z = Y$, if Y does not contain A or the point were the walk from Y to A differs from the one from Y to U, if $Y \supseteq A$.

Compare Z and A. By 1.1(9,10) and 3.1(2(f)), there is $\mu \in A \cap \mathrm{dom}(a_n)$ such that

$$Z \cap A = A \cap \mu.$$

Also there is $S \in Z \cup \{Z\} \cap \mathrm{dom}(a_n)$ such that $otpS = otpA$ and $Z \cap A = S \cap A$ (remember that A is isomorphic to A_i which is a minimal including X and U is the image of X). If S is the final model of the walk from Y to U then we are done. Suppose otherwise. Let $T \neq S$ be such model.

Note that then necessary $otpT = otpA$, since A is a minimal including U the only other possibility is $otpT > otpA$. But if this happens then there will be $T' \in T$ isomorphic to A and with U inside by 1.1(10). Which is impossible, since then T' must be one of the immediate predecessors of T or a member of them by 1.1(11). So it is possible to continue the walk contradicting to the choice of T as the final model.

Now $T \in \mathrm{dom}(a_n)$. The rest of the argument repeats completely those of Claim B2.1.

Subcase B3.8.2 U is not an image of X.

Suppose for simplicity that $A = A_i$. Then U will be in X and it will be obtained by moving from $A_{i-1} \cap \mathrm{dom}(a_n)$ to X.

If the walk from Y to U goes via X, then using the previous case this walk will be in q and we are done. Suppose that this does not happen.

If $otpX > otpY$, then by 3.1(2(f)) there is $X' \in A$ in q with $otpY = otpX'$ and $Y \cap X = Y \cap X'$. Clearly, $U \in X'$. Note that $Y \neq X'$ unless Y is in X. If $Y \in X$, then both $U, Y \in X$. Hence the argument of the previous case applies. Assume so, that $Y \neq X'$.

Set $K = X$, if $otpX \leq otpY$ and $K = X'$, if $otpX > otpY$.

Let Z be the first model of the walk from Y to U which does not contain K. Compare Z and K. By 1.1(9,10) and 3.1(2(f)), there is $S \in Z \cup \{Z\}$ in q such that $otpS = otpK$ and $Z \cap K = S \cap K$. If the walk from Y (or Z which is the same) to U goes through S, then use π_{KS} to copy to S the walk from K to U. The walk from K to U is in q, so the one copied (from S to U) must be in q as well by 3.1(2(e)). But the walk from Y to U is the combination of the walks from Y to S with the walk from S to U and both are in q. So we are done. Just note that the walk from Y to S is in q, since we have either

(a) $K = X$ and then $otpS = otpX$.

So S is an image of X and hence, B3.8.1 applies

or

(b) $K = X'$ and then $otpY = otpX' = otpS$.

In this case we must have $Y = S$ and so the walk is trivial.

Suppose now that the walk from Y (or Z which is the same) to U does not go through S.

Pick Y_0 to be the last member of the common part of the walks from Y to S and to U. Then it should be a successor model. Let Y_{00}, Y_{01} be its immediate predecessors with $Y_{00} \in C^{\kappa^+}(Y_0)$. Then Y_{01} should include S, since otherwise the walk to a common point U of both models S, Y_0 must go into direction of S, which is not the case. Now, Y_0 must be in q. It follows from the definition of the walk and 3.1(2(h)) applied to Y, S in q. But then both Y_{00}, Y_{01} are in q by 3.1(2(h)) and (l). Hence also $S_0 = \pi_{Y_{00},Y_{01}}[S]$ is in q, by 3.1(2(e)). We have still $U \in S_0$. If S_0 is on the walk from Y to U, then we are done exactly as above. Just as above use $\pi_{S_0 K}$ to copy the walk from S_0 to U to the one from K to U. Otherwise, pick Y_1 to be the last member of the common part of the walks from Y to S_0 and to U. It should be a successor model below Y_{00}. Let Y_{10}, Y_{11} be its immediate predecessors with $Y_{10} \in C^{\kappa^+}(Y_1)$. Then Y_{11} should include S_0, since otherwise the walk to a common point U of both models S_0, Y_{10} must go into direction of S_0, which is not the case. Now, Y_1 must be in q. It follows from the definition of the walk and 3.1(2(h)) applied to $Y, S_0 \in q$. But then both Y_{10}, Y_{11} are in q by 3.1(2(h)) and (l). Hence also $S_1 = \pi_{Y_{10},Y_{11}}[S_0]$ is in q, by 3.1(2(e)). We have still $U \in S_1$. If S_1 is on the walk from Y to U, then we are done. Otherwise, pick Y_2 to be the last member of the common part of the walks from Y to S_1 and to U. It should be a successor model below Y_{10}. Continue as above and define $S_2 \in q$. If $S_2 \neq T$, then we can continue to go down and to define Y_3 etc. After finitely many stages a model S^* which is on the walk from Y to U will be reached.

This completes the checking of 3.1(2(h)).

The checking of 3.1(2(k)) repeats those of Claim B2.1. The rest of the conditions hold trivially.

□ of the claim.

It remains only to deal with the case when X is an ordinal. Recall that $i > 0$ is the least with $X \in A_i$. We can assume now that all the immediate predecessors of A_i (and there are at most two and at least one) are already in $\mathrm{dom}(a_n)$. Let us denote $X = \alpha$ and let A'_{i-1} denotes the immediate predecessor different from A_{i-1}, if such exists.

Split now into two cases.

Case B4.1 There is no $Z \in A_i \cap \mathrm{dom}(a_n)$ with $\alpha < \sup(\kappa^{+3} \cap Z)$.

We proceed as in Case A. We assume that $a_n(A_i) \prec H(\chi^{+k})$ for $k \gg 2$. Pick X^* to be an element of $a_n(A_i)$ such that

(1) $X^* \prec H(\chi^{+k-1})$

(2) $X^* \cap \kappa^{+n+3}$ is an ordinal of cofinality κ^{+n+2}

(3) for every $Z \in A_i \cap \mathrm{dom}(a_n)$ we have $a_n(Z) \cap H(\chi^{+k-1}) \in X^*$

We take X^* to be the image of $X = \alpha$. Move this setting now to all the elements of $\mathrm{dom}(a_n)$ isomorphic to A_i (if any). The arguments of Cases A,B apply in order to show the result is a condition.

Case B4.1 There is $Z \in A_i \cap \mathrm{dom}(a_n)$ with $\alpha < \sup(\kappa^{+3} \cap Z)$.

Then we pick $Z \in A_i \cap \mathrm{dom}(a_n)$ with $\min(\kappa^{+3} \cap Z \backslash \alpha)$ as small as possible. Let β be $\min(\kappa^{+3} \cap Z \backslash \alpha)$, for such Z. Using induction we first add β to $\mathrm{dom}(a_n)$. Assume without loss of generality that both $a_n(A_i), a_n(\beta) \prec H(\chi^{+k})$ for $k \gg 2$. Pick X^* to be an element of $a_n(A_i) \cap a_n(\beta)$ such that

(1) $X^* \prec H(\chi^{+k-1})$

(2) $X^* \cap \kappa^{+n+3}$ is an ordinal of cofinality κ^{+n+2}

(3) for every $Z' \in A_i \cap \mathrm{dom}(a_n)$ with $\sup(Z') \cap \kappa^{+3} < \beta$ we have $a_n(Z') \cap H(\chi^{+k-1}) \in X^*$

We take X^* to be the image of $X = \alpha$. Move this setting now to all the elements of $\mathrm{dom}(a_n)$ isomorphic to A_i (if any). The arguments of Cases A,B apply in order to show the result is a condition. The only new possibility that was not considered in the checking of 3.1(2(k)) above is the following: Y a model in $\mathrm{dom}(a_n)$ with $\sup Y > \alpha$, $\alpha \notin Y$ and $otp(Y) < otp(A_i)$. Then use Subclaim B1.1.3.1 to find $W \in \mathrm{dom}(a_n)$ such that $W \supseteq Y$ and $otp(W) = otp(A_i)$. Suppose first that $\alpha \notin W$. Now compare W and A_i. There is $\xi \in W \cap \mathrm{dom}(a_n)$ such that $A_i \cap W = W \cap \xi$. By 1.1(10(c)), then $\xi = \min(W \backslash \sup(W \cap A_i)) > \sup(A_i) > \alpha$. Hence $\min(Y \backslash \alpha) = \min(Y \backslash \xi)$. But both Y and ξ are in $\mathrm{dom}(a_n)$. So we are done.

Suppose now that $\alpha \in W$. Then $\alpha = \pi_{WA_i}[\alpha]$. Set $Y' = \pi_{WA_i}[Y]$. Then $\sup Y' > \alpha$ since π_{WA_i} is order preserving. Induction on wl can be applied now to Y'. So, $\tau' = \min(Y' \backslash \alpha) \in \mathrm{dom}(a_n)$. Set $\tau = \pi_{A_i W}(\tau')$. Then $\tau \in \mathrm{dom}(a_n)$ and $\tau = \min(Y \backslash \alpha)$ by the elementarity of π_{WA_i}.

This completes the proof of the lemma.

$\Box$ of 3.3

Remark 3.4 The proof of 3.3 provides a bit more information. Thus, if X is a model that we like to add to $\mathrm{dom}(a_n)$, then

(1) if $X \in C^{\kappa^+}(A)$, for some $A \in \mathrm{dom}(a_n)$, and either

    (a) X is a limit model with $cof(otp_{\kappa^+}(X) - 1) > \kappa$
    or
    (b) X is a successor model, the immediate successor of X in $C^{\kappa^+}(A)$ is not in $\mathrm{dom}(a_n)$ and the same for the immediate predecessor (if it exists at all)
    or
    (c) X is a successor model, the immediate successor of X in $C^{\kappa^+}(A)$ is in $\mathrm{dom}(a_n)$, but X is its unique immediate predecessor, the immediate predecessor (if it exists) is not in $\mathrm{dom}(a_n)$ or it is, but then it is the unique immediate predecessor of X,

    then X can be added without adding other additional models or ordinals except the images of X under isomorphisms. I.e. if $A' \in \mathrm{dom}(a_n)$ is the least model including X and $B \in \mathrm{dom}(a_n)$ has the same otp as those of A', then $\pi_{A'B}[X]$ is added also. Note that necessarily $A' \in C^{\kappa^+}(A)$, otherwise just use 1.1(10),(4(a)).

(2) if X is a successor model, $X \in C^{\kappa^+}(A)$, for some $A \in \mathrm{dom}(a_n)$, and (b), (c) of the previous case fail, then adding X requires adding another of its immediate predecessor of the immediate successor of X in $C^{\kappa^+}(A)$, by 3.1(2(l)) or of another its immediate predecessor. Which in turn requires further additions which will be specified in (3) below.

(3) If $X \notin C^{\kappa^+}(A)$, for a smallest $A \in \mathrm{dom}(a_n)$ including X (note that it is possible to have many such A's but all of them will have the same otp and will agree about X not being on the sequence C^{κ^+}, just due to isomorphism. Also its possible in general to have them immediate successors of X. If one likes to fix one, then the one of the least *rank* can be used), then in order to add X we need to add finitely many models of the walk from A to X. Adding a model B_1 which is an immediate predecessor of a model B not in $C^{\kappa^+}(B)$ for B that was in $\mathrm{dom}(a_n)$ or was added during the walk requires adding the immediate predecessor B_0 of B inside $C^{\kappa^+}(B)$, as well as an ordinal $\alpha_0 \in B_0$ such that

$$B_1 \cap B_0 = B_0 \cap \alpha_0.$$

Adding α_0 may require further adding of the immediate predecessors of B_0 or of some model in $C^{\kappa^+}(B_0)$. This in turn requires adding of new ordinal and so on. The rank (or wl) of the models and the ordinals involved is decreasing. Hence after finitely many additions the process will terminate. Again after each addition we need to take all the isomorphic images.

The ordering $\leq^*$ on $\mathcal{P}$ and $\leq_n$ on Q_{n0} is not closed in the present situation. Thus it is possible to find an increasing sequence of $\aleph_0$ conditions $\langle\langle a_{ni}, A_{ni}, f_{ni}\rangle \mid i < \omega\rangle$ in Q_{n0} with no upper bound. The reason is that the union of maximal models of these conditions, i.e. $\bigcup_{i<\omega} \max(\operatorname{dom} a_{ni})$ need not be in A_{11} for any A_{11} in $G(\mathcal{P}')$. The next lemma shows that still $\leq_n$ and so also $\leq^*$ share a kind of strategic closure.

Lemma 3.5 *Let $n < \omega$. Then $\langle Q_{n0}, \leq_n \rangle$ does not add new sequences of ordinals of the length $< \kappa_n$, i.e. it is (κ_n, ∞)-distributive.*

Proof. Let $\delta < \kappa_n$ and $\underset{\sim}{f}$ be a Q_{n0}-name of a function from δ to ordinals. Using genericity of $G(\mathcal{P}')$ (or stationarity of the set $\{A^{0\kappa^+} | A^{0\kappa^+}$ appears in an element of $G(\mathcal{P}')\}$) it is not hard to find elementary submodel M of some $H(\nu)$ for ν big enough so that

(a) $Q_{n0}, \underset{\sim}{f}, \mathcal{P}' \in M$

(b) $|M| = \kappa^+$

(c) $M^* = M \cap H(\kappa^{+3})$ appears in $A^{1\kappa^{++}}$ of a condition of $G(\mathcal{P}')$.

(d) $cf(M^* \cap \kappa^{++}) = \delta$

(e) $^{\delta>}M \subseteq M$.

Note that for such M, $M^* = M \cap H(\kappa^{+3})$ must be a limit model, since by 1.1(12) successor models are closed under κ sequences, but M^* is not by (d) above.

We have $C^{\kappa^+}(M^*)\backslash\{M^*\} \subseteq M^*$ and, by elementarity of M, $C^{\kappa^+}(B) \in M$ for each $B \in C^{\kappa^+}(M^*)\backslash\{M^*\}$. Also the cofinality of $C^{\kappa^+}(M^*)\backslash\{M^*\}$ under the inclusion must be δ, since it is an $\in$-increasing continuous sequence of elements of M^* with limit M^* and by (d) above $cf(M^* \cap \kappa^{++}) = \delta$. Fix an increasing continuous sequence $\langle A_i \mid i < \delta\rangle$ of elements of $C^{\kappa^+}(M^*)\backslash\{M^*\}$ such that $\bigcup_{i<\delta} A_i = M^*$, A_0 is a successor model and for each limit model A_i in the sequence A_{i+1} is its immediate successor in

$C^{\kappa^+}(M^*)$. By (e), each initial segment of it will be in M. Now we decide inside M one by one values of f and put models from $\langle A_i \mid i < \delta \rangle$ to be maximal models of conditions used. This way we insure that unions of such conditions is a condition.

We define by induction an increasing sequence of conditions

$$\langle \langle a(i), A(i), f(i) \rangle | i < \delta \rangle$$

and an increasing continuous subsequence

$$\langle A_{k_i} | i < \delta \rangle \text{ of } \langle A_i | i < \delta \rangle$$

such that for each $i < \delta$

(1) $\langle a(i), A(i), f(i) \rangle \in M$,

(2) $\langle a(i+1), A(i+1), f(i+1) \rangle$ decides $\underset{\sim}{f(i)}$

(3) $A_{k_i}, A_{k_{i+1}} \in \text{dom}(a(i))$ and $A_{k_{i+1}}$ is the maximal model of $\text{dom}(a(i))$

There is no problem with $A(i)$'s and $f(i)$'s in this construction. Thus we have enough completeness to take intersections of $A(i)$'s and unions of $f(i)$'s. The only problematic part is $a(i)$. So let us concentrate only on building of $a(i)$'s.

i=0

Then let us pick some $Y_0 \prec Y_1 \prec H(\chi^\omega)$ of cardinality κ_n^{+n+2}, closed under κ_n^{+n+1}-sequences of its elements and $Y_0 \in Y_1$. Set $a(0) = \langle \langle A_0, Y_0 \rangle, \langle A_1, Y_1 \rangle \rangle$.

i+1

Then we first extend $\langle a(i), A(i), f(i) \rangle$ to a condition $\langle a(i)', A(i)', f(i)' \rangle$ deciding $\underset{\sim}{f(i)}$. Then perform swt (see 1.3) to turn $\langle a(i)', A(i)', f(i)' \rangle$ into an equivalent condition $\langle a(i)'', A(i)', f(i)' \rangle$ with $A_{k_i} \in C^{\kappa^+}(\max \text{dom}(a(i)''))$. Pick a successor model A_j (from the cofinal sequence $\langle A_i \mid i < \delta \rangle$) including $\max \text{dom}(a(i)'')$. Set $k_{i+1} = j$ and add it to $\text{dom}(a(i)'')$, using 3.3 and swt inside A_j if necessary. Finally we add A_{j+1}, using 3.3.

i is a limit ordinal

Then we need to turn $a = \bigcup_{j<i} a(j)$ into condition. For this we will need to add to $\text{dom}(a)$ models and ordinals which are limits of elements of $\text{dom}(a)$. First we extend a by adding to it $\langle A_{k_i}, \bigcup_{j<i} a(A_{k_j}) \rangle$, where $k_i = \bigcup_{j<i} k_j$. Then for each non decreasing sequence $\langle \alpha_j | j < i \rangle$ of ordinals in $\text{dom}(a)$ we add the pair $\langle \bigcup_{j<i} \alpha_j, \bigcup_{j<i} (a(\alpha_j) \cap H(\chi^{+\ell})) \rangle$, if it is not

already in the dom(a), where $\ell \leq \omega$ the maximal such that for unboundedly many j's in i $a(\alpha_j) \prec H(\chi^{+\ell})$, if the maximum exists or $\ell \gg n$ otherwise. Finally, for each model $B \in$ dom(a) if there is a nondecreasing sequence $\langle B_j | j < i \rangle$ of elements of $C^{\kappa^+}(B)$ in dom(a) and B is the least possible (under inclusion or with least sup) including the sequence, then we add the pair $\langle \cup_{j<i} B_j, \cup_{j<i}(a(B_j) \cap H(\chi^{+\ell})) \rangle$, if it is not already in the dom(a), where $\ell \leq \omega$ is the minimum between the least k such that $a(B) \subseteq H(\chi^{+k})$ and the maximal ℓ' such that for unboundedly many j's in i $a(B_j) \prec H(\chi^{+\ell'})$, if the maximum exists

or

it is k, if the maximum does not exist and $k < \omega$,

or

$\ell \gg n$, if the maximum does not exist and $k = \omega$.

Denote the result by b.

Claim 3.5.1 b satisfies 3.1(2).

Proof. Let start with 3.1(2(e)). Suppose that $A, B \in$ dom(b) are different and $otp(A) = otp(B)$. Pick $A', B' \in$ dom(a) to be the smallest possible (under inclusion and *rank*, but actually any choice of a smallest under inclusion alone will do) including A, B respectively.

Subclaim 3.5.1.1 $otp(A') = otp(B')$.

Proof. Suppose otherwise. Let, for example $otp(A') < otp(B')$. Then by 3.1(2(f)), there will be $B'' \in B \cap$ dom(a) with $otp(A') = otp(B'')$. Note that i is limit so for some $j < i$ big enough both A', B' are in dom(a_j) which is a part of a condition and so satisfies 3.1. Now, $\pi_{A'B''}[A]$ will be in dom(b) as well, just all the models from the increasing sequence converging to A are moved by $\pi_{A'B''}$ to form such sequence for $\pi_{A'B''}[A]$. We can assume that $A' \subset B'$ just replacing A by $\pi_{A'B''}[A]$ and A' by B'' if necessary. Considering the walk from B' to A' it is not hard to see that either $A' \in C^{\kappa^+}(B')$ or there is $A'' \in C^{\kappa^+}(B) \cap$ dom(a) of the same otp as A'. Suppose for simplicity that $A' \in C^{\kappa^+}(B')$. Otherwise just use $\pi_{A'A''}$ to move to A''. But $A \in C^{\kappa^+}(A')$. Hence $A \in C^{\kappa^+}(B')$. Also $B \in C^{\kappa^+}(B')$, by the choice of B'. $otp(A) = otp(B)$ implies then $A = B$. This contradicts the minimality of B', since we have now

$$B = A \subset A' \subset B'.$$

$\square$ of the subclaim.

Once we have $otp(A') = otp(B')$, the isomorphism $\pi_{A'B'}$ between A', B' and those between $a(A') \cap H(\chi^{+k}), a(B') \cap H(\chi^{+k})$, k is minimal such that $a(A') \subseteq H(\chi^{+k}), a(B') \subseteq H(\chi^{+k})$, can be used to induce isomorphisms between A and B, $b(A) \cap H(\chi^{+m})$ and $b(B) \cap H(\chi^{+m})$, where m is minimal such that $b(A) \subseteq H(\chi^{+m}), b(B) \subseteq H(\chi^{+m})$. Note that by the definition of b we must have $m \leq k$. Such isomorphisms will respect those of the members of the sequences converging to A and B, since isomorphisms between members of the sequences are induced in the same way.

Let turn now to 3.1(2(f)). Suppose that $A, B \in \mathrm{dom}(b)$, $A \not\subseteq B$ and $B \not\subseteq A$. We may assume that at least one of them is new. Again pick the smallest models A', B' with $A \in A', B \in B'$. Now, as in 3.3, Claim B3, we can use induction on wl - the walk length. Thus basically we need only to consider a situation when $A' \in C^{\kappa^+}(B')$. But then A, B just extend one another and we are done.

Let us check that 3.1(2(g)) holds. Suppose that $\alpha, A \in \mathrm{dom}(b)$ and $\alpha \in A$. If A is new model then α belongs to one of the members of the sequence of models converging to A. So we can assume that A is old one, i.e. in $\mathrm{dom}(a)$. Let $\langle \alpha_j | j < i \rangle$ be a nondecreasing sequence from $\mathrm{dom}(a)$ converging to α. By 3.1(2(k)), $\gamma_j = \min(A \backslash \alpha_j) \in \mathrm{dom}(a)$. Then $\langle \gamma_j | j < i \rangle$ will be also a sequence converging to α. Apply 3.1(2(g)) to A and its members. Then either there will be the model $B \in C^{\kappa^+}(A) \cap \mathrm{dom}(a)$ which satisfy 3.1(2(g)) for a final segment of γ_j's, or we will have an increasing sequence of such models. In the former case B must be a successor model and so closed under κ sequences with $a(B)$ closed under $< -\kappa_n$ sequences. Hence α will be in B, $b(\alpha) \in a(B)$ and we are done. In the later case the union of the sequence of models will be in $\mathrm{dom}(b)$ and it will be as desired.

Let us check 3.1(2(f)).

Suppose that $A, B \in \mathrm{dom}(b)$ and $B \in A$. We need to show that the walk from A to B is in $\mathrm{dom}(b)$. If A is new, then it is limit. So B will belong to a member of the sequence converging to A consisting of elements of $C^{\kappa^+}(A) \cap \mathrm{dom}(a)$. So we can assume without loss of generality that already $A \in \mathrm{dom}(a)$. Pick $B' \in \mathrm{dom}(a)$ to be the smallest model with $B \in B'$. Now, as in 3.3, Claim B3, we can use induction on wl - the walk length. Thus basically we need only to consider a situation when $B' \in C^{\kappa^+}(A)$. But then everything is trivial.

Turn now to 3.1(2(k)) Thus let $A, \alpha \in \mathrm{dom}(a)$ and $\sup A > \alpha$. If α is old then sup of one of the models converging to A will be above α. 3.1(2(k)) applies and we are done. Suppose so that α is new. If also A is new, then sup of one of the models converging to A will be above α. We can replace then A by one of such models. So without loss of generality, we can assume that $A \in \text{dom}(a)$. Let $\langle \alpha_j | j < i \rangle$ be a nondecreasing sequence from $\text{dom}(a)$ converging to α. By 3.1(2(k)), $\gamma_j = \min(A \backslash \alpha_j) \in \text{dom}(a)$. If $\langle \gamma_j | j < i \rangle$ is eventually constant, then the constant value will be as desired. Suppose otherwise. Then $\langle \gamma_j | j < i \rangle$ will be also a converging to α sequence. Apply 3.1(2(g)) to A and its members. It follow that $\alpha \in A$. So $\min(A \backslash \alpha) = \alpha \in \text{dom}(b)$ and we are done.

The condition 3.1(2(l)) is satisfied since all new models that were added are limit models. By 1.1(11(b)), such models are unique immediate predecessors of their immediate successor models. Hence even if some model in a_n got the immediate predecessor it must be the unique one.

The rest of the conditions hold trivially.
□ of the claim.

Now it remains only to add A_{k_i+1} as the top model to $\text{dom}(b)$ which can be done easily using 3.3.
□ of the lemma.

Lemma 3.6 $\langle \mathcal{P}, \leq^* \rangle$ *does not add new sequences of ordinals of the length* $< \kappa_0$.

Proof. Repeat the argument of 3.5 with $\mathcal{P}$ replacing Q_{n0}. Then use 2.10 of [1] to insure 3.2(4).   □

The argument of 3.5 can be used in a standard fashion to show the Prikry condition (i.e. the standard argument runs inside elementary submodel M with δ replaced by κ^+).

Lemma 3.7 $\langle \mathcal{P}, \leq^* \rangle$ *satisfies the Prikry condition.*

Finally we define $\rightarrow$ on $\mathcal{P}$ similar to those of [1] or [2].
Using 3.4, the arguments of [2, 3.19] can be used to derive the following.

Lemma 3.8 $\langle \mathcal{P}, \rightarrow \rangle$ *satisfies* κ^{++}-*c.c.*

Proof. Suppose otherwise. Work in V. Let $\langle \underset{\sim}{p}_\alpha \mid \alpha < \kappa^{++} \rangle$ be a name of an antichain of the length κ^{++}. Using 1.7 we find an increasing sequence $\langle \langle \langle A_\alpha^{0\kappa^+}, A_\alpha^{1\kappa^+}, C_\alpha^{\kappa^+} \rangle, A_\alpha^{1\kappa^{++}} \rangle \mid \alpha < \kappa^{++} \rangle$ of elements of $\mathcal{P}'$ and a sequence $\langle p_\alpha \mid \alpha < \kappa^{++} \rangle$ so that for every $\alpha < \kappa^{++}$ the following holds:

(a) $\langle\langle A_{\alpha+1}^{0\kappa^+}, A_{\alpha+1}^{1\kappa^+}, C_{\alpha+1}^{\kappa^+}\rangle, A_{\alpha+1}^{1\kappa^+}\rangle \Vdash \underset{\sim}{p}_\alpha = \check{p}_\alpha$

(b) $\bigcup_{\beta<\alpha} A_\beta^{0\kappa^+} = A_\alpha^{0\kappa^+}$

(c) $^\kappa A_{\alpha+1}^{0\kappa^+} \subseteq A_{\alpha+1}^{0\kappa^+}$

(d) $A_{\alpha+1}^{0\kappa^+}$ is a successor model

(e) $\langle \cup A_\beta^{1\kappa^+} \mid \beta < \alpha \rangle \in A_{\alpha+1}^{0\kappa^+}$

(f) for every $\alpha \le \beta < \kappa^{++}$ we have

$$A_\alpha^{0\kappa^+} \in C^\beta(A_\beta^{0\kappa^+})$$

(g) $A_{\alpha+2}^{0\kappa^+}$ is not an immediate successor model of $A_{\alpha+1}^{0\kappa^+}$, for every $\alpha < \kappa^{++}$

(h) $p_\alpha = \langle p_{\alpha n} \mid n < \omega \rangle$

(i) for every $n \ge \ell(p_\alpha)$ $A_{\alpha+1}^{0\kappa^+}$ is the maximal model of $\mathrm{dom}(a_{\alpha n})$ where $p_{\alpha n} = \langle a_{\alpha n}, A_{\alpha n}, f_{\alpha n} \rangle$

Let $p_{\alpha n} = \langle a_{\alpha n}, A_{\alpha n}, f_{\alpha n} \rangle$ for every $\alpha < \kappa^{++}$ and $n \ge \ell(p_\alpha)$. Extending by 3.3 if necessary, let us assume that $A_\alpha^{0\kappa^+} \in \mathrm{dom}(a_{\alpha n})$, for every $n \ge \ell(p_\alpha)$. Shrinking if necessary, we assume that for all $\alpha, \beta < \kappa^+$ the following holds:

(1) $\ell = \ell(p_\alpha) = \ell(p_\beta)$

(2) for every $n < \ell$ $p_{\alpha n}$ and $p_{\beta n}$ are compatible in Q_{n1} i.e. $p_{\alpha n} \cup p_{\beta n}$ is a function.

(3) for every $n, \ell \le n < \omega$ $\langle \mathrm{dom}(a_{\alpha n}), \mathrm{dom}(f_{\alpha n}) \mid \alpha < \kappa^{++} \rangle$ form a Δ-system with the kernel contained in $A_0^{0\kappa^+}$

(4) for every $n, \omega > n \ge \ell$ $\mathrm{rng}(a_{\alpha n}) = \mathrm{rng}(a_{\beta n})$.

Shrink now to the set S consisting of all the ordinals below κ^{++} of cofinality κ^+. Let α be in S. For each $n, \ell \le n < \omega$, there will be $\beta(\alpha) < \alpha$ such that

$$\mathrm{dom}(a_{\alpha n}) \cap A_\alpha^{0\kappa^+} \subseteq A_{\beta(\alpha,n)}^{0\kappa^+}.$$

Just recall that $|a_{\alpha n}| < \kappa_n$. Shrink S to a stationary subset S^* so that for some $\alpha^* < \min S^*$ of cofinality κ^+ we will have $\beta(\alpha, n) < \alpha^*$, whenever $\alpha \in S^*, \ell \le n < \omega$. Now, the cardinality of $A_{\alpha^*}^{0\kappa^+}$ is κ^+. Hence, shrinking S^* if necessary, we can assume that for each $\alpha, \beta \in S^*, \ell \le n < \omega$

$$\mathrm{dom}(a_{\alpha n}) \cap A_\alpha^{0\kappa^+} = \mathrm{dom}(a_{\beta n}) \cap A_\beta^{0\kappa^+}.$$

Let us add $A_{\alpha^*}^{0\kappa^+}$ to each $p_\alpha, \alpha \in S^*$. By 3.3, 3.4(1(a)) it is possible to do this without adding other additional models or ordinals except the images of $A_{\alpha^*}^{0\kappa^+}$ under isomorphisms. Denote the result for simplicity by p_α as well. Note that (again by 3.3, 3.4(1(a))) any $A_\gamma^{0\kappa^+}$ for $\gamma \in S^* \cap (\alpha^*, \alpha)$ or, actually any other successor or limit model $X \in C^{\kappa^+}(A_\alpha^{0\kappa})$ with $cof(otp_{\kappa^+}(X)) = \kappa^+$, which is between $A_{\alpha^*}^{0\kappa^+}$ and $A_\alpha^{0\kappa^+}$ can be added without adding other additional models or ordinals except the images of it under isomorphisms.

Let now $\beta < \alpha$ be ordinals in S^*. We claim that p_β and p_α are compatible in $\langle \mathcal{P}, \rightarrow \rangle$.

First extend p_α by adding $A_{\beta+2}^{0\kappa^+}$ As it was remarked above this will not add other additional models or ordinals except the images of $A_{\beta+2}^{0\kappa^+}$ under isomorphisms to p_α.

Let p be the resulting extension. Assume that $\ell(q) = \ell(p)$. Otherwise just extend q in an appropriate manner to achieve this. Let $n \geq \ell(p)$ and $p_n = \langle a_n, A_n, f_n \rangle$. Let $q_n = \langle b_n, B_n, g_n \rangle$. Without loss of generality we may assume that $a_n(A_{\beta+2}^{0\kappa^+})$ is an elementary submodel of $\mathfrak{A}_{n,k_n}$ with $k_n \geq 5$. Just increase n if necessary. Now, we can realize the $k_n - 1$-type of $rng(b_n)$ inside $a_n(A_{\beta+2}^{0\kappa^+})$ over the common parts $dom(b_n)$ and $dom(a_n)$. This will produce $q_n' = \langle b_n', B_n, g_n \rangle$ which is $k_n - 1$-equivalent to q_n and with $rng(b_n') \subseteq a_n(A_{\beta+2}^{0\kappa^+})$. Doing the above for all $n \geq \ell(p)$ we will obtain $q' = \langle q_n' \mid n < \omega \rangle$ equivalent to q (i.e. $q' \longleftrightarrow q$).

Extend q' to q'' by adding to it $\langle A_{\beta+2}^{0\kappa^+}, a_n(A_{\beta+2}^{0\kappa^+}) \rangle$ as the maximal set for every $n \geq \ell(p)$. Recall that $A_{\beta+1}^{0\kappa^+}$ was its maximal model. So we are adding a top model, also, by the condition (g) above $A_{\beta+2}^{0\kappa^+}$ is not an immediate successor of $A_{\beta+1}^{0\kappa^+}$. Hence no additional models or ordinals are added at all. Let $q_n'' = \langle b_n'', B_n, g_n \rangle$, for every $n \geq \ell(p)$.

Combine now p and q'' together. Thus for each $n \geq \ell(p)$ we add b_n'' to a_n as well as all of its isomorphic images by $\pi_{A_{\beta+2}^{0\kappa^+} X}$, for every X in $dom(a_n)$ which is isomorphic to $A_{\beta+2}^{0\kappa^+}$. The rest of the parts are combined in the obvious fashion (we put together the functions and intersect sets of measure one moving first to the same measure). Add if necessary a new top model to insure 3.1(2(d)). Let $r = \langle r_n \mid n < \omega \rangle$ be the result, where $r_n = \langle c_n, C_n, h_n \rangle$, for $n \geq \ell(p)$.

Claim 3.8.1 $r \in \mathcal{P}$ and $r \geq p$.

Proof. Fix $n \geq \ell(p)$. The main points here are that b_n'' and a_n agree on the common part and adding of b_n'' to a_n does not require other additions of models or of ordinals except the images of b_n'' under isomorphisms.

Thus let $A \in \operatorname{dom}(b_n'') \backslash \operatorname{dom}(a_n)$ be a model. Let $B \in \operatorname{dom}(a_n) \backslash \operatorname{dom}(b_n'')$. Note that it is the main possibility. Once we know how to handle it, dealing with isomorphic images can be reduced to the present case as it was done in 3.3. Suppose first that $B \not\supseteq A$. This implies that $B \not\supseteq A_\alpha^{0\kappa^+}$ and $B \not\subseteq A_\alpha^{0\kappa^+}$ (just if $B \subset A_\alpha^{0\kappa^+}$, then $B \in A_{\alpha^*}^{0\kappa^+}$ and so it is in $\operatorname{dom}(b_n)$). Then

$$A \cap B = A \cap A_{\beta+2}^{0\kappa^+} \cap B = A \cap A_\alpha^{0\kappa^+} \cap B = A \cap \rho,$$

for some $\rho \in A_\alpha^{0\kappa^+} \cap A_\alpha^{1\kappa^{++}} \cap \operatorname{dom}(a_n)$, since both $B, A_\alpha^{0\kappa^+} \in \operatorname{dom}(a_n)$ and 3.1(2(f)) holds. But now we must have this ρ in $A_{\alpha^*}^{0\kappa^+}$ and then in $\operatorname{dom}(b_n)$. So, $a_n(\rho) = b_n(\rho) = b_n''(\rho)$. Hence,

$$b_n''(A) \cap a_n(B) = b_n''(A) \cap a_n(A_\alpha^{0\kappa^+}) \cap a_n(B) = b_n''(A) \cap a_n(\rho) = b_n''(A) \cap b_n''(\rho),$$

due to the choice of the type of $\operatorname{rng}(b_n'')$.

Consider now the side of B of the intersection. So now B is a model. Compare it with $A_{\beta+2}^{0\kappa^+}$. If $otp(B) < otp(A_{\beta+2}^{0\kappa^+})$ then there is $D \in \operatorname{dom}(a_n) \cap A_{\beta+2}^{0\kappa^+}$ of the order type of B and such that

$$B \cap A_{\beta+2}^{0\kappa^+} = B \cap D.$$

But then, again $D \in A_{\alpha^*}^{0\kappa^+}$ and so in $\operatorname{dom}(b_n)$. Hence $D \cap A$ can be handled on the side of D, i.e. $D \cap A = E \cap \xi$, for some $E \in D \cup \{D\} \cap \operatorname{dom}(b_n)$ and an ordinal $\xi \in D \cap \operatorname{dom}(b_n)$. But $D \in A_{\alpha^*}^{0\kappa^+}$, hence $E, \xi \in A_{\alpha^*}^{0\kappa^+}$. This implies that $E, \xi \in \operatorname{dom}(a_n)$. So $a_n(E) = b_n(E), a_n(\xi) = b_n(\xi)$.

If $otp(B) \geq otp(A_{\beta+2}^{0\kappa^+})$ then there is $B' \in (B \cup \{B\}) \cap \operatorname{dom}(a_n)$ such that $otp(B') = otp(A_{\beta+2}^{0\kappa^+})$ and

$$B \cap A_{\beta+2}^{0\kappa^+} = B' \cap A_{\beta+2}^{0\kappa^+}.$$

Now we move to the B- side using $\pi_{A_{\beta+2}^{0\kappa^+} B'}[A]$.

The same argument works once $B \supseteq A$, since then necessarily, $otp(B) \geq otp(A_{\beta+2}^{0\kappa^+})$.

The above shows 3.1(2(f)).

Let us check 3.1(2(h)). Suppose that $A \in \operatorname{dom}(b_n''), B \in \operatorname{dom}(a_n)$. Again, it is the main possibility. Once we know how to handle it, dealing with isomorphic images can be reduced to the present case as it was done in 3.3.

Case 3.8.1.1 $A \supset B$.

Then $B \in A_{\beta+2}^{0\kappa^+} \subseteq A_{\alpha}^{0\kappa^+}$ and hence $B \in A_{\alpha*}^{0\kappa^+} \cap \mathrm{dom}(a_n)$. Which implies that $B \in \mathrm{dom}(b_n)$. But then the walk from A to B in $\mathrm{dom}(b_n'')$ and we are done.

Case 3.8.1.2 $A \subset B$.

If $otp(B) < otp(A_{\beta+2}^{0\kappa^+})$ then there is $D \in \mathrm{dom}(a_n) \cap A_{\beta+2}^{0\kappa^+}$ of the order type of B and such that

$$B \cap A_{\beta+2}^{0\kappa^+} = B \cap D.$$

So $D \supset A$ and hence $A \in D$. But D must be in $A_{\alpha*}^{0\kappa^+} \cap \mathrm{dom}(a_n)$. Then A as an element of D must be in this intersection as well. So, both A, B are in $\mathrm{dom}(a_n)$. Hence the walk from B to A is in $\mathrm{dom}(a_n)$ and we are done.

If $otp(B) \geq otp(A_{\beta+2}^{0\kappa^+})$ then there is $B' \in (B \cup \{B\}) \cap \mathrm{dom}(a_n)$ such that $otp(B') = otp(A_{\beta+2}^{0\kappa^+})$ and

$$B \cap A_{\beta+2}^{0\kappa^+} = B' \cap A_{\beta+2}^{0\kappa^+}.$$

Now we can use $\pi_{A_{\beta+2}^{0\kappa^+} B'}$. It is identity on the common part of $A_{\beta+2}^{0\kappa^+}, B'$ and so does not move A. The walk from $A_{\beta+2}^{0\kappa^+}$ to A will be copied to those from B' to A. Once on the B-side we can run induction on wl as it was done in 3.3.

Now let us turn to 3.1(2(k)). Suppose that $A, \xi \in \mathrm{dom}(a_n) \cup \mathrm{dom}(b_n'')$ and $\sup A > \xi$.

Case 3.8.1.3 $A \in \mathrm{dom}(b_n'')$.

If $\xi \in \mathrm{dom}(b_n'')$ then we are done. Suppose otherwise. Then, $\xi \notin A_{\beta+2}^{0\kappa^+}$. Consider $\rho = \min(A_{\beta+2}^{0\kappa^+} \setminus \xi)$. By 3.1(2(k)), $\rho \in \mathrm{dom}(a_n)$. But then $\rho \in A_{\alpha*}^{0\kappa^+}$ and, so it is in $\mathrm{dom}(b_n)$. Now, the least $\mu \in A \setminus \rho$ will be in $\mathrm{dom}(b_n'')$ and will be as desired.

Case 3.8.1.4 $A \in \mathrm{dom}(a_n) \setminus \mathrm{dom}(b_n'')$.

Assume that $\xi \in \mathrm{dom}(b_n'') \setminus \mathrm{dom}(a_n)$. Compare A with $A_{\beta+2}^{0\kappa^+}$.

Subcase 3.8.1.4.1 $otp(A) < otp(A_{\beta+2}^{0\kappa^+})$.

There is $C \in \mathrm{dom}(a_n)$ such that $A \subseteq C$ and $otp(C) = otp(A_{\beta+2}^{0\kappa^+})$. Note that $A_{\beta+2}^{0\kappa^+} \in C^{\kappa^+}(A_{\alpha+1}^{0\kappa^+})$ (by (f) above) and $A_{\beta+2}^{0\kappa^+} \not\supseteq A$, so the walk from $A_{\alpha+1}^{0\kappa^+}$ to A splits from $C^{\kappa^+}(A_{\alpha+1}^{0\kappa^+})$ above $A_{\beta+2}^{0\kappa^+}$. Now it is not hard to find such C. Move now to $A_{\beta+2}^{0\kappa^+}$. Set $A' = \pi_{CA_{\beta+2}^{0\kappa^+}}[A]$. Then A' is in

$\text{dom}(a_n) \cap A_\alpha^{0\kappa^+}$. Hence it is in $\text{dom}(b_n)$. So $\rho = \min(A' \backslash \xi) \in \text{dom}(b_n'')$. If $\xi \in C$, then $\pi_{A_{\beta+2}^{0\kappa^+} C}(\rho)$ will be as desired.

Suppose that $\xi \notin C$. Compare C with $A_{\beta+2}^{0\kappa^+}$. There will be $\mu \in C \cap \text{dom}(a_n)$ such that

$$C \cap A_{\beta+2}^{0\kappa^+} = C \cap \mu.$$

By 1.1(8), $C, A_{\beta+2}^{0\kappa^+}$ are isomorphic over $C \cap A_{\beta+2}^{0\kappa^+}$. So

$$C \cap \mu = C \cap \sup(C \cap A_{\beta+2}^{0\kappa^+}) = A_{\beta+2}^{0\kappa^+} \cap \sup(C \cap A_{\beta+2}^{0\kappa^+}).$$

Then ξ must be above $\sup(C \cap A_{\beta+2}^{0\kappa^+})$ since $\xi \notin C$. But $\xi < \sup(A)$, so by 1.1(10), $\sup(A_{\beta+2}^{0\kappa^+}) < \mu$. Hence $\mu = \min(C \backslash \xi)$. Now, clearly, $\min(A \backslash \xi) = \min(A \backslash \mu)$. But $A, \mu \in \text{dom}(a_n)$, so we are done.

Subcase 3.8.1.4.2 $otp(A) \geq otp(A_{\beta+2}^{0\kappa^+})$.

Pick then $C \in C^{\kappa^+}(A_{\alpha+1}^{0\kappa^+})$ of the otp equal to $otp(A)$. As in 3.3 (induction on wl), $C \in \text{dom}(a_n)$. By (f) above and the coherence of C^{κ^+}, we have $A_{\beta+2}^{0\kappa^+} \in C^{\kappa^+}(C)$. So, $\xi \in C$. There is $\delta \in A \cap \text{dom}(a_n)$ such that

$$A \cap C = A \cap \delta.$$

But $\xi \in C \backslash A$ and $\xi < \sup A$, so by 1.1(10), we must have $\sup C < \delta$. Hence $\delta = \min(A \backslash \xi)$ and we are done.

The rest of the conditions hold trivially.

$\square$ of the claim.

Now we have $r \geq p, q''$. Hence, $p \to r$ and $q \to r$. Contradiction. $\square$

Acknowledgment

The results of this paper were presented at the conference Computational Prospects of Infinity, at the Institute of Mathematical Sciences, Singapore. We would like thank the organizers and specially to Qi Feng for their hospitality.

We are grateful to Assaf Sharon for encouraging us to write this paper, for discussions on the subject, for his corrections and eagerness to have all the details. We would like to thank to Assaf Rinot for the help with T_EX and to the referee of the paper for his helpful remarks.

References

[1] M. Gitik, Blowing up power of a singular cardinal, Annals of Pure and Applied Logic 80 (1996) 349-369.

[2] M. Gitik, Blowing up power of a singular cardinal-wider gaps, Annals of Pure and Applied Logic 116 (2002) 1-38.

[3] D. Velleman, Simplified morasses with linear limits, J. of Symbolic Logic, 49 (1984), 1001-1021.

LIMIT COMPUTABILITY AND
CONSTRUCTIVE MEASURE

Denis R. Hirschfeldt

Department of Mathematics
University of Chicago
Chicago, IL 60637, USA
E-mail: drh@math.uchicago.edu

Sebastiaan A. Terwijn

Institute of Discrete Mathematics and Geometry
Technical University of Vienna
Wiedner Hauptstrasse 8-10/E104, 1040 Vienna, Austria
E-mail: terwijn@logic.at

In this paper we study constructive measure and dimension in the class Δ_2^0 of limit computable sets. We prove that the lower cone of any Turing-incomplete set in Δ_2^0 has Δ_2^0-dimension 0, and in contrast, that although the upper cone of a noncomputable set in Δ_2^0 always has Δ_2^0-*measure* 0, upper cones in Δ_2^0 have nonzero Δ_2^0-*dimension*. In particular the Δ_2^0-dimension of the Turing degree of $\emptyset'$ (the Halting Problem) is 1. Finally, it is proven that the low sets do not have Δ_2^0-measure 0, which means that they do not form a small subset of Δ_2^0. This result has consequences for the existence of bi-immune sets.

1. Introduction

In his study of randomness [27], Schnorr introduced the notion of a Schnorr null set as a more constructive version of Martin-Löf's [21] notion of null set. We briefly review the relevant definitions. For motivation and discussion of these notions we refer the reader to Schnorr's book [27], the monograph by Li and Vitányi [17], and the recent surveys [5, 8, 31].

For $\sigma \in 2^{<\omega}$ and $X \in 2^\omega$, we write $\sigma \sqsubset X$ to mean that σ is an initial segment of X. A class $\mathcal{A} \subseteq 2^\omega$ is a Σ_1^0-*class* if there is a c.e. set $A \subseteq 2^{<\omega}$ such that $\mathcal{A} = \bigcup_{\sigma \in A}[\sigma]$, where $[\sigma] = \{X \in 2^\omega : \sigma \sqsubset X\}$. Whenever we mention a Σ_1^0-class $\mathcal{A}$, we assume we have fixed such a set of generators A, and identify $\mathcal{A}$ with A. Note that we can assume that A is *prefix-free*, that is, if $\sigma \in A$ and $\sigma \prec \tau$ then $\tau \notin A$.

Let μ be the usual Lebesgue measure on 2^ω. A set $\mathcal{A} \subseteq 2^\omega$ is called *Martin-Löf null* (or Σ_1^0-*null*) if there is a uniformly c.e. sequence $\{\mathcal{U}_i\}_{i \in \omega}$ of Σ_1^0-classes (called a *test*) such that $\mu(\mathcal{U}_i) \leqslant 2^{-i}$ and $\mathcal{A} \subseteq \bigcap_i \mathcal{U}_i$. The set $\mathcal{A}$ is *Schnorr null* if in addition the measures $\mu(\mathcal{U}_i)$ are uniformly computable reals. A test with this extra property is called a *total test* or a *Schnorr test*. Equivalently, $\mathcal{A}$ is Schnorr null if there is a test $\{\mathcal{U}_i\}_{i \in \omega}$ such that $\mu(\mathcal{U}_i) = 2^{-i}$ and $\mathcal{A} \subseteq \bigcap_i \mathcal{U}_i$.

The corresponding randomness notions are defined by saying that $A \in 2^\omega$ is Σ_1^0-*random* (or 1-*random* or *Martin-Löf random*) if $\{A\}$ is not Σ_1^0-null, and A is *Schnorr random* if $\{A\}$ is not Schnorr null.

A different treatment of measure is the one of Ville [32] using martingales. A *martingale* is a function $d : 2^{<\omega} \to \mathbb{Q}^+$ that satisfies for every $\sigma \in 2^{<\omega}$ the averaging condition $2d(\sigma) = d(\sigma 0) + d(\sigma 1)$, and d is a *supermartingale* if merely $2d(\sigma) \geqslant d(\sigma 0) + d(\sigma 1)$. A (super)martingale d *succeeds on* a set A if $\limsup_{n \to \infty} d(A \upharpoonright n) = \infty$. We say that d succeeds on, or *covers*, a class $\mathcal{A} \subseteq 2^\omega$ if d succeeds on every $A \in \mathcal{A}$. The *success set* $S[d]$ of d is the class of all sets on which d succeeds. Ville proved that the class of null sets of the form $S[d]$, with d of arbitrary complexity, coincides with the class of classical (Lebesgue) null sets.

Schnorr gave characterizations of the above notions of effectively null set in terms of martingales. In particular, he introduced null sets of the form

$$S_h[d] = \Big\{ X : \limsup_{n \to \infty} \frac{d(X \upharpoonright n)}{h(n)} = \infty \Big\},$$

where d is a martingale and h is a nondecreasing unbounded function (called an *order*), and proved the following theorem.

Theorem 1.1. (Schnorr [27], Sätze 9.4, 9.5) *A set $\mathcal{A} \subseteq 2^\omega$ is Schnorr null if and only if there are a computable martingale d and a computable order h such that $\mathcal{A} \subseteq S_h[d]$.*

Schnorr also addressed null sets of *exponential* order, that is, of the form $S_h[d]$ with $h(n) = 2^{\varepsilon n}$ and $\varepsilon \in (0, 1]$. Although he did not make an explicit reference to Hausdorff dimension, it turns out that the theory of Hausdorff dimension can be cast precisely in terms of such null sets of exponential order, so that Schnorr's notion of effective measure in a natural way leads us into the theory of dimension.

Lutz [18] used effective martingales to develop his theory of resource bounded measure. He defined $A \in 2^\omega$ to be *computably random* if there is no computable martingale d such that $A \in S[d]$.[a] This framework for studying measure and randomness at the level of complexity classes can be used to constructivize Hausdorff dimension along the same lines:

Definition 1.2. For a complexity class $\mathcal{C}$, a set $\mathcal{A} \subseteq 2^\omega$ has $\mathcal{C}$-*dimension* α if

$$\alpha = \inf \left\{ s \in \mathbb{Q} : \exists d \in \mathcal{C} \, (\, d \text{ is a supermartingale and } \mathcal{A} \subseteq S_{2^{(1-s)n}}[d] \,) \right\}.$$

Lutz [19, 20] used a variant of martingales called *gales* in his presentation. In this paper we stick to martingales and the null sets of the form $S_h[d]$ used by Schnorr in our treatment of Hausdorff dimension. That this makes no difference was pointed out by several authors, including those of [2, 3, 31]. If $\mathcal{C}$ consists of all functions, then the notion of $\mathcal{C}$-dimension is equivalent to classical Hausdorff dimension. We say that a function $2^{<\omega} \to \mathbb{Q}^+$ is in Σ_1^0 if it is approximable from below by a nondecreasing computable function. If $\mathcal{C} = \Sigma_1^0$ then the notion of $\mathcal{C}$-dimension is equivalent to Lutz's definition [20] of constructive Σ_1^0-dimension.

In this paper we are interested in the quantitative structure of Δ_2^0. The appropriate measures to use in this context are those for which Δ_2^0 itself does not have measure 0, but for which every element of Δ_2^0 does have measure 0. Since there are Σ_1^0-random sets in Δ_2^0, Martin-Löf's Σ_1^0-measure is too weak for our purposes. For Σ_2^0-measure, obtained by relativizing Σ_1^0-measure to the halting set $\emptyset'$, the class Δ_2^0 has measure 0, so this measure is too strong. However, relativizing the notions of Schnorr null and computably null to $\emptyset'$ gives measures that meet our requirements:

Definition 1.3. A set $\mathcal{A} \subseteq 2^\omega$ has Δ_2^0-*measure* 0 (or is Δ_2^0-*null*) if there is a $\emptyset'$-computable martingale that succeeds on $\mathcal{A}$.

A set $\mathcal{A} \subseteq 2^\omega$ has *Schnorr* Δ_2^0-*measure* 0 (or is *Schnorr* Δ_2^0-*null*) if there is a $\emptyset'$-computable Schnorr test that covers $\mathcal{A}$.

A first study of the quantitative structure of Δ_2^0 using these measures was made in Terwijn [29, 30].

Relativizing computable randomness yields Δ_2^0-*randomness*, and relativizing Schnorr randomness yields *Schnorr* Δ_2^0-*randomness*. The relations

---

[a]Schnorr also considered this definition in relativized form [27], p. 55.

between the various notions are as follows:

$$\Delta_2^0\text{-random}$$
$$\Downarrow$$
Schnorr Δ_2^0-random $\implies$ Δ_2^0-dimension 1
$$\Downarrow \qquad\qquad\qquad\qquad \Downarrow$$
Σ_1^0-random $\implies$ Σ_1^0-dimension 1
$$\Downarrow$$
computably random $\qquad\qquad\qquad \Downarrow$
$$\Downarrow$$
Schnorr random $\implies$ computable dimension 1

No other implications hold than the ones indicated. That there are Schnorr random sets that are not computably random was proved by Wang [33]. (See Nies, Stephan, and Terwijn [22] for more information on the separation between the various randomness notions.) The strictness of the other implications in the first column follows from elementary observations and results in Schnorr [27], and is discussed in [8, 30]. That there are no more implications between the first and the second column follows from the next proposition. The strictness of the two implications in the second column follows by similar means.[b]

Proposition 1.4. *There are sets A such that A is not Schnorr random and A has Δ_2^0-dimension 1.*

Proof. Let R be Δ_2^0-random, and let $D = \{2^x : x \in \omega\}$ be an exponentially sparse computable domain. Then $A = R \cup D$ is not Schnorr random, since no Schnorr random set contains an infinite computable subset, but no Δ_2^0-martingale can succeed on A exponentially fast. □

Clearly, the "Δ_2^0-dimension 1" in Proposition 1.4 can be improved to "Δ_n^0-dimension 1" by the same proof, if one is considering higher orders of randomness.

The rest of this paper is organized as follows. Ambos-Spies, Merkle, Reimann, and Stephan [2] investigated resource bounded dimension in the exponential time class E. Among other things, they proved that under polynomial time many-one reducibility the complete degree in E has dimension

---

[b]It is easy to see (cf. [20]) that the class of computable sets has Σ_1^0-dimension 0, but is not computably null, so in particular this class has computable dimension 1. Also, Lutz [20] has shown that there are sets in Δ_2^0 of any given rational Σ_1^0-dimension, but it is obvious that every set in Δ_2^0 has Δ_2^0-dimension 0.

1, and that the set of possible dimensions of p-m-degrees in E is dense in $[0, 1]$. In Section 2 we show that under Turing reducibility in Δ_2^0 the complete degree has Δ_2^0-dimension 1, and all other degrees have Δ_2^0-dimension 0. In Section 3 we present a proof that the low sets do not have Δ_2^0-measure 0 by showing that for every $\emptyset'$-computable martingale there is a low set that is not covered by it. This means that the low sets do not form a small subset of Δ_2^0.

Our notation generally follows Odifreddi [23, 24] and Soare [28]. We write $^{\leqslant_T}A$ for the lower cone $\{B : B \leqslant_T A\}$ and $A^{\leqslant_T}$ for the upper cone $\{B : A \leqslant_T B\}$.

2. Δ_2^0-dimension

The next theorem is a strengthening of Theorem 5.5 in [30], which states that the lower cone of every $A <_T \emptyset'$ has Schnorr Δ_2^0-measure 0. We will make use of the following definition and lemma.

Definition 2.1. For functions f and g and rational $q \in (0, 1]$, we say that f is *q-dominated* by g if

$$\liminf_{n \to \infty} \frac{|\{i \leqslant n : g(i) \geqslant f(i)\}|}{n} \geqslant q. \tag{2.1}$$

Lemma 2.2. *For every $q \in (0, 1]$ there is a function $f \leqslant_T \emptyset'$ such that f is not q-dominated by any function $g <_T \emptyset'$.*

Proof. Let $h \leqslant_T \emptyset'$ be a function not dominated by any function $g <_T \emptyset'$, for example, $h(x) = \mu s(\emptyset'_s \upharpoonright x = \emptyset' \upharpoonright x)$ (the smallest s such that all the $y \in \emptyset'$ smaller than x are enumerated into $\emptyset'$ within s steps). Without loss of generality, $q = \frac{1}{c}$ for some $c \in \omega$. Define $f(x) = h(\lfloor \log_c x \rfloor)$ (where $\log_c$ is the logarithm with base c). If g satisfies (2.1) then for almost every k there is a natural number $x \in [c^k, c^{k+1})$ such that $g(x) \geqslant f(x)$. But then the function $\widehat{g}$ defined by $\widehat{g}(k) = \max\{g(x) : x \in [c^k, c^{k+1})\}$ dominates h, a contradiction. $\qquad\square$

Theorem 2.3. *Let $A \in \Delta_2^0$ be any Turing-incomplete set. Then the Δ_2^0-dimension of the lower cone $^{\leqslant_T}A$ is 0.*

Proof. Let $q \in (0, 1]$ be rational and suppose that $A <_T \emptyset'$. We define uniformly in $\emptyset'$ for every $e \in \omega$ a martingale d_e such that

$$R_e : \qquad \Phi_e^A \text{ total and } \{0, 1\}\text{-valued} \implies \Phi_e^A \in S_{2^{(1-q)n}}[d_e].$$

By the usual sum trick this suffices to prove the theorem: The sum $d(\sigma) = \sum_{e \in \omega} 2^{-e} d_e(\sigma)$ is again a $\emptyset'$-computable martingale, and if $X \in S_{2^{(1-q)n}}[d_e]$ then for all $q' > q$ we have $X \in S_{2^{(1-q')n}}[d]$, which shows that $\{B : B \leqslant_T A\}$ has Δ_2^0-dimension $\leqslant q$. Since $q > 0$ was arbitrary the theorem follows.

By Lemma 2.2, let $f \leqslant_T \emptyset'$ be a function that is not q-dominated by any function $g <_T \emptyset'$.

We now define d_e in stages s. At stage s we define d_e on all strings $\sigma \in 2^{<\omega}$ of length s. The value $d_e(\sigma)$ will depend only on $|\sigma|$. (Ambos-Spies, Mayordomo, Wang, and Zheng [1] called martingales with this property 'oblivious'.)

Stage $s = 0$. Define $d_e(\lambda) = 1$, where λ is the empty string.

Stage $s + 1$. Given $d_e(\sigma)$ with $|\sigma| = s$, use the oracle $\emptyset'$ to search for a string $\tau \sqsubset A$ with $|\tau| \leqslant f(s)$ such that $\Phi_{e,|\tau|}^{\tau}(s) \downarrow$. If such τ does not exist, or if $\Phi_{e,|\tau|}^{\tau}(s) \downarrow \notin \{0,1\}$, do not make a bet; that is, let $d_e(\sigma i) = d_e(\sigma)$ for $i \in \{0,1\}$. If τ exists and $\Phi_{e,|\tau|}^{\tau}(s) \downarrow = i \in \{0,1\}$, define $d_e(\sigma i) = 2d_e(\sigma)$; that is, bet all our capital on $\Phi_e^A(|\sigma|) = i$. This concludes the definition of d_e.

It is clear that d_e is defined on all strings for every e, uniformly in $\emptyset'$. We check that R_e is satisfied. Suppose that Φ_e^A is total and computes a set. Then the function

$$g_e(n) = \mu t \Big((\exists \tau \sqsubset A) \big[|\tau| = t \wedge \Phi_{e,t}^{\tau}(n) \downarrow \big] \Big)$$

is A-computable. By the choice of f, there are infinitely many N such that for more than $(1-q)N$ many $n < N$ we have $f(n) \geqslant g_e(n)$. For these n, in the definition of d_e the string τ is found and a bet is placed successfully. (Note that we never make a wrong bet.) Hence $\Phi_e^A \in S_{2^{(1-q)n}}[d_e]$.   □

Theorem 2.4. *For every noncomputable $A \in \Delta_2^0$, the upper cone $A^{\leqslant_T}$ has Schnorr Δ_2^0-measure 0.*

Proof. This theorem is an effectivization of the well-known result of de Leeuw, Moore, Shannon, and Shapiro [16] and Sacks [26] that the upper cone of a noncomputable set has Lebesgue measure 0.[c] Lutz and Terwijn [29] showed that there is a $\emptyset'$-computable martingale that succeeds on $A^{\leqslant_T}$ when $A \in \Delta_2^0$ is noncomputable. We give here a direct proof using total

---

[c]For another approach to effectivizing this result, see Hirschfeldt, Nies, and Stephan [10].

$\emptyset'$-computable tests, which gives the stronger result of the theorem. Fix a noncomputable $A \in \Delta_2^0$, and define for every i and n the open sets

$$\mathcal{E}_{i,n} = \{B : A \upharpoonright n = \Phi_i^B \upharpoonright n\}.$$

For every i we have $\{B : A = \Phi_i^B\} = \bigcap_n \mathcal{E}_{i,n}$, so $\mu(\bigcap_n \mathcal{E}_{i,n}) = 0$. Furthermore, the $\mathcal{E}_{i,n}$ are uniformly $\emptyset'$-computable because A is $\emptyset'$-computable, and the $\mu(\mathcal{E}_{i,n})$ are uniformly $\emptyset'$-computable reals. So if we let $f(k)$ be the least n such that $\mu(\mathcal{E}_{i,n}) \leqslant 2^{-k}$ and define $\mathcal{F}_{i,k} = \mathcal{E}_{i,f(n)}$, then $\mathcal{F}_{i,0}, \mathcal{F}_{i,1}, \ldots$ is a total $\emptyset'$-computable test, and we still have $\{B : A = \Phi_i^B\} = \bigcap_k \mathcal{F}_{i,k}$.

Because the tests $\bigcap_k \mathcal{F}_{i,k}$ are $\emptyset'$-uniform in i, it follows from an easily proved effective union lemma that $\bigcup_i \bigcap_k \mathcal{F}_{i,k} = A^{\leqslant T}$ is also of Schnorr measure 0 relative to $\emptyset'$. $\qquad\square$

The next theorem shows that for every $A \in \Delta_2^0$ the Δ_2^0-dimension of the upper cone of A is maximal. In particular, although the Schnorr Δ_2^0-measure of $\deg_T(\emptyset')$ is 0, there is no Δ_2^0-martingale that succeeds on this Turing degree exponentially fast.

Theorem 2.5. *The Δ_2^0-dimension of $\deg_T(\emptyset')$ is* 1.

Proof.[d] Given a martingale $d \in \Delta_2^0$ and a rational $\varepsilon > 0$, we build a $B \equiv_T \emptyset'$ such that $B \notin S_{2^{\varepsilon n}}[d]$. The idea is simple: We code $\emptyset'$ on an exponentially sparse computable domain D, and define B by $\emptyset'$-effectively diagonalizing against d outside D and taking the coded version of $\emptyset'$ on D. Since D is exponentially sparse, d cannot succeed fast on B, and we have $B \leqslant_T \emptyset' \leqslant_T B \oplus D \leqslant_T B$ since D is computable. Note that this idea works for every computably sparse domain D, so that in fact $\deg_T(\emptyset')$ is not included in any null set of the form $S_h[d]$ for a Δ_2^0-martingale d and a computable order h. Theorem 2.4 shows that the same is not true for all $\emptyset'$-computable orders h. For the theorem as stated it suffices to take $D = \{2^m - 1 : m \in \omega\}$ and define

$$B(x) = \begin{cases} \emptyset'(n) & \text{if } x = 2^m - 1 \\ 0 & \text{if } x \notin D \text{ and } d((B \upharpoonright x)0) < d((B \upharpoonright x)1) \\ 1 & \text{otherwise.} \end{cases}$$

---

[d]This proof is a few years old. Coding techniques similar to the one used in it have meanwhile been used in the context of Hausdorff dimension independently by several authors, cf. e.g. Reimann [25].

Then $d(B \restriction n - 1) \leqslant d(B \restriction n)$ except possibly when $n = 2^m - 1$ for some n, so

$$\limsup_{n \to \infty} \frac{d(B \restriction n)}{2^{\varepsilon n}} \cdot \leqslant \limsup_{m \to \infty} \frac{d(B \restriction 2^m)}{2^{\varepsilon 2^m}} \leqslant \limsup_{m \to \infty} \frac{2^m}{2^{\varepsilon 2^m}} < 1,$$

and hence $B \notin S_{2^{\varepsilon n}}[d]$. □

It follows from Theorems 2.3 and 2.5 that the only possibilities for the Δ_2^0-dimension of a Turing degree are 0 or 1:

Corollary 2.6. *For $A \in \Delta_2^0$, the Δ_2^0-dimension of $\deg_T(A)$ is 1 if A is Turing complete, and 0 otherwise.*

3. The measure of the low sets

It is known that the class of sets that are bounded by a 1-generic set has Σ_1^0-measure 0 (by effectivizing Theorem 4.2 in Kurtz [14], cf. [30], or by Demuth and Kučera [4]). In particular the subclass of the low sets consisting of the Δ_2^0 1-generic sets has Σ_1^0-measure 0. In this section we prove that the low sets do not form a small subset of Δ_2^0, that is, that they do not have Δ_2^0-measure 0. It is easily verified that the computable sets have Δ_2^0-measure 0, and that most sets in Δ_2^0 are bi-immune for the computable sets.[e] Although the low sets do not have Δ_2^0-measure 0, Downey, Hirschfeldt, Lempp, and Solomon [6] were able to construct a Δ_2^0 set A that is bi-immune for the low sets (i.e. there is no infinite low subset of either A or its complement). The set they constructed in fact truth-table reduces to $\emptyset'$. It is not difficult to see that the sets that tt-reduce to $\emptyset'$ have Schnorr Δ_2^0-measure 0, i.e. there is a total $\emptyset'$-computable test covering them. So in this sense the set constructed in [6] does not exhibit the typical behavior of a Δ_2^0-set. Theorem 3.1 shows that indeed it is not the case that almost every set in Δ_2^0 is bi-immune for the low sets.

Theorem 3.1. *The low sets do not have Δ_2^0-measure 0.*

Proof. Let M be a universal Σ_1^0-martingale and let N be an arbitrary Δ_2^0-martingale. We will exhibit a low set B on which N does not succeed.

---

[e]Indeed, this is even true for Σ_1^0-measure: The class of sets that are not bi-immune for the computable sets has Σ_1^0-measure 0.

Define a new martingale L by

$$L(\emptyset) = \tfrac{1}{2}\big(M(\emptyset) + N(\emptyset)\big),$$

$$L(\sigma) = \tfrac{1}{2}\big(M(\sigma) + N(\sigma(0)\sigma(2)\ldots\sigma(2n))\big) \text{ if } |\sigma| = 2n+1 \text{ or } 2n+2.$$

So L is essentially a sum of the behaviour of M and the behaviour of N restricted to the even bits. We leave it to the reader to check that L is indeed a martingale.

Now let $A \in \Delta^0_2$ be such that L does not succeed on A. (Such an A exists because Δ^0_2 does not have Δ^0_2-measure 0.) Then M is bounded on A, and hence A is Σ^0_1-random. Also, N is bounded on the set B defined by $B(n) = A(2n)$ for every n. We claim that B is low, being half of a Σ^0_1-random set below $\emptyset'$. Thus we have exhibited a low set on which the arbitrary Δ^0_2-martingale N does not succeed.

· To prove the claim that B is low, suppose that C is the odd part of A, i.e. the unique set with $A = B \oplus C$. Since A is Σ^0_1-random, by a result of van Lambalgen [15] the set C is Σ^0_1-random relative to B. Nies and Stephan (see Theorem 3.4 in [7]) showed that this implies that B is low. For completeness, we include a proof of this fact.

Let $\mathrm{use}(\Phi^B_e(e))$ be the partial function that for every e measures the number of computation steps of $\Phi^B_e(e)$, if this is defined. Since C is a Δ^0_2 set, it has a computable approximation C_s such that $\lim_s C_s(n) = C(n)$ for every n. Let the convergence modulus of this approximation be the function $m(n) = \mu t(\forall s \geqslant t)[C_s(n) = C_t(n)]$. Now if there were infinitely many e such that $\mathrm{use}(\Phi^B_e(e)) \geqslant m(e)$ then B could compute infinitely many points of C, contradicting the fact that C is Σ^0_1-random relative to B. Hence, for almost all e, whenever $\Phi^B_e(e)$ is defined we have that $\mathrm{use}(\Phi^B_e(e)) < m(e)$. Since $m \leqslant_T \emptyset'$ we obtain that $B' \leqslant_T \emptyset'$. $\square$

Corollary 3.2. *The Δ^0_2-Hausdorff dimension of the low sets is* 1.

Acknowledgments

The first author was supported by National Science Foundation Grants DMS-02-00465 and DMS-05-00590. The second author was supported by the Austrian Research Fund FWF under grant P17503-N12. The second author thanks Bjørn Kjos-Hanssen for several discussions [12], and in particular for pointing out a mistake in [30].[f] The present paper has greatly benefitted from the comments and suggestions of an anonymous referee, who in particular shortened and improved the proof of Theorem 3.1.

References

1. K. Ambos-Spies, E. Mayordomo, Y. Wang, and X. Zheng, *Resource-bounded balanced genericity, stochasticity and weak randomness*, Proc. 13th Symp. on Theoret. Aspects of Comp. Sci., Lect. Notes in Comp. Sci. 1046 (1996) 63–74, Springer-Verlag.
2. K. Ambos-Spies, W. Merkle, J. Reimann, and F. Stephan, *Hausdorff dimension in exponential time*, Computational Complexity 2001, 210-217, IEEE Computer Society, 2001.
3. C. S. Calude, L. Staiger, and S. A. Terwijn, *On partial randomness*, Ann. Pure Appl. Logic 138 (2006) 20–30.
4. O. Demuth and A. Kučera, *Remarks on 1-genericity, semigenericity and related concepts*, Comment. Math. Univ. Carolinae, 28 (1987), 85–94.
5. R. G. Downey, *Some computability-theoretic aspects of reals and randomness*, in *The Notre Dame Lectures* (P. Cholak, ed.), Lect. Notes Log., 18, Assoc. Symbol. Logic, Urbana, IL, 2005, 97–147.
6. R. G. Downey, D. R. Hirschfeldt, S. Lempp, and R. Solomon, *A Δ_2^0 set with no infinite low subset in either it or its complement*, Journal of Symbolic Logic 66 (2001) 1371–1381.
7. R. G. Downey, D. R. Hirschfeldt, J. S. Miller, and A. Nies, *Relativizing Chaitin's halting probability*, Journal of Mathematical Logic 5 (2005) 167–192.
8. R. G. Downey, D. R. Hirschfeldt, A. Nies, and S. A. Terwijn, *Calibrating randomness*, Bulletin of Symbolic Logic 12 (2006) 411–491.
9. K. Falconer, *Fractal Geometry, Mathematical Foundations & Applications*, Wiley & Sons, 1992.
10. D. R. Hirschfeldt, A. Nies, and F. Stephan, *Using random sets as oracles*, Journal of the London Mathematical Society 75 (2007) 610–622.
11. J. M. Hitchcock, J. H. Lutz and S. A. Terwijn, *The arithmetical complexity of dimension and randomness*, ACM Trans. Comput. Log. 8 (2007), no. 2, Art. 13, 22pp.
12. B. Kjos-Hanssen, personal communication, July 2003.
13. A. Kučera, *An alternative, priority-free solution to Post's Problem*, in: Lecture Notes in Computer Science 233, 493–500, Springer, 1986.

---

[f] In Theorem 4.1 in [30] it was proven that the lower cone of every incomplete c.e. set has Δ_2^0-measure 0. Also, the upper cone of every noncomputable set in Δ_2^0 has Δ_2^0-measure 0 (Theorem 2.3). Now in Corollary 4.2 it was falsely claimed that from the proofs of these theorems it follows that almost every set in Δ_2^0 is incomparable with every noncomputable and incomplete c.e. set. Kjos-Hanssen pointed out that this contradicts Kučera's result that every 1-random set in Δ_2^0 bounds a noncomputable c.e. set, cf. [13].

14. S. A. Kurtz, *Randomness and genericity in the degrees of unsolvability*, PhD Thesis, University of Illinois, 1981.
15. M. van Lambalgen, *Random sequences*, PhD Thesis, University of Amsterdam, 1987.
16. K. de Leeuw, E. F. Moore, C. F. Shannon, and N. Shapiro, *Computability by probabilistic machines*, in *Automata Studies*, Annals of Mathematics Studies 34, Princeton University Press, 1956, 183–212.
17. M. Li and P. Vitányi, *An Introduction to Kolmogorov Complexity and its Applications*, 2nd edition, Springer-Verlag, 1997.
18. J. H. Lutz, *Category and measure in complexity classes*, SIAM J. Comput. 19 (1990) 1100–1131.
19. J. H. Lutz, *Dimension in complexity classes*, SIAM J. Comput. 32 (2003), 1236–1259 (extended abstract in *15th Annual IEEE Conference on Computational Complexity (Florence, 2000)* (F. Titsworth, ed.), IEEE Computer Soc., Los Alamitos, CA, 2000, 158–169).
20. J. H. Lutz, *The dimensions of individual strings and sequences*, Inform. and Comput. 187 (2003), 49–79 (preliminary version: *Gales and the constructive dimension of individual sequences*, in *Proc. 27th International Colloquium on Automata, Languages, and Programming* (U. Montanari, J. D. P. Rolim, E. Welzl, eds.), Springer-Verlag, 2000, 902–913).
21. P. Martin-Löf, *The definition of random sequences*, Information and Control 9 (1966) 602–619.
22. A. Nies, F. Stephan, and S. A. Terwijn, *Randomness, relativization and Turing Degrees*, Journal of Symbolic Logic 70(2) (2005) 515–535.
23. P. G. Odifreddi, *Classical Recursion Theory*, Vol. 1, Studies in logic and the foundations of mathematics Vol. 125, North-Holland, 1989.
24. P. G. Odifreddi, *Classical Recursion Theory*, Vol. 2, Studies in logic and the foundations of mathematics Vol. 143, North-Holland, 1999.
25. J. Reimann, *Computability and fractal dimension*, PhD Thesis, University of Heidelberg, 2004.
26. G. E. Sacks, *Degrees of Unsolvability*, Annals of Mathematics Studies 55, Princeton University Press, 1963.
27. C.-P. Schnorr, *Zufälligkeit und Wahrscheinlichkeit*, Lect. Notes in Math. 218, Springer-Verlag, 1971.
28. R. Soare, *Recursively Enumerable Sets and Degrees*, Springer, 1987.
29. S. A. Terwijn, *Computability and Measure*, PhD Thesis, University of Amsterdam/ILLC, 1998.
30. S. A. Terwijn, *On the quantitative structure of Δ_2^0*, in: U. Berger, H. Osswald, and P. Schuster (eds.), *Reuniting the Antipodes — Constructive and Nonstandard Views of the Continuum*, Synthèse Library, Kluwer, 2000, 271–284.
31. S. A. Terwijn, *Complexity and Randomness*, Rendiconti del Seminario Matematico di Torino 62 (2004) 1–38 (preliminary version: Research Report CDMTCS-212, University of Auckland, March 2003).
32. J. Ville, *Étude Critique de la Notion de Collectif*, Gauthier-Villars, 1939.
33. Y. Wang, *A separation of two randomness concepts*, Information Processing Letters 69(3) (1999) 115–118.

THE STRENGTH OF SOME COMBINATORIAL PRINCIPLES RELATED TO RAMSEY'S THEOREM FOR PAIRS

Denis R. Hirschfeldt

Department of Mathematics, The University of Chicago
Chicago, IL 60637-1514, USA
E-mail: drh@math.uchicago.edu

Carl G. Jockusch, Jr.

Department of Mathematics, University of Illinois at Urbana-Champaign
1409 W. Green St., Urbana, IL 61801-2975, USA
E-mail: jockusch@math.uiuc.edu

Bjørn Kjos-Hanssen

Department of Mathematics, University of Hawai'i at Mānoa
2565 McCarthy Mall
Honolulu HI 96822, USA
E-mail: bjoern@math.hawaii.edu

Steffen Lempp

Department of Mathematics, University of Wisconsin
Madison, WI 53706-1388, USA
E-mail: lempp@math.wisc.edu

Theodore A. Slaman

Department of Mathematics The University of California
Berkeley, CA 94720-3840, USA
E-mail: slaman@math.berkeley.edu

We study the reverse mathematics and computability-theoretic strength of (stable) Ramsey's Theorem for pairs and the related principles COH and DNR. We show that SRT_2^2 implies DNR over RCA_0 but COH does not, and answer a question of Mileti by showing that every computable stable 2-coloring of pairs has an incomplete Δ_2^0 infinite homogeneous set. We also give some extensions of the latter result, and relate it to potential approaches to showing that SRT_2^2 does not imply RT_2^2.

1. Introduction

In this paper we establish some results on the reverse mathematics and computability-theoretic strength of combinatorial principles related to Ramsey's Theorem for pairs. This topic has attracted a large amount of recent research (see for instance [2, 4, 9, 10]), but certain basic questions still remain open.

For a set X, let $[X]^2 = \{Y \subset X \mid |Y| = 2\}$. A 2-*coloring* of $[\mathbb{N}]^2$ is a function from $[\mathbb{N}]^2$ into $\{0, 1\}$. A set $H \subseteq \mathbb{N}$ is *homogeneous* for a 2-coloring C of $[\mathbb{N}]^2$ if C is constant on $[H]^2$. Ramsey's Theorem for pairs (RT_2^2) is the statement in the language of second-order arithmetic that every 2-coloring of $[\mathbb{N}]^2$ has an infinite homogeneous set. A 2-coloring C of $[\mathbb{N}]^2$ is *stable* if for each $x \in \mathbb{N}$ there exists a $y \in \mathbb{N}$ and a $c < 2$ such that $C(\{x, z\}) = c$ for all $z > y$. Stable Ramsey's Theorem for pairs (SRT_2^2) is RT_2^2 restricted to stable colorings.

It follows from work of Jockusch (Theorem 5.7 in [5]) that if $n > 2$ then Ramsey's Theorem for n-tuples is equivalent to arithmetical comprehension (ACA_0), but Seetapun [11] showed that RT_2^2 does not imply ACA_0. (All implications and nonimplications discussed here are over the standard base theory RCA_0 of reverse mathematics. For background on reverse mathematics and discussions of many of the techniques used below, see Simpson [12].)

A long-standing open question in reverse mathematics is whether RT_2^2 implies Weak König's Lemma (WKL_0), the statement that every computable infinite binary tree has an infinite path. (That WKL_0 does not imply RT_2^2 follows from a result of Jockusch (Theorem 3.1 in [5]) discussed below.) As is well-known, WKL_0 is equivalent to the statement that for each set A, there is a $0, 1$-valued function function f that is diagonally noncomputable relative to A (where a total function f is *diagonally noncomputable* if $\forall e\, (f(e) \neq \Phi_e(e))$.) A natural way to weaken this statement is to drop the requirement that f be $0, 1$-valued, and allow it to take arbitrary values in ω; the corresponding axiom system has been named DNR. In Section 2 we show that RT_2^2 implies DNR over RCA_0. In other words, whereas we do not know whether RT_2^2 implies WKL_0, we have a partial result toward this implication. In fact, we show that the possibly weaker system SRT_2^2 already implies DNR. It is not known whether SRT_2^2 is strictly weaker than RT_2^2; we will discuss this question further below.

An infinite set X is *cohesive* for a family $R_0, R_1, \ldots$ of sets if for each i, one of $X \cap R_i$ or $X \cap \overline{R}_i$ is finite. COH is the principle stating that every family of sets has a cohesive set. Having seen that $RCA_0 + SRT_2^2 \vdash DNR$, and recalling that RT_2^2 is equivalent over RCA_0 to $SRT_2^2 + COH$ (see Lemma 7.11 in [2] and Corollary A.1.4 in [9]), we proceed to compare COH and DNR. As noted by Cholak, Jockusch, and Slaman (Lemma 9.14 in [2]), even WKL_0 does not imply COH, so certainly DNR does not imply COH. We establish that COH does not imply DNR in Section 3. This result was independently and simultaneously obtained by Hirschfeldt and Shore (Corollary 2.21 in [4]), and as we will see, the main ideas of the proof were already present in [2].

Jockusch (Theorem 3.1 in [5]) constructed a computable 2-coloring of $[\mathbb{N}]^2$ with no Δ_2^0 infinite homogeneous set. On the other hand, computable stable 2-colorings always have Δ_2^0 infinite homogeneous sets. Indeed, the problem of finding an infinite homogeneous set for a computable stable 2-coloring is essentially the same as the problem of finding an infinite subset of either A or $\overline{A}$ for a Δ_2^0 set A. More precisely, we have the following. If A is Δ_2^0 then there is a computable stable 2-coloring C of $[\mathbb{N}]^2$ such that if H is homogeneous for C then $H \subseteq A$ or $H \subseteq \overline{A}$. Conversely, if C is a computable stable 2-coloring of $[\mathbb{N}]^2$ then there is a Δ_2^0 set A such that any infinite set B with $B \subseteq A$ or $B \subseteq \overline{A}$ computes an infinite homogeneous set for C. (See Proposition 2.1 in [5] and Lemma 3.5 in [2], or Claim 5.1.3 in [9].)

Cholak, Jockusch, and Slaman (Theorem 3.1 in [2]) showed that every computable 2-coloring of $[\mathbb{N}]^2$ has a low_2 infinite homogeneous set, and suggested the possibility of separating SRT_2^2 and RT_2^2 by showing that every computable *stable* 2-coloring of $[\mathbb{N}]^2$ has a *low* infinite homogeneous set. Such a result, if relativizable, would allow us to build an ω-model of SRT_2^2 consisting entirely of low sets, which would therefore not be a model of RT_2^2. (An ω-model of second-order arithmetic is one whose first-order part is standard, and such a model is identified with its second-order part.) However, Downey, Hirschfeldt, Lempp, and Solomon [3] constructed a computable stable 2-coloring of $[\mathbb{N}]^2$ with no low infinite homogeneous set.

Mileti (Theorem 5.3.7 in [9]) showed that for each $X <_T 0'$ there is a computable stable 2-coloring of $[\mathbb{N}]^2$ with no X-computable infinite homogeneous set. (He also showed that this is true for any low_2 set X.)

In light of these results, Mileti (Question 5.3.8 in [9]) asked whether there is an infinite Δ_2^0 set A such that every infinite Δ_2^0 subset of A or $\overline{A}$ is complete (i.e., has degree $0'$); in other words, whether there is a computable stable 2-coloring of $[\mathbb{N}]^2$ such that any Δ_2^0 infinite homogeneous set is complete. Hirschfeldt gave a negative answer to this question; this previously unpublished result appears as Corollary 4.10 below. In Theorem 4.5, we modify the proof of this result to show that, in fact, if $C_0, C_1, \ldots >_T 0$ are uniformly Δ_2^0, then for every Δ_2^0 set A there is a Δ_2^0 subset X of either A or $\overline{A}$ such that $\forall i\, (C_i \not\leq_T X)$. In proving that RT_2^2 does not imply ACA_0, Seetapun [11] showed that if $C_0, C_1, \ldots >_T 0$ then every 2-coloring of $[\mathbb{N}]^2$ has an infinite homogeneous set that does not compute any of the C_i. Our result can be seen as a Δ_2^0 analogue of this theorem. The restriction to stable colorings is of course necessary in this case, since as mentioned above, there are 2-colorings of pairs with no Δ_2^0 infinite homogeneous set.

There is still a large gap between the negative answer to Mileti's question and the result of Downey, Hirschfeldt, Lempp, and Solomon [3] mentioned above. In particular, we would like to know the answer to the following question.

Question 1.1. Let A be Δ_2^0. Must there be an infinite subset of either A or $\overline{A}$ that is both Δ_2^0 and low_2?

A relativizable positive answer to this question would lead to a separation between SRT_2^2 and RT_2^2, since it would allow us to build an ω-model of $\mathrm{RCA}_0 + \mathrm{SRT}_2^2$ that is not a model of RT_2^2, as we now explain. We begin with the ω-model $\mathcal{M}_0$ consisting of the computable sets. Let C_0 be a stable 2-coloring of $[\mathbb{N}]^2$ in $\mathcal{M}_0$. Assuming a positive answer to Question 1.1, we have an infinite homogeneous set H_0 for C_0 that is both Δ_2^0 and low_2. Note that H_0' is low over $0'$ and c.e. over $0'$.

Now let $\mathcal{M}_1$ be the ω-model consisting of the H_0-computable sets, and let C_1 be a stable 2-coloring of $[\mathbb{N}]^2$ in $\mathcal{M}_1$. Again assuming a (relativizable) positive answer to Question 1.1, we have an infinite homogeneous set H_1 for C_1 such that $H_0 \oplus H_1$ is both Δ_2^0 in H_0 and low_2. As before, $(H_0 \oplus H_1)'$ is low over $0'$. It may no longer be c.e. over $0'$, but it is 2-CEA over $0'$ (that is, it is c.e. in and above a set that is itself c.e. in and above $0'$).

Now let $\mathcal{M}_2$ be the ω-model consisting of the $H_0 \oplus H_1$-computable sets, and continue in this way, making sure that for every i and every stable 2-coloring C of $[\mathbb{N}]^2$ in $\mathcal{M}_i$, we have $C_j = C$ for some j. Let $\mathcal{M} = \bigcup_i \mathcal{M}_i$. By construction, $\mathcal{M}$ is an ω-model of $\mathrm{RCA}_0 + \mathrm{SRT}_2^2$, and for every set X in $\mathcal{M}$, we have that X' is low over $0'$ and m-CEA over $0'$ for some m. By the extension of Arslanov's Completeness Criterion given by Jockusch, Lerman, Soare, and Solovay [6], no such X can have PA degree over $0'$ (that is, X cannot be the degree of a nonstandard model of arithmetic with an extra predicate for $0'$). However, Jockusch and Stephan (Theorem 2.1 in [8]) showed that a degree contains a p-cohesive set (that is, a set that is cohesive for the collection of primitive recursive sets) if and only if its jump is PA over $0'$. Thus $\mathcal{M}$ is not a model of COH, and hence not a model of RT_2^2.

Note that to achieve the separation described above, it would be enough to show (in a relativizable way) that every Δ_2^0 set A has a subset of either it or its complement that is both Δ_2^0 and low_n for some n (which may depend on A). However, we do not even know whether every Δ_2^0 set has a subset of either it or its complement that is both Δ_2^0 and nonhigh.

The ultimate refutation of this approach to separating SRT_2^2 and RT_2^2 would be to build a computable *stable* 2-coloring of $[\mathbb{N}]^2$ for which the jump of every infinite homogeneous set has PA degree over $0'$. (Without the condition of stability, such a coloring was built by Cholak, Jockusch, and Slaman (Theorem 12.5 in [2]).) Indeed, such a construction (if relativizable) would show that every ω-model of $RCA_0 + SRT_2^2$ is a model of RT_2^2, as we now explain. Suppose that such stable colorings exist, and let $\mathcal{M}$ be an ω-model of $RCA_0 + SRT_2^2$. Relativizing the result of Jockusch and Stephan (Theorem 2.1 in [8]) on p-cohesive sets mentioned above, we can show that $\mathcal{M}$ is a model of COH. But as mentioned above, $SRT_2^2 + COH$ is equivalent to RT_2^2 over RCA_0, so $\mathcal{M}$ is a model of RT_2^2.

2. SRT_2^2 implies DNR

The proof that SRT_2^2 implies DNR over RCA_0 is naturally given in two parts: first we show that each ω-model of SRT_2^2 is a model of DNR, and then that we can in fact carry out the proof of this implication in RCA_0, that is, using only Σ_1^0-induction.

2.1. *The argument for ω-models*

A set A is *effectively bi-immune* if there is a computable function f such that for each e, if $W_e \subseteq A$ or $W_e \subseteq \overline{A}$, then $|W_e| < f(e)$.

Lemma 2.1. *There is an effectively bi-immune set $A \leqslant_T 0'$. In fact, we can choose the function f witnessing the bi-immunity of A to be defined by $f(e) = 3e + 2$.*

Proof. We build A in stages, via a $0'$-computable construction. At each stage we decide the value of $A(n)$ for at most three n's. At stage e, we check whether W_e has at least $3e + 2$ many elements. If so, then there are at least two elements $n_0, n_1 \in W_e$ at which we have not yet decided the value of A. Let $A(n_0) = 0$ and $A(n_1) = 1$. In any case, if $A(e)$ is still undefined then let $A(e) = 0$. □

We also need the following lemma, which follows immediately from the equivalence mentioned above between finding homogeneous sets for computable stable colorings and finding subsets of Δ_2^0 sets or their complements. A *Turing ideal* is a subset of 2^ω closed under Turing reduction and join. A subset of 2^ω is a Turing ideal if and only if it is an ω-model of RCA_0.

Lemma 2.2. *A Turing ideal $\mathcal{I}$ is an ω-model of SRT_2^2 if and only if for each set A, if $A \leqslant_T C'$ for some $C \in \mathcal{I}$, then there is an infinite $B \in \mathcal{I}$ such that either $B \subseteq A$ or $B \subseteq \overline{A}$.*

We can now prove the implication between SRT_2^2 and DNR for ω-models.

Theorem 2.3. *Each ω-model of SRT_2^2 is a model of DNR.*

Proof. Let $\mathcal{I}$ be a Turing ideal that is an ω-model of SRT_2^2. We show that $\mathcal{I}$ contains a diagonally noncomputable function. The proof clearly relativizes to get a function that is diagonally noncomputable relative to X for any $X \in \mathcal{I}$.

Let A be as in Lemma 2.1. By Lemma 2.2, there is an infinite $B \in \mathcal{I}$ such that B is a subset of A or $\overline{A}$. By the choice of A, for all e, if $W_e \subseteq B$ then $|W_e| < 3e + 2$.

Let g be such that $W_{g(e)}$ is the set consisting of the first $3e + 2$ many elements of B (in the usual ordering of ω). For any e, if $W_e = W_{g(e)}$ then $W_e \subseteq B$, and so $|W_e| < 3e+2$. But $|W_{g(e)}| = 3e+2$, so this is a contradiction. Thus $\forall e\, (W_e \neq W_{g(e)})$.

Now let f be a computable function such that $W_{f(e)} = W_{\Phi_e(e)}$ if $\Phi_e(e)\downarrow$, and $W_{f(e)} = \emptyset$ otherwise. Then $h = g \circ f$ is diagonally noncomputable, since it is total and for each e, if $\Phi_e(e)\downarrow$ then $W_{h(e)} \neq W_{f(e)} = W_{\Phi_e(e)}$. But h is also computable in B, and hence belongs to $\mathcal{I}$. □

2.2. *The proof-theoretic argument*

We now simply need to analyze the above proof to ensure that Σ_1^0-induction suffices to carry it out. The formal analog of Lemma 2.2 is the statement that SRT_2^2 is equivalent to the following principle, called D_2^2: For every $0, 1$-valued function $d(x, s)$, if $\lim_s d(x, s)$ exists for all x, then there is an infinite set B and a $j < 2$ such that $\lim_s d(x, s) = j$ for all $x \in B$. The equivalence of SRT_2^2 and D_2^2 over RCA_0 is claimed in Lemma 7.10 of [2]. However, the argument indicated there for the $D_2^2 \rightarrow SRT_2^2$ direction appears to require Π_1^0-bounding, which is not provable in RCA_0. It is unknown whether $D_2^2 \rightarrow SRT_2^2$ is provable in RCA_0. Fortunately, we need only the other direction, since we are starting with the assumption that SRT_2^2 holds. This direction is proved as in Lemma 7.10 of [2], and we reproduce the proof here for the reader's convenience. Work in $RCA_0 + SRT_2^2$. Let a function $d(x, s)$ be given that satisfies the hypothesis of D_2^2. Give the pair $\{x, s\}$ with $x < s$ the color $d(x, s)$. The infinite homogeneous set produced by SRT_2^2 for this stable coloring satisfies the conclusion of D_2^2.

Theorem 2.4. $RCA_0 \vdash SRT_2^2 \rightarrow DNR$.

Proof. Given the existence of a set A as in Lemma 2.1 (or more precisely, of a function $d(x, s)$ such that $A(x) = \lim_s d(x, s)$), the definition of the diagonally noncomputable function h given in the proof of Theorem 2.3 can clearly be carried out using D_2^2 and Σ_1^0-induction.

So the only part of the proof of Theorem 2.3 we need to consider more carefully is the construction of A and the satisfaction of all bi-immunity requirements. More precisely, fix a model $\mathcal{M}$ of $RCA_0 + SRT_2^2$. Within that model, we have an enumeration of the $\mathcal{M}$-c.e. sets $W_0, W_1, \ldots$ (where the indices range over all elements of the first-order part of $\mathcal{M}$). We need to show the existence of a function $d(x, s)$ in $\mathcal{M}$ such that $\lim_s d(x, s)$ exists for all x, and for every W_e, if there is a $j < 2$ such that $\forall x \in W_e$ ($\lim_s d(x, s) = j$), then $|W_e| < 3e + 2$. (We will actually be able to use $2e + 2$ instead of $3e + 2$.)

We can build d in much the same way as we built A, but we need to be more careful because we no longer have access to an oracle for $0'$. So we need a computable construction to replace the $0'$-computable construction in the proof of Lemma 2.2. Let R_e be the eth bi-immunity requirement.

In this construction, R_e may control up to two numbers n_e^0 and n_e^1 at any point in the construction. At stage $t = \langle e, s \rangle$, if $|W_{e,s}| \geqslant 2e + 2$, then for each $i < 2$ such that n_e^i is undefined, define n_e^i to be different from each $n_{e'}^j$ for $e' \leqslant e$, and undefine all $n_{e'}^j$ for $e' > e$. In any case, for each n, if $n = n_k^j$ for some j and k, then let $d(n, t) = j$, and otherwise let $d(n, t) = 0$.

It is now easy to check (in RCA_0) that $\lim_t d(n, t)$ exists for all n, since for each n, either n is never controlled by a requirement, in which case $d(n, t) = 0$ for all t, or there is a stage t at which n is controlled by R_e for some e. In the latter case, since control of a number can only pass to stronger requirements, there are at most e many $u \geqslant t$ such that $d(n, u + 1) \neq d(n, u)$.

The last thing we need to check is that each R_e is satisfied. It follows by induction that for each e, there are at most $2e$ many numbers that are ever controlled by any $R_{e'}$ with $e' < e$, and thus there is a stage v_e by which all such numbers have been controlled by such requirements. (This is an instance of Π_1^0-induction, which holds in RCA$_0$ (see Lemma 3.10 in Simpson [12]), using a formula saying that for all finite sequences of size $2e+1$ of distinct elements and for all t, it is not the case that each element of the sequence has been controlled by some $R_{e'}$ with $e' < e$ by stage t.) So if $|W_e| \geqslant 2e+2$, then picking a stage $t = \langle e, s \rangle \geqslant v_e$ such that $|W_{e,s}| \geqslant 2e+2$, the n_e^i must be defined at stage t, and will never be undefined at a later stage, so $\lim_u d(n_e^i, u) = i$. Thus R_e is satisfied. $\square$

3. COH does not imply DNR

In this section we show that COH does not imply DNR over RCA$_0$. We first recall a connection between diagonally noncomputable functions and special Π_1^0 classes.

Definition 3.1. For $n \geqslant 1$ and $A \in 2^\omega$, a Π_n^0 subclass of 2^ω is A-*special* if it has no A-computable members. A class is *special* if it is $\emptyset$-special.

Theorem 3.2. (Jockusch and Soare (Corollary 1.3 in [7])) *If A computes an element of a special Π_2^0 class, then A computes an element of a special Π_1^0 class.*

Corollary 3.3. *Any diagonally noncomputable function computes an element of a special Π_1^0 class.*

Proof. Consider the special Π_2^0 class

$$\{A \mid \forall x, t \, \exists y \, \exists s > t \, [\langle x, y \rangle \in A \wedge \neg(\Phi_{x,s}(x) \downarrow = y)] \wedge$$
$$\forall x, a, b \, [(\langle x, a \rangle \in A \wedge \langle x, b \rangle \in A) \to a = b]\}.$$

It is easy to check that any diagonally noncomputable function computes an element of this class. The corollary now follows from Theorem 3.2. $\square$

We now consider the relationship between cohesiveness and special Π_1^0 classes.

Lemma 3.4. (Cholak, Jockusch and Slaman (Lemma 9.16 in [2])) *Let* $A \in 2^\omega$, *let* P *be an* A-*special* Π_1^0 *class, and let* $R_0, R_1, \ldots \leqslant_T A$. *Then there is an* $\vec{R}$-*cohesive set* G *that does not compute any element of* P.

This lemma is proved using Mathias forcing with A-computable conditions. We will use two results about Mathias forcing, but since we will not work with this notion directly, we refer to Section 9 of [2], Section 6 of [1], and Section 2 of [4] for the relevant definitions. Analyzing the proof of Lemma 3.4, we immediately obtain the following result.

Corollary 3.5. (to the proof of Lemma 3.4) *There is an* $m \in \omega$ *such that if* G *is* m-A-*generic for Mathias forcing with* A-*computable conditions, then* G *is cohesive with respect to any collection of sets* $\vec{R} \leqslant_T A$.

It is clear that Lemma 3.4 generalizes to deal with all Π_1^0 classes at once; this is proved directly in Lemma 6.3 of [1].

Lemma 3.6. (Binns, Kjos-Hanssen, Lerman, and Solomon [1]) *Let* P *be a* Π_1^0 *class and let* A *be a set. Let* G *be* 3-A-*generic for Mathias forcing with* A-*computable conditions. If* P *is* A-*special, then* P *is* $(G \oplus A)$-*special.*

We are now ready to establish the result in the section heading.

Theorem 3.7. *There is an* ω-*model of* $RCA_0 + COH$ *that is not a model of DNR.*

Proof. Let $m \geqslant 3$ be as in Corollary 3.5. Let $A_0 = \emptyset$, and inductively let A_{n+1} be $A_n \oplus G_n$, where G_n is m-A_n-generic for Mathias forcing with A_n-computable conditions. Let $\mathcal{I}$ be the Turing ideal generated by $\{A_n \mid n \in \omega\}$.

Let $\mathcal{M}$ be the ω-model determined by $\mathcal{I}$. If $\vec{R} \in \mathcal{I}$ is a collection of sets then $\vec{R} \leqslant_T A_n$ for some n. By Corollary 3.5, G_n is $\vec{R}$-cohesive. Since $G_n \in \mathcal{I}$, it follows that $\mathcal{M}$ is a model of COH.

On the other hand, if B computes a diagonally noncomputable function, then by Corollary 3.3, there is a special Π_1^0 class P such that B computes an element of P. In other words, P is not B-special. However, if $B \in \mathcal{I}$ then $B \leqslant_T A_n$ for some n. By Lemma 3.6 and induction, P is A_n-special, and hence P is B-special. So if B computes a diagonally noncomputable function, then $B \notin \mathcal{I}$. Thus $\mathcal{M}$ is not a model of DNR. $\square$

So DNR separates SRT_2^2 from COH. That is, SRT_2^2 implies DNR, whereas COH does not.

4. Degrees of homogeneous sets for stable colorings

In this section we give our negative answer to Mileti's question mentioned in the introduction. We will need two auxiliary results. One is an extension of the low basis theorem noted by Linda Lawton (unpublished).

Theorem 4.1. (Lawton) *Let T be an infinite, computable, computably bounded tree, and let $C_0, C_1, \ldots >_T 0$ be uniformly Δ_2^0. Then T has an infinite low path P such that $\forall i (C_i \not\leq_T P)$, and an index of such a P can be $0'$-computed from an index of T.*

This theorem is proved by forcing with Π_1^0 classes, and lowness is achieved just as in the usual proof of the low basis theorem. Steps are interspersed to guarantee cone avoidance, which is possible by the following lemma.

Lemma 4.2. *Let C be a noncomputable set and let Q be a nonempty computably bounded Π_1^0 class. Let Φ be a Turing reduction. Then Q has a nonempty Π_1^0 subclass R such that $\Phi^f \neq C$ for all $f \in R$. Furthermore, there is a fixed procedure that computes an index of R from indices of Q and Φ and an oracle for $C \oplus 0'$.*

Proof. Let U be a computable tree with $Q = [U]$. For each n, let U_n be the set of strings σ in U such that $\Phi^\sigma(n)$ is either undefined or has a value other than $C(n)$. (Here we use the convention that computations with string oracles σ run for at most $|\sigma|$ steps.) Then U_n is a computable tree, and an index of it can be computed from a C-oracle. Note that U_n is infinite for some n, since otherwise C is computable. Furthermore, $\{n \mid U_n \text{ is infinite }\} \leq_T C \oplus 0'$, since C can compute an index of U_n as a computable tree, and then $0'$ can determine whether U_n is infinite by asking whether it contains a string of every length. Let $R = [U_n]$ for the least n with U_n infinite. $\square$

Below, we will use the following relativized form of Theorem 4.1, which can be proved in the same way: Let L be a low set. Let T be an infinite, L-computable, L-computably bounded tree, and let $C_0, C_1, \ldots \not\leq_T L$ be uniformly Δ_2^0. Then T has an infinite low path P such that $\forall i (C_i \not\leq_T P)$, and an index of such a P can be $0'$-computed from indices of L and T.

The other result we will use below is that if $C_0, C_1, \ldots >_T 0$ are uniformly Δ_2^0 and the complement $\overline{A}$ of the Δ_2^0 set A has no infinite Δ_2^0 subset Y such that $\forall i (C_i \not\leq_T Y)$, then A cannot be too sparse.

Definition 4.3. An infinite set Z is *hyperimmune* if for every computable increasing function f, there is an n such that the interval $[f(n), f(n + 1))$ contains no element of Z.

If Z is not hyperimmune, then a computable f such that $[f(n), f(n + 1)) \cap Z \neq \emptyset$ is said to *witness the non-hyperimmunity of Z*.

Proposition 4.4. *Let A be Δ_2^0. Let $C_0, C_1, \ldots$ be uniformly Δ_2^0 and let L be an infinite Δ_2^0 set such that $C_i \not\leq_T L$ for all i. If $A \cap L$ is L-hyperimmune, then there is an infinite Δ_2^0 set $Y \subseteq \overline{A}$ such that $\forall i\,(C_i \not\leq_T Y)$.*

Proof. We build Y by finite extensions; that is, we define $\gamma_0 \prec \gamma_1 \prec \cdots$ and let $Y = \bigcup_i \gamma_i$.

For a string σ and a set X, we write $\sigma \sqsubset X$ to mean that $\{n < |\sigma| \mid \sigma(n) = 1\} \subseteq X$.

Begin with γ_0 defined as the empty sequence. At stage $s = \langle e, i \rangle$, given the finite binary sequence $\gamma_s \sqsubset \overline{A} \cap L$, we $0'$-computably search for either

(1) an m and extensions $\gamma_s \sigma_0$ and $\gamma_s \sigma_1$ such that $\Phi_e^{\gamma_s \sigma_0}(m) \downarrow \neq \Phi_e^{\gamma_s \sigma_1}(m) \downarrow$ and $\gamma_s \sigma_k \sqsubset \overline{A} \cap L$ for $k = 0, 1$; or

(2) an m such that for all extensions $\gamma_s 0^m \sigma \sqsubset L$, either $\Phi_e^{\gamma_s 0^m \sigma}(m) \uparrow$ or $\Phi_e^{\gamma_s 0^m \sigma}(m) \downarrow \neq C_i(m)$.

We claim one of these must be found. Suppose not. Then for every m we can find an extension $\gamma_s 0^m \sigma_0 \sqsubset L$ such that $\Phi_e^{\gamma_s 0^m \sigma_0}(m) \downarrow = C_i(m)$. Since $C_i \not\leq_T L$, there must be infinitely many m for which there is also an extension $\gamma_s 0^m \sigma_1 \sqsubset L$ such that $\Phi_e^{\gamma_s 0^m \sigma_1}(m) \downarrow \neq C_i(m)$. So we can L-computably enumerate an infinite set M such that for each $m \in M$, there are $\gamma_s 0^m \sigma_k \sqsubset L$ for $k = 0, 1$ such that $\Phi_e^{\gamma_s 0^m \sigma_0}(m) \downarrow \neq \Phi_e^{\gamma_s 0^m \sigma_1}(m) \downarrow$. Let $m \in M$. Since we are assuming that case 1 above does not hold, there must be a k such that $\gamma_s 0^m \sigma_k \not\sqsubset \overline{A} \cap L$. So letting l_m be the maximum of $|\gamma_s 0^m \sigma_k|$ for $k = 0, 1$, we are guaranteed the existence of an element of $A \cap L$ in the interval $[m, l_m)$. Now we can find $m_0, m_1, \ldots \in M$ such that $m_{j+1} > l_{m_j}$, and define $f(j) = m_j$. Then f is a witness to the non-L-hyperimmunity of $A \cap L$, contrary to hypothesis.

So one of the two cases above must eventually hold. If case 1 holds, let k be such that $\Phi_e^{\gamma_s \sigma_k}(m) \neq C_i(m)$ and define $\gamma_s' = \gamma_s \sigma_k$. If case 2 holds, define $\gamma_s' = \gamma_s 0^m$. In either case, let $\gamma_{s+1} \sqsubset \overline{A} \cap L$ be an extension of γ_s' such that $\gamma_{s+1}(j) = 1$ for some $j > |\gamma_s|$. Such a string must exist since $\gamma_s' \sqsubset \overline{A} \cap L$ and $\overline{A} \cap L$ is infinite (as otherwise $A \cap L$ would be cofinite within L, and hence not L-hyperimmune). This definition ensures that $\Phi_e^Y \neq C_i$. $\qquad \square$

We are now ready to prove the main result of this section.

Theorem 4.5. *Let A be Δ_2^0 and let $C_0, C_1, \ldots >_T 0$ be uniformly Δ_2^0. Then either A or $\overline{A}$ has an infinite Δ_2^0 subset X such that $C_i \not\leq_T X$ for all i.*

Proof. Assume that $\overline{A}$ has no infinite Δ_2^0 subset Y such that $C_i \not\leq_T Y$ for all i. We use Proposition 4.4 to build an infinite Δ_2^0 set X such that $C_i \not\leq_T X$ for all i, via a $0'$-computable construction satisfying the following requirements:

$$R_{e,i} : \Phi_e^X \text{ total } \Rightarrow \exists n \, (\Phi_e^X(n) \neq C_i(n)).$$

We first discuss how to satisfy the single requirement $R_{0,0}$. By Proposition 4.4 (with $L = \omega$), A is not hyperimmune. Suppose we have a computable function f witnessing the non-hyperimmunity of A. Let the computable, computably bounded tree $\widehat{T}$ consist of the nodes $(m_0, \ldots, m_{k-1})$ with $f(j) \leqslant m_j < f(j+1)$ for all $j < k$. Such a node represents a guess that $m_j \in A$ for each $j < k$. Note that the choice of f ensures that $\widehat{T}$ has at least one path along which all such guesses are correct.

Now prune $\widehat{T}$ as follows. For each node $\sigma = (m_0, \ldots, m_{k-1})$, if there are nonempty $F_0, F_1 \subseteq \text{rng}(\sigma)$ and an n such that $\Phi_0^{F_0}(n) \downarrow \neq \Phi_0^{F_1}(n) \downarrow$ with uses bounded by the largest element of $F_0 \cup F_1$, then prune $\widehat{T}$ to ensure that σ is not extendible to an infinite path. Note that we can do this pruning in such a way as to end up with a computable tree T.

Now $0'$ can determine whether T is finite. If so, then we can find a leaf σ of T such that $\text{rng}(\sigma) \subset A$. There are nonempty $F_0, F_1 \subseteq \text{rng}(\sigma)$ and an n such that $\Phi_0^{F_0}(n) \downarrow \neq \Phi_0^{F_1}(n) \downarrow$ with uses bounded by the largest element z of $F_0 \cup F_1$, so if we let k be such that $\Phi_0^{F_k}(n) \neq C_0(n)$ and define X so that $X \restriction z + 1 = F_k \restriction z + 1$, then we ensure that $\Phi_0^X(n) \neq C_0(n)$.

On the other hand, if T is infinite then by Theorem 4.1, $0'$ can find a low path P of T such that $C_i \not\leq_T P$ for all i. There must be an n such that either $\Phi_0^Y(n) \uparrow$ for every $Y \subseteq \text{rng}(P)$ or there is a $Y \subseteq \text{rng}(P)$ such that $\Phi_0^Y(n) \downarrow \neq C_0(n)$, since otherwise we could P-compute $C_0(n)$ for each n by searching for a finite $F \subset \text{rng}(P)$ such that $\Phi_0^F(n) \downarrow$. But by the construction of T, this means that there is an n such that for every infinite $Y \subseteq \text{rng}(P)$, either $\Phi_0^Y(n) \uparrow$ or $\Phi_0^Y(n) \downarrow \neq C_0(n)$. So if we now promise to make $X \subseteq \text{rng}(P)$, we ensure that $\Phi_0^X \neq C_0$. Notice that we can make such a promise because $C_i \not\leq_T P$ for all i, and hence $C_i \not\leq_T \text{rng}(P)$ for all i (since P is an increasing sequence), which implies that $A \cap \text{rng}(P)$ is infinite.

Let us now consider how to satisfy another requirement, say $R_{0,1}$. The action taken to satisfy $R_{0,0}$ results in either a finite initial segment of X

being determined, or a promise being made to keep X within a given infinite low set that does not compute any of the C_i. We can handle both cases at once by assuming that we have a number r_1 and an infinite low set L_1 containing the finite set F_1 of numbers less than r_1 currently in X, such that $C_i \not\leq_T L_1$ for all i. We want $X \restriction r_1 = F_1$ and $X \subseteq L_1$.

Suppose that we have an L_1-computable function g witnessing the non-L_1-hyperimmunity of $A \cap L_1$. We can then proceed much as we did for $R_{0,0}$, but taking r_1 and L_1 into account, in the following way. We can assume that $g(0) \geqslant r_1$. Define $\widehat{T}$ to consist of the nodes $(m_0, \ldots, m_{k-1})$ with $g(j) \leqslant m_j < g(j+1)$ and $m_j \in L_1$ for all $j < k$. For each node $\sigma = (m_0, \ldots, m_{k-1})$, if there are nonempty $G_0, G_1 \subseteq \mathrm{rng}(\sigma)$ and an n such that $\Phi_1^{F_1 \cup G_0}(n) {\downarrow} \neq \Phi_1^{F_1 \cup G_1}(n) {\downarrow}$ with uses bounded by the largest element of $G_0 \cup G_1$, then prune $\widehat{T}$ to ensure that σ is not extendible to an infinite path, thus obtaining a new L_1-computable tree T.

If T is finite then find a leaf σ of T such that $\mathrm{rng}(\sigma) \subset A \cap L_1$. Then there is a nonempty $G \subseteq \mathrm{rng}(\sigma)$ and an n such that $\Phi_1^{F_1 \cup G}(n) {\downarrow} \neq C_0(n)$ with use bounded by the largest element z of G, so if we define X such that $X \restriction z + 1 = (F_1 \cup G) \restriction z + 1$ then we ensure that $\Phi_1^X(n) \neq C_0(n)$.

If T is infinite then $0'$ can find a low path P of T. If we now promise that all future elements of X will be in $\mathrm{rng}(P)$, we ensure that $\Phi_1^X \neq C_0$ as before. Notice that we can make such a promise because $\mathrm{rng}(P) \subseteq L_1$ and, as before, $A \cap \mathrm{rng}(P)$ is infinite.

Thus we can satisfy $R_{0,1}$, and the action we take results in a number r_2 and an infinite low set L_2 that does not compute any of the C_i (and contains the finite set F_2 of numbers less than r_2 currently in X) such that we want $X \restriction r_2 = F_2$ and $X \subseteq L_2$. In other words, we are in the same situation we were in after satisfying $R_{0,0}$, and we could now proceed to satisfy another requirement as we did $R_{0,1}$.

However, there is a crucial problem with proceeding in this way for all the $R_{e,i}$ at once, which is that we know no $0'$-computable way to determine the witnesses to non-hyperimmunity required by the construction. The best we can do is guess at them. That is, we have a $0''$-partial computable function w such that if l is a lowness index for an infinite set L (that is, $\Phi_l^{0'} = L'$) then $\Phi_{w(l)}^L$ witnesses the non-L-hyperimmunity of $A \cap L$.

We are now ready to describe our construction. We give our requirements a priority ordering by saying that $R_{e,i}$ is stronger than $R_{e',i'}$ if $\langle e, i \rangle < \langle e', i' \rangle$. All numbers added to X at a stage s of our construction will be greater than s, thus ensuring that $X \leqslant_T \emptyset'$. Let X_s be the set of numbers added to X by the beginning of stage s.

Throughout the construction, we run a $0'$-approximation to w. Associated with each $R_{e,i}$ are a number $r_{\langle e,i \rangle}$ and a low set $L_{\langle e,i \rangle}$ with lowness index $l_{\langle e,i \rangle}$ (all of which might change during the construction). If the approximation to $w(l_{\langle e,i \rangle})$ changes, then for all $\langle e',i' \rangle \geqslant \langle e,i \rangle$ the strategy for $R_{e',i'}$ is immediately canceled, $R_{e',i'}$ is declared to be unsatisfied, and $r_{\langle e',i' \rangle}$, $L_{\langle e',i' \rangle}$, and $l_{\langle e',i' \rangle}$ are reset to the current values of $r_{\langle e,i \rangle}$, $L_{\langle e,i \rangle}$, and $l_{\langle e,i \rangle}$, respectively. It is important to note that the approximation to w continues to run during the action of a strategy at a fixed stage. That is, we may find a change in the approximation to some $w(l_{\langle e,i \rangle})$ with $\langle e,i \rangle \leqslant \langle e',i' \rangle$ in the middle of a stage s at which we are trying to satisfy $R_{e',i'}$. If this happens then we immediately end the stage and cancel strategies as described above.

Initially, all requirements are unsatisfied. At the beginning of stage 0, for every e,i, let $r_{\langle e,i \rangle} = 0$ and $L_{\langle e,i \rangle} = \omega$, and let $l_{\langle e,i \rangle}$ be a fixed lowness index for ω.

At stage s, let $R_{e,i}$ be the strongest unsatisfied requirement and proceed as follows.

We have a number $r_{\langle e,i \rangle}$ and a low set $L_{\langle e,i \rangle}$ with lowness index $l_{\langle e,i \rangle}$, such that $L_{\langle e,i \rangle}$ contains $X_s \upharpoonright r_{\langle e,i \rangle}$, and $C_j \not\leqslant_{\mathrm{T}} L_{\langle e,i \rangle}$ for all j. As before, we want to ensure that $X \upharpoonright r_{\langle e,i \rangle} = X_s \upharpoonright r_{\langle e,i \rangle}$ and $X \subseteq L_{\langle e,i \rangle}$. Let v be the current approximation to $w(l_i)$ and let $g = \Phi_v^{L_{\langle e,i \rangle}}$. By shifting the values of g if necessary, we can assume that $g(0) \geqslant \max(r_{\langle e,i \rangle}, s)$. Define $\widehat{T}$ to consist of the nodes $(m_0, \ldots, m_{k-1})$ with $g(j) \leqslant m_j < g(j+1)$ and $m_j \in L_{\langle e,i \rangle}$ for all $j < k$. Note that g may not be total, in which case $\widehat{T}$ is finite.

For each node $\sigma = (m_0, \ldots, m_{k-1})$, if there are nonempty $G_0, G_1 \subseteq \sigma$ and an n such that $\Phi_e^{X \upharpoonright r_i \cup G_0}(n) \downarrow \neq \Phi_e^{X \upharpoonright r_i \cup G_1}(n) \downarrow$ with uses bounded by the largest element of $G_0 \cup G_1$, then prune $\widehat{T}$ to ensure that σ is not extendible to an infinite path, thus obtaining a new $L_{\langle e,i \rangle}$-computable tree T.

We want to $0'$-effectively determine whether T is finite. More precisely, the question we ask $0'$ is whether the pruning process described above ever results in all the nodes at some level of $\widehat{T}$ becoming non-extendible. A positive answer means T is finite. If g is total then a negative answer means T is infinite. However, if g is not total, so that $\widehat{T}$ is finite, we may still get a negative answer, because the pruning process may get stuck waiting forever for a level of $\widehat{T}$ to become defined.

If the answer to our question is positive, then look for a leaf σ of T such that $\mathrm{rng}(\sigma) \subset A \cap L_{\langle e,i \rangle}$. If no such leaf exists, then either $L_{\langle e,i \rangle}$ is finite or $v \neq w(l_{\langle e,i \rangle})$, so end the stage and cancel the strategies for $R_{\langle e',i' \rangle}$ with $\langle e',i' \rangle \geqslant \langle e,i \rangle$ as described above. (That is, declare $R_{\langle e',i' \rangle}$ to be unsatisfied, and reset $r_{\langle e',i'\rangle}$, $L_{\langle e',i'\rangle}$, and $l_{\langle e',i'\rangle}$ to the current values of $r_{\langle e,i\rangle}$, $L_{\langle e,i\rangle}$, and $l_{\langle e,i\rangle}$, respectively.) Otherwise, there are a nonempty $G \subseteq \text{rng}(\sigma)$ and an n such that $\Phi_e^{X\upharpoonright r_i \cup G}(n) \downarrow \neq C_i(n)$ with use bounded by the largest element of G. Let $r_{\langle e,i\rangle+1}$ be the largest element of G, let $L_{\langle e,i\rangle+1} = L_{\langle e,i\rangle}$, and let $l_{\langle e,i\rangle+1} = l_{\langle e,i\rangle}$. Put every element of G into X.

If the answer to our question is negative, then use the relativized form of Theorem 4.1 to $0'$-effectively obtain a low path P of T such that $C_j \not\leq_T P$ for all j, and a lowness index $l_{\langle e,i\rangle+1}$ for $L_{\langle e,i\rangle+1} = X \upharpoonright r_i \cup \text{rng}(P)$. If g is not total, then the construction in the proof of Theorem 4.1 will still produce such an $L_{\langle e,i\rangle+1}$ and $l_{\langle e,i\rangle+1}$, but $L_{\langle e,i\rangle+1}$ may be finite. (Which will of course be a problem for weaker priority requirements, but in this case the strategy for $R_{e,i}$ will eventually be canceled, and hence $L_{\langle e,i\rangle+1}$ will eventually be redefined.) Let $r_{\langle e,i\rangle+1} = r_{\langle e,i\rangle}$. Search for an element of $A \cap L_{\langle e,i\rangle+1}$ greater than $\max\{r_{\langle e',i'\rangle} \mid \langle e', i'\rangle \leqslant \langle e,i\rangle\}$ not already in X and put this number into X. If $L_{\langle e,i\rangle+1}$ is infinite, such a number must be found. Otherwise, such a number may not exist, but this situation can only happen if the approximation to $w(l_{\langle e',i'\rangle})$ at the beginning of stage s is incorrect for some $\langle e', i'\rangle \leqslant \langle e,i\rangle$, in which case the strategy for $R_{e,i}$ will be canceled, and the stage ended as described above.

In either case, if the action of the strategy for $R_{e,i}$ has not been canceled, then declare $R_{e,i}$ to be satisfied, and for $\langle e', i'\rangle > \langle e,i\rangle$, declare $R_{e',i'}$ to be unsatisfied, let $r_{\langle e',i'\rangle+1} = r_{\langle e,i\rangle+1}$, let $L_{\langle e',i'\rangle+1} = L_{\langle e,i\rangle+1}$, and let $l_{\langle e',i'\rangle+1} = l_{\langle e,i\rangle+1}$.

This completes the construction. Since every element entering X at stage s is in A and is greater than s, we have that X is a Δ_2^0 subset of A. Furthermore, at each stage a number is added to X unless the strategy acting at that stage is canceled, so once we show that every requirement is eventually permanently satisfied, we will have shown that X is infinite.

Assume by induction that for all $\langle e', i'\rangle < \langle e,i\rangle$, the requirement $R_{e',i'}$ is eventually permanently satisfied, and that $r_{\langle e,i\rangle}$, $L_{\langle e,i\rangle}$, and $l_{\langle e,i\rangle}$ eventually reach a final value, for which $L_{\langle e,i\rangle}$ is infinite. Let s be the least stage by which this situation obtains and the approximation to $w(l_{\langle e,i\rangle})$ has settled to a final value v. Note that at stage $s - 1$, either the strategy for some $R_{\langle e',i'\rangle}$ with $\langle e', i'\rangle < \langle e,i\rangle$ acted, or the approximation to $w(l_{\langle e,i\rangle})$ changed, so at the beginning of stage s, it must be the case that $R_{e,i}$ is the strongest unsatisfied requirement. Thus at that stage the strategy for $R_{e,i}$ acts, and the function $g = \Phi_v^{L_{\langle e,i\rangle}}$ it works with at that stage is in fact a witness to the non-$L_{\langle e,i\rangle}$-hyperimmunity of $A \cap L_{\langle e,i\rangle}$. Thus $R_{e,i}$ will become satisfied at the end of the stage, and $r_{\langle e,i\rangle+1}$, $L_{\langle e,i\rangle+1}$, and $l_{\langle e,i\rangle+1}$ will not be redefined after the end of the stage.

If the tree T built at stage s is finite, then a leaf σ of T is found such that $\mathrm{rng}(\sigma) \subset A \cap L_{\langle e,i\rangle}$, and there are a nonempty $G \subseteq \mathrm{rng}(\sigma)$ and an n such that $\Phi_e^{X \upharpoonright r_{\langle e,i\rangle} \cup G}(n)\downarrow \neq C_i(n)$ with use bounded by the largest element $r_{\langle e,i\rangle+1}$ of G. Since $r_{\langle e,i\rangle+1}$ is never again redefined, $\Phi_e^X(n) \neq C_i(n)$, and thus the requirement $R_{e,i}$ is satisfied. Furthermore, $L_{\langle e,i\rangle+1}$ is defined to be $L_{\langle e,i\rangle}$ at this stage, and hence is infinite.

If T is infinite, then $L_{\langle e,i\rangle+1}$ is defined to contain the range of a path of T, and hence is infinite. Furthermore, $L_{\langle e,i\rangle+1}$ is never redefined, and by the way X is defined, $X \subseteq A \cap L_{\langle e,i\rangle+1}$. There must be an n such that either $\Phi_e^Y(n)\uparrow$ for every $Y \subseteq L_{\langle e,i\rangle+1}$ or there is a $Y \subseteq L_{\langle e,i\rangle+1}$ such that $\Phi_0^Y(n)\downarrow \neq C_i(n)$, since otherwise we could $L_{\langle e,i\rangle+1}$-compute $C_i(n)$ for each n by searching for a finite $F \subset L_{\langle e,i\rangle+1}$ such that $\Phi_e^F(n) \downarrow$. But by the definition of T, this means that there is an n such that for every infinite $Y \subseteq L_{\langle e,i\rangle+1}$, either $\Phi_e^Y(n)\uparrow$ or $\Phi_e^Y(n)\downarrow \neq C_i(n)$. So since $X \subseteq L_{\langle e,i\rangle+1}$, we have $\Phi_e^X \neq C_i$, and hence the requirement $R_{e,i}$ is satisfied. $\qquad\square$

Theorem 4.5 gives the negative answer to Mileti's question mentioned above.

Corollary 4.6. (Hirschfeldt) *Every Δ_2^0 set has an incomplete infinite Δ_2^0 subset of either it or its complement. In other words, every computable stable 2-coloring of $[\mathbb{N}]^2$ has an incomplete Δ_2^0 infinite homogeneous set.*

We can improve on this result by using the following unpublished result due to Jockusch.

Proposition 4.7. (Jockusch) *Let Z be hyperimmune. Then there is a 1-generic $G \leqslant_T Z \oplus 0'$ such that $Z \subseteq G$.*

Proof. We build G by finite extensions; that is, we define $\gamma_0 \prec \gamma_1 \prec \cdots$ and let $G = \bigcup_i \gamma_i$. Let $S_0, S_1, \ldots$ be an effective listing of all c.e. sets of finite binary sequences.

Begin with γ_0 defined as the empty sequence. At stage i, given the finite binary sequence γ_i, search for an extension $\alpha \in S_i$ of $\gamma_i 1^{f(n)}$. If one is found then let $f(n+1) = |\alpha|$.

If f is total then, since Z is hyperimmune, there is an n such that the interval $[|\gamma_i| + f(n), |\gamma_i| + f(n+1))$ contains no element of Z. So $Z \oplus 0'$-computably search for either such an interval or for an n such that $f(n+1)$

is undefined. In the first case, let α be as above and let $\gamma_{i+1} = \alpha$. In the second case, let $\gamma_{i+1} = \gamma_i 1^{f(n)}$.

It is now easy to check by induction that $Z \subseteq G$ and that G meets or avoids each S_i. $\qquad\Box$

Corollary 4.8. *Let $X \subset Y$ be such that X is Y-hyperimmune. Then there are $G, H \leqslant_T X \oplus Y'$ such that*

(1) $H \leqslant_T G \oplus Y$,
(2) G is 1-generic relative to Y,
(3) $X \subseteq H \subset Y$, and
(4) $Y \setminus H$ is infinite.

Proof. Let $h(0) < h(1) < \cdots$ be the elements of Y, and let $Z = h^{-1}(X)$. By Proposition 4.7 relativized to Y, there is a $G \leqslant_T Z \oplus Y'$ such that G is 1-generic relative to Y and $Z \subseteq G$. Let $H = h(G)$. Since $h \leqslant_T Y$ and h is increasing, we have $H \leqslant_T G \oplus Y$, and $X \subseteq H \subset Y$ by the definition of h. Finally, $Y \setminus H = h(\overline{G})$, and hence is infinite. $\qquad\Box$

Corollary 4.9. *Let A be a Δ_2^0 set such that $\overline{A}$ has no infinite low subset, and let L be low. Then $A \cap L$ is not L-hyperimmune.*

Proof. Suppose that $A \cap L$ is L-hyperimmune. We can apply Corollary 4.8 to $X = A \cap L$ and $Y = L$ to obtain G and H as above. Since $A \cap L$ and L' are both Δ_2^0, so is G. Since G is also 1-generic relative to L, and L is low, $G \oplus L$ is low. But $H \oplus L \leqslant_T G \oplus L$, and hence $H \oplus L$ is low. Thus $L \setminus H$ is an infinite low subset of $\overline{A}$, which is a contradiction. $\qquad\Box$

Corollary 4.10. (Hirschfeldt) *Let A be a Δ_2^0 set such that $\overline{A}$ has no infinite low subset. Then A has an incomplete infinite Δ_2^0 subset.*

Proof. (Sketch) The proof is similar to that of Theorem 4.5. Instead of working with the given sets C_i, we build a Δ_2^0 set C while satisfying the requirements

$$R_e : \Phi_e^X \text{ total } \Rightarrow \exists n \, (\Phi_e^X(n) \neq C(n)).$$

At stage s, we work with the least unsatisfied requirement R_i. We have a number r_i and a low set L_i with lowness index l_i, such that L_i contains $X_s \restriction r_i$. We define $\widehat{T}$ as before. For each node $\sigma = (m_0, \ldots, m_{k-1})$, if there is a nonempty $G \subseteq \sigma$ such that $\Phi_e^{X \restriction r_i \cup G}(s) \downarrow$ with use bounded by the largest element of G, then we prune $\widehat{T}$ to ensure that σ is not extendible to an infinite path, thus obtaining a new L_i-computable tree T.

If T is finite, then we look for a leaf σ of T and a G as above, let r_{i+1} be the largest element of G, define $C(s) \neq \Phi_i^{X \restriction r_i \cup G}(s)$, let $L_{i+1} = L_i$, let $l_{i+1} = l_i$, and put every element of G into X.

If T is infinite, we $0'$-effectively obtain a low path P of T and a lowness index l_{i+1} for $L_{i+1} = X \restriction r_i \cup \text{rng}(P)$. We then let $r_{i+1} = r_i$ and $C(s) = 0$, search for an element of $A \cap L_{i+1}$ greater than $\max\{r_j \mid j \leqslant i\}$ not already in X, and put this number into X.

The further details of the construction are as before, and the verification that it succeeds in satisfying all the requirements is similar.　　　　□

Acknowledgments

All of the authors thank the Institute for Mathematical Sciences of the National University of Singapore for its generous support during the period July 18 – August 15, 2005, when part of this research was carried out. The first author was partially supported by NSF grants DMS-0200465 and DMS-0500590. The second author was supported by NSFC Grand International Joint Project 'New Directions in Theory and Applications of Models of Computation', No. 60310213. The fourth author was partially supported by NSF grant DMS-0140120. The fifth author was partially supported by NSF grant DMS-0501167.

References

1. S. Binns, B. Kjos-Hanssen, M. Lerman and D. R. Solomon, On a conjecture of Dobrinen and Simpson concerning almost everywhere domination, J. Symbolic Logic 71 (2006) 119–136.
2. P. A. Cholak, C. G. Jockusch, Jr., and T. A. Slaman, On the strength of Ramsey's theorem for pairs, J. Symbolic Logic 66 (2001), 1–55.
3. R. Downey, D. R. Hirschfeldt, S. Lempp, and R. Solomon, A Δ_2^0 set with no infinite low subset in either it or its complement, J. Symbolic Logic 66 (2001), 1371–1381.
4. D. R. Hirschfeldt and R. A. Shore, Combinatorial principles weaker than Ramsey's Theorem for Pairs, J. Symbolic Logic 72 (2007) 171–206.
5. C. G. Jockusch, Jr., Ramsey's theorem and recursion theory, J. Symbolic Logic 37 (1972), 268–280.
6. C. G. Jockusch, Jr., M. Lerman, R. I. Soare, and R. M. Solovay, Recursively enumerable sets modulo iterated jumps and extensions of Arslanov's completeness criterion, J. Symbolic Logic 54 (1989), 1288–1323.
7. C. G. Jockusch, Jr. and R. I. Soare, Degrees of members of Π_1^0 classes, Pacific J. Math. 40 (1972), 605–616.

8. C. G. Jockusch, Jr. and F. Stephan, A cohesive set which is not high, Math. Log. Quart. 39 (1993) 515–530.

9. J. R. Mileti, Partition Theorems and Computability Theory, PhD Dissertation, University of Illinois, 2004.

10. J. R. Mileti, Partition theorems and computability theory, Bull. Symbolic Logic 11 (2005), 411–427.

11. D. Seetapun and T. A. Slaman, On the strength of Ramsey's Theorem, Notre Dame J. Formal Logic 36 (1995), 570–582.

12. S. G. Simpson, Subsystems of Second Order Arithmetic, Springer-Verlag, Berlin, 1999.

ABSOLUTENESS FOR UNIVERSALLY BAIRE SETS AND THE UNCOUNTABLE II

Ilijas Farah

Department of Mathematics and Statistics, York University
4700 Keele Street, Toronto, Canada M3J 1P3
and
Matematicki Institut, Kneza Mihaila 35
11 000 Beograd, Serbia
E-mail: ifarah@mathstat.yorku.ca
URL: http://www.mathstat.yorku.ca/~ifarah

Richard Ketchersid[*] and Paul Larson[†]

Department of Mathematics and Statistics, Miami University
Oxford, Ohio 45056, USA
E-mails: [*]richard.ketchersid@gmail.com
[†]larsonpb@muohio.edu
[†]URL: http://www.users.muohio.edu/larsonpb/

Menachem Magidor

Mathematics Institute, Hebrew University
Givat Ram, 91904 Jerusalem, Israel
E-mail: menachem@math.huji.ac.il

Using $\lozenge$ and large cardinals we extend results of Magidor–Malitz and Farah–Larson to obtain models correct for the existence of uncountable homogeneous sets for finite-dimensional partitions and universally Baire sets. Furthermore, we show that the constructions in this paper and its predecessor can be modified to produce a family of 2^{ω_1}-many such models so that no two have a stationary, costationary subset of ω_1 in common. Finally, we extend a result of Steel to show that trees on reals of height ω_1 which are coded by universally Baire sets have either an uncountable path or an absolute impediment preventing one.

The first author is partially supported by NSERC. The third author was supported in part by NSF grant DMS-0401603 and a summer research grant from Miami University. Part of this research was conducted while the third and fourth authors were in residence at the Institute for Mathematical Sciences at the National University of Singapore, and we thank the institute for its hospitality.

163

In [4] it was shown (using large cardinals) that if a model of a theory T satisfying a certain second-order property P can be forced to exist, then a model of T satisfying P exists already. The properties P considered in [4] included the following.

(1) Containing any specified set of $\aleph_1$-many reals.
(2) Correctness about NS_{ω_1}.
(3) Correctness about any given universally Baire set of reals (with a predicate for this set added to the language).

In this paper we add the following properties, all proved under the assumption of Jensen's $\Diamond$ principle.

(4) Correctness about Magidor–Malitz quantifiers (and even about the existence of uncountable homogeneous sets for subsets of $[\omega_1]^{<\omega}$ and any $[\kappa]^{<\omega}$).
(5) Correctness about the countable chain condition for partial orders.
(6) Correctness about uncountable chains through (some) trees of height and cardinality ω_1.
(7) Containing a function on ω_1 dominating any such given function on a club.

These results are obtained using two main tools (both due to Woodin):

(a) iterable models (also called $\mathbb{P}_{max}$-preconditions), introduced in [22],
(b) stationary-tower forcing ([11]), or more specifically, Woodin's proof of Σ_1^2-absoluteness ([21]) .

While (b) requires higher large cardinal strength than (a), it allows one to assure (1). Aside from (1) and (7), we can obtain all of these properties simultaneously using the method (a) (with "some" being "all" for (6)). Aside from (1) and (4) we can prove all of these properties simultaneously using the method (b). Property (4) subsumes the next two properties in the list, but we do not see how to obtain it by stationary tower constructions. As a matter of fact, simultaneously obtaining (1) and (4), and even (1) and (6), would imply Σ_2^2-absoluteness conditioned on $\Diamond$ (see Conjecture 3.1 and Theorem 3.8). That $\Diamond$ implies that one can iterate a $\mathbb{P}_{max}$ pre-condition to be correct about the countable chain condition for partial orders on ω_1 is due to Larson and Yorioka [13].

In the fourth section we show abstractly that these arguments can be modified to build a family of 2^{ω_1} many models, no two having a stationary-costationary subset of ω_1 in common. In the final section we prove a generalized version of a theorem of Steel which can be used to show that the existence of a model of a given sentence which is correct about a given universally Baire set is absolute, given a proper class of Woodin cardinals.

1. Magidor–Malitz logic

The language $L(Q^{<\omega})$ is formed by adding to the language of set theory quantifiers Q^n for each n in ω. In this paper we restrict our attention to the so-called ω_1-interpretation of this language. That is, a formula of the form

$$Q^n x_1, x_2, \ldots, x_n \phi(x_1, \ldots, x_n)$$

is interpreted as saying that there is an uncountable subset of ω_1 such that every n-tuple from this set satisfies ϕ. The expressive power of this language is not diminished by requiring ϕ to be *symmetric*, i.e., invariant under permuting its free variables $x_1, \ldots, x_n$. Recall that $[Z]^n$ is the set of all n-element subsets of Z and $[Z]^{<\omega}$ is the set of all finite subsets of Z. Given $K \subseteq [\omega_1]^n$, we say (following [5]) that $X \subseteq \omega_1$ is a K-*cube* if and only if $[X]^n \subseteq K$ (or just a *cube* if the corresponding partition is clear). Since an interpretation of a symmetric formula is a subset of $[\omega_1]^n$, correctness for Magidor–Malitz logic is equivalent to correctness for the existence of uncountable cubes (note that the existence of countable cubes of any given order type is absolute between transitive models.) We therefore say that a model M is *correct* for Magidor–Malitz logic (or for Ramsey quantifiers) if ω_1^M is uncountable and, for every $n \in \omega$ and every $K \subset ([\omega_1^M]^n)^M$ definable in M from parameters in M (note that we do not assume here that M satisfies ZFC), there is an uncountable K-cube in V if and only if one exists in M. The following theorem was proved in [14].

Theorem 1.1 ($\Diamond$). *If T is a theory in the language $L(Q^{<\omega})$ and it is consistent with ZFC that T has a model which is correct for Magidor–Malitz logic, then T has such a model.* □

Requiring T from Theorem 1.1 to contain a large enough fragment of ZFC is not a loss of generality. Here (and throughout this paper) "large enough fragment of ZFC" means large enough to make ultrapower embeddings for generic ultrafilters on ω_1 elementary. This requires some form of the Axiom of Choice, but the theory ZFC° from [12] suffices. In this case Theorem 1.1 can be equivalently reformulated as follows:

Theorem 1.2 ($\Diamond$). *If a theory T extends a large enough fragment of ZFC and it is consistent, then there exists a model M for T such that ω_1^M is* *uncountable and M is correct about the existence of uncountable cubes for partitions of $[\omega_1^M]^n$ for all $n \in \mathbb{N}$.* □

The model M guaranteed by this result is an ω-model but it is not necessarily well-founded. As a matter of fact, asserting well-foundedness of M requires some large cardinal strength (see [4, Proposition 8.9]). A model M of a large enough fragment of ZFC is *correct for partitions of* $[\omega_1^M]^{<\omega}$ if it is correct about the existence of uncountable cubes for partitions $K \subseteq [\omega_1^M]^{<\omega}$ in M. This assertion implies ω_1^M is uncountable, but note that we do not require it to be well-founded. Our first result, proved at the end of this section as Theorem 1.12, is a strengthening of Theorem 1.2.

Theorem 1.3 ($\Diamond$). *If a theory T extends a large enough fragment of ZFC and it is ω-consistent, then there exists a model M for T such that ω_1^M is uncountable and M is correct about the existence of uncountable cubes for partitions of $[\omega_1^M]^{<\omega}$ that belong to M, for each $n \in \omega$.*

The difference between Theorem 1.2 and Theorem 1.3 is that in the latter the dimension of K is not bounded. In Proposition 2.8 we show that the conclusion of Theorem 1.3 is stronger than the conclusion of Theorem 1.1.

Continuing along the lines of [4], we also show that in the presence of large cardinals correctness for partitions of $[\omega_1]^{<\omega}$ can be combined with correctness for any given universally Baire set of reals, with respect to the logic of forceability. Analogously to [4, §5], given a set of reals A let $L(A)$ be the language of set theory with an additional unary predicate for A. We say that a model M is *correct for A and partitions of* $[\omega_1^M]^{<\omega}$ (in short, $L(Q^{\leq\omega}, A)$-*correct*) if $\omega_1^M = \omega_1$, M interprets the additional unary symbol as $A \cap M$ and it is correct for partitions of $[\omega_1^M]^{<\omega}$. Since correctness for Π_1^1-sets already implies well-foundedness of ω_1^M, assuming $\omega_1^M = \omega_1$ is not a loss of generality in this context. The symbol $\leq \omega$ in the term '$L(Q^{\leq\omega}, A)$-correct' is perhaps misleading, but it was chosen in order to emphasize the difference between partitions of $[\omega_1]^{<\omega}$ and of $[\omega_1]^n$ for a fixed n, since $L(Q^{<\omega})$ is an established notation for Magidor–Malitz logic. The reader may wish to compare the following theorem with the results in [1].

Theorem 1.4. *Suppose that there exist proper class many Woodin cardinals, let A be a universally Baire set of reals, and let T be a set of sentences in $L(A)$. Suppose that there exists an $L(Q^{\leq\omega}, A)$-correct model of T in some set forcing extension. Then there exists an $L(Q^{\leq\omega}, A)$-correct model of T in every set forcing extension satisfying $\Diamond$.*

Proof. Immediate from Lemma 1.5, Lemma 1.6, and Theorem 1.7 below.   □

The logic $L_{\omega_1\omega}(Q^{<\omega})$ allows countable disjunctions in addition to quantifiers Q^n ($n \in \mathbb{N}$). It is well-known that an analogue of Theorem 1.1 for this logic can be proved using the methods of [14]; for a proof see e.g., [5]. By standard methods (see e.g., [3] for the case of $L_{\omega_1\omega}(Q)$), the case of Theorem 1.4 when A is a Borel set follows. This cannot be extended even to analytic sets unless large cardinals are assumed ([4, Proposition 8.7]). An alternative way for proving these results using iterated generic ultrapowers is outlined in our proofs of Theorem 1.3 and Theorem 1.12. Note that this semantical result does not recover the full strength of Keisler or Magidor–Malitz theorems. This is because these results provide completeness theorems for logics $L_{\omega_1\omega}(Q)$ and $L_{\omega_1\omega}(Q^{<\omega})$. We do not know whether this can be achieved for the logic with the quantifier corresponding to the existence of uncountable cubes of $[\omega_1]^{<\omega}$.

1.1. Proofs

Continuing in the vein of [4], the proofs of this section employ $\mathbb{P}_{max}$ preconditions, also known as *iterable pairs*. ZFC° is a fragment of ZFC that holds in $H(\theta)$ for a regular $\theta \geq \aleph_2$ (the precise definition will not be needed here; see [12, §1]). If N_0 is a transitive model of ZFC° and I_0 is a normal ideal on $\omega_1^{N_0}$ in N_0, then an *iteration of (N_0, I_0) of length γ* is $\langle (N_\eta, I_\eta), j_{\xi\eta}, G_\eta, \xi < \eta \leq \gamma \rangle$, where $j_{\xi\eta} \colon (N_\xi, I_\xi) \to (N_\eta, I_\eta)$ is a commuting family of elementary embeddings, $G_\eta \subseteq (\mathcal{P}(\omega_1)/I_\eta)^{N_\eta}$ is a generic filter, $j_{\eta\eta+1}$ is the corresponding generic ultrapower embedding, and for a limit η and $\xi < \eta$, $j_{\xi\eta}$ and N_η are the direct limit of $j_{\xi\zeta}$ and N_ζ for $\xi < \zeta < \eta$. An iteration is *well-founded* if all the models occurring in it are well-founded. A pair is *iterable* if all of its iterations are well-founded. If $A \in N_0$ is a universally Baire set then a pair (N_0, I_0) is *A-iterable* if it is iterable and its iterations compute A correctly. We shall follow the standard convention and identify an *iteration* of length γ with the final model together with the embedding $j_{0\gamma} \colon (N_0, I_0) \to (N_\gamma, I_\gamma)$. For more information we refer the reader to [22, 12].

Lemma 1.5 below is proved in [4, Lemma 3.3]. In the presence of a proper class of Woodin cardinals, universally Baire sets of reals are δ^+-weakly homogeneously Suslin for all δ.

Lemma 1.5. *Assume that $\delta < \lambda$ are a Woodin and a measurable cardinal respectively, A and $\omega^\omega \setminus A$ are δ^+-weakly homogeneously Suslin sets of reals, and ϕ is a sentence whose truth is preserved by σ-closed forcing. If there* *exists a partial order in V_δ that forces that ϕ holds in $H(\theta)$ for some $\theta \geq$ $(2^\lambda)^+$, then there exists an A-iterable model $(N, \mathrm{NS}_{\omega_1}^N)$ that satisfies ϕ.* □

A forcing $\mathbb{P}$ has *property K_n* if for each family p_α $(\alpha < \omega_1)$ of conditions there is an uncountable $I \subseteq \omega_1$ such that $p_{\alpha(1)}, \ldots, p_{\alpha(n)}$ has a lower bound for all $\alpha(1), \ldots, \alpha(n)$ in I. It has *precaliber $\aleph_1$* if for each family p_α $(\alpha < \omega_1)$ of conditions there is an uncountable $I \subseteq \omega_1$ such that every finite subset of p_α $(\alpha \in I)$ has a lower bound. The following is well-known.

Lemma 1.6. *Assume $K \subseteq [Z]^{<\omega}$. The statement 'there are no uncountable K-cubes' is absolute for σ-closed forcing extensions and for precaliber $\aleph_1$ forcing extensions. If furthermore $K \subseteq [Z]^n$ then the statement 'there are no uncountable K-cubes' is absolute for property K_n forcing extensions.*

Proof. In all three cases the forcing preserves $\aleph_1$, and therefore we only need to check that it does not add an uncountable cube. Assume $\mathbb{P}$ is σ-closed and it forces the existence of an uncountable cube, and let $\dot{H}$ be its name. Pick a decreasing ω_1-sequence of conditions p_α such that p_α decides the first α elements of $\dot{H}$. Then the decided set is uncountable and a K-cube. If $\mathbb{P}$ has precaliber $\aleph_1$ and it forces an existence of an uncountable cube $\dot{H}$, pick p_α that decides αth element of $\dot{H}$ is ξ_α. If every finite subset of $\{p_\alpha \mid \alpha \in I\}$ has a lower bound, then $\{\xi_\alpha \mid \alpha \in I\}$ is a cube. The proof of the case when $\mathbb{P}$ has the property K_n is very similar. □

Theorem 1.7 ($\Diamond$). *If (M, I) is an iterable pair then there is an iteration $j : (M, I) \to (M^*, I^*)$ of length ω_1 such that M^* is correct for partitions of $[\omega_1]^{<\omega}$.*

We give two proofs of Theorem 1.7. The first uses the presentation of Magidor–Malitz logic given in [5] and its modularity makes it more susceptible to generalizations. The second is shorter and more straightforward.

1.2. First proof of Theorem 1.7

We use standard model-theoretic terminology as in [5] or any standard model theory text. For a transitive model N of ZFC° let L_N be the language of set theory extended by adding the constants for elements of N (and all universally Baire sets and NS_{ω_1}). Let $(\forall^{\aleph_0} x \in z)\phi(x)$ be the shortcut for 'z is uncountable and $\phi(x)$ holds for all but countably many $x \in z$.' For a 1-type Φ in L_N let

$$\partial\Phi(x) = \{(\forall^{\aleph_0} z \in x)\phi(z) \mid \phi(z) \in \Phi(z)\}.$$

Also let $\partial^\circ \Phi = \Phi$ and $\partial^{n+1}\Phi = \partial(\partial^n \Phi)$. A type Φ is *totally unsupported* in N if $\partial^n \Phi$ is not realized in N for all $n \geq 0$.

If $j\colon N \to N^*$ is an elementary embedding and Φ is an N-type then $j\Phi$ is a type defined in the natural way:

$$j\Phi(x) = \{\phi(x, j(\vec{a})) \mid \phi(x, \vec{a}) \in \Phi(x)\}$$

(here $\vec{a}$ stands for an arbitrary n-tuple of parameters and $j(\vec{a})$ has the natural interpretation). We emphasize that in the following lemma the types Φ_i are not required to belong to N.

Lemma 1.8. *Assume (N, I) is an iterable pair and types Φ_i $(i < \omega)$ are totally unsupported in N. Then there is N-generic $G \subseteq I^+$ such that each $j\Phi_i$ is totally unsupported in N^*, where $j\colon N \to N^*$ is the corresponding generic embedding.*

Proof. Let $\nu = \omega_1^N$. Enumerate all pairs $(f, \partial^n \Phi_i)$ for $f\colon \nu \to \nu$ in N. Pick G recursively, by finding a decreasing sequence A_k $(k \geq 0)$ in I^+. Assure that A_{2k} is in k-th dense subset of I^+ in N. To find A_{2k+1}, consider the k-th pair $(f, \partial^n \Phi_i)$. If there is $\phi \in \partial^n \Phi_i$ such that the set

$$B_\phi = \{\alpha \in A_{2k} \mid N \models \neg\phi(f(\alpha))\}$$

is I-positive then let $A_{2k+1} = B_\phi$.

We claim that such a ϕ has to exist. Otherwise let $D = \nabla\{B_\phi \mid B_\phi \in \mathrm{NS}_{\omega_1}\}$. Then $D \in N$ and $C = A_{2k} \setminus D$ is equal to A_{2k} modulo I. Also, for every $\phi \in \partial^n \Phi_i$ we have that $N \models (\forall^{\aleph_0}\alpha \in C)\phi(f(\alpha))$. We consider two possibilities. First, if $C' = f[C]$ is uncountable, then C' realizes $\partial^{n+1}\Phi_i$, contradicting our assumption that this type is totally unsupported in N. Otherwise there is $\alpha \in N$ such that $f^{-1}(\{\alpha\}) \cap C$ is uncountable. Therefore $N \models \phi(\alpha)$ for all $\phi \in \partial^n \Phi_i$, contradicting the assumption that $\partial^n \Phi_i$ is totally unsupported in N.

The construction clearly satisfies the requirements. We need to check that N^* does not realize any one of $j\partial^n \Phi_i$. Fix a name $\dot{x}$ for an element of N^*, $n \in \mathbb{N}$, and $i \in \mathbb{N}$. Then $\mathrm{Int}_G(\dot{x}) = [f]_G$ for some $f \in N$. Let k be such that $(f, \partial^n \Phi_i)$ appears as the kth pair. Then $A_{2k+1} \subseteq \{\alpha \in \omega_1 \mid N \models \neg\phi(f(\alpha), \vec{a})\}$ for some $\phi \in \partial^n \Phi_i$, hence $A_{2k+1} \Vdash \neg\phi(\dot{x}, j(\vec{a}))$ and therefore $\dot{x}$ does not realize $\partial^n \Phi_i$. $\square$

Let N be a model of ZFC°, let $X \subseteq \omega_1^N$ (not necessarily in N), and let $\psi(x)$ be a formula. We write

$(\mathrm{aa}\, x \in X)\psi(x)$ for 'the set of $x \in X$ such that $\neg\psi(x)$ holds (in V)

is bounded in ω_1^N,'

$(\text{aa}\,\vec{x} \in X^n)\psi(\vec{x})$ for $(\text{aa}\,x_1 \in X)(\text{aa}\,x_2 \in X)\dots(\text{aa}\,x_n \in X)\psi(\vec{x})$,

where $\vec{x}$ is an n-tuple of variables.

For a type $\Phi = \Phi(x_0, x_1, \dots)$ in N and $X \subseteq \omega_1^N$ write ($\vec{x}$ is assumed to be of appropriate length, this length being ω in the definition of $\Phi_X^{<\omega}$)

$$\Phi_X(x) = \{\phi(x,\vec{a}) \in \Phi \mid \vec{a} \in N, (\text{aa}\,x \in X)N \models \phi(x,\vec{a})\}. \qquad (*)$$
$$\Phi_X^n(\vec{x}) = \{\phi(\vec{x},\vec{a}) \in \Phi \mid \vec{a} \in N, (\text{aa}\,\vec{x} \in X^n)N \models \phi(\vec{x},\vec{a})\}.$$
$$\Phi_X^{<\omega}(\vec{x}) = \bigcup_n \Phi_X^n(\vec{x} \restriction n).$$

Assume $K \subseteq [\omega_1^N]^{<\omega}$ is in N and X is a K-cube. Then every finite set realizing $\Phi_X^{<\omega}$ is in K. Also, since we are allowing parameters from N in the definitions of Φ_X, the set $\{a \in \omega_1^N \mid a \text{ realizes } \Phi_X\}$ is in this situation automatically a K-cube. Finally, note that if $Z \in N$ realizes $\partial^1 \Phi_X$ then 'almost all' finite subsets of Z are in K.

We suppress writing parameters $\vec{a} \in N$ from now on, with the understanding that ϕ is a formula in the language extended by adding constants for all elements of the model N. The proof of the following lemma is modeled on [5, Lemma 7.3.4].

Lemma 1.9. *Assume that N is a model of ZFC° and $X \subseteq \omega_1^N$. If $\partial^d \Phi_X$ is realized in N for some $d \geq 1$ then there is an uncountable $Y \in N$ such that $\Phi_Y^{<\omega} \supseteq \Phi_X^{<\omega}$.*

In particular, if X is an uncountable K-cube and $\partial^d \Phi_X$ is realized in N for some $d \geq 1$, then in N there exists an uncountable K-cube.

Proof. For $n \in \mathbb{N}$ and $d \in \mathbb{N}$ write

$(\forall^{\aleph_0} x \in^d Z)\phi(x)$ for $(\forall^{\aleph_0} x_1 \in Z)(\forall^{\aleph_0} x_2 \in x_1)\dots(\forall^{\aleph_0} x_d \in x_{d-1})\phi(x)$,

$(\forall^{\aleph_0} \vec{x} \in Z^n)\phi(x)$ for $(\forall^{\aleph_0} x_1 \in Z)(\forall^{\aleph_0} x_2 \in Z)\dots(\forall^{\aleph_0} x_n \in Z)\phi(\vec{x})$,

$(\forall^{\aleph_0} \vec{x} \in^d Z^n)\phi(x)$ for $(\forall^{\aleph_0} x_1 \in^d Z)(\forall^{\aleph_0} x_2 \in^d Z)\dots(\forall^{\aleph_0} x_n \in^d Z)\phi(\vec{x})$.

Hence Z realizes $\partial^d \Phi$ in N if and only if $N \models (\forall^{\aleph_0} \vec{x} \in^d Z)\phi(\vec{x})$ for each $\phi \in \Phi$.

Claim. *Assume H realizes $\partial^d \Phi_X$ for some $d \geq 1$. Then for all $m \geq 0$ and $n \geq 1$ we have $(\forall \phi \in \Phi_X^{m+n})(\text{aa}\,\vec{a} \in X^m)N \models (\forall^{\aleph_0} \vec{x} \in^d H^n)\phi(\vec{a}, \vec{x})$.*

Proof. Induction on n, for all m simultaneously. Assume $n = 1$ and pick $\phi \in \Phi_X^{m+1}$ (with parameters suppressed) so that $(\text{aa}\,\vec{x} \in X^{m+n})N \models \phi(\vec{x})$.

For $\vec{y} \in X^m$ let $\psi_{\vec{y}}(x)$ be $\phi(\vec{y}, x)$. Since $(\text{aa}\, \vec{y} \in X^m)\psi_{\vec{y}}(x) \in \Phi_X(x)$, by the assumption on H we have

$$(\text{aa}\, \vec{y} \in X^m)N \models (\forall^{\aleph_0} x \in^d H)\psi_{\vec{y}}(x).$$

Now assume the assertion holds for n and fix $\phi \in \Phi_X^{m+n+1}$. By the inductive assumption, $(\text{aa}\, \vec{w} \in X^m)(\text{aa}\, z \in X)N \models (\forall^{\aleph_0} \vec{x} \in^d H^n)\phi(\vec{w}, z, \vec{x})$. Fix $\vec{w} \in X^m$ such that

$$(\text{aa}\, z \in X)N \models (\forall^{\aleph_0} \vec{x} \in^d H^n)\phi(\vec{w}, z, \vec{x}).$$

If we let $\psi_{\vec{w}}(y)$ be $(\forall^{\aleph_0} \vec{x} \in^d H^n)\phi(\vec{w}, y, \vec{x})$, then $\psi_{\vec{w}}(y) \in \Phi_X(y)$ and therefore

$$N \models (\forall^{\aleph_0} y \in^d H)(\forall^{\aleph_0} \vec{x} \in^d H)\phi(\vec{w}, y, \vec{x}),$$

equivalently $N \models (\forall^{\aleph_0} \vec{x} \in^d H^{n+1})\phi(\vec{w}, \vec{x})$. $\qquad\square$

Assume $\partial^d \Phi_X$ is realized in N for some d. By the claim, for every n and $\phi \in \Phi_X^n$ we have $M \models (\forall^{\aleph_0} \vec{x} \in^d H^n)\phi(\vec{x})$. Write $x \in^d Z$ for $x \in \bigcup \cdots \bigcup Z$, where $\bigcup$ occurs $d-1$ times, and $A \subseteq^d B$ if $A \subseteq \bigcup \cdots \bigcup B$, where $\bigcup$ occurs $d-1$ times. Note that the quantifier $(\forall^{\aleph_0} x \in^d z)$ introduced earlier agrees with these conventions.

A set $E \subseteq^d H$ is *solid* if for all $m, n \in \mathbb{N}$, every $\vec{e} \in E^m$, and every $\phi \in \Phi_X^{m+n}$ we have $M \models (\forall^{\aleph_0} \vec{x} \in^d H^n)\phi(\vec{e}, \vec{x})$. Since E is solid if and only if each of its finite subsets is solid, by Zorn's Lemma we can find a maximal solid $E \subseteq X$. We claim E is uncountable. Assume otherwise. For all $m, n \in \mathbb{N}$, $\vec{e} \in E^n$ and $\phi \in \Phi_X^{m+n}$ we have $N \models (\forall^{\aleph_0} \vec{x} \in^d H^n)\phi(\vec{e}, \vec{x})$. Since there are only countably many such quadruples $(m, n, \vec{e}, \phi)$, we can find $a \in^d H$ such that $a \notin E$ and $E \cup \{a\}$ is still solid, contradicting the maximality of E.

Let $Y \subseteq^d H$ be uncountable and solid. Then for every $n \geq 1$ and $\phi \in \Phi_X^n$ we have $N \models \phi(\vec{b})$ for all $\vec{b} \in Y^n$, therefore Y is as required. $\qquad\square$

The following consequence of Lemma 1.9 is an extension of [5, Lemma 7.3.4].

Lemma 1.10. *Assume N is a model of ZFC° and $K \in N$ is such that N models '$K \subseteq [\omega_1]^{<\omega}$ and there are no uncountable K-cubes.' If $X \subseteq \omega_1^N$ is a maximal K-cube, then Φ_X is totally unsupported in N.*

Proof. If Φ_X is realized by some $b \in N$, then $b \neq a$ for all $a \in X$ and $X \cup \{b\}$ is still a K-cube, contradicting the maximality of X. Now assume $\partial^n \Phi_X$ for some $n \geq 1$ is realized by some H. By Lemma 1.9 there is an uncountable $Y \in N$ such that every $\vec{a} \in Y^{<\omega}$ satisfies $\vec{a} \in K$, contradicting our assumption. $\qquad\square$

First Proof of Theorem 1.7. It will suffice to construct $M^* = M_{\omega_1}$ with correct ω_1 and such that for every $K \subseteq [\omega_1]^{<\omega}$ in M, if there are no uncountable K-cubes in M then there are no uncountable K-cubes in V.

Let $\langle \sigma_\alpha : \alpha < \omega_1 \rangle$ be a $\Diamond$-sequence. We recursively build an iteration

$$\langle (M_\alpha, I_\alpha), G_\beta, j_{\alpha\gamma} : \alpha \leq \gamma \leq \omega_1, \beta < \omega_1 \rangle$$

of (M, I) and a set $U \subset \omega_1$ as follows. For each $\alpha < \omega_1$, let Φ^α be Φ_{σ_α} as defined in (*) using M_α for N, and put $\alpha \in U$ if and only if Φ^α is totally unsupported in M_α. When constructing G_α, we apply Lemma 1.8 to ensure that each $j_{\beta(\alpha+1)}\Phi^\beta$ $(\beta \in U \cap (\alpha + 1))$ is totally unsupported in $M_{\alpha+1}$.

Having completed the construction of the iteration, fix $K \subset [\omega_1]^{<\omega}$ in M_{ω_1} such that in M_{ω_1} there exists no uncountable K-cube $Y \subset \omega_1$. Let X be a maximal K-cube of ω_1, i.e., such that $[X]^{<\omega} \subset K$ but $[X \cup \{\xi\}]^{<\omega} \not\subset K$ for any $\xi \in \omega_1 \setminus X$. Let $\alpha < \omega_1$ and $k \in M_\alpha$ be such that $K = j_{\alpha\omega_1}(k)$, and let $\beta \in [\alpha, \omega_1)$ be such that

(1) $\omega_1^{M_\beta} = \beta$;

(2) $\sigma_\beta = X \cap \beta$;

(3) $j_{\beta\omega_1}\Phi^\beta$ is contained in Φ_X as computed over M_{ω_1}; $(M_\beta, X \cap \beta)$ is an elementary submodel of (M_{ω_1}, X), in particular σ_β is $j_{\alpha\beta}(k)$-maximal over M_β.

Then Lemma 1.10 implies Φ_β is totally unsupported in M_β, hence $\beta \in U$. Then $j_{\beta\omega_1}\Phi^\beta$ is totally unsupported in M_{ω_1}. If $\xi \in X \setminus \beta$, then (3) implies $j_{\beta\omega_1}\Phi_{X\cap\beta}$ is realized by ξ in M_{ω_1}, a contradiction. Therefore $X \subseteq \beta$, and we conclude that there are no uncountable K-cubes in M_{ω_1}. $\qquad\square$

1.3. Second proof of Theorem 1.7

This proof is similar and uses the following notion from [9]: a subset of $[\omega_1]^{<\omega}$ is *stationary* if it contains a subset of every club subset of ω_1. More generally, given a normal uniform ideal I on ω_1 we say that a subset of $[\omega_1]^{<\omega}$ is I-positive if it contains a subset disjoint from each member of I. We also let $a < b$ mean $\sup(a) < \inf(b)$, when a and b are sets of ordinals.

Let $\langle \sigma_\delta : \delta < \omega_1 \rangle$ be a $\Diamond$-sequence. We construct an iteration

$$\langle M_\alpha, I_\alpha, G_\beta, j_{\alpha,\gamma} : \beta < \omega_1, \alpha \leq \gamma \leq \omega_1 \rangle$$

in the usual way, with the following modifications. We allow the ordinary construction to determine cofinally many members of each G_β, including the first one, and fill in the intervening steps ourselves. For each $\beta < \omega_1$, let Φ_β be the set of unary formulas with constants in M_β satisfied by every member of σ_β, and for $\gamma \in [\beta, \omega_1]$, let Φ_β^γ be $j_{\beta\gamma}\Phi_\beta$, the set of ϕ such that for some $\phi' \in \Phi_\beta$, ϕ is ϕ' with its constants replaced by their $j_{\beta\gamma}$-images.

While constructing G_β, we include a stage for each tuple (B, f, ξ) of the following type:

- B is a stationary subset of $[\omega_1^{M_\beta}]^p$ in M_β for some nonzero $p \in \omega$;
- $f \colon B \to \omega_1^{M_\beta}$ is a function in M_β with $f(b) \geq \max(b)$ for all $b \in dom(f)$;
- $\xi \leq \beta$.

When we come to the stage for a given (B, f, ξ), we have some $A \in I_\beta^+$ which we have decided to put into G_β. If A has stationary intersection with the complement of the first-coordinate projection of B, then we put this intersection in G_β. Otherwise, if possible, we find some $\phi \in \Phi_\xi^\beta$ such that the following set A' is in I_β^+.

- if $p = 1$, $A' = \{\alpha \in A \mid M_\beta \models \neg\phi(f(\alpha))\}$;
- if $p = m + 1$, A' is the set of $\alpha \in A$ for which for I_β^+-many $a \in [\omega_1^{M_\beta}]^m$, $(\alpha, a) \in B$ and $M_\beta \models \neg\phi(f(\alpha, a))$.

Then we put A' in G_β. If there is no such ϕ, we do nothing at this stage.

Now, at the end of our construction, consider some $K \subset [\omega_1]^{<\omega}$ in M^* and suppose that $X \subset \omega_1$ is an uncountable K-cube. Fix α and K' such that $K = j_{\alpha\omega_1}(K')$. We will derive a contradiction from the assumption that for no $\beta \in (\alpha, \omega_1)$ is there an uncountable $j_{\alpha\beta}(K')$-cube in M_β.

Let Φ be the set of unary formulas with constants from M that are satisfied in M by every member of X. Note then that the set of countable ordinals satisfying all the members of Φ in M is uncountable. Fix $\xi \in [\alpha, \omega_1)$ such that $\omega_1^{M_\xi} = \xi$, $\sigma_\xi = X \cap \xi$ and $\Phi_\xi^{\omega_1}$ is the set of formulas in Φ with constants in the $j_{\xi\omega_1}$-image of M_ξ.

Now suppose that $\beta \geq \xi$, $p \in \omega \setminus \{0\}$, $A \in G_\beta$, $B \in M_\beta$ is a stationary subset of $[\omega_1^{M_\beta}]^p$ with first-coordinate projection containing A modulo I_β, and $f \colon B \to \omega_1^{M_\beta}$ is a function in M_β. Then there exist a (possibly 0) $k \in \omega$, a k-tuple b contained in the critical sequence of $j_{\xi\beta}$, an I_ξ-positive $B' \subset [\omega_1^{M_\xi}]^{k+p}$ in M_ξ and a function $f' \colon B' \to \omega_1^{M_\xi}$ in M_ξ such that $B \cap [\omega_1^{M_\xi} \setminus (\max(b) + 1)]^p = \{a \mid (b \cup a) \in j_{\xi\beta}(B') \wedge \max(b) < \min(a \setminus b)\}$ and $f(a) = j_{\xi\beta}(f')(b \cup a)$ for all $a \in B \cap [\omega_1^{M_\xi} \setminus (\max(b) + 1)]^p$. By induction on k, we show, under the assumption that $f(b) \geq \max(b)$ for all $b \in dom(f)$, that there exists a $\phi \in \Phi_\xi^\beta$ such that the following set A_ϕ' is in I_β^+.

- if $p = 1$, $A_\phi' = \{\alpha \in A \mid M_\beta \models \neg\phi(f(\alpha))\}$;
- if $p = m + 1$, A_ϕ' is the set of $\alpha \in A$ for which for I_β^+-many $a \in [\omega_1^{M_\beta}]^m$, $(\alpha, a) \in B$ and $M_\beta \models \neg\phi(f(\alpha, a))$.

By our construction, this shows that $X \subset \xi$.

In the case where $k = 0$ and $p = 1$, if there is no ϕ as desired then $A \in \mathcal{P}(\omega_1)^{M_\beta} \setminus I_\beta$ and $f \in (\omega_1^{\omega_1})^{M_\beta}$ are such that A forces $[f]_{G_\beta}$ to satisfy each member of $\Phi_\xi^{\beta+1}$. For each $n \in \omega$, we show that there is a club $E_n \subset \omega_1^{M_\beta}$ in M_β such that for all increasing n-tuples $\langle \nu_0, \ldots, \nu_{n-1} \rangle$ from $A \cap C$, $\langle f(\nu_0), \ldots, f(\nu_{n-1}) \rangle$ is an increasing sequence and $\{f(\nu_0), \ldots, f(\nu_{n-1})\}$ is in $j_{\alpha\beta}(K')$." Then in M_β there exists a sequence of clubs $\langle E_n' : n < \omega \rangle$ such that each E_n' satisfies this statement for n, and their intersection is the desired set.

Note first of all that since $f(\alpha) \geq \alpha$ for all $\alpha \in dom(f)$, we may assume by shrinking if necessary that for all finite sequences $\langle \nu_0, \ldots, \nu_n \rangle$ from $A \cap C$, $\langle f(\nu_0), \ldots, f(\nu_n) \rangle$ is an increasing sequence.

For each n, by reverse (finite) induction starting at $i = n - 1$ and ending at $i = 0$ we show that the following holds for each i: for each i-tuple a from σ_ξ, M_β satisfies the sentence "there is a club subset $C \subset \omega_1$ such that for all $(n - i)$-tuples $\langle \nu_0, \ldots, \nu_{n-i-1} \rangle$ from $A \cap C$, if $\langle f(\nu_0), \ldots, f(\nu_{n-i-1}) \rangle$ is an increasing sequence above $\sup(a)$, then $a \cup \{f(\nu_0), \ldots, f(\nu_{n-i-1})\}$ is in $j_{\alpha\beta}(K')$." Since σ_ξ is a $j_{\alpha\xi}(K')$-cube, this holds for $i = n - 1$. It if holds for $i = j + 1$, then for each j-tuple a from σ_ξ there is a club set $D_a \in \mathcal{P}(\omega_1)^{M_\beta}$ such that in M_β, for each $\chi \in D_a \cap A$ there is a club $C_{a,\xi}$ such that for all $(n-i)$-tuples $\langle \nu_0, \ldots, \nu_{n-i-1} \rangle$ from $A \cap C_{a,\chi}$, if $\langle f(\chi), f(\nu_0), \ldots, f(\nu_{n-i-1}) \rangle$ is an increasing sequence above $\sup(a)$, then $a \cup \{f(\chi), f(\nu_0), \ldots, f(\nu_{n-i-1})\}$ is in $j_{\alpha\beta}(K')$. Then letting $E_n^a = D_a \cap \triangle\{C_{a,\chi} : \chi \in D \cap A\}$, we have the desired statement for i and a, and $E_n^\emptyset$ is the desired club E_n.

The case where $k = 0$ and $p = m + 1$ is similar. Suppose that for every $\phi \in \Phi_\beta$ the set A_ϕ' is nonstationary. Let B' be the set of members of B whose least members are in A. For each $n \in \omega$ we find a club $E_n \subset \omega_1^{M_\beta}$ in M_β such that for all increasing n-tuples $\langle b_0, \ldots, b_{n-1} \rangle$ from $B' \cap [E_n]^{m+1}$, $\langle f(b_0), \ldots, f(b_{n-1}) \rangle$ is an increasing sequence and $\{f(b_0), \ldots, f(b_{n-1})\}$ is in $j_{\alpha\beta}(K')$. Then M_β has an intersection of such clubs as above. Again, we may assume by shrinking if necessary that $f(b) < \alpha$ for all $\alpha \in A$ and $b \in B' \cap [\alpha]^{<\omega}$.

Again, by reverse finite induction starting at $i = n - 1$ and ending at $i = 0$ we show that the following holds for each i: for each i-tuple a from σ_ξ, M_β satisfies the sentence "there is a club subset $C \subset \omega_1$ such that for all $(n - i)$-tuples $\langle b_0, \ldots, b_{n-i-1} \rangle$ from $B' \cap \mathcal{P}(C)$, if $\langle f(b_0), \ldots, f(b_{n-i-1}) \rangle$ is an increasing sequence above $\sup(a)$, then $a \cup \{f(b_0), \ldots, f(b_{n-i-1})\}$ is in $j_{\alpha\beta}(K')$." Since σ_ξ is a $j_{\alpha\xi}(K')$-cube, this holds for $i = n - 1$. It if holds for $i = j + 1$, then for each j-tuple a from σ_ξ there is a club set $D_a \in \mathcal{P}(\omega_1)^{M_\beta}$ such that in M_β, for each $b \in B' \cap \mathcal{P}(D)$ there is a club $C_{a,b}$ such that for all $(n - i)$-tuples $\langle b_0, \ldots, b_{n-i-1} \rangle$ from $B' \cap \mathcal{P}(C_{a,b})$, if $\langle f(b), f(b_0), \ldots, f(b_{n-i-1}) \rangle$ is an increasing sequence above $\sup(a)$, then $a \cup \{f(b), f(b_0), \ldots, f(b_{n-i-1})\}$ is in $j_{\alpha\beta}(K')$. Then letting

$$E_n^a = D_a \cap \triangle\{C_{a,b} : b \in B' \cap \mathcal{P}(D)\},$$

we have the desired statement for i and a, and $E_n^\emptyset$ is the desired club E_n.

If $k = j + 1$, let $\eta = \max(b)$ and let $b^- = b \setminus \{\eta\}$. Then by our induction hypothesis there is a $\phi \in \Phi_\xi$ such that the set of $\alpha < \omega_1^{M_\eta}$ for which for stationarily many $a \in [\omega_1^{M_\eta}]^n$, $b^- \cup \{\alpha\} \cup a \in j_{\xi\eta}(B')$ and $M_\eta \models \neg\phi(j_{\xi\eta}(f)(b^- \cup \{\alpha\} \cup a))$ is in G_η. Then ϕ is as desired.

Remark. If M is a model of a sufficient fragment of ZFC which is correct about ω_1 we say that M is *correct about* NS_{ω_1} if $\mathrm{NS}_{\omega_1} \cap M = \mathrm{NS}_{\omega_1}^M$. We note that either of the above proofs of Theorem 1.3 allows one to easily add correctness about NS_{ω_1} to M^* to the conclusion. To see this, note the the proofs of Theorem 1.7 do not require putting any specific set into the generic filter at a given stage. The standard $\mathbb{P}_{max}$ bookkeeping argument then allows putting the images of each stationary subset of ω_1 in each model of the iteration into the generic filter stationarily often, thus assuring NS_{ω_1}-correctness (see the game-theoretic formulation of the basic iteration lemma for $\mathbb{P}_{max}$ in [12]).

In [19], to $S \subseteq \omega_1$ Todorcevic associates $K_S \subseteq [\omega_1]^2$ such that if there is an uncountable K-cube then S contains a club and if S contains a club then a proper forcing notion adds an uncountable K-cube. Hence one may ask whether correctness about partitions of $[\omega_1]^2$, or for partitions of $[\omega_1]^{<\omega}$, implies correctness about NS_{ω_1}. However, the above proof can easily be adapted to make M^* incorrect about NS_{ω_1}, showing that correctness about partitions of $[\omega_1]^{<\omega}$ does not imply correctness about NS_{ω_1}. To see this, take some costationary subset of ω_1 in some model and keep it and its images out of all the generic filters, thus assuring that the image of this set will be nonstationary in V even though it is stationary in the final

model. With these observations one gets the following strengthening of
Theorem 1.4.

Theorem 1.11. *Suppose that there exist proper class many Woodin cardinals, let A be a universally Baire set of reals, and let T be a set of sentences in $L(A)$. Suppose that there exists an $L(Q^{\leq \omega}, A)$-correct model of T in some set forcing extension. Then in every set forcing extension satisfying $\Diamond$ there exist $L(Q^{\leq \omega}, A)$-correct models M, M' of T such that M is correct about NS_{ω_1} and M' is not.* □

1.4. Proof of Theorem 1.3

Theorem 1.3 follows from the proof of Theorem 1.7 once we notice that the proof of correctness for partitions of $[\omega_1]^{<\omega}$ did not require iterability (i.e., we did not use the fact that the models produced were wellfounded). One could rephrase Theorem 1.7 as follows.

Theorem 1.12 ($\Diamond$). *Assume M is a countable model of a large enough fragment of ZFC. Then M has an elementary extension M^* whose ω_1 is uncountable and which is correct about partitions of $[\omega_1^{M^*}]^n$ for each $n \in \omega$. If M is an ω-model, then it has an elementary extension M^* whose ω_1 is uncountable and which is correct about partitions of $[\omega_1^{M^*}]^{<\omega}$. Moreover, M^* is correct about all Borel sets with codes in the well-founded part of M.*

Proof. The proof of this is largely the same as the proofs of Theorem 1.7. Let $\langle \sigma_\alpha : \alpha < \omega_1 \rangle$ be a $\Diamond$-sequence. We recursively build a sequence

$$\langle (M_\alpha, I_\alpha), G_\beta, j_{\alpha\gamma} : \alpha \leq \gamma \leq \omega_1, \beta < \omega_1 \rangle$$

such that each G_β an M_β-normal ultrafilter of $\mathcal{P}(\omega_1^{M_\beta})^{M_\beta}$ and $M_{\beta+1}$ is the G_β-ultrapower of M_β. We don't require the ultrafilters G_β to be generic over the models M_β.

Note that if $X \in M_\beta$ is countable in M_β and $f : \omega_1^{M_\beta} \to X$ is a function in M_β, then the M_β-normality of G_β implies that f is constant on a set in G_β. Conversely, if X is uncountable and f is injective, then f represents a new element of $j(X)$ in the ultrapower (these facts are standard; the point is just that they don't depend on wellfoundedness). It follows that elements of M_{ω_1} will have uncountable extent if and only if they are uncountable in M_{ω_1}.

One can likewise construct the iteration following the construction in the proof of Theorem 1.7. The argument goes through without change except for one point. If n is a nonstandard integer of M_β, then clearly we cannot argue by finite reverse induction on n. If the integers of M are nonstandard, then we have to settle for correctness about partitions of $[\omega_1^{M^*}]^n$ for each standard integer n. □

While M^* constructed in the above proof of Theorem 1.12 need not be well-founded, the wellfounded part of its ω_1 contains the well-founded part of ω_1^M. Therefore, assuming M is well-founded, M^* is correct about $L_{\omega_1\omega}$ sentences belonging to M. Note that the method of the proof gives proof of the following consequence of Keisler's completeness theorem for $L_{\omega_1\omega}(Q)$: For any $L_{\omega_1\omega}(Q)$ sentence ϕ the statement 'ϕ has a correct model' is forcing-absolute.

In [14] Magidor and Malitz provide an axiomatization for $L(Q^{<\omega})$ and, using $\diamondsuit$, prove the corresponding completeness theorem. Their axiomatization involves schemata of arbitrarily high complexity (and necessarily so; see [15]), Our result is purely semantic and we do not know whether there is a reasonable axiomatization for the logic of 'correctness for partitions of $[\omega_1]^{<\omega}$.' Note that we have completely avoided the problem of defining the syntax for this logic by embedding T into ZFC.

2. More on correctness for partitions of finite sets

2.1. Partitions of $[\kappa]^{<\omega}$ for $\kappa > \omega_1$

If (M, I) is an iterable pair and $j : (M, I) \to (M^*, I^*)$ is an iteration, then M^* is equal to the collection of all sets of the form $j(f)(a)$, where f is a function in M and a is a finite subset of the critical sequence corresponding to j. It follows that if M is countable and j is an iteration of length ω_1, then M^* is the union of countably many sets each having cardinality $\aleph_1$ in M^*. The results of the previous section then give the following.

Theorem 2.1 ($\diamondsuit$). *If (M, I) is an iterable pair, then there is an iteration $j : (M, I) \to (M^*, I^*)$ of length ω_1 such that M^* is correct about the existence of uncountable cubes for partitions of $[\kappa]^{<\omega}$ for every $\kappa \in M$.* □

Theorem 2.2. *Suppose that there exist proper class many Woodin cardinals, let A be a universally Baire set of reals, and let T be a set of sentences in $L(A)$. Suppose that there exists an A-correct model of T that is correct about the existence of uncountable cubes for partitions of any $[\kappa]^{<\omega}$ in some set forcing extension. Then there exists such model in every set forcing extension satisfying $\diamondsuit$.* □

Correctness about the existence of uncountable cubes for partitions of pairs implies the following.

Corollary 2.3 ($\Diamond$). *Suppose that there exist proper class many Woodin cardinals. Let T be a large enough fragment of ZFC that holds in some forcing extension. Then there is an uncountable transitive model M of T that is correct about the countable chain condition of all partial orders in M. We can also assure M is A-correct for any given universally Baire set A.*   □

Analogously to Theorem 1.3 we obtain the following.

Theorem 2.4 ($\Diamond$). *If a theory T extends a large enough fragment of ZFC and it is consistent, then there exists a model M for T such that ω_1^M is uncountable and M is correct about the existence of uncountable cubes for partitions of $[\kappa]^{<\omega}$ that belong to M for every $\kappa \in M$.*   □

2.2. $[\omega_1]^n$ vs. $[\omega_1]^{<\omega}$

By the results of §1, the existence of class many Woodin cardinals and $\Diamond$ imply the following.

$R^{<\omega}$ If A is universally Baire and ϕ is a sentence of $L(Q^{<\omega}, A)$ that has a correct model in some forcing extension, then ϕ has a correct model.

$R^{\leq\omega}$ If A is universally Baire and ϕ is a sentence of $L(Q^{\leq\omega}, A)$ that has a correct model in some forcing extension, then ϕ has a correct model.

The case of $R^{<\omega}$ ($R^{\leq\omega}$, respectively) when A is a Borel set easily follows from the method of [14] (Theorem 1.12, respectively) and it does not require large cardinals. The general case of $R^{<\omega}$ (and therefore of $R^{\leq\omega}$) requires large cardinals; this follows from [4, §8.2], where the weaker logic $L(Q, A)$ was considered.

In this section we shall show that $R^{\leq\omega}$ is a genuine strengthening of $R^{<\omega}$ already in the case when A is Borel. For a universally Baire set A consider the following two assertions.

$R^{<\omega}(A)$ If ϕ is a sentence of $L(Q^{<\omega}, A)$ that has a correct model in some forcing extension, then ϕ has a correct model.

$R^{\leq\omega}(A)$ If ϕ is a sentence of $L(Q^{\leq\omega}, A)$ that has a correct model in some forcing extension, then ϕ has a correct model.

For a collection Γ consisting of universally Baire sets of reals, $R^{<\omega}(\Gamma)$ and $R^{\leq\omega}(\Gamma)$ assert respectively that $R^{<\omega}(A)$ and $R^{\leq\omega}(A)$ hold for each $A \in \Gamma$. Along with proving that $\Diamond$ implies $R^{<\omega}(\text{Borel})$, Magidor and Malitz showed that forcing with a Cohen algebra preserves $R^{<\omega}(\text{Borel})$ ([14, p. 257]). This shows that $R^{<\omega}$(Borel) does not imply CH. Below we dwell on their ideas and further investigate in which models $R^{<\omega}(A)$ and $R^{\leq\omega}(A)$ hold.

Lemma 2.5. *Assume $R^{<\omega}$ (Borel). Then there exists a Suslin tree, a ccc-destructible (ω_1, ω_1)-gap in $P(\omega)/Fin$ and an entangled set of reals.*

Proof. This is immediate; for the definitions see e.g., [18]. □

Before stating a less trivial consequence of $R^{\leq\omega}$, let us record an immediate consequence of Lemma 1.6.

Lemma 2.6. *Assume A is universally Baire.*

(1) *If $R^{<\omega}(A)$ holds then it holds in every forcing extension by a forcing that has property K_n for all n.*

(2) *If $R^{\leq\omega}(A)$ holds then it holds in every forcing extension by a forcing that has precaliber $\aleph_1$.* □

The content of (1) of Lemma 2.7 below is in the well-known equivalence of statements 'the real line is not covered by $\aleph_1$ many Lebesgue null sets' and 'the Lebesgue measure algebra has precaliber $\aleph_1$.' We reproduce its proof for the convenience of the reader and to ensure that the former assertion's expressibility in $L_{\omega_1\omega}(Q^{\leq\omega})$ is transparent. Clause (2) is essentially given in [14, p. 257], where it was shown that $R^{<\omega}$ is preserved by the forcing for adding any number of Cohen reals. We don't state the obvious variations for $R^{<\omega}(A)$ or $R^{\leq\omega}(A)$ of two propositions below.

Proposition 2.7. (1) *Assume $R^{\leq\omega}$ (Borel). Then the real line can be covered by $\aleph_1$ Lebesgue null sets.*

(2) *Every model of ZFC has a forcing extension in which $R^{\leq\omega}$ (Borel) holds but the real line cannot be covered by $\aleph_1$ meager sets.*

(3) *Every model of ZFC has a forcing extension in which $R^{<\omega}$ (Borel) holds but the real line cannot be covered by $\aleph_1$ Lebesgue null sets.*

Proof. (1) We shall find a sentence ϕ of $L(Q^{\leq\omega},\text{Borel})$ that has a correct model if and only if the real line can be covered by $\aleph_1$ null sets.

Assume for a moment there is an increasing sequence of null G_δ sets N_α ($\alpha < \omega_1$) such that $\bigcup_{\alpha<\omega_1} N_\alpha = \mathbb{R}$. Let $F_\alpha \subseteq \mathbb{R}$ be a compact set of positive measure disjoint from N_α, and define $K \subseteq [\omega_1]^{<\omega}$ by $s \in K$ if and only if $\bigcap_{\alpha\in s} F_\alpha \neq \emptyset$. An uncountable K-cube gives a family of compact sets with a finite intersection property, and the intersection of this family is disjoint from $\bigcup_\alpha N_\alpha$. Therefore a sentence ϕ asserting enough ZFC plus 'There exist compact sets of positive measure F_α ($\alpha < \omega_1$) such that the partition K defined by $s \in K$ if and only if $\bigcap_{\alpha\in s} F_\alpha \neq \emptyset$ has no uncountable cube' has a correct model in every extension in which the real line can be covered by $\aleph_1$ many null sets. (Note that we only need correctness for a rather simple Borel set.)

We claim that the converse is also true. Assume otherwise. Let M be a model correct for ϕ and assume the real line cannot be covered by $\aleph_1$ many null sets. Let F_α ($\alpha < \omega_1$) be compact positive sets witnessing ϕ in M. By downward Löwenheim–Skolem theorem we may assume M is of size $\aleph_1$. By the ccc-ness of Lebesgue measure algebra there is a compact positive set $F \in M$ forcing that the generic filter contains uncountably many of the F_α's. Let $r \in F$ be a real that avoids all null sets coded in M. Then r is a random real over M, hence $H = \{\alpha < \omega_1 \mid r \in F_\alpha\}$ is uncountable. Then H is an uncountable cube, contradicting the assumption on M.

(2) A model of $\Diamond$ satisfies $R^{<\omega}(\text{Borel})$ by the $L_{\omega_1\omega}$ variant of Theorem 1.1, e.g., Theorem 1.12. The standard forcing for adding $\aleph_2$ Cohen reals has precaliber $\aleph_1$ and it forces that the real line cannot be covered by fewer than $\aleph_2$ meager sets . By Lemma 2.6 the extension obtained by adding Cohen reals to a model of $\Diamond$ is as required.

(3) This is similar to the proof of (2), using the well-known fact that every measure algebra has property K_n for all n.   □

The model of (3) of Proposition 2.7 gives the following.

Proposition 2.8. *Every model of ZFC has a forcing extension in which $R^{<\omega}(\text{Borel})$ holds, but $R^{\leq\omega}(\text{Borel})$ fails.*   □

3. Extensions of the Σ_1^2-absoluteness argument

Let us recall a conjecture of John R. Steel presented in [20].

Conjecture 3.1. *Assuming sufficient large cardinals, every Σ_2^2 sentence ϕ that holds in some forcing extension satisfying $\Diamond$ holds in all forcing extensions satisfying $\Diamond$.*

Since ¬CH is Σ_2^2, the requirement that $\Diamond$ holds in the forcing extension in which ϕ holds cannot be dropped in Conjecture 3.1. Note the resemblance to the following result of Woodin ([21], [11], [2]).

Theorem 3.2. *Assume there are class many measurable Woodin cardinals. Then every Σ_1^2 sentence ϕ that holds in some forcing extension holds in all forcing extensions satisfying CH.*   □

This was one of the starting points to the first part of this paper ([4]). By standard facts about Woodin cardinals ([11, Theorem 2.5.10]), Conjecture 3.1 is equivalent to its consequence stating that $\Diamond$ (together with appropriate large cardinals) implies every Σ_2^2 statement true in some forcing extension satisfying $\lozenge$. Results of §1 can be interpreted as confirmation of Conjecture 3.1 in the case when the Σ_2^2 sentence ϕ states the existence of a partition of $[\omega_1]^{<\omega}$ satisfying some first-order properties with no uncountable cubes. However, Conjecture 3.1 is not likely to be proved by iterating $\mathbb{P}_{max}$ preconditions as in §1. A major obstacle is that for each $\mathbb{P}_{max}$ precondition (N, I) there exists a real number not belonging to any of the iterates of (N, I) (take e.g., the real coding (N, I)). At this point we do not see how to prove a version of absoluteness for Magidor–Malitz logic using the stationary tower or Todorcevic's method of using a saturated ideal in a Lévy collapse of a large cardinal to $\aleph_2$ (see [2]). In this section we solve some other technical problems related to Conjecture 3.1. Assuming $\lozenge$ and using stationary tower, we find a model containing all reals and satisfying the following

(1) Correctness about the countable chain condition for partial orders on ω_1 (Theorem 3.6).
(2) Correctness about uncountable chains through (some) trees of height and cardinality ω_1 (Theorem 3.6).
(3) Containing a function on ω_1 dominating any such given function on a club (Proposition 3.10).

While both (1) and (2) are consequences of correctness for the existence of uncountable cubes for partitions of $[\omega_1]^2$, (3) cannot be obtained using $\mathbb{P}_{max}$ preconditions. The following fundamental fact ([22, 12]) about $\mathbb{P}_{max}$ iterations shows that for every iterable pair (M, I) there is a function $f\colon \omega_1 \to \omega_1$ such that for every iteration $j\colon (M, I) \to (M^*, I^*)$ of length ω_1, f dominates every member of $\omega_1^{\omega_1} \cap M^*$ on a club: if (M, I) is an iterable pair coded by a real x such that M is countable and $x^{\#}$ exists, then for every countable ordinal β and every iteration $j\colon (M, I) \to (M^*, I^*)$ of length β, the ordinal height of M^* is less than the least x-indiscernible above β. This is one of the points in Woodin's proof that the saturation of NS_{ω_1} together with the existence of $H(\aleph_2)^{\#}$ implies CH fails ([22, §3.1]).

The version for correctness about the countable chain condition was proved in [13] before the work in this paper and its predecessor. The version for trees on ω_1 is left to the reader.

3.1. The setup

Definitions of the stationary towers $\mathbb{P}_{<\delta}$ and $\mathbb{Q}_{<\delta}$ can be found e.g., in [22] or [11]. We work with the terms from Section 4 of [4]. There, $V[h]$ is a forcing extension of V, and M is a model whose ω_1 (which we also call λ) is a Woodin cardinal in $V[h]$, which sees a club $C \subset \lambda$ contained in the Woodin cardinals of $V[h]$ whose limit points β have the property that $C \cap \beta$ is contained modulo a tail in each club subset of β in $V[h]$, and such that $V_\zeta[h] \in M$ for some strongly inaccessible cardinal $\zeta > \lambda$ of $V[h]$. Inside the model M, then, one can construct $V[h]$-generics for $\mathbb{Q}_{<\lambda}^{V[h]}$. The following theorem (due to Woodin, see [11, 4]) summarizes the situation. As discussed in [4], the assumption of a measurable Woodin cardinal can be replaced with a weaker, so-called *full*, Woodin cardinal.

Theorem 3.3. *Suppose that δ is a measurable Woodin cardinal and $\kappa > \delta$ is a Woodin cardinal. Then there is a condition $a \in \mathbb{P}_{<\kappa}$ such that if $G \subset \mathbb{P}_{<\kappa}$ is a V-generic and $a \in G$, then $G \cap V_\delta$ is a V-generic filter for $\mathbb{Q}_{<\delta}$ and, letting $j \colon V \to M$ be the generic ultrapower induced by G,*

- $j(\omega_1^V) = \delta$;
- *κ is a Woodin cardinal in $V[G]$;*
- *M is closed under sequences of length less than κ in $V[G]$;*
- *there exists in M a club set $C \subset \delta$ contained in the Woodin cardinals of V such that for each limit point β of C, $(C \cap \beta) \setminus D$ is a bounded subset of β for each club $D \subset \beta$ in V.*   □

Note that in this context, since δ is strongly inaccessible in V and ω_1 is in M, in M there exist V-generic filters for each partial order in V_δ. In Theorem 3.6 we show that if $\Diamond$ holds in M then M can build the generic so that the final image model is correct about the countable chain condition for its partial orders on ω_1, and correct about whether its trees of height and cardinality ω_1 have uncountable paths. In each case the argument involves inserting cofinally many steps into the construction of each generic filter H_α (or just stationarily many), in order to ensure that a set given by a fixed $\Diamond$-sequence is not an initial segment of an uncountable path or antichain.

Lemma 3.4. *Suppose that $\delta < \lambda$ are Woodin cardinals, $G \subset \mathbb{Q}_{<\delta}$ is V-generic, $a \in \mathbb{Q}_{<\lambda}$ is such that $\mathbb{Q}_{<\delta}$ regularly embeds into the restriction $\mathbb{Q}_{<\lambda}(a)$ of $\mathbb{Q}_{<\lambda}$ to a. Let $j \colon V \to N$ be the embedding induced by G. Let $T \in N$ be a tree on ω_1^N of height ω_1^N with no uncountable branches in N. Let p be a cofinal branch of T in some outer model of $V[G]$, let b be a condition in $\mathbb{Q}_{<\lambda}(a)/\mathbb{Q}_{<\delta}$ and let f be a function in V from b to ω_1^V. Then there is a $b' \leq b$ forcing that $[f]_H$ does not extend p, where H is the induced $\mathbb{Q}_{<\lambda}$-generic.*

Proof. It is a standard fact that in this situation N and $V[G]$ agree about the existence or nonexistence of cofinal paths through T. More generally, they agree about Σ_1 sentences with parameters in $\mathcal{P}(\delta)^N$. This follows from the fact that there is in some outer model an elementary embedding with critical point above δ from N into a model containing $\mathcal{P}(\delta)^{V[G]}$; as an example of this, see the relationship between M_λ and M on page 95 of [11]. Consider then the set of nodes in T which b forces $[f]_H$ to extend. This set cannot be p, but it must be a pairwise compatible set, so it cannot contain p, either. So extend b to b' forcing that $[f]_H$ does not extend some fixed member of p. □

Lemma 3.5. *Suppose that $\delta < \lambda$ are Woodin cardinals, $G \subset \mathbb{Q}_{<\delta}$ is V-generic, $a \in \mathbb{Q}_{<\lambda}$ is such that $\mathbb{Q}_{<\delta}$ regularly embeds into the restriction $\mathbb{Q}_{<\lambda}(a)$ of $\mathbb{Q}_{<\lambda}$ to a. Let $j: V \to N$ be the embedding induced by G. Let $P \in M$ be a partial order on ω_1^N which is c.c.c. in M. Let A be a predense subset of P in some outer model of $V[G]$, let b be a condition in $\mathbb{Q}_{<\lambda}(a)/\mathbb{Q}_{<\delta}$ and let f be a function in V from b to ω_1^V. Then there is a $b' \leq b$ forcing that $[f]_H$ is compatible with some member of A, where H is the induced $\mathbb{Q}_{<\lambda}$-generic.*

Proof. By the same standard fact as in the proof of Lemma 3.4, N and $V[G]$ agree about the existence or nonexistence of uncountable antichains of P. Consider then the set X of elements of P which b forces $[f]_H$ to be incompatible with. If X does not contain A then the lemma clearly holds, so assume otherwise. In $V[G]$, and thus in M there is a countable $X' \subset X$ such that every element of P is compatible with an element of X if and only if it is compatible with an element of X'. Since A is predense and $A \subset X$, this means that X' is predense, so every element of P is compatible with some member of X'. Since X' is countable, it will continue to have this property in the $\mathbb{Q}_{<\lambda}$-ultrapower, contradicting that b forces that $[f]_H$ will be incompatible with every member of X'. □

We say that a model N is correct about the countable chain condition on partial orders on ω_1 if $\omega_1^N = \omega_1$ and for every partial order P on ω_1 in N, P has an uncountable antichain in N if and only if it has one in V. We say that a model N is correct about uncountable paths through trees of height and cardinality ω_1 if $\omega_1^N = \omega_1$ and for every tree of height and cardinality ω_1 in N, P has an uncountable branch in N if and only if it has one in V.

Theorem 3.6. *Suppose that κ is a measurable Woodin cardinal. Let A be a κ-universally Baire set of reals and let ϕ be a sentence in the language of set theory with one additional unary predicate. Then the following hold, where the models are taken to be over a language with an additional unary predicate for the interpretation of A in the corresponding model.*

(1) *Suppose that some partial order $P \in V_\kappa$ forces the existence of a model N of ϕ which is correct about uncountable paths through trees of height and cardinality ω_1. Then in every set forcing extension of V by a forcing in V_κ which satisfies $\Diamond$ there exists a model M of ϕ which is correct about uncountable paths through trees of height and cardinality ω_1.*

(2) *Suppose that some partial order $P \in V_\kappa$ forces the existence of a model N of ϕ which is correct about the ccc on partial orders on ω_1. Then in every set forcing extension of V by a forcing in V_κ which satisfies $\Diamond$ there exists a model M of ϕ which is correct about the ccc on partial orders on ω_1.*

Correctness about NS_{ω_1} can be added to conclusion of Theorem 3.6 and Theorem 3.7 below. The proof of each theorem involves adding a few steps to the construction of each H_α in the proof of the corresponding theorem in [4]. The point is that the model M from that proof constructs a collection of $V[h]$-generic filters H_α ($\alpha < \lambda$), and if at a given stage a $\Diamond$-sequence in M guesses a cofinal branch in a given tree in the current model $((V[h][H_\alpha])_\varsigma)$, Lemma 3.4 says that we can extend our construction in such a way that that branch is not extended in the extension of the tree. Similarly, if at a given stage a $\Diamond$-sequence in M guesses a maximal antichain in a given partial order in the current model, Lemma 3.5 says that we can extend our construction in such a way that that antichain is not extended in the extension of the partial order. The new elements of the construction discussed here require only cofinally many stages of the construction of each H_α, and so do not interfere with the original argument. They do not interfere with each other, either: one can combine these two arguments to obtain both correctness properties. However, they do interfere with the argument that allows the construction in M to put any given real in the model it is constructing, as adding a given real to a model requires control over the entire construction of the generic filter at that stage. If we restrict to the set of ω_1-trees, however, then we can obtain correctness about paths while picking up all the reals. The point here is that for each level of each ω_1-tree in the construction, there is only one stage where nodes on that level are created. So once Lemma 3.4 has been applied to make sure that a given path is not extended, that path can never be extended accidentally later in the construction, while picking up a given real, say. Combining this observation with the arguments from Section 4 of [4], we have the following.

Theorem 3.7. *Suppose that κ is a measurable Woodin cardinal. Let A be a κ-universally Baire set of reals and let ϕ be a sentence in the language of set theory with one additional unary predicate. Suppose that some partial order $P \in V_\kappa$ forces the existence of a model N of ϕ (with the additional symbol interpreted as A^{V^P}) which is correct about uncountable paths through ω_1-trees and which contains all the reals. Then in every set forcing extension of V by a forcing in V_κ which satisfies $\Diamond$ there exists a model M of ϕ (with the additional symbol interpreted as $A \cap M$) which is correct about uncountable paths through ω_1-trees and contains any given $\aleph_1$-many reals.*   □

The following well-known observation shows that a version of this construction which obtained correctness about uncountable paths through trees of height and cardinality ω_1 while picking up all the reals would show that $\Diamond$ decides all Σ_2^2 sentences with respect to models obtained by set forcing.

Theorem 3.8. *Suppose that M is a transitive model of ZFC + CH which contains the reals, and for every tree T of height and cardinality ω_1 in M, T has an uncountable path in M if and only if it has one in V. Suppose that M satisfies a sentence ϕ of the form $\exists A \subset \mathbb{R} \forall B \subset \mathbb{R} \psi(A, B)$, where the quantifiers of ψ range over the reals. Then ϕ holds in V.*

Proof. Let $A \subset \mathbb{R}$ be such that $\forall B \subset \mathbb{R} \psi(A, B)$ holds in M, let $\langle x_\alpha : \alpha < \omega_1 \rangle$ be a listing of the reals in M, and for each $\alpha < \omega_1$ and any set of reals X let $X \upharpoonright \alpha$ denote $X \cap \{x_\beta : \beta < \alpha\}$. Then for any $X \subset \mathbb{R}$ and any formula θ whose quantifiers range only over reals, $\theta(A, B)$ holds if and only if there is a club $C \subset \omega_1$ such that for all $\alpha \in C$, $\theta_\alpha(X \upharpoonright \alpha)$ holds, where θ_α is the formula θ with its quantifiers restricted to $\{x_\alpha : \alpha < \omega_1\}$. Since CH holds in M, there is a natural tree in M of height and cardinality ω_1 giving the initial segments of a supposed club $C \subset \omega_1$ and set $B \subset \mathbb{R}$ such that for all $\alpha \in C$, $\neg\psi_\alpha(A \upharpoonright \alpha, B \upharpoonright \alpha)$ holds. Since A witnesses ϕ in M, there is no uncountable path through this tree in M, and thus by the assumption of the theorem, there is none in V, which means that A witnesses ϕ in V.   □

As we noted above, the following theorem follows easily from Theorem 1.7, though it can be proved much more easily using the approach from Lemmas 3.4 and 3.5.

Theorem 3.9 ($\Diamond$). *If (M, I) is an iterable pair, then there is an iteration $j : (M, I) \to (M^*, I^*)$ of (M, I) such that the model M^* is correct about the ccc on partial orders on ω_1 and about the existence of uncountable paths through trees of height and cardinality ω_1.* □

Proposition 3.10. *In the situation of Theorem 3.3, assuming $\Diamond$ holds in $V[h]$ then there is a function in $V[h]$ whose image under the $\mathbb{Q}_{<\lambda}^{V[h]}$-generic embedding can be made to dominate any function from λ to λ in M on a club.*

The proof given below uses the following standard fact about the stationary tower $\mathbb{Q}_{<\lambda}$ (see [11]): for any ordinal $\gamma < \lambda$, the function on $\mathcal{P}_{\aleph_1}(\gamma)$ which takes each $X \subset \gamma$ to the ordertype of X represents γ in the generic ultrapower. The stationary set defined in Lemma 3.11 below then forces that the image of g will take the value γ at δ. This contrasts with the situation when *canonical function bounding* (see [10], for instance) holds; then, no function in $\omega_1^{\omega_1}$ can represent any ordinal above the ω_2 of the ground model.

Lemma 3.11. *Let δ be a Woodin cardinal and let $\langle \sigma_\alpha : \alpha < \omega_1 \rangle$ be a sequence witnessing that $\Diamond$ holds. Define $g : \omega_1 \to \omega_1$ by letting $g(\alpha)$ be the corresponding ordertype if σ_α codes a wellordering of α, and 0 otherwise. Then for any $\gamma > \delta$ the set of countable $X \subset V_\gamma$ such that $\mathrm{o.\,t.}(X \cap \gamma) = g(\mathrm{o.\,t.}(X \cap \delta))$ and X captures every predense subset of $\mathbb{Q}_{<\delta}$ in X is compatible with every condition in $\mathbb{Q}_{<\delta}$.*

Proof. Pick $a \in \mathbb{Q}_{<\delta}$ and $F : [V_\gamma]^{<\omega} \to V_\gamma$. By a standard argument (see Corollary 2.7.12 of [11]), there exists a continuous increasing $\subset$-chain $\langle X_\alpha : \alpha \leq \omega_1 \rangle$ of countable subsets of V_γ such that

- $X_0 \cap \cup a \in a$;
- each X_α is closed under F;
- each $X_{\alpha+1}$ end-extends X_α below δ;
- each X_α captures every predense subset of $\mathbb{Q}_{<\delta}$ in X_α.

Let $f : \omega_1 \to (X_{\omega_1} \cap \gamma)$ be a bijection, and let S be the set of $(\alpha, \beta) \in \omega_1^2$ such that $f(\alpha) \leq f(\beta)$. For club many $\alpha < \omega_1$, the ordertype of $X_\alpha \cap \delta$ is α and $f[\alpha] = X_\alpha \cap \gamma$. For some such α, σ_α codes $S \cap \alpha^2$, and this α is as desired. □

Proof of Proposition 3.10. Suppose $H \subset \mathbb{Q}_{<\eta_\alpha}$ is a $V[h]$-generic filter as in the Σ_1^2-absoluteness proof in [4]. Suppose that γ is less than $\eta_{\alpha+1}$ (which itself can be chosen to arbitrarily large below λ). Let a be the stationary set of countable subsets of V_γ given by Lemma 3.11. Then $\mathbb{Q}_{<\eta_\alpha}$ regularly embeds into the restriction of $\mathbb{Q}_{<\eta_{\alpha+1}}$ to conditions below a, and a forces that the image of g under the induced $\mathbb{Q}_{<\eta_{\alpha+1}}$-embedding will take value γ at η_α. In this way, Lemma 3.11 can be used in M to ensure that the image of the function g dominates any given function in M on a club. $\qquad\square$

Another warm-up problem towards proving Σ_2^2-absoluteness from $\Diamond$ is the question of whether the model M^* constructed in the Σ_1^2 absoluteness proof can be made to contain a sequence which is a $\Diamond$-sequence in M. If M had contained a canonical function which necessarily dominated every function in N on a stationary set then this would have shown that M^* could not contain a $\Diamond$-sequence of M. The following observation relates preserving $\Diamond$-sequences to the Σ_2^2 absoluteness problem. The idea is very similar to the proof of Theorem 3.8.

Lemma 3.12. *Suppose that M is a model of ZFC $+\ \Diamond$. Then for every Σ_2^2 sentence ϕ which holds in M there is a $\Diamond$-sequence Σ_ϕ in M such that ϕ holds in any outer model of M in which Σ_ϕ remains a $\Diamond$-sequence.*

Proof. Let $\langle \sigma_\alpha : \alpha < \omega_1^M \rangle$ be a $\Diamond$ sequence in M. Fix a Σ_2^2 sentence

$$\phi \equiv \forall X \subset \mathbb{R} \exists Y \subset \mathbb{R}\, \psi(X, Y),$$

where all quantifiers in ψ range over reals and integers, and suppose that $A \subset \mathbb{R}^M$ witnesses ϕ in M. Let $\langle a_\alpha : \alpha < \omega_1^M \rangle$ be a wellordering of $H(\omega_1)^M$ in M, and for each $\alpha < \omega_1$, let ψ_α be the formula obtained by restricting all the real quantifiers of ψ to range only over $\mathbb{R}_\alpha = \mathbb{R} \cap \{a_\beta : \beta < \alpha\}$. For each $B \in \mathcal{P}(\mathbb{R})^M$ there is a club $C \in \mathcal{P}(\omega_1)^M$ such that for all $\alpha \in C$, the structure

$$\langle \{a_\beta : \beta < \alpha\}, A \cap \mathbb{R}_\alpha, B \cap \mathbb{R}_\alpha, \in \rangle$$

is an elementary submodel of $\langle H(\omega_1), A, B, \in \rangle$. It follows that, letting Z be the set of $\alpha < \omega_1^M$ such that $\psi_\alpha(A \cap \mathbb{R}_\alpha, \sigma_\alpha)$ holds, $\Sigma_\phi = \langle \sigma_\alpha : \alpha \in Z \rangle$ is a $\Diamond$-sequence in M. The same argument shows that if Σ_ϕ remains a $\Diamond$-sequence in some outer model of M, then A witnesses ϕ in this outer model. $\qquad\square$

3.2. Trees of models

Let $S(\alpha, \beta, \gamma)$ $(\alpha, \beta, \gamma < \omega_1)$ be pairwise disjoint stationary subsets of ω_1. Let N_x $(x \in 2^{\leq \omega_1})$ be a collection of transitive models of ZFC such that $\mathcal{P}(\omega_1)^{N_x}$ is countable for each N_x, and for each pair $x \subset y$ in $2^{\leq \omega_1}$ let $j_{xy} \colon N_x \to N_y$ be an elementary embedding with critical point $\omega_1^{N_x}$. Suppose that for each $x \in 2^{\leq \omega_1}$ of limit length the model N_x is the direct limit of the models N_y $(y \subsetneq x)$ under these embeddings. For each $x \in 2^{<\omega_1}$ of limit length we let $\langle A_\alpha^x : \alpha \in dom(x) \rangle$ list the stationary, costationary subsets of $\omega_1^{N_x}$ in N_x, in such that a way that $x \subset y$ implies that $A_\alpha^y = j_{xy}(A_\alpha^x)$ for each $\alpha \in dom(x)$.

Suppose further that

- whenever $x(\gamma) = 0$ and $\omega_1^{N_x} \in S(\alpha, \beta, \gamma)$ for some β and some $\alpha < \omega_1^{N_x}$, then $\omega_1^{N_x}$ is in A_α^y for all $y \supset x$, and that
- whenever $x(\gamma) = 1$ and $\omega_1^{N_x} \in S(\alpha, \beta, \gamma)$ for some α and some $\beta < \omega_1^{N_x}$, then $\omega_1^{N_x}$ is not in A_β^y for any $y \supset x$.

Now let x, y be any two distinct elements of 2^{ω_1}, and suppose that $\gamma < \omega_1$ is such that $x(\gamma) = 0$ and $y(\gamma) = 1$. Let $C_x = \{\omega_1^{N_{x'}} : x' \subsetneq x\}$ and let $C_y = \{\omega_1^{N_{y'}} : y' \subsetneq y\}$, and let B_x and B_y be two stationary, costationary subsets of ω_1 in N_x and N_y respectively. Fix $\alpha, \beta < \omega_1$ such that $B_x = A_\alpha^x$ and $B_y = A_\beta^y$, and let η be the maximum of $\min(C_x \setminus (\alpha + 1))$ and $\min(C_y \setminus (\beta + 1))$. Then $B_x \triangle B_y$ contains $C_x \cap C_y \cap S(\alpha, \beta, \gamma) \cap (\omega_1 \setminus \eta)$, and so is stationary.

The argument just given shows that the constructions given in this section can be modified to produce a $2^{<\omega_1}$-tree of models whose paths produce models with no stationary, costationary subsets of ω_1 in common. During the construction of a $\mathbb{P}_{max}$ iteration or a sequence of stationary tower generic as in the Σ_1^2 absoluteness argument, one can take any given stationary, costationary set in the current model and choose whether to put the current ω_1 in the image of this set (for $\mathbb{P}_{max}$ this is standard, for the Σ_1^2 argument this was shown in [4]). The tree-of-models construction above is an attempt to capture the idea that if CH implies some Σ_1^2 statement ϕ (which doesn't follow from ZFC), then it implies that there are 2^{ω_1} many distinct witnesses ϕ. Undoubtedly this can be made more precise.

4. Special trees on reals

In [16], Steel shows that in the presence of large cardinals, trees on reals in $L(\mathbb{R})$ without uncountable branches in V have an absolute impediment preventing such a branch from being added by forcing. In this section we generalize this result to trees coded by arbitrary universally Baire sets, using results of Woodin on the inner model HOD (the class model consisting of all hereditarily ordinal definable sets, see [6, 8]) in place of inner model theory.

Given a tree T, we let T^+ denote the set of sequences whose proper initial segments are all in T. We think of the trees on reals in this section as sets of reals.

Theorem 4.1 ([16]). *Assume that there exist infinitely many Woodin cardinals below a measurable cardinal. Let $T \subset \mathbb{R}^{<\omega_1}$ be a tree in $L(\mathbb{R})$. Then exactly one of the following holds.*

- *There is an uncountable branch of T in V.*
- *There is a function $f \colon T^+ \to \omega^\omega$ in $L(\mathbb{R})$ such that for each $p \in T^+$, $f(p)$ codes a wellordering of ω in ordertype $dom(p)$.*

In our generalization of this result, we can prove one of the two directions in a slightly more general context than the other.

Recall that HOD_x is the class of all sets hereditarily ordinal-definable with x as a parameter. The two following theorems are due to Woodin and appear in [7].

A cone of Turing degrees is a set of the form $\{x \subset \omega \mid y \text{ is Turing-reducible to } x\}$, for some $y \subset \omega$.

Theorem 4.2. *Assume $ZF + AD$. Suppose that Y is a set and $a \in H(\omega_1)$. Then for a Turing cone of x,*

$$\mathrm{HOD}_{Y,a,[x]_{Y,a}} \models \omega_2^{\mathrm{HOD}_{Y,a,x}} \text{ is a Woodin cardinal,}$$

where $[x]_{Y,a} = \{z \in \omega^\omega \mid \mathrm{HOD}_{Y,a,z} = \mathrm{HOD}_{Y,a,x}\}$. $\qquad\square$

Given a sets Y, a, the (Y, a)-cone of reals above a given real x is the set of all reals z such that $x \in \mathrm{HOD}_{Y,a,z}$.

Theorem 4.3. *Assume $ZF+AD$. Suppose that Y is a set, $a \in H(\omega_1)$ and $\alpha < \omega_1$. Then for a (Y, a)-cone of x,*

$$\mathcal{P}(\alpha) \cap \mathrm{HOD}_{Y,a,[x]_{Y,a}} \subset \mathcal{P}(\alpha) \cap \mathrm{HOD}_{Y,a}.$$

Theorem 4.4. *Let $T \subset \mathbb{R}^{<\omega_1}$ be a tree, let S be a set of ordinals coding trees on the ordinals projecting to T and its complement, and suppose that $L(S, \mathbb{R}) \models AD$. Then at least one of the following two statements is true.*

(1) There is an uncountable branch of T in V.
(2) There is a function $f \colon T^+ \to \omega^\omega$ in $L(S, \mathbb{R})$ such that for each $p \in T^+$, $f(p)$ codes a wellordering of ω in ordertype $dom(p)$.

Furthermore, if there exists a Woodin cadinal δ and every set of reals in $L(S, \mathbb{R})$ is δ^+ weakly homogeneously Suslin in V, at least one of (1) and (2) is false.

Proof. We work in $L(S, \mathbb{R})$. First suppose that (2) fails. We show that (1) holds. Since there are wellorderings of $\mathcal{P}(\omega)^{\mathrm{HOD}_{S,p}}$ uniformly definable from p, there must be a $p \in T^+$ which is uncountable in $\mathrm{HOD}_{S,p}$. Letting Y be S, a be p and α be ω, we have from Theorems 4.3 and 4.2 that there is a real x such that p is uncountable in $M = \mathrm{HOD}_{S,p,[x]_{S,p}}$ and $\delta = \omega_2^{\mathrm{HOD}_{S,p,x}}$ is a Woodin cardinal in M.

Since δ is countable in V, we can choose an M-generic filter g for $\mathrm{Coll}(\omega_1, <\delta)^M$. Then the nonstationary ideal is presaturated in $M[g]$. Furthermore, since $S \in M$, there are trees in $M[g]$ projecting in V to T and its complement. This means that $M[g]$ is T-iterable [22, 12]. Stepping outside of $L(S, \mathbb{R})$ to a model of Choice and taking any iteration j of $M[g]$ of length ω_1, then, $j(p)$ is an uncountable member of T^+.

To see the last part of the Theorem, suppose that T and f are coded by δ^+-weakly homogeneously Suslin sets of reals, and suppose that p is an uncountable path through T. Then there is a countable elementary submodel X of some large enough initial segment of the universe containing δ, T, f and p whose transitive collapse M has the property that (letting $\bar{\delta}$ be the image of δ under the collapse), if $M[g]$ is a forcing extension of M by $\mathrm{Coll}(\omega_1, \bar{\delta})^M$, then $M[g]$ is (T, f)-iterable ([22, 12, 4]). Letting $\bar{p}$ be the image of p under the collapse, then, every forcing extension of $M[g]$ by $(\mathcal{P}(\omega_1)/NS_{\omega_1})^{M[g]}$ has $f(\bar{p})$ as an element, which means that $M[g]$ has $f(\bar{p})$ as an element, giving a contradiction, since $f(\bar{p})$ codes a wellordering of ω of the same length as $\bar{p}$, and this length is $\omega_1^{M[g]}$. $\square$

The following theorems can be used to show that if there exists a proper class of Woodin cardinals and the tree T is a weakly homogeneously Suslin set of reals, then there is a model of the form $L(S, \mathbb{R})$ satisfying AD, where S is a set of ordinals coding trees projecting to T and its complement. In this context, then, exactly one of (1) and (2) above hold.

Theorem 4.5 (Steel [17, 11]). *Suppose that there exist proper class many Woodin cardinals. Then universally Baire sets of reals have universally Baire scales.* $\square$

Theorem 4.6 (Woodin [17]). *Suppose that there exist proper class many Woodin cardinals, and let A be a weakly homogeneously Suslin set of reals. Then A is universally Baire and $L(A, \mathbb{R}) \models AD$.* $\square$

References

[1] U. Abraham and S. Shelah. A Δ_2^2 well-order of the reals and incompactness of $L(Q^{\mathrm{MM}})$. *Ann. Pure Appl. Logic*, 59(1):1–32, 1993.

[2] I. Farah. A proof of the Σ_1^2-absoluteness theorem. In S. Gao et al., editor, *Advances in Logic*, Contemporary Mathematics. Amer. Math. Soc., Rhode Island, 2007, 9–22.

[3] I. Farah, M. Hrušák, and C.A. Martínez Ranero. A countable dense homogeneous set of size $\aleph_1$. *Fundamenta Mathematicae*, 186:71–77, 2005.

[4] I. Farah and P.B. Larson. Absoluteness for universally Baire sets and the uncountable I. *Quaderni di Matematica*, 17:47–92, 2006.

[5] W. Hodges. *Building models by games*, volume 2 of *London Mathematical Society Student Texts*. Cambridge University Press, Cambridge, 1985.

[6] T. Jech. *Set Theory*. Academic Press, 1978.

[7] P. Koellner and W.H. Woodin. Large cardinals from determinacy. In M. Foreman and A. Kanamori, editors, *Handbook of Set Theory*. to appear.

[8] K. Kunen. *An Introduction to Independence Proofs*. North–Holland, 1980.

[9] P. Larson. A uniqueness theorem for iterations. *J. Symbolic Logic*, 67:1344–1350, 2002.

[10] P. Larson and S. Shelah. Bounding by canonical functions, with CH. *J. Math Logic*, 3:193–215, 2003.

[11] P.B. Larson. *The stationary tower*, volume 32 of *University Lecture Series*. American Mathematical Society, Providence, RI, 2004. Notes on a course by W. Hugh Woodin.

[12] P.B. Larson. Forcing over models of determinacy. In M. Foreman and A. Kanamori, editors, *Handbook of Set Theory*. to appear.

[13] P.B. Larson and T. Yorioka. $\mathbb{P}_{max}$ variations for destructible gaps. in preparation, 2005.

[14] M. Magidor and J. Malitz. Compact extensions of $L(Q)$. Ia. *Ann. Math. Logic*, 11(2):217–261, 1977.

[15] S. Shelah and C. Steinhorn. The nonaxiomatizability of $L(Q^2_{\aleph_1})$ by finitely many schemata. *Notre Dame J. Formal Logic*, 31(1):1–13, 1990.

[16] J. Steel. Note to Jouko Väänänen. unpublished, 2000.

[17] J. Steel. The derived model theorem. preprint, 2003.

[18] S. Todorcevic. *Partition Problems in Topology*, volume 84 of *Contemporary mathematics*. American Mathematical Society, Providence, Rhode Island, 1989.

[19] S. Todorcevic. Stationary sets and partitions of $[\omega_1]^2$. Handwritten note, May 1997.

[20] W. H. Woodin. Beyond Σ^2_1 absoluteness. In *Proceedings of the International Congress of Mathematicians, Vol. I (Beijing, 2002)*, pages 515–524, Beijing, 2002. Higher Ed. Press.

[21] W.H. Woodin. Σ^2_1-absoluteness. Handwritten note of May 1985.

[22] W.H. Woodin. *The Axiom of Determinacy, forcing axioms and the nonstationary ideal*, volume 1 of *de Gruyter Series in Logic and Its Applications*. de Gruyter, 1999.

MONADIC DEFINABILITY OF ORDINALS

Itay Neeman

Department of Mathematics
University of California Los Angeles
Los Angeles, CA 90095-1555, USA
E-mail: ineeman@math.ucla.edu

We identify precisely which singular ordinals are definable by monadic second order formulae over the ordinals, assuming knowledge of the definable regular cardinals.

In the paper Neeman [4] the author defines a class of finite state automata acting on transfinite sequences, connects these automata with monadic second order truth over the ordinals, and uses the connection to show that $\aleph_\omega$ is not definable by a monadic second order formula over $(ON; <)$, and in fact no singular cardinal is definable. This can be viewed as a "negative" result, but it turns out that the same tools can be used to produce some positive results. Here we use the connection between automata and monadic truth to show that if an ordinal $\theta > 0$ is definable then:

(1) $\mathrm{cof}(\theta)$ is definable.
(2) There are definable ordinals $\delta, \gamma < \theta$ so that $\theta = \delta + \gamma \cdot \mathrm{cof}(\theta)$.

This, and the observation that the set of definable ordinals is closed under ordinal addition and multiplication, leads to a complete characterization of the definable singular ordinals in term of the definable regular cardinals: an ordinal is definable by a monadic formula iff it can be obtained from definable regular cardinals using ordinal addition and multiplication. (The results of Magidor [3] strongly suggest that the question of which regular cardinals are definable is independent of ZFC.)

In light of the fact that the definability of θ implies the definability of $\mathrm{cof}(\theta)$, it is tempting to imagine that, for all θ, $\mathrm{cof}(\theta)$ is definable with

---

This material is based upon work supported by the National Science Foundation under Grant No. DMS-0094174.

parameter θ. Similarly, in light of the closure of the set of definable ordinals under addition and multiplication, it is tempting to imagine that $\alpha + \beta$ and $\alpha \cdot \beta$ are definable with parameters α and β. We end the paper with a precise analysis of monadic definability with parameters, showing that both these fantasies are false.

1. Preliminaries

For a function $t \colon \delta \to S$, where δ is an ordinal and S a set, define $\mathrm{cf}(t) = \{b \in S \mid \text{the set } \{\xi \mid t(\xi) = b\} \text{ is cofinal in } \delta\}$.

Definition 1.1. Let S be a set. The language $\mathcal{L}_S^*$, used to describe structures of the form $(\gamma; s, r)$ where $\gamma \in \mathrm{ON}$, $s \colon \gamma \to S$, and $r \colon \gamma \rightharpoonup S$ (partial), is the second order language generated through the following clauses:

(1) $\alpha \in A$, $s(\alpha) = b$, $r(\alpha) = b$, $b \in \mathrm{cf}(s)$, and $b \in \mathrm{cf}(s \restriction \alpha)$ are atomic formulae of $\mathcal{L}_S^*$, where α is a first order variables, A a second order variables, and b an element of S.

(2) If φ and ψ are formulae in $\mathcal{L}_S^*$ then so are $\neg \varphi$ and $(\varphi \wedge \psi)$.

(3) If φ is a formula in $\mathcal{L}_S^*$ then so is $(\exists A)\varphi$, where A is a second order variable.

(4) If φ is a formula in $\mathcal{L}_S^*$ then so are $(\forall^* \alpha < \beta)\varphi$ and $(\forall^* \alpha)\varphi$, where α and β are first order variables.

When a formula φ in the language $\mathcal{L}_S^*$ is interpreted over the structure $(\gamma; s, r)$, its first order variables range over elements of γ, and its second order variables range over subsets of γ. Truth value over $(\gamma; s, r)$ is defined in the obvious way for formulae generated through conditions (1)–(3) of Definition 1.1. As for formulae generated through condition (4): $(\gamma; s, r) \models (\forall^* \alpha < \beta)\varphi$ just in case that:

(1) β is a limit ordinal of cofinality at least ω_1, and

(2) there exists a club $C \subset \beta$ so that $(\gamma; s, r) \models \varphi[\alpha]$ for all $\alpha \in C$.

$(\gamma; s, r) \models (\forall^* \alpha)\varphi$ just in case that the same conditions hold, but with β replaced by γ.

Claim 1.2. *Let φ be a sentence in $\mathcal{L}_S^*$. Then the truth value of φ in a structure $(\gamma; s, r)$ with γ of cofinality ω (or a successor) depends only on $\mathrm{cf}(s)$.*

Definition 1.3. Two structures $(\gamma; s, r)$ and $(\gamma^*; s^*, r^*)$ are **similar**, denoted $(\gamma; s, r) \sim (\gamma^*; s^*, r^*)$, if:

(1) $\mathrm{cf}(s) = \mathrm{cf}(s^*)$.

(2) There are clubs C in γ and C^* in γ^*, and an order preserving bijection $f\colon C \to C^*$, so that $s^*(f(\xi)) = s(\xi)$ and $r^*(f(\xi)) = r(\xi)$ for all $\xi \in C$.

Claim 1.4. *Let φ be a sentence in $\mathcal{L}_S^*$. Let $(\gamma; s, r)$ and $(\gamma^*; s^*, r^*)$ be similar. Then $(\gamma; s, r) \models \varphi$ iff $(\gamma^*; s^*, r^*) \models \varphi$.*

For proofs of Claims 1.2 and 1.4 see Neeman [4]. Using Claim 1.2 define $D \models \psi$, where $D \subset S$ and ψ is a sentence of $\mathcal{L}_S^*$ to hold iff $(\gamma; s, r) \models \psi$ for some (and hence all) structures $(\gamma; s, r)$ with $\mathrm{cof}(\gamma) = \omega$ and $\mathrm{cf}(s) = D$.

Definition 1.5. Let Σ be a finite non-empty set. A Σ-**automaton** is a tuple $\mathcal{A} = \langle S, P, T, \vec{\varphi}, \Psi, h, u \rangle$ where:

(1) S and P are finite non-empty sets.
(2) $T \subset S \times \Sigma \times S$.
(3) $\vec{\varphi} = \langle \varphi_1, \ldots, \varphi_k \rangle$ is a finite tuple of sentences in $\mathcal{L}_S^*$.
(4) Ψ is a function from 2^k into S, where $k = \mathrm{lh}(\vec{\varphi})$.
(5) u is a function from S into $\{U \mid U \subsetneq P\}$.
(6) h is a function from S into P with the property that $h(b) \in P - u(b)$ for each $b \in S$.

$\mathcal{A}$ is called **deterministic** if T is a function from $S \times \Sigma$ into S, meaning that for each pair $\langle b, \sigma \rangle \in S \times \Sigma$ there is precisely one $b^* \in S$ so that $\langle b, \sigma, b^* \rangle \in T$.

We refer to Σ as the **alphabet,** to S as the set of **states** of $\mathcal{A}$, and to P as the set of **pebbles.** T is the **successor transition table.** $\vec{\varphi}$ and Ψ determine limit transitions in a way that we explain below. h and u determine the placement and maintenance of pebbles.

Definition 1.6. Let $\vec{\varphi}$ and Ψ be as in conditions (3) and (4) above. Given a domain $(\gamma; s, r)$ with $\gamma \in \mathrm{ON}$, $s\colon \gamma \to S$, and $r\colon \gamma \to S$, define $t^{\vec{\varphi}}_{(\gamma;s,r)}\colon k \to 2$ by setting $t^{\vec{\varphi}}_{(\gamma;s,r)}(i) = 1$ if $(\gamma; s, r) \models \varphi_i$ and $t^{\vec{\varphi}}_{(\gamma;s,r)}(i) = 0$ otherwise for each $i \leq k$. Define a function $\Psi \oplus \vec{\varphi}$, acting on domains $(\gamma; s, r)$ as above, by setting $(\Psi \oplus \vec{\varphi})(\gamma; s, r) = \Psi(t^{\vec{\varphi}}_{(\gamma;s,r)})$.

Remark 1.7. For $D \subset S$ set $t^{\vec{\varphi}}_D(i) = 1$ if $D \models \varphi_i$ and $t^{\vec{\varphi}}_D(i) = 0$ otherwise. Set $(\Psi \oplus \vec{\varphi})(D) = \Psi(t^{\vec{\varphi}}_D)$. For γ of cofinality ω then, $(\Psi \oplus \vec{\varphi})(\gamma; s, r) = (\Psi \oplus \vec{\varphi})(\mathrm{cf}(s))$.

Definition 1.8. Let α be an ordinal and let $X \colon \alpha \to \Sigma$. A pair $\langle s, r \rangle$ where $s \colon \alpha + 1 \to S$ and $r \colon \alpha \rightharpoonup S$ is called a **run** of $\mathcal{A}$ on X just in case that it satisfies the following conditions:

(S) $\langle s(\xi), X(\xi), s(\xi + 1) \rangle \in T$ for each $\xi < \alpha$.

(L) $s(\lambda) = (\Psi \oplus \bar{\varphi})(\lambda; s {\restriction} \lambda, r {\restriction} \lambda)$ for each limit $\lambda \leq \alpha$.

(R) If there exists some $\gamma > \xi$ so that $h(s(\xi)) \notin u(s(\gamma))$ then $r(\xi) = s(\gamma)$ for the least such γ, and otherwise $r(\xi)$ is undefined.

Condition (S) governs successor transitions, condition (L) governs limit transitions, and condition (R) determines values for r.

The Σ-automaton $\mathcal{A}$ should be viewed as running over the input $X \colon \alpha \to \Sigma$ and producing a run $\langle s, r \rangle$ through a transfinite sequence of stages. In each stage β the automaton determines $s(\beta)$ through either condition (S) or condition (L), depending on whether β is a successor or a limit. In the case of successor $\xi + 1$, the automaton determines the state $s(\xi+1)$ based on the previous state $s(\xi)$ and the input $X(\xi)$. The transition table T dictates the possible choices, as $s(\xi+1)$ must be picked to that $\langle s(\xi), X(\xi), s(\xi+1) \rangle \in T$. In the case of a limit λ, the automaton determines $s(\lambda)$ based on a bounded fragment of the almost-all theory of the run $(\lambda; s {\restriction} \lambda, r {\restriction} \lambda)$ produced so far. The fragment consulted is the restriction of the theory to the sentences in $\bar{\varphi}$. The function Ψ tells the automaton how to set the state $s(\lambda)$ based on this fragment.

Having determined $s(\beta)$, the automaton places the pebble $p = h(s(\beta))$ on the ordinal β. The pebble p remains placed on β until a later stage β^* is reached with $p \notin u(s(\beta^*))$. At the first such stage β^* the automaton removes the pebble from β, and sets $r(\beta) = s(\beta^*)$. This is expressed precisely in condition (R). $r(\beta)$ remains undefined until the pebble placed on β is removed, and may indeed remain undefined throughout, if the pebble is not removed at all during the run. The use of pebbles therefore introduces a delay into part of the construction of a run. This delay is essential in the proof of Theorems 1.10 and 1.11 below.

Notice that the value of $r {\restriction} \lambda$ known by stage λ—call it $(r {\restriction} \lambda)^{\mathrm{local}}$—is not the same as the final value $r {\restriction} \lambda$ known by the end of the run, after stage α, as there may be ordinals $\xi < \lambda$ so that the pebble $h(s(\xi))$ placed on ξ is removed at a stage $\gamma \geq \lambda$. But there may only be finitely many such ordinals, since the number of pebbles is finite and since no pebble is ever located on two ordinals at the same stage (to see this use the restriction $h(b) \notin u(b)$ in condition (6) of Definition 1.5). Thus $(r {\restriction} \lambda)^{\mathrm{local}}$ and $r {\restriction} \lambda$ may only differ on a finite set.

When reaching a limit stage λ the automaton looks at the value of $r{\restriction}\lambda$ known by stage λ, setting $s(\lambda)$ equal to $(\Psi \oplus \vec{\varphi})(\lambda; s{\restriction}\lambda, (r{\restriction}\lambda)^{\text{local}})$. This assignment satisfies condition (L) in Definition 1.8 since, by Claim 1.4 and the fact that $(r{\restriction}\lambda)^{\text{local}}$ and $r{\restriction}\lambda$ differ only on a finite set, the structures $(\lambda; s{\restriction}\lambda, (r{\restriction}\lambda)^{\text{local}})$ and $(\lambda; s{\restriction}\lambda, r{\restriction}\lambda)$ satisfy precisely the same sentences.

Remark 1.9. Coding runs and inputs by sets of ordinals, one can express the existence of a run starting at a given state b and ending at a given state b^* in the monadic second order language. To be precise, for $X\colon \theta \to \Sigma$ and $\sigma \in \Sigma$ set $A_{X,\sigma} = \{\xi \mid X(\xi) = \sigma\}$. Let $\sigma_1, \ldots, \sigma_l$ enumerate Σ. Then for every $b, b^* \in S$ there is a formula φ_{b,b^*} so that (for all θ and all $X\colon \theta \to \Sigma$) $(\theta; <) \models \varphi_{b,b^*}[A_{X,\sigma_1}, \ldots, A_{X,\sigma_l}]$ iff there is a run $\langle s, r \rangle$ of $\mathcal{A}$ on X with $s(0) = b$ and $s(\theta) = b^*$.

An **accepting condition** for an automaton $\mathcal{A}$ is a pair $\langle I, F \rangle$ where $I \in S$ and $F \subset S$. $\langle \mathcal{A}, I, F \rangle$ is said to **accept** $X\colon \alpha \to \Sigma$ just in case that there exists a run $\langle s, r \rangle$ of $\mathcal{A}$ on X so that $s(0) = I$ and $s(\alpha) \in F$. Notice that if $\mathcal{A}$ is deterministic then it has exactly one run $\langle s, r \rangle$ on X with $s(0) = I$, so that $\langle \mathcal{A}, I, F \rangle$ accepts X iff $s(\text{lh}(X)) \in F$ for s taken from this unique run.

Call $\langle \mathcal{A}, I, F \rangle$ and $\langle \mathcal{A}^*, I^*, F^* \rangle$ **equivalent** if for every ordinal α and every $X\colon \alpha \to \Sigma$, $\langle \mathcal{A}, I, F \rangle$ accepts X iff $\langle \mathcal{A}^*, I^*, F^* \rangle$ accepts X.

Theorem 1.10 (Neeman [4]). *For any automaton $\mathcal{A}$ and accepting condition $\langle I, F \rangle$, there is a deterministic automaton $\mathcal{A}^*$ with accepting condition $\langle I^*, F^* \rangle$ so that $\langle \mathcal{A}, I, F \rangle$ and $\langle \mathcal{A}^*, I^*, F^* \rangle$ are equivalent.*

Theorem 1.10 extends the work of Büchi [1] and Büchi–Zaiontz [2] to automata acting on inputs of lengths ω_2 and greater. The specific details of the definition of automata above are of course important to the proof of the theorem.

For $a \in \text{ON}$ define $\chi_{\mathsf{f}}(a)\colon \text{ON} \to 2$ through the condition $\chi_{\mathsf{f}}(a)(\gamma) = 1$ if $\gamma = a$ and $\chi_{\mathsf{f}}(a)(\gamma) = 0$ otherwise. For $a \subset \text{ON}$ define $\chi_{\mathsf{s}}(a)\colon \text{ON} \to 2$ through the condition $\chi_{\mathsf{s}}(a)(\gamma) = 1$ if $\gamma \in a$ and $\chi_{\mathsf{s}}(a)(\gamma) = 0$ otherwise. (f and s here stand for "first order" and "second order.") Given a monadic second order formula φ with free variables $x_1, \ldots, x_k$ let $\text{sig}(\varphi)\colon k \to \{\mathsf{s}, \mathsf{f}\}$ be the function defined by the condition $\text{sig}(\varphi)(i) = \mathsf{s}$ if x_i is a second order variable, and $\text{sig}(\varphi)(i) = \mathsf{f}$ if x_i is a first order variable. A sequence $\langle a_1, \ldots, a_k \rangle$ **fits the signature** of φ if a_i is an ordinal for i such that $\text{sig}(\varphi)(i) = \mathsf{f}$, and a set or class of ordinals for i such that $\text{sig}(\varphi)(i) = \mathsf{s}$. Given a sequence $\langle a_1, \ldots, a_k \rangle$ which fits the signature of

φ define $\chi(a_1, \ldots, a_k) \colon \mathrm{ON} \to 2^k$ through the condition $\chi(a_1, \ldots, a_k)(\gamma) = \langle \chi_{\mathrm{sig}(\varphi)(1)}(a_1)(\gamma), \ldots, \chi_{\mathrm{sig}(\varphi)(k)}(a_k)(\gamma) \rangle$. This is the **characteristic function** of $\langle a_1, \ldots, a_k \rangle$.

The following theorem, essentially a converse to Remark 1.9, completes the connection between monadic second order formulae over the ordinals and finite state automata. The theorem is proved by induction on the complexity of φ. The case of existential quantification is easy when working with non-deterministic automata, and the case of negation is easy when working with deterministic automata. Thus the bulk of the work in reaching Theorem 1.11 is securing Theorem 1.10, namely the equivalence between deterministic and non-deterministic automata.

Theorem 1.11 (Neeman [4]). *Let φ be a monadic second order formula in the language of order, with k free variables say. Then there is a deterministic finite state automaton $\mathcal{A}$, with accepting condition $\langle I, F \rangle$, so that: for every ordinal θ (also for $\theta = \mathrm{ON}$), and for every sequence $a_1, \ldots, a_k$ which fits the signature of φ, $(\theta; <) \models \varphi[a_1, \ldots, a_k]$ iff $\langle \mathcal{A}, I, F \rangle$ accepts $\chi(a_1, \ldots, a_k) \restriction \theta$.*

Call an ordinal θ **definable** if there is a monadic formula $\varphi(v)$ so that $(\mathrm{ON}; <) \models \varphi[\alpha]$ iff $\alpha = \theta$, and **definable with parameters** $x_1, \ldots, x_k$ if there is a formula $\varphi(v_1, \ldots, v_k, v)$ so that $(\mathrm{ON}; <) \models \varphi[x_1, \ldots, x_k, \alpha]$ iff $\alpha = \theta$.

An ordinal θ can be **pinpointed** iff there is a sentence ψ so that $(\theta; <) \models \psi$ and for all $\alpha < \theta$, $(\alpha; <) \not\models \psi$. ψ is said to **pinpoint** θ.

It is clear that ordinals which can be pinpointed are definable. Using Theorem 1.11 one can prove the converse:

Lemma 1.12 (Neeman [4]). *If θ is definable then it can be pinpointed.*

For $\alpha \in \mathrm{ON}$ let 0^α denote the input $X \colon \alpha \to \{0\}$ defined by $X(\xi) = 0$ for all ξ. If θ can be pinpointed then by Theorem 1.11 there is a deterministic automaton $\mathcal{A}$ and accepting condition $\langle I, F \rangle$ so that $\langle \mathcal{A}, I, F \rangle$ accepts 0^θ and does not accept 0^α for any $\alpha < \theta$. This makes the following claim useful in the study of monadic definability of ordinals:

Claim 1.13 (Neeman [4]). *Let $\mathcal{A}$ be a deterministic automaton and let $\langle s, r \rangle$ be a run of $\mathcal{A}$ on 0^θ. Let $D = \mathrm{cf}(s \restriction \theta)$. Let $\delta < \theta$ be least so that $\{s(\xi) \mid \delta \leq \xi < \theta\} = D$. Let γ be least so that $\{s(\xi) \mid \delta \leq \xi < \delta + \gamma\} = D$ and $s(\delta + \gamma) = s(\delta)$. Let $C = \{\alpha \in (\delta, \theta] \mid \alpha$ is closed under addition of $\gamma\}$. Then $\mathrm{cf}(s \restriction \alpha) = D$ for every $\alpha \in C$, and, for $\alpha, \beta \in C$, if $\mathrm{cof}(\alpha) = \mathrm{cof}(\beta)$ then $s(\alpha) = s(\beta)$.*

Notice that, in the notation of the last claim, θ is closed under addition of γ (else $\mathrm{cf}(s{\restriction}\theta)$ would not be equal to D). Thus it follows from the claim that either $\theta = \delta + \gamma \cdot \mathrm{cof}(\theta)$, or else there is $\alpha < \theta$, namely $\alpha = \delta + \gamma \cdot \mathrm{cof}(\theta)$, so that $s(\alpha) = s(\theta)$. If $\mathcal{A}$ and $\langle I, F \rangle$ are obtained from Theorem 1.11 using a formula which pinpoints θ, and $\langle s, r \rangle$ is an accepting run of $\langle \mathcal{A}, I, F \rangle$, then the latter is impossible. Thus:

Claim 1.14 (Neeman [4]). *Suppose that θ is definable. Then there is a deterministic automaton $\mathcal{A}$ and a run $\langle s, r \rangle$ of $\mathcal{A}$ on 0^θ so that $\theta = \delta + \gamma \cdot \mathrm{cof}(\theta)$ where δ and γ are defined from s as in Claim 1.13.*

In particular, no singular cardinal can be definable. Neeman [4] concludes with this result.

2. Definability, forward

In this section we make the simple observation that the set of definable ordinals is closed under ordinal addition and multiplication. This is true even though neither addition nor multiplication is a definable operation, and $\alpha + \beta$ need not in general be definable from α and β, as we shall see in Section 4. (Recall that definability here is in the monadic language.)

Given a monadic formula $\varphi(v_1, \ldots, v_k)$, let $\varphi^{\mathrm{rel}}(v_1, \ldots, v_k, A)$ be the formula obtained from φ by replacing all first order quantifiers $(\exists v_i)$ and $(\forall v_i)$ in φ with (the formal equivalent of) $(\exists v_i \in A)$ and $(\forall v_i \in A)$.

Claim 2.1. *For a sentence ψ, $(\mathrm{ON}; <) \models \psi^{\mathrm{rel}}[A]$ iff $(\mathrm{ot}(A); <) \models \psi$, where $\mathrm{ot}(A)$ is the order type of A.*

Proof. Fix A. Let $f \colon \mathrm{ot}(A) \to A$ be the unique order preserving bijection. Given a formula $\varphi(v_1, \ldots, v_k)$, call $x_1, \ldots, x_k$ and $x_1^*, \ldots, x_k^*$ **similar** if $x_i \in \mathrm{ot}(A)$ and $x_i^* = f(x_i)$ for i such that v_i is first order, and $x_i \subset \mathrm{ot}(A)$ and $x_i^* \cap A = f''(x_i)$ for i such that v_i is second order. Then for every formula $\varphi(v_1, \ldots, v_k)$, $(\mathrm{ot}(A); <) \models \varphi[x_1, \ldots, x_k]$ iff $(\mathrm{ON}; <) \models \varphi^{\mathrm{rel}}[x_1^*, \ldots, x_k^*, A]$ whenever $x_1, \ldots, x_k$ and $x_1^*, \ldots, x_k^*$ are similar. This statement is easily proved by induction on the complexity of φ, and the case of $k = 0$ yields the claim. $\qquad\square$

Claim 2.2. *Let α and β be ordinals. Suppose that α is definable with parameters $x_1, \ldots, x_k$ and that β is definable. Then $\alpha + \beta$ is definable with parameters $x_1, \ldots, x_k$.*

Proof. Let ψ_β pinpoint β. Let $\psi_\alpha(v_1, \ldots, v_k, v)$ witness that α is definable with parameters $x_1, \ldots, x_k$. Let $\varphi(v_1, \ldots, v_k, y)$ be the formula

$$(\exists \eta)(\exists A)(\psi_\alpha(v_1, \ldots, v_k, \eta) \wedge (\xi \in A \iff \eta \leq \xi < y) \wedge$$
$$\psi_\beta^{\mathrm{rel}}(A) \wedge (\forall u < y) \neg \psi_\beta^{\mathrm{rel}}(A \cap u))$$

Then $(\mathrm{ON}; <) \models \varphi[x_1, \ldots, x_k, \theta]$ iff $\theta = \alpha + \beta$.   □

Claim 2.3. *Let α and β be definable ordinals. Then $\alpha \cdot \beta$ is definable.*

Proof. Fix ψ_α and ψ_β which pinpoint α and β respectively. Then $\theta = \alpha \cdot \beta$ iff there exists A so that:

- A is a closed unbounded subset of $\theta + 1$ and 0 and θ are both in A.
- $\psi_\beta^{\mathrm{rel}}[A \cap \theta]$ holds, and for every $u < \theta$, $\psi_\beta^{\mathrm{rel}}[A \cap u]$ fails. So $\mathrm{ot}(A \cap \theta) = \beta$.
- $\psi_\alpha^{\mathrm{rel}}[C]$ holds whenever $C = [\zeta, \zeta^*)$ with $\zeta \in A$ and ζ^* equal to the first element of A above ζ, and $\psi_\alpha^{\mathrm{rel}}[C \cap u]$ fails for all $u < \zeta^*$. So $\mathrm{ot}(C) = \alpha$.

These conditions can be phrased in the monadic language, providing a formula that defines $\alpha \cdot \beta$.   □

An ordinal is a **multiple** of γ if it has the form $\gamma \cdot \nu$ for some ordinal ν.

Claim 2.4. *Suppose that γ is definable. Then there is a formula $\varphi(v)$ so that $(\mathrm{ON}; <) \models \varphi[\theta]$ iff θ is a multiple of γ.*

Proof. Similar to the proof of the previous claim.   □

Given ordinals γ and α define $\mathrm{trunc}_\gamma(\alpha)$ to be the largest multiple of γ which is $\leq \alpha$. To give just two examples, $\mathrm{trunc}_1(\alpha) = \alpha$ and $\mathrm{trunc}_0(\alpha) = 0$.

Claim 2.5. *Suppose that γ is definable. Then $\mathrm{trunc}_\gamma(\alpha)$ is definable with parameter α.*

Proof. Immediate from the last claim.   □

Claims 2.2 and 2.5 show that given a class $P \subset \mathrm{ON}$, every ordinal of the form $\mathrm{trunc}_\gamma(\alpha) + \beta$, where $\alpha \in P$ and γ, β are definable with no parameters, is definable with parameters from P. We shall see later that in fact these are *all* the ordinals definable with parameters from P.

3. Definability, backward

Let θ be a definable limit ordinal. Let ψ pinpoint θ. Using Theorem 1.11 fix a deterministic automaton $\mathcal{A}$ and an accepting condition $\langle I, F \rangle$ so that $\langle \mathcal{A}, I, F \rangle$ accepts 0^α iff $(\alpha; <) \models \psi$. Let $\langle s, r \rangle$ be the unique run of $\mathcal{A}$ on 0^θ with $s(0) = I$. As in Claim 1.13 set:

  (i) $D = \mathrm{cf}(s \!\restriction\! \theta)$.
  (ii) δ is least so that $\{ s(\xi) \mid \delta \leq \xi < \theta \} = D$.
  (iii) γ is least so that $\{ s(\xi) \mid \delta \leq \xi < \delta + \gamma \} = D$ and $s(\delta + \gamma) = s(\delta)$.

Let $b^* = s(\theta)$ and notice that $s(\alpha) \neq b^*$ for $\alpha < \theta$ since $\langle \mathcal{A}, I, F \rangle$ does not accept 0^α. Let $C = \{ \delta + \gamma \cdot \omega \cdot \xi \mid \xi \geq 1 \wedge \delta + \gamma \cdot \omega \cdot \xi \leq \theta \}$. θ belongs to C and by Claim 1.13 there is no $\alpha < \theta$ in C with $\mathrm{cof}(\alpha) = \mathrm{cof}(\theta)$. Hence $\theta = \delta + \gamma \cdot \mathrm{cof}(\theta)$.

Claim 3.1. δ *and* γ *are definable.*

Proof. By Remark 1.9, and using the fact that $\mathcal{A}$ is deterministic, there is for each $b \in S$ a monadic formula $\varphi_b(v)$ so that $(\mathrm{ON}; <) \models \varphi_b[\xi]$ iff $s(\xi) = b$. Conditions (ii) and (iii) can thus be phrased in the monadic language, providing a definition of γ, and a definition of δ from θ. Since θ is definable, both δ and γ are definable. $\qquad\qquad\square$

Lemma 3.2. $\mathrm{cof}(\theta)$ *is definable.*

Proof. We may assume that $\mathrm{cof}(\theta) > \omega$, since otherwise $\mathrm{cof}(\theta)$ is clearly definable. Let X denote 0^θ, so that $\langle s, r \rangle$ is a run of $\mathcal{A}$ on X. Note to begin with that for every $\alpha, \beta \in (\delta, \theta)$, if $s(\alpha) = s(\beta)$ then $r(\alpha) = r(\beta)$. In fact, if $s(\alpha) = s(\beta)$ then $s(\alpha + \xi) = s(\beta + \xi)$ for all ξ so that $\alpha + \xi, \beta + \xi < \theta$, since $\mathcal{A}$ is deterministic and $X(\alpha + \xi) = X(\beta + \xi)$. From this and the fact that $\{ s(\zeta) \mid \alpha < \zeta < \theta \} = \{ s(\zeta) \mid \beta < \zeta < \theta \}$ it follows that $r(\alpha) = r(\beta)$.

Thus there is a function $R \colon S \to S$ so that, for $\alpha \in (\delta, \theta)$, $r(\alpha) = R(s(\alpha))$.

Let $\langle S, P, T, \vec{\varphi}, \Psi, h, u \rangle$ constitute the automaton $\mathcal{A}$. For each $i < \mathrm{lh}(\vec{\varphi})$ let φ_i^* be obtained from φ_i by replacing each occurrence of $r(\xi) = b$ with $\bigvee_{\bar{b} \in R^{-1}(b)} s(\xi) = \bar{b}$. Then:

  (iv) For each limit $\lambda \in (\delta, \theta]$, $(\Psi \oplus \vec{\varphi}^*)(\lambda; s \!\restriction\! \lambda, r \!\restriction\! \lambda) = (\Psi \oplus \vec{\varphi})$
    $(\lambda; s \!\restriction\! \lambda, r \!\restriction\! \lambda)$.
  (v) The sentences in $\vec{\varphi}^*$ make no mention of r.

Let $b_1, \ldots, b_j$ enumerate the states in D. Set $T^* = \{ \langle b_i, 0, b_{i+1} \rangle \mid i < j \} \cup \{ \langle b_j, 0, b_1 \rangle \}$. Let $\mathcal{A}^*$ be the automaton $\langle S, P, T^*, \vec{\varphi}^*, \Psi, h, u \rangle$. Let $\tau = \mathrm{cof}(\theta)$ and let $\langle s^*, r^* \rangle$ be the unique run of $\mathcal{A}^*$ on 0^τ with $s^*(0) = s(\delta)$.

Define $r' : \theta \rightharpoonup S$ by setting $r'(\delta + \gamma \cdot \omega \cdot \xi) = r^*(\omega \cdot \xi)$ and leaving $r'(\zeta)$ undefined on ζ not covered by this clause.

Claim 3.3. *For every* $\xi \leq \tau$:

    (1) $s(\delta + \gamma \cdot \omega \cdot \xi) = s^*(\omega \cdot \xi)$.

    (2) *Let* $\zeta = \delta + \gamma \cdot \omega \cdot \xi$. *Then for* $\xi \geq 1$, $(\zeta; s{\restriction}\zeta, r'{\restriction}\zeta)$ *and* $(\omega \cdot \xi; s^*{\restriction}\omega \cdot \xi, r^*{\restriction}\omega \cdot \xi)$ *are similar.*

Proof. The proof is by induction on ξ. Condition (1) for $\xi = 0$ follows from the definition, as $s^*(0)$ was set equal to $s(\delta)$. Condition (1) for $\xi \geq 1$ follows from condition (2) for ξ together with conditions (iv) and (v) above and Claim 1.4. Condition (2) for ξ of cofinality $\leq \omega$ follows simply from the fact that $\mathrm{cf}(s{\restriction}\zeta) = \mathrm{cf}(s^*{\restriction}\omega \cdot \xi)$. (Both are equal to D, by Claim 1.13 in the case of $\mathrm{cf}(s{\restriction}\zeta)$ and by the definition of T^* in the case of $\mathrm{cf}(s^*{\restriction}\omega \cdot \xi)$.) For ξ of cofinality $> \omega$ condition (2) follows from the fact that $\mathrm{cf}(s{\restriction}\zeta) = \mathrm{cf}(s^*{\restriction}\omega \cdot \xi)$, and from condition (1) below ξ, which gives clubs below ζ and $\omega \cdot \xi$ on which $s{\restriction}\zeta$ and $s^*{\restriction}\omega \cdot \xi$ are equal. □

Since $s(\theta) = b^*$ and $s(\alpha) \neq b^*$ for $\alpha < \theta$, it follows from the last claim that $\tau = \mathrm{cof}(\theta)$ is least so that $s^*(\tau) = b^*$. Using Remark 1.9 this can be turned into a definition of τ in the monadic second order language. □ (Lemma 3.2)

Theorem 3.4. *Let* θ *be an ordinal. Then* θ *is definable iff* $\theta = 0$ *or the following conditions hold:*

    (1) $\mathrm{cof}(\theta)$ *is definable.*

    (2) *There are definable ordinals* $\delta, \gamma < \theta$ *so that* $\theta = \delta + \gamma \cdot \mathrm{cof}(\theta)$.

Proof. The right-to-left direction is immediate from the results in Section 2. The left-to-right direction is clear for 0 and for successor θ, and follows from Claim 3.1 and Lemma 3.2 for limit θ. □

Theorem 3.5. *An ordinal is definable iff it can be obtained from definable regular cardinals using ordinal addition and multiplication.*

Proof. Again the right-to-left direction is immediate from the results in Section 2. The left-to-right direction is proved by induction using Theorem 3.4. □

4. Parameters

Let $a_1 < \cdots < a_l$ be ordinals, and let $\theta \notin \{a_1, \ldots, a_l\}$ be a limit ordinal definable with parameters $a_1, \ldots, a_l$. Let φ be a monadic formula such that $(\text{ON}; <) \models \varphi[a_1, \ldots, a_l, \alpha]$ iff $\alpha = \theta$.

Adding a bogus parameter if needed we may assume that there is $k < l$ so that $a_k < \theta < a_{k+1}$. For $\alpha \in (a_k, a_{k+1})$ define $X_\alpha \colon \text{ON} \to 2$ through the conditions $X(a_i) = 1$ for each i, $X(\alpha) = 1$, and $X(\xi) = 0$ for $\xi \notin \{\alpha, a_1, \ldots, a_l\}$.

Using Theorem 1.11 find a deterministic automaton $\mathcal{A}$ and an accepting condition $\langle I, F \rangle$ so that, for each $\alpha \in (a_k, a_{k+1})$, $\langle \mathcal{A}, I, F \rangle$ accepts X_α iff $(\text{ON}; <) \models \varphi[a_1, \ldots, a_l, \alpha]$.

Let $\langle s, r \rangle$ be the unique run of $\mathcal{A}$ on X_θ with $s(0) = I$. Set $D = \text{cf}(s{\restriction}\theta)$, $\delta > a_k$ least so that $\{s(\xi) \mid \delta \le \xi < \theta\} = D$, and γ least so that $\{s(\xi) \mid \delta \le \xi < \delta + \gamma\} = D$ and $s(\delta + \gamma) = s(\delta)$.

Let $b^* = s(\theta)$. Let $C = \{\delta + \gamma \cdot \omega \cdot \xi \mid \xi \ge 1\}$. Note that θ is closed under addition of γ, so $\theta \in C$. Let τ be least so that $s(\delta + \gamma \cdot \omega \cdot \tau) = b^*$. By Claim 1.13, $s(\alpha) = s(\beta)$ for $\alpha, \beta \in C$ of the same cofinality, so τ is a regular cardinal, and $s(\delta + \gamma \cdot \xi) = b^*$ for all ξ of cofinality τ.

Claim 4.1. γ *is definable with no parameters,* τ *is definable with no parameters, and* δ *is definable with parameters* $a_1, \ldots, a_l$.

Proof. Similar to the proofs of Claim 3.1 and Lemma 3.2. $\square$

Claim 4.2. *Suppose that* θ *is not the first ordinal of cofinality* τ *in* C. *Then* $a_{k+1} < \theta + \gamma \cdot \tau \cdot \omega$.

Proof. Suppose otherwise. Let $\alpha = \theta + \gamma \cdot \tau$. We show that $\langle \mathcal{A}, I, F \rangle$ accepts X_α. Since $\alpha < a_{k+1}$ this implies that $(\text{ON}; <) \models \varphi[a_1, \ldots, a_l, \alpha]$, contradicting the fact that φ defines θ from $a_1, \ldots, a_l$.

Let $\langle s^*, r^* \rangle$ be the unique run of $\mathcal{A}$ on X_α with $s^*(0) = I$. Since $X_\alpha {\restriction} \theta = X_\theta {\restriction} \theta$, s and s^* are the same up to an including θ. Thus $s^*(\theta) = b^*$.

For ζ in the interval $[\theta, \alpha)$, that is the interval $[\theta, \theta + \gamma \cdot \tau)$, $X_\alpha(\zeta)$ is equal to 0. By assumption θ is not the least ordinal of cofinality τ in C, hence $\theta \ge \delta + \gamma \cdot \tau \cdot 2$, so that $X_\theta(\zeta) = 0$ for ζ in the interval $[\delta + \gamma \cdot \tau, \delta + \gamma \cdot \tau \cdot 2)$.

Thus $X_\alpha(\theta + \xi) = X_\theta(\delta + \gamma \cdot \tau + \xi)$ for all $\xi < \gamma \cdot \tau$. It follows from this, the fact that $s^*(\theta) = s(\delta + \gamma \cdot \tau)$ (both are equal to b^*), and the determinism of $\mathcal{A}$, that $s^*(\theta + \xi) = s(\delta + \gamma \cdot \tau + \xi)$ for each $\xi \le \gamma \cdot \tau$. In particular then $s^*(\theta + \gamma \cdot \tau)$ is equal to $s(\delta + \gamma \cdot \tau \cdot 2)$, and the latter is equal to b^* since $\delta + \gamma \cdot \tau \cdot 2$ is in C and of cofinality τ.

We established so far that $s^*(\alpha) = s(\theta)$ (both are equal to b^*). Note that $X_\alpha(\alpha + \xi) = X_\theta(\theta + \xi)$ for every ξ: for $\xi = 0$ both are equal to 1, for $\xi \in (0, \gamma \cdot \tau \cdot \omega)$ both are equal to 0 as $a_{l+1} \geq \theta + \gamma \cdot \tau \cdot \omega$ by assumption, and for $\xi \geq \gamma \cdot \tau \cdot \omega$, $\alpha + \xi$ is equal to $\theta + \xi$. From the fact that $X_\alpha(\alpha + \xi) = X_\theta(\theta + \xi)$ for all ξ, the fact that $s^*(\alpha) = s(\theta)$, and the determinism of $\mathcal{A}$ it follows that $s^*(\alpha + \xi) = s(\theta + \xi)$ for all ξ.

s^* and s are thus the same on a tail-end of ON. Since $\langle s, r \rangle$ is an accepting run of $\langle \mathcal{A}, I, F \rangle$, it follows that so is $\langle s^*, r^* \rangle$. $\langle \mathcal{A}, I, F \rangle$ therefore accepts X_α, meaning that $(\text{ON}; <) \models \varphi[a_1, \ldots, a_l, \alpha]$, contradicting the fact that φ defines θ. □

Claim 4.3. *Either there is $n < \omega$ so that $\theta = \delta + \gamma \cdot \tau \cdot n$, or else there is $n < \omega$ so that $\theta = \text{trunc}_{\gamma \cdot \tau \cdot \omega}(a_{k+1}) + \gamma \cdot \tau \cdot n$.*

Proof. The choice of τ above is such that θ has the form $\delta + \gamma \cdot \xi$ for some ξ of cofinality $\geq \tau$. Suppose that θ is not equal to $\delta + \gamma \cdot \tau \cdot n$ for any n. Then θ must have the form $\delta + \gamma \cdot \tau \cdot \xi$ for $\xi \geq \omega$, and since $\delta + \gamma \cdot \tau \cdot \omega$ is a multiple of $\gamma \cdot \tau$ we may drop δ, concluding that θ has the form $\gamma \cdot \tau \cdot \xi$ for some ξ.

Recall that $\text{trunc}_{\gamma \cdot \tau \cdot \omega}(a_{k+1})$ is the largest multiple of $\gamma \cdot \tau \cdot \omega$ which is $\leq a_{k+1}$. Since θ is not equal to $\delta + \gamma \cdot \tau$, it follows from the previous claim that $\theta \geq \text{trunc}_{\gamma \cdot \tau \cdot \omega}(a_{k+1})$. $\theta < a_{k+1}$ and a_{k+1} of course is smaller than $\text{trunc}_{\gamma \cdot \tau \cdot \omega}(a_{k+1}) + \gamma \cdot \tau \cdot \omega$. Thus θ belongs to the interval $[\text{trunc}_{\gamma \cdot \tau \cdot \omega}(a_{k+1}), \text{trunc}_{\gamma \cdot \tau \cdot \omega}(a_{k+1}) + \gamma \cdot \tau \cdot \omega)$. From this and the fact that θ is a multiple of $\gamma \cdot \tau$ it follows that θ has the form $\text{trunc}_{\gamma \cdot \tau \cdot \omega}(a_{k+1}) + \gamma \cdot \tau \cdot n$ for some $n < \omega$. □

Corollary 4.4. *Let θ be definable with parameters $a_1, \ldots, a_l$. Then at least one of the following conditions holds:*

(1) *There is $\delta < \theta$ definable with parameters $a_1, \ldots, a_l$, and an ordinal β definable with no parameters, so that $\theta = \delta + \beta$.*

(2) *There is $i \leq l$ and ordinals α, β definable with no parameters, so that $\theta = \text{trunc}_\alpha(a_i) + \beta$.*

Proof. If θ is a successor ordinal then condition (1) holds with $\delta = \theta - 1$. If $\theta = 0$ then condition (2) holds with $\alpha = \beta = 0$. If $\theta \in \{a_1, \ldots, a_l\}$ then condition (2) holds with $\alpha = 1$ and $\beta = 0$. Finally, for θ a limit ordinal not in $\{a_1, \ldots, a_l\}$, the corollary follows from the previous claim using the fact that γ and τ (an hence also $\gamma \cdot \tau$, $\gamma \cdot \tau \cdot n$ for each $n < \omega$, and $\gamma \cdot \tau \cdot \omega$) are definable with no parameters. □

Theorem 4.5. *Let $P \subset \text{ON}$ be a non-empty class. Then θ is definable with parameters from P iff it belongs to the class $\{\text{trunc}_\gamma(\alpha) + \beta \mid \alpha \in P$ and γ, β are definable with no parameters$\}$.*

Proof. The right-to-left direction follows from the results in Section 2. The left-to-right follows from Corollary 4.4 by induction on θ. □

Theorem 4.5 shows that very little can be gained from parameters in the case of monadic definability over $(\text{ON}; <)$. It has the following immediate consequences:

Claim 4.6. *There are ordinals α, β so that $\alpha + \beta$ is not definable with parameters α, β. There are ordinals α, β so that $\alpha \cdot \beta$ is not definable with parameters α, β. There is an ordinal α so that $\text{cof}(\alpha)$ is not definable with parameter α.*

Proof. The ordinal ω^ω is (by Theorem 3.5) a multiple of all the definable countable ordinals, and not itself definable. It follows from Theorem 4.5 that $\omega^\omega + \omega^\omega$ is not definable with parameter ω^ω (and, equivalently, $\omega^\omega \cdot 2$ is not definable with parameters ω^ω and 2).

As for the cofinality function, let τ be the first regular cardinal which is not definable. Let $\alpha = \tau + \tau$. All the definable $\gamma \leq \alpha$ are smaller than τ by Theorem 3.5, and therefore $\text{trunc}_\gamma(\alpha)$ is always either α or 0 for definable γ. By Theorem 4.5, τ is not definable with parameter α. □

Claim 4.6 is not surprising in the case of monadic definability. But notice that each of its clauses *fails* with the added condition that α and β are definable, by Claims 2.2 and 2.3 and Lemma 3.2.

References

[1] J. Richard Büchi. Decision methods in the theory of ordinals. *Bull. Amer. Math. Soc.*, 71:767–770, 1965.

[2] J. Richard Büchi and Charles Zaiontz. Deterministic automata and the monadic theory of ordinals $< \omega_2$. *Z. Math. Logik Grundlag. Math.*, 29(4):313–336, 1983.

[3] Menachem Magidor. Reflecting stationary sets. *J. Symbolic Logic*, 47(4):755–771, 1982.

[4] Itay Neeman. Finite state automata and monadic definability of singular cardinals. *J. Symbolic Logic*, 73(2): 412–438, 2008.

A CUPPABLE NON-BOUNDING DEGREE

Keng Meng Ng

School of Mathematics, Statistics and Computer Science
Victoria University of Wellington
P.O. Box 600, Wellington, New Zealand
E-mail: Keng.Meng.Ng@mcs.vuw.ac.nz

The classical examples of naturally occuring definable Turing ideals in $\mathcal{R}$ are the ideals generated by the cappable, noncuppable, and non-bounding degrees. We will provide a proof for the conjecture put forward by Nies in [18], that there is a cuppable, non-bounding r.e. degree. This implies that the ideals generated by the non-bounding and/or noncuppable degrees are new, and different from the known ones.

1. Introduction

In recent years, a major area of research in computability theory has been the study of definability in the Turing degree structure. One would be interested to ask which relations and properties of the Turing degrees are expressible in the first order language of degrees, with the partial ordering $\leq_T$. Nies, Shore and Slaman [21], Nerode, Jockusch, Simpson, Woodin and many others have contributed to this end. A particularly interesting result is the characterization of the definable relations in the r.e. degrees (similarly, in all Turing degrees) which are invariant under the double jump, by looking at their definability in first order (second order) arithmetic.

Much work has also been done on lattice-theoretic related problems of $\mathcal{R}$ (the upper semi-lattice of r.e. degrees), such as lattice embedding, the study of automorphisms, and attempts to give a full algebraic breakdown of $\mathcal{R}$. Various natural (definable) subsets in $\mathcal{R}$ arose, such as the cappable, cuppable, and promptly simple degrees. The cappable and noncappable degrees gave the first algebraic decomposition of $\mathcal{R}$, into a proper ideal and a strong filter. It soon became apparent that more attention had to be paid to how fast an element is enumerated into an r.e. set (relative to other r.e. sets), instead of whether or not an element is eventually enumerated. Classes such as the h-simple degrees, hh-simple degrees and promptly simple degrees were introduced, with the promptly simple degrees surprisingly coinciding with various other unrelated classes.

In this paper, we will construct a cuppable, non-bounding r.e. degree. This settles the conjecture by Nies in [18], and shows that the two ideals generated by the union and intersection of the non-cuppable and non-bounding ideals are new. We state this here as our main theorem:

Main Theorem. *There is an r.e. degree $a > 0$ that is cuppable, and non-bounding.*

We begin with a few definitions. A coinfinite r.e. set A is said to be *promptly simple*, if there is a recursive function p, and an enumeration $\{A_s\}_{s \in \omega}$ of A, such that for every infinite r.e. set W_e, there is some s, x where x is enumerated into W_e at stage s, and $x \in A_{p(s)}$. An r.e. degree is said to be promptly simple, if it contains a promptly simple r.e. set.

   (i) Let **M** be the set of all cappable r.e. degrees. Let **NCAP** be $\mathcal{R} \backslash \mathbf{M}$, the set of non-cappable r.e. degrees, and **ENCAP** be the set of effectively non-cappable r.e. degrees.
   (ii) Let **LCUP** be the set of all low cuppable r.e. degrees, and **NCUP** be the set of non-cuppable r.e. degrees.
   (iii) **PS** is the set of promptly simple r.e. degrees.
   (iv) **NB** is the set of non-bounding r.e. degrees, i.e. those r.e. degrees which do not bound a minimal pair.
   (v) **SPH̄** is the set of r.e. sets definable in $\mathcal{R}$ as the non-hh-simple r.e. sets with the splitting property.

Theorem 1.1.

   (i) (*Ambos-Spies et al.* [1]) **ENCAP=NCAP=LCUP=PS=SPH̄**.
   (ii) $\mathcal{R} = \mathbf{M} \cup \mathbf{NCAP}$, where **M** is a proper ideal in $\mathcal{R}$, and **NCAP** is a strong filter in $\mathcal{R}$.

As pointed out by Slaman, **NCUP** trivially forms an ideal in $\mathcal{R}$. The question thus arose as to whether there are any other naturally occuring, definable ideals in $\mathcal{R}$ (Shore [22]). Nies answered the question by proving that the ideal generated by a definable subset of $\mathcal{R}$, is itself also definable.

Thus, any naturally defined class of r.e. degrees will generate a definable ideal, the only concern being whether or not it is new.

Theorem 1.2.

(i) (*Nies* [18]) *If $D \subseteq \mathcal{R}$ is definable, then so is the ideal in $\mathcal{R}$ generated by D:*

$$[D]_{id} := \{x \in \mathcal{R} \mid \exists C \subseteq D \ \wedge \ C \text{ is finite } \wedge \ x \leq \sup C\}.$$

(ii) *There is a cappable degree not in $[\textbf{NB}]_{id}$, as well as a r.e. degree which is both noncuppable, and non-bounding.*

Theorem 1.3.

(i) (*Yang, Yu* [26]) *There is a noncuppable r.e. degree which is not below the join of finitely many non-bounding degrees.*

(ii) (*Yang, Yu* [26]) *There is a cappable degree which is not below the join of a noncuppable degree and finitely many non-bounding degrees.*

Therefore the ideals $\textbf{M}$, $\textbf{NCUP}$, and $[\textbf{NB}]_{id}$ are related in the following way :

$$\textbf{NCUP} \subsetneq \textbf{M}, \qquad [\textbf{NB}]_{id} \subsetneq \textbf{M}, \qquad \textbf{NCUP} \neq [\textbf{NB}]_{id},$$

with all three being pairwise distinct. Furthermore, we also have

$$[\textbf{NCUP} \cap \textbf{NB}]_{id} \neq \emptyset, \qquad [\textbf{NCUP} \cup \textbf{NB}]_{id} \subsetneq \textbf{M}.$$

These are the classical examples of elementarily definable Turing ideals. Even with the result by Nies in Theorem 1.2, it is still very difficult to find new examples of such ideals, since one still has the task of obtaining an elementary (lattice-theoretic) characterization of the subsets of $\mathcal{R}$. We will give examples below, of some of the other ideals identified so far.

- Chaitin [3] and Solovay [25] studied the class of K-trivial sets. Recall that a set A is K-trivial if the prefix complexity of each initial segment of A is as low as it can be. That is,

$$\forall n \ \left(K(A{\restriction}n) \leq K(n) + b \right)$$

for some constant b. This class is, in some sense, far away from the notion of being random, since a set A is Martin-Löf random iff $\forall n \ K(A{\restriction}n) \geq n - b$ for some constant b. Thus, the K-trivial sets behave in the same way as the recursive sets when we examine their complexity, yet Solovay [25], and Downey, Hirschfeldt, Nies and Stephan [9] gave constructions of a non-recursive K-trivial set. This interesting class has been studied extensively, and Chaitin [3] has shown that every K-trivial set is Δ_2^0. Nies [19] showed that the class of K-trivial sets are the same as the sets which are low for 1-randomness, and Downey, Hirschfeldt, Nies and Stephan [9] also showed that the K-trivial sets form a natural solution to Post's Problem (i.e. every K-trivial set is Turing incomplete), and also constructed a promptly simple K-trivial set.

It is also interesting to note that the r.e. K-trivial reals are closed under join [9], and downwards closed [19]. It is also easy to see that $\{e \mid \Phi_e \text{ total} \wedge \Phi_e^{\emptyset'} \text{ is } K\text{-trivial} \} \in \Sigma_3^0$, hence the ideal of the r.e. K-trivial degrees is a Σ_3^0 ideal. Furthermore, Nies [20] and Downey, Hirschfeldt [8] also showed that every non-trivial Σ_3^0 ideal in $\mathcal{R}$ is bounded above by a low$_2$ r.e. set.

- Bickford, Mills [2] defined an r.e. degree a to be deep, if for all other r.e. degrees b, we have $(a \cup b)' = b'$. The deep degrees forms a definable ideal in $\mathcal{R}$, but Lempp, Slaman [15] showed that this ideal is trivial, i.e. there is no deep degree other than 0. In [4], Cholak, Groszek and Slaman weakened the requirement for a degree to be deep, by introducing the notion of an almost deep degree. That is, instead of requiring that for every b, when b is joined with a, the jump is preserved, we will instead look at what happens if only lowness was preserved. Say that an r.e. degree a is almost deep, if for every low r.e. degree b, the join $a \cup b$ is also low. [4] also contains the construction of a non recursive, almost deep r.e. degree. This shows that the ideal of the almost deep degrees is non-trivial, and is contained in $\mathbf{M}$.

- An r.e. degree is said to be contiguous, if it contains a single r.e. wtt degree. The concept of a contiguous degree was first introduced by Ladner [13], and Ladner, Sasso [14] and used in the study of transfer techniques (i.e. results which transfer from wtt degrees to the r.e. degrees). This was taken up by Downey [5] who introduced the notion of a strongly contiguous degree (i.e. an r.e. degree with only a single wtt degree), and showed by transfer techniques, that there is an r.e. (Turing) degree with the strong anti-cupping property.

The most significant difference between $\mathcal{R}_{wtt}$ (r.e. wtt degrees) and $\mathcal{R}$ is probably the fact that $\mathcal{R}_{wtt}$, but not $\mathcal{R}$ forms a distributive upper semi-lattice. Even so, the distributivity of $\mathcal{R}_{wtt}$ transfer locally to the contiguous degrees in $\mathcal{R}$. Downey, Lempp in [11] proved that an r.e. degree a is contiguous if and only if it is locally distributive, i.e.

$$\forall a_0 \forall a_1 \forall b \; \Big((a_0 \cup a_1 = a) \wedge (b \leq a) \Rightarrow$$
$$\exists b_0 \exists b_1 \; (b_0 \cup b_1 = b \wedge b_0 \leq a_0 \wedge b_1 \leq a_1) \Big).$$

Hence, the r.e. contiguous degrees are elementarily definable in $\mathcal{R}$, and generates a definable ideal. In unpublished work, Downey and the author showed that $0'$ is the join of two contiguous degrees, and hence the generated ideal is non-proper.

Other related work includes the study of totally ω-r.e. and totally ω^ω-r.e. degrees, the array recursive degrees, and some lattice embedding results by Downey, Greenberg, Walk, and Weber. In [7], Downey, Greenberg, Walk and Weber showed that the totally ω-r.e. degrees forms a definable subset of $\mathcal{R}$ (with a non-proper generated ideal, since every contiguous degree was also totally ω-r.e.), while it is still not known if the array recursive degrees are definable. For more details, the reader is referred to [6] and [10].

2. Overview of the construction

We shall build r.e. sets A, B, such that A is non-bounding, and $K \leq_T A \oplus B$. Our notation is standard, and follows Soare [24]. We fix an effective listing $\{(\Phi_i, \Psi_i, X_i, Y_i)\}_{i \in \omega}$ of all 4–tuples where Φ_i, Ψ_i are p.r. functionals, and X_i, Y_i are r.e. sets. We will build r.e. sets A, B, C, and Turing functional Γ to satisfy the requirements

$\mathcal{P}_e : A \neq \overline{W_e}$ (A is non-recursive)

$\mathcal{R}_e : (\Phi_e^A = X_e) \wedge (\Psi_e^A = Y_e) \wedge (X_e, Y_e$ are both non-recursive)

    $\Rightarrow (\exists$ an r.e. $D_e)(D_e \leq_T X_e \wedge D_e \leq_T Y_e \wedge D_e$ is non-recursive)

    (non-bounding strategy)

$\mathcal{N}_e : \Phi_e^B \neq C$ (incompleteness of B),

and such that $\Gamma(A \oplus B) = K$.

For any r.e. set Z, we write $x \searrow Z_s$ to mean that $x \in Z_s - Z_{s-1}$. We will assume that for all x, e and s,

$$x \searrow X_{e,s} \Rightarrow \Phi_{e,s}^{A_s}(x) = 1 \qquad \text{and} \qquad x \searrow Y_{e,s} \Rightarrow \Psi_{e,s}^{A_s}(x) = 1,$$

since we are only interested in the $\mathcal{R}_e$'s where the premise holds. Fix an enumeration $\{K_s\}$ of K, in which $\forall s \left| \{x \mid x \searrow K_s\} \right| \le 1$, and $\exists^\infty s(K_s = K_{s-1})$. We also define the length agreement

$$l^{\Phi_e}(s) = \max\{x \mid (\forall y < x)(X_{e,s}(y) = \Phi_{e,s}^{A_s}(y))\},$$
$$l^{\Psi_e}(s) = \max\{x \mid (\forall y < x)(Y_{e,s}(y) = \Psi_{e,s}^{A_s}(y))\},$$
$$l^e(s) = \min\{l^{\Phi_e}(s), l^{\Psi_e}(s)\}.$$

3. Strategy of a single requirement

For the requirements $\mathcal{P}_e$ and $\mathcal{N}_e$, the individual strategy is the standard one. In the case of $\mathcal{P}_e$ (*positive requirement*), we wait for an element $x \searrow W_{e,s}$, then we will put $x \searrow A_s$, otherwise we keep x out of A. In the case of $\mathcal{N}_e$ (*negative requirement*), we wait for $\Phi_{e,s}^{B_s}(x) \downarrow= 0$, then we will put x into C, and preserve the computation $\Phi_{e,s}^{B_s}(x)$ by keeping elements out of B. At the end of every stage s in the construction, we will move one step closer to building $\Gamma(A \oplus B) = K$, by either extending the definition of Γ_s, or by correcting any wrong approximation made by Γ_s, due to changes in K (i.e. some $x \searrow K_s$). To this end, we will maintain a set of markers $\{\gamma(n,s)\}$, and $\gamma(n)$ will be enumerated into B for the sake of the correctness of $\Gamma(A \oplus B; n)$. Finally, for the requirement $\mathcal{R}_i$, we will subdivide it into infinitely many sub-requirements

$$\mathcal{R}_{i,j} : D_i \ne \overline{W_j}$$

The requirement $\mathcal{R}_i$ will construct an r.e. set D_i below X_i and Y_i, and attempt to satisfy all sub-requirements $\mathcal{R}_{i,j}$ for every j. Either it will succeed in doing so, or else some $\mathcal{R}_{i,j}$ fails, and we can make use of this to compute either X_i or Y_i. Fix a j, and we will now describe the strategy of a single sub-requirement $\mathcal{R}_{i,j}$. Let x denote the current witness that $\mathcal{R}_{i,j}$ is working on, and $r_1(s), r_2(s)$ be the restraint imposed at stage s by X_i and Y_i respectively. Let $r(s) := \max\{r_1(s), r_2(s)\}$. The action of $\mathcal{R}_{i,j}$ consists of the following steps:

(Step 1): Wait for a stage s such that $x \searrow W_{j,s}$. If no such stage exists, then $\mathcal{R}_{i,j}$ is satisfied. Otherwise, set $r_1(s+1) = 0$, and go to step 2. This action is also called *opening of an X-gap*.

(Step 2): Wait for the next stage $t > s$, where $l^{\Phi_i}(t) > x$. This will happen if the premise in $\mathcal{R}_i$ holds. If $X_{i,s}{\restriction}_x \ne X_{i,t}{\restriction}_x$, then we *close the X-gap successfully* by performing the following : Set $r_2(t+1) = 0$ (*open a Y-gap*), and go to step 3. Otherwise if $X_{i,s}{\restriction}_x = X_{i,t}{\restriction}_x$, then we *close the X-gap unsuccessfully* by defining $r_1(t+1) = t$, reset x by choosing another witness $> t + 1$, and return back to step 1.

(Step 3): Wait for the next stage $u > t$, where $l^{\Psi_i}(u) > x$. If $Y_{i,t}\lceil x \neq Y_{i,u}\lceil x$, then we *close the Y-gap successfully* by enumerating x into D_i, and then halt. This satisfies $\mathcal{R}_{i,j}$. Otherwise if $Y_{i,t}\lceil x = Y_{i,u}\lceil x$, then we *close the Y-gap unsuccessfully* by defining $r_2(u+1) = u$, reset x by choosing another witness $> u + 1$, and return back to step 1.

Note that if there are only finitely many opening of X-gaps and Y-gaps, then $\mathcal{R}_{i,j}$ is satisfied, and $\lim_s r(s) < \infty$. Furthermore, any element x that enters D_i can only do so if it obtains permission from both X_i and Y_i, and hence $D_i \leq_T X_i$ and $D_i \leq_T Y_i$. Hence $\mathcal{R}_i$ will be satisfied, if $\mathcal{R}_{i,j}$ is satisfied for all j. Suppose that there are infinitely many Y-gaps, then $\liminf_s r(s) = 0$ and for any y, $Y_{i,s}\lceil y = Y_i\lceil y$ for the least stage s such that the witness $x(s) > y$ and a Y-gap is open. Hence, Y_i is recursive. If there are finitely many Y-gaps but infinitely many X-gaps, then $\liminf_s r(s) = \lim_s r_2(s) < \infty$, and X_i is recursive. Therefore, $\mathcal{R}_i$ is satisfied even when $\mathcal{R}_{i,j}$ fails for some j.

4. Interaction of strategies

Now we look at the possible conflicts amongst the different requirements. The non-bounding strategy may impose an unbounded restraint on the nodes below it if X or Y is recursive, thus there is a need for the strategy to enter the gap stages, in which the nodes (on the true path) below it has only a finite restraint (on A) to work with. These gap stages provide an opportunity for the strategies below the node working on the non-bounding strategy to enumerate numbers into A.

The end of stage action will maintain the correctness of the functional $\Gamma(A \oplus B)$, and whenever a correction needs to be taken, we will put $\gamma(n,s)$ into B to make $\Gamma_s(A \oplus B; n) \uparrow$, in preparation for it to receive a new definition later. The problem is that if we *always* put numbers into B whenever a K-change is observed, we might end up coding K into B, and hence there is no chance for the $\mathcal{N}$-strategies (incompleteness strategies) to work. To prevent this situation, we will require both A and B to be involved in carrying the information contained in K. The important point here is that the end of stage actions taken to construct Γ has the highest priority over any strategy on the tree. So, whenever an incompleteness strategy observes a B-computation and wants to preserve it, the strategy can only try to limit the damage done to itself by the end of stage actions. This is done as follow:

Each incompleteness node $\alpha = \mathcal{N}_e$ will choose a number $n(\alpha, s)$, in which it believes that $K_s\lceil n(\alpha,s) = K\lceil n(\alpha,s)$. Whenever α witnesses $\Phi_{e,s}^{B_s}(z)\downarrow$, it will attempt a diagonalization by enumerating z into C, move all markers $\gamma(p, s)$ for $p \geq n(\alpha, s)$ to values larger than $u := $ the use of $\Phi_{e,s}^{B_s}(z)$. In order to do this, α will enumerate $\gamma(n(\alpha, s), s)$ into A, and trust that $K_s\lceil n(\alpha,s) = K\lceil n(\alpha,s)$. In future, this diagonalization will be preserved, unless the end of stage action puts in some number $< u$, which can only happen if $K_s\lceil n(\alpha,s) \neq K\lceil n(\alpha,s)$. At the next visit to α, we will give up the diagonalization attempt, and start on a new one by choosing a new witness z', and wait for $\Phi_{e,t}^{B_t}(z')\downarrow$ again. This new attempt can once again be ruined, but only if $K_t\lceil n(\alpha,s) \neq K\lceil n(\alpha,s)$. Hence, there will be at most $n(\alpha, s)$ many failed attempts.

Note that the incompleteness strategies will make enumerations into A in an attempt to guarantee their success, so that in future, the enumerations made by the end of stage action can only ruin them finitely often. Hence, for each x, the final coding location $\lim_{s\to\infty} \gamma(x, s)$ depends on *both* A and B.

5. Priority tree layout

Our requirements will be laid out on the priority tree, where the nodes are ordered by the standard ordering $<_L$:

- For a node α on the tree, $|\alpha| = 4e$, we assign the requirement $\mathcal{P}_e$, with possible outcomes $succ$ (success) $<_L w$ (waiting).
- If $|\alpha| = 4e+1$, we have the requirement $\mathcal{N}_e$ with possible outcomes $0 <_L 1 <_L \cdots$ indicating the restraint imposed by the negative requirement $\mathcal{N}_e$ on B.
- For $|\alpha| = 4e+2$, we will assign it to some $\mathcal{R}_{i(\alpha)}$ which attempts to guess if $(\Phi_{i(\alpha)}^A = X_{i(\alpha)}) \wedge (\Psi_{i(\alpha)}^A = Y_{i(\alpha)})$ holds. The outcomes are 0 (stands for infinitely many expansionary stages) $<_L 1$ (finitely many expansionary stages).
- For $|\alpha| = 4e + 3$, we assign it some sub-requirement $\mathcal{R}_{i(\alpha),j(\alpha)}$. The possible outcomes are : $succ$ (success in putting some x in $D_{i(\alpha)}$) $<_L g_2$ (infinitely many Y-gaps) $<_L g_1$ (infinitely many X-gaps) $<_L w$ (waiting for $x \searrow W_{j(\alpha)}$).

To complete the description of the priority tree, we need to define the functions i, j. We use the functions L_0, L_1 (where for all $\alpha \in dom(L_0)$, $L_0(\alpha), L_1(\alpha) \subseteq \omega$) to help us. For $|\alpha| = 2$, define $i(\alpha) = 0$, and let $L_0(\alpha) =$

$L_1(\alpha) = \omega$. If $|\alpha| = 4n$ or $4n + 1(n > 0)$, we let $i(\alpha) \uparrow, j(\alpha) \uparrow, L_0(\alpha) \uparrow$, and $L_1(\alpha) \uparrow$. If $|\alpha| = 4n + 2$ or $4n + 3$, we let β be the maximal node such that $\beta \prec \alpha$ and $|\beta| = 4n' + 2$ or $4n' + 3$ for some n'. Let $\beta^\wedge a \preceq \alpha$, and we define $L_0(\alpha), L_1(\alpha)$ as follow:

- $a \in \{1, g_1, g_2\}$:

$$L_0(\alpha) = \Big(L_0(\beta) \setminus \{i(\beta)\}\Big) \cup \Big\{k \mid k > i(\beta)\Big\},$$

$$L_1(\alpha) = \Big(L_1(\beta) \setminus \{\langle i(\beta), m \rangle \mid m \in \omega\}\Big) \cup \Big\{\langle k, m \rangle \mid k > i(\beta), m \in \omega\Big\}.$$

- $a = 0$:

$$L_0(\alpha) = L_0(\beta) \setminus \{i(\beta)\},$$
$$L_1(\alpha) = L_1(\beta).$$

- $a \in \{succ, w\}$:

$$L_0(\alpha) = L_0(\beta),$$
$$L_1(\alpha) = L_1(\beta) \setminus \{\langle i(\beta), j(\beta) \rangle\}.$$

If $|\alpha| = 4n + 2$, then define $i(\alpha) = \min L_0(\alpha)$ and $j(\alpha) \uparrow$. If $|\alpha| = 4n + 3$, then define $i(\alpha) = i', j(\alpha) = j'$, where $\langle i', j' \rangle = \min L_1(\alpha)$.

For convenience of notation, we will say that a node $\alpha = \mathcal{P}_e$ if $|\alpha| = 4e$, and $\alpha = \mathcal{N}_e$ if $|\alpha| = 4e + 1$. Similarly, we say that $\alpha = \mathcal{R}_i$ if $|\alpha| = 4e + 2$ for some e, and $i(\alpha) = i$. Also, say that $\alpha = \mathcal{R}_{i,j}$ if $|\alpha| = 4e + 3$ for some e, $i(\alpha) = i$, and $j(\alpha) = j$. We say that $\alpha \neq \mathcal{P}_e$ (similarly for $\mathcal{N}_e, \mathcal{R}_i, \mathcal{R}_{i,j}$), if $\forall k(|\alpha| \neq 4k)$. We let

$$\tau(\alpha) = (\mu\beta \preceq \alpha)\Big(\beta = \mathcal{R}_{i(\alpha)} \wedge$$

$$(\neg\exists\gamma)(\beta \preceq \gamma \prec \alpha)\Big[i(\gamma) < i(\alpha) \ \wedge \ \alpha(|\gamma|) \in \{1, g_1, g_2\}\Big]\Big),$$

and if no such β exists, let $\tau(\alpha) \uparrow$. For any infinite path h, and $i \in \omega$, let $\tau(h, i) = $ maximal node α such that $\alpha \prec h$ and $\alpha = \mathcal{R}_i$, and let $E(h, i) = \{\beta \mid \beta \succeq \tau(h, i) \ \wedge \ \tau(\beta) \downarrow = \tau(h, i)\}$.

It follows by a simple induction that for every infinite path h and every $i \in \omega$, $(\exists^{<\infty}\alpha \prec h)(\alpha = \mathcal{R}_i)$. Thus it follows that either $\forall n(h(n) \neq \mathcal{R}_i)$, or else $\tau(h, i) \downarrow$. If $\alpha = \mathcal{R}_j$, then $\tau(\alpha) = \alpha$, and it follows that $\alpha = \mathcal{R}_j \ \wedge \ \alpha \in E(h, i) \Rightarrow \alpha = \tau(h, i)$.

6. The construction

We shall construct r.e. sets A, B, C and Turing functional Γ, with A_s, B_s, C_s, Γ_s to denote the finite sets of elements enumerated so far at stage s. For a node α on the tree, let $\langle \alpha \rangle$ be the number assigned to α under some effective coding of the tree. $\gamma(x, s)$ will denote the approximation at stage s, of the final use $u(A \oplus B; x)$, and the current state of the module α is denoted by $F(\alpha, s)$. We will now state down the rest of the parameters used in the construction.

- For a node $\alpha = \mathcal{P}_e$, let $r(\alpha, s) = 0$ (restraint on A contributed by α).
- For each node $\alpha = \mathcal{N}_e$, $n(\alpha, s)$ will mark the location of x in which α believes that $K_s \restriction_x = K \restriction_x$, and $z(\alpha, s)$ denotes the witness z at stage s that attempts to make $\Phi_e^B(z) \neq C(z)$. We let $r(\alpha, s) = 0$ (since $\mathcal{N}_e$ only keeps elements out of B.
- At each node $\alpha = \mathcal{R}_i$, the r.e. set D_α will be formed by elements contributed by the nodes $\{\beta \mid \beta \succeq \alpha \ \wedge \ \tau(\beta) \downarrow = \alpha\}$. We always define $r(\alpha, s) = 0$.
- For a node $\alpha = \mathcal{R}_{i,j}$, we let $r_1(\alpha, s)$ and $r_2(\alpha, s)$ denote the restraint put up at stage s by X_i and Y_i respectively, in the attempt to make X_i and Y_i computable. We let $r(\alpha, s) = \max\{r_1(\alpha, s), r_2(\alpha, s)\}$, and let $x(\alpha, s)$ be the current witness of the basic module $\mathcal{R}_{i,j}$.

All parameters will remain in force until re-assigned (or initialized). Hence, we may drop s from the notation without ambiguity, and refer to the parameters as $\gamma(x), p(\alpha), n(\alpha), A, B, C$, etc. Define the restraint function $\bar{r}(\alpha, s) = \max\{r(\beta, s) \mid \beta \leq_L \alpha\}$.

To *reset the witness* $x(\alpha)$ (and $z(\alpha)$) at stage s, is to re-define $x(\alpha, s)$ (similarly $z(\alpha, s)$) to be the least $x \in \omega^{[\alpha]}, x > s$, and $x > x(\alpha, s - 1)$.

To *reset* $\gamma(x)$ *above* y at stage s is to do the following. Cancel the existing value of $\gamma(x)$, and redefine $\gamma(x)$ to be the least value in $\omega^{[x]}$ larger than $y, \max A_s, \max B_s$ and all the previous values of $\gamma(x)$. Next, for each $z > x$, cancel the existing value of $\gamma(z)$ and redefine $\gamma(z)$ to be the least value in $\omega^{[z]}$ larger than $\max\{y, \max A_s, \max B_s, \gamma(x), \cdots, \gamma(z - 1)\}$ and all previous values of $\gamma(z)$.

To *initialize a node* α at stage s means the following. If $\alpha = \mathcal{P}_e$ and $s = 0$, we set $F(\alpha) = w$. If $\alpha = \mathcal{N}_e$, we set $n(\alpha)$ to be the least $x > n(\alpha, s - 1)$, $x \in \omega^{[\alpha]}$, such that $\gamma(x) > \bar{r}(\alpha)$, reset $z(\alpha)$, and set $F(\alpha) = 0$. For $\alpha = \mathcal{R}_i$, we will remove any link with top α. Lastly, if $\alpha = \mathcal{R}_{i,j}$, we will reset $x(\alpha)$, set $r_1(\alpha) = r_2(\alpha) = 0$, remove any links with bottom α, and if $F(\alpha) \neq succ$ we set $F(\alpha) = w$.

The construction will proceed by induction on stage s. If $s = 0$, we initialize all nodes on the tree, reset $\gamma(0)$ above 0, and do nothing else. For each stage $s > 0$, we will define $\delta_{s,t}$ at each substage $t < s$, and state the action of the node $\delta_{s,t}$. We will have $\delta_{s,0} \prec \cdots \prec \delta_{s,s-1} = \delta_s$. We sometimes refer to a substage t of a stage s by (s,t), and order the substages (s,t) lexicographically. We will say that a node α is *visited* at substage (s,t), if $\delta_{s,t} = \alpha$, and that α is visited at stage s, if it is visited at some substage of s. For a node $\alpha = \mathcal{R}_i$, we say that a stage $s > 0$ is α-*expansionary*, if $l^i(s) > \max\{l^i(t) \mid t < s \wedge \alpha$ is visited at stage $t\}$. Let $s > 0$, and proceed with the construction as follow.

(*Substage $t = 0$*): Define $\delta_{s,0} = \mathcal{P}_0$, and check if $F(\mathcal{P}_0) = w$, and there is some $y > \gamma(\langle \mathcal{P}_0 \rangle)$ such that $y \in W_0 \cap \omega^{[\mathcal{P}_0]}$. If there is such y, enumerate the least such into A, initialize all $\beta >_L \mathcal{P}_0$, set $F(\mathcal{P}_0) = succ$, reset $\gamma(z)$ (where z is the least such that $y \leq \gamma(z)$) above 0, and go to the next substage. If there is no such y, or $F(\alpha) = succ$, proceed to the next substage with no action needed.

(*Substage $0 < t < s$*): Assume that $\delta_{s,t-1}$ has been defined, and its action taken. Define $\alpha = \delta_{s,t}$ by the following : if $\delta_{s,t-1} = \mathcal{R}_i, F(\delta_{s,t-1}) = 0$, and there exists a link $(\delta_{s,t-1}, \beta)$ for some β, then let $\alpha = \beta$, otherwise let $\alpha = \delta_{s,t-1} {}^\wedge F(\delta_{s,t-1})$. The corresponding action to be taken by α is listed below.

($\alpha = \mathcal{P}_e$): If $F(\alpha) = w$, and there is some y such that $y \in W_e \cap \omega^{[\alpha]}$ with $y > \max\{\bar{r}(\alpha), \gamma(\langle \alpha \rangle)\}$, enumerate the least such y into A, initialize all $\beta >_L \alpha$, set $F(\alpha) = succ$, reset $\gamma(z)$ (where z is the least such that $y \leq \gamma(z)$) above 0, and go to the next substage. Otherwise, go to the next substage with no action needed.

($\alpha = \mathcal{N}_e$): There are four possibilities, and their corresponding actions are listed below.

($\mathcal{N}.1$): If $F(\alpha) = 0$ and $\Phi_{e,s}^{B_s}(z(\alpha)) \downarrow = 0$, then we will perform the following actions: enumerate $z(\alpha)$ into C, make $\Gamma_s(A_s \oplus B_s; y) \uparrow$ (for all $y \geq n(\alpha)$) by enumerating $\gamma(n(\alpha))$ into A, reset $\gamma(n(\alpha))$ above u, where u is the use of $\Phi_{e,s}^{B_s}(z(\alpha))$, set $F(\alpha) = u$, and initialize all $\beta >_L \alpha$.

($\mathcal{N}.2$): If we have $F(\alpha) = 0$ and $\Phi_{e,s}^{B_s}(z(\alpha)) \neq 0$, we proceed to the next substage with no action.

($\mathcal{N}$.3): Suppose that $F(\alpha) > 0$, and there is some $y < n(\alpha)$ such that $y \searrow K_{u'}$ for some $u \le u' < s$, and u is the previous stage in which α is visited. Then, reset $z(\alpha)$ and set $F(\alpha) = 0$.

($\mathcal{N}$.4): Suppose that $F(\alpha) > 0$, and $K_{u-1}\!\restriction_{n(\alpha)} = K_{s-1}\!\restriction_{n(\alpha)}$. Then, proceed to the next substage with no action.

($\alpha = \mathcal{R}_i$): If stage s is α-expansionary, set $F(\alpha) = 0$, otherwise set $F(\alpha) = 1$.

($\alpha = \mathcal{R}_{i,j}$): Choose the first clause from the following list ($\mathcal{R}$.1) – ($\mathcal{R}$.3) that applies, and perform the action stated.

($\mathcal{R}$.1): α is ready to open an X-gap : that is, $\tau(\alpha) \downarrow$, $F(\alpha) = w$, $x(\alpha) \in W_{j,s}$, and $x(\alpha) < l^i(s)$. The action to be taken is to open an X-gap by setting $r_1(\alpha) = 0$, $F(\alpha) = g_1$, initialize all $\beta \ge_L \alpha^\wedge w$, and create a link $(\tau(\alpha), \alpha)$.

($\mathcal{R}$.2): α is ready to close an X-gap : that is, $F(\alpha) = g_1$ and s is $\tau(\alpha)$-expansionary. Let $u < s$ be the stage where the current X-gap was opened. If $X_{i,u}\!\restriction_{x(\alpha)} \ne X_{i,s}\!\restriction_{x(\alpha)}$ we close the X-gap successfully and open a Y-gap by setting $r_2(\alpha) = 0$, $F(\alpha) = g_2$, and initializing all nodes $\beta \ge_L \alpha^\wedge g_1$. Otherwise if $X_{i,u}\!\restriction_{x(\alpha)} = X_{i,s}\!\restriction_{x(\alpha)}$, we close the X-gap unsuccessfully by setting $r_1(\alpha) = s$, $F(\alpha) = w$, resetting $x(\alpha)$, initializing all nodes $\beta \ge_L \alpha^\wedge w$, and removing the link $(\tau(\alpha), \alpha)$. If the X-gap was closed unsuccessfully, go directly to substage s, and set $\delta_s = \delta_{s,t}$.

($\mathcal{R}$.3): α is ready to close a Y-gap : $F(\alpha) = g_2$ and s is $\tau(\alpha)$-expansionary. As above, let $u < s$ be the stage where the current Y-gap was opened. If $Y_{i,u}\!\restriction_{x(\alpha)} \ne Y_{i,s}\!\restriction_{x(\alpha)}$ we close the Y-gap successfully by enumerating $x(\alpha)$ into $D_{\tau(\alpha)}$, setting $F(\alpha) = succ$, $r_1(\alpha) = r_2(\alpha) = 0$, and initializing all $\beta \ge_L \alpha^\wedge g_2$. Otherwise if $Y_{i,u}\!\restriction_{x(\alpha)} = Y_{i,s}\!\restriction_{x(\alpha)}$, we close the Y-gap unsuccessfully by setting $F(\alpha) = w$, $r_2(\alpha) = s$, resetting $x(\alpha)$, and initializing all nodes $\beta \ge_L \alpha^\wedge g_1$. In either case, we remove the link $(\tau(\alpha), \alpha)$, set $\delta_s = \delta_{s,t}$, and go directly to substage s.

If none of the clauses ($\mathcal{R}$.1) – ($\mathcal{R}$.3) holds, do nothing and proceed to the next substage.

(*Substage* $t = s$): If there is some $y \searrow K_s$, we enumerate $\gamma(y)$ into B and
reset $\gamma(y)$ above 0. Otherwise if $K_s = K_{s-1}$, we pick the least y
such that $\Gamma_s(A_s \oplus B_s; y) \uparrow$, and set $\Gamma_s(A_s \oplus B_s; y) \downarrow = K_s(y)$ with
use $2\gamma(y) + 1$.

This ends the construction.

7. Verification

We will now state some straightforward properties exhibited by the links,
which can be shown by induction.

- If (ρ, γ) is a link that had been formed in the construction, then
  necessarily we must have $\gamma \succeq \rho^\wedge 0$, and also

$$\nexists \sigma \Big(\rho \preceq \sigma \preceq \gamma \ \wedge \ \sigma = \mathcal{R}_j \text{ for some } j < i(\rho) \Big), \qquad (7.1)$$

$$\forall \sigma \Big((\sigma \succeq \gamma^\wedge g_1 \vee \sigma \succeq \gamma^\wedge g_2) \wedge i(\sigma) \geq i(\rho) \wedge \tau(\sigma) \downarrow \ \Rightarrow \ \tau(\sigma) \succ \gamma \Big). \tag{7.2}$$

- It is also clear that for any node $\sigma \preceq \delta_s$, a link (ρ, γ) exists, and
  is travelled during stage s for some $\rho \prec \sigma \prec \gamma \Leftrightarrow \sigma$ is not visited
  during stage s.
- Combining the above fact with (7.1) and (7.2), we see that any links
  that exist simultaneously may be nested, but never crossing.
- Therefore for any node $\sigma \preceq \delta_s$, σ is visited at stage $s \Leftrightarrow \nexists$ a link
  (ρ, γ) at substage $(s, 0)$ with $\rho^\wedge 0 \preceq \sigma$ and $\gamma \npreceq \sigma$. That is, any link
  (ρ, γ) existing at the beginning of stage s with $\rho^\wedge 0 \preceq \sigma$, must have
  bottom $\gamma \preceq \sigma$.
- Any link that is created, can be travelled at most twice before it
  is removed. Also, any link that is removed must be travelled in the
  same stage.
- Nested links are removed from outermost inwards.
- No link may be formed and travelled in the same stage.
- If a link (ρ, γ) is formed during the construction, then every node α
  such that $\rho \prec \alpha^\wedge g_1 \preceq \gamma$ or $\rho \prec \alpha^\wedge g_2 \preceq \gamma$ must have a gap open at
  the instance of formation, and the gap will remain open until after
  the link (ρ, γ) is removed. Therefore at any stage s, every node α
  such that $\alpha^\wedge g_1 \preceq \delta_s$ or $\alpha^\wedge g_2 \preceq \delta_s$ must have a (X- or Y-gap,
  respectively) open at the end of stage s.

The true path (TP) of the construction is defined to be $TP(n) = \liminf_s F(TP\!\upharpoonright_n)$, for all n. The $\liminf_s F(\alpha)$ always exists for any node α, because α only has finitely many possible outcomes, with the exception of $\alpha = \mathcal{N}_e$, where we require that $F(\alpha, t) < F(\alpha, s) \Rightarrow \exists u (t \le u < s \wedge F(\alpha, u) = 0)$.

Lemma 7.1. *The true path is the leftmost path visited infinitely often. That is for any n, there are infinitely many stages s, where $TP\!\upharpoonright_n$ is visited, and $\exists^{<\infty} s (\delta_s <_L TP\!\upharpoonright_n)$.*

Proof. The lemma obviously holds when $n = 0$. Let $n > 0$, and assume the lemma holds for $\alpha = TP\!\upharpoonright_{n-1}$. Let $a = TP(n-1)$, and choose a stage s_0 such that $\forall s > s_0 (\delta_s \not<_L \alpha \wedge F(\alpha, s) \not<_L a)$. For any $s_1 > s_0$, if $\delta_{s_1} <_L \alpha^\wedge a$, then there must be a link (ρ, γ) with $\rho \prec \alpha$ and $\alpha^\wedge b \preceq \gamma$ for some $b <_L a$, which is travelled during stage s. Since α is visited infinitely often, there will be a $s_2 \ge s_1$ where α is visited. Hence by induction, we have for all $s \ge s_2$, we have $\delta_s \not<_L \alpha^\wedge a$, and there will be no link (ρ', γ') with $\rho' \prec \alpha$ and $\alpha^\wedge b \preceq \gamma'$ for any $b <_L a$ existing at the end of stage s.

To show that $\alpha^\wedge a$ is visited infinitely often, we first show that α can only be initialized finitely often. To see this, consider a stage $s_3 > s_0$, in which no $\beta \prec \alpha$ is initialized after stage s_3. After stage s_3, α can only be initialized by α^- where $\alpha = (\alpha^-)^\wedge b$ for some b. The case $\alpha^- = \mathcal{P}_e$ or $\mathcal{R}_e$ is trivial, and if $\alpha^- = \mathcal{N}_e$, we have $n(\alpha^-, t) = n(\alpha^-, s_3)$ for all $t \ge s_3$, and subsequently α can only be initialized by α^- at most $n(\alpha^-, s_3) + 1$ many times. If $\alpha^- = \mathcal{R}_{i,j}$ initializes α at infinitely many stages $t \ge s_3$, then there must be infinitely many stages $t' \ge s_3$ in which $F(\alpha^-, t') <_L b$. Therefore, we can let $s_4 > s_0$ be a stage after which α is never initialized, α is visited at stage s_4, where $F(\alpha, s_4)$ is set to a due to the action of α. We may assume that there is some link (α, γ) existing at stage s_4, and that $a = 0$, because otherwise $\alpha^\wedge a$ will be visited at stage s_4. This link will be travelled at most twice before it is removed at some stage $s_5 \ge s_4$. Let $s_6 > s_5$ be the least stage such that α is visited, and $F(\alpha, s_6)$ is set to 0. Then for all $s_5 < u < s_6$, we have $\delta_u \not\succeq \alpha^\wedge 0$, and there is no link (ρ', γ') with $\rho' \prec \alpha^\wedge 0 \preceq \gamma'$ existing at the end of stage u. Thus, $\alpha^\wedge 0$ will be visited at stage s_6.   □

Lemma 7.2.

(i) *For each node $\alpha \ne \mathcal{R}_{i,j}$ on the true path, we have*

$$\lim_{s \to \infty} \{\bar{r}(\alpha, s) \mid \alpha \text{ is visited at stage } s\} < \infty.$$

(ii) *For all* x, $\lim\limits_{s\to\infty} \gamma(x,s) < \infty$.

Proof.

(i) Firstly we note that for any $\beta = \mathcal{R}_{i,j}$ on the true path, we have

$$TP(|\beta|) \in \{succ, w\} \Rightarrow \exists r_0 \forall^\infty s\big(r(\alpha,s) = r_0\big), \tag{7.3}$$
$$TP(|\beta|) = g_1 \Rightarrow \exists r_1 \forall^\infty s\big(r(\alpha,s) = r_1 \text{ when an } X\text{-gap is open}\big), \tag{7.4}$$
$$TP(|\beta|) = g_2 \Rightarrow \forall^\infty s\big(r(\alpha,s) = 0 \text{ whenever a } Y\text{-gap is open}\big). \tag{7.5}$$

Now fix $\alpha \neq \mathcal{R}_{i,j}$ on the true path. Let s_0 be a stage such that $\forall s > s_0$, $\delta_s \not\prec_L \alpha$ and the relevant clauses within the brackets on the right side of the implications in $(7.3), (7.4)$ and (7.5) hold for every $\beta \prec \alpha$. For any $s > s_0$ in which α is visited, we have $r(\alpha, s) = 0$, and every $\beta \prec \alpha$ such that $TP(|\beta|) = g_1$ or g_2 will have a X- or respectively Y-gap open. Hence $\lim\limits_{s\to\infty} \{\bar{r}(\alpha, s) \mid \alpha \text{ is visited at stage } s\}$ exists.

(ii) During the construction, only the action taken by some node $\alpha = \mathcal{P}_e$ or $\mathcal{N}_e$, or the action at the end of a stage can move $\gamma(x)$. For a particular x, we choose s_0 so that $K_{s_0}\!\restriction_{x+1} = K\!\restriction_{x+1}$, and for all $s > s_0$, if $\alpha = \mathcal{P}_e$ and $\langle\alpha\rangle < x$, then α does not enumerate any element into A_s. Therefore after stage s_0, only the action of finitely many nodes α (where $\alpha = \mathcal{N}_e$) can move $\gamma(x)$. Let α be such a node, and let $n(\alpha, s) \leq x$ for some $s > s_0$. If α is never initialized after stage s, then α will reset $\gamma(n(\alpha))$ at most $n(\alpha) + 1$ many times. $\qquad\qquad\square$

Lemma 7.3. *Along the true path, the requirements succeed.*

Proof. First consider the requirement $\mathcal{P}_e : A \neq \overline{W_e}$: let $\alpha = TP\!\restriction_{4e}$, and if $TP(4e) = succ$ then $A \cap W_e \neq \emptyset$. Thus, we suppose that $TP(4e) = w$. At all times in the construction, we have

$\gamma(x)$ value is reassigned $\Leftrightarrow$

$$\exists y \leq \gamma(x) \ (y \text{ is enumerated into } A \text{ or } B). \tag{7.6}$$

We claim that $A^{[\alpha]}$ is finite, since the only nodes that can enumerate elements into $A^{[\alpha]}$ is the node α itself, as well as the node β such that $\beta = \mathcal{N}_i$ and $\langle\alpha\rangle \in \omega^{[\beta]}$. Now, α contributes no elements to $\omega^{[\alpha]}$, while β can only put finitely many elements into $\omega^{[\alpha]}$ (by Lemma 7.2(ii), and (7.6)). On the other hand, $W_e^{[\alpha]}$ is also finite because of Lemma 7.2. Hence, $A \neq \overline{W_e}$.

$\mathcal{R}_e$ - *Non-bounding Strategy* : assume that $\Phi_e^A = X_e$, $\Psi_e^A = Y_e$, and X_e, Y_e are both non-recursive. We first show that there is no node α such that $\exists j(\alpha = \mathcal{R}_{e,j})$ and $\alpha^\wedge g_1 \prec TP$. Suppose such α exists, then $\tau(\alpha) \downarrow$ and we let s_0 be a stage after which α is never initialized, and $\forall s > s_0 (\delta_s \not\prec_L \alpha)$. To compute $X_e(y)$, we wait for a stage $s > s_0$ such that $x(\alpha, s) > y$, α is visited at stage s, where a X-gap is opened. Then, $X_{e,s}\lceil_{y+1} = X_e\lceil_{y+1}$ because during any stage $t > s$ where α has no open X-gap, we must have $\delta_t \not\succ \alpha^\wedge g_1$. The same argument also shows that there is no node α such that $\exists j(\alpha = \mathcal{R}_{e,j})$ and $\alpha^\wedge g_2 \prec TP$. Therefore $\tau := \tau(TP, e) \downarrow$, and there is no β such that $\tau \prec \beta \prec TP$, $i(\beta) < e$ and $TP(|\beta|) \in \{1, g_1, g_2\}$. We also have $\tau^\wedge 0 \prec TP$, and for almost all j, there is some α such that $\alpha = \mathcal{R}_{e,j}$, $\alpha \prec TP$, and $\alpha \in E(TP, e)$. Therefore, we can conclude that the r.e. set D_τ is non-recursive. It only remains to show that $D_\tau \leq_T X_e$ and $D_\tau \leq_T Y_e$. A X_e-recursive test of whether $y \in D_\tau$ is the following. Wait for a stage s_0 such that $x(\alpha, s_0) = y$ for some $\alpha \in E(TP, e)$, and $X_{e,s_0}\lceil_y = X_e\lceil_y$. If no X-gap is open at stage s_0 then $x \notin D_\tau$, otherwise we wait for stage $s_1 > s_0$ where all gaps (X or Y) are closed, and we have $D_\tau(y) = D_{\tau,s_1}(y)$. A similar argument is used for $D_\tau \leq_T Y_e$.

$\mathcal{N}_e$ - $\Phi_e^B \neq C$: let $\alpha = TP\lceil_{4e+1}$, and let s_0 be a stage after which α is never initialized, $\forall s > s_0 (\delta_s \not\prec_L \alpha)$, $K_{s_0}\lceil_{n(\alpha, s_0)} = K\lceil_{n(\alpha, s_0)}$, and α is visited at stage $s_0 + 1$. Let $s_1 > s_0 + 1$ be the next stage where α is visited, and let $z := z(\alpha, s_1) = \lim_{t \to \infty} z(\alpha, t)$. Thus we have

$$TP(|\alpha|) = 0 \Rightarrow z \notin C \wedge \Phi_e^B(z) \neq 0,$$
$$TP(|\alpha|) > 0 \Rightarrow z \in C \wedge \Phi_e^B(z) = 0,$$

which satisfies the requirement $\mathcal{N}_e$.

$\Gamma(A \oplus B) = K$: fix an x, and by Lemma 7.2 and (7.6) there is a stage s_0 where we set $\Gamma_{s_0}(A_{s_0} \oplus B_{s_0}; x) \downarrow= K_{s_0}(x)$ with use $2\gamma(x, s_0) + 1$, and $(\forall y \leq x)(\forall s > s_0)\left(\gamma(y, s) = \lim_{t \to \infty} \gamma(y, t)\right)$. We have $(A_{s_0} \oplus B_{s_0})\lceil_{2\gamma(x, s_0)+2} = (A \oplus B)\lceil_{2\gamma(x, s_0)+2}$, and thus $\Gamma(A \oplus B; x) = K_{s_0}(x) = K(x)$. $\square$

References

1. K. Ambos-Spies, C. Jockusch Jr., R. Shore, R. Soare, (1984). An algebraic decomposition of the recursively enumerable degrees and the coincidence of several degree classes with the promptly simple degrees. *Trans Amer. Math Soc.* **281**, 109-128.
2. M. Bickford, C. Mills, (1983). Lowness properties of r.e. sets. *Typescript, unpublished.*
3. G. Chaitin, (1976). Information-theoretical characterizations of recursive infinite strings. *Theoretical Computer Science* **2**, 45-48.
4. P. Cholak, M. Groszek, T. Slaman, (2001). An almost deep degree. *J. Symb Logic* **66**(2), 881-901.

5. R. Downey, (1987). Δ_2^0 degrees and transfer theorems. *Illinois J. Mathematics* **31**, 419-427.

6. R. Downey, (1990). Lattice nonembeddings and initial segments of the recursively enumerable degrees. *Ann. Pure and Appl. Logic* **49**, 97-119.

7. R. Downey, N. Greenberg, R. Weber. Totally ω computably enumerable degrees I : bounding critical triples. *Submitted.*

8. R. Downey, D. Hirschfeldt, (2006). Algorithmic Randomness and Complexity. To appear.

9. R. Downey, D. Hirschfeldt, A. Nies, F. Stephan, (2003). Trivial reals. *Proc. 7th and 8th Asian Logic Conf.*, 103-131.

10. R. Downey, C. Jockusch, M. Stob, (1990). Array nonrecursive sets and multiple permitting arguments. *Recursion Theory Week, Lecture Notes in Mathematics, Springer-Verlag, Heidelberg* **1432**, 141-174.

11. R. Downey, S. Lempp, (1997). Contiguity and distributivity in the enumerable degrees. *J. Symb Logic* **62**, 1215-1240.

12. H. Enderton, H. Putnam, (1970). A note on the hyperarithmetical hierarchy. *J. Symb Logic* **35**, 429-430.

13. R. Ladner, (1973). A completely mitotic nonrecursive recursively enumerable degree. *Trans. Amer. Math. Soc.* **184**, 479-507.

14. R. Ladner, L. Sasso, (1975). The weak truth table degrees of r.e. sets. *Ann. Math. Logic* **4**, 429-448.

15. S. Lempp, T. Slaman, (1989). A limit on relative genericity in the recursively enumerable sets. *J. Symb Logic* **54**, 376-395.

16. A. Nerode, R. Shore, (1980). Second order logic and first order theories of reducibility orderings. *The Kleene Symposium*, 181-200.

17. A. Nerode, R. Shore, (1980). Reducibility orderings : theories, definability and automorphisms. *Ann. Math. Logic* **18**, 61-89.

18. A. Nies, (2000). Definability in the c.e. degrees : questions and results. *Contemporary Mathematics* **257**, 207-213.

19. A. Nies, (2005). Lowness properties and randomness. *Adv. Math.* **197**(1), 274-305.

20. A. Nies, (2006). Ideals in the recursively enumerable degrees. To appear.

21. A. Nies, R. Shore, T. Slaman, (1998). Interpretability and definability in the recursively enumerable degrees. *Proc. London Math. Soc. (3)* **77**, 241-291.

22. R. Shore, (2000). Natural definability in degree structures. *Contemporary Mathematics* **257**, 255-271.

23. T. Slaman, H. Woodin, (1986). Definability in the turing degrees. *Illinois J. Math* **30**, 320-334.

24. R. Soare, (1987). Recursively enumerable sets and degrees. *Springer-Verlag, Berlin.*

25. R. Solovay, (1975). Draft of a paper (or series of papers) on Chaitin's work. *IBM Thomas J. Watson Research Center, Yorktown Heights, NY*, 215 pages.

26. Y. Yang, L. Yu, (2005). On the definable ideal generated by nonbounding c.e. degrees. *J. Symb Logic* **70**, 252-270.

ELIMINATING CONCEPTS

André Nies*

Department of Computer Science, University of Auckland
Private Bag 92019, Auckland, New Zealand
E-mails: andre@cs.auckland.ac.nz
andrenies@gmail.com

Four classes of sets have been introduced independently by various researchers: low for K, low for ML-randomness, basis for ML-randomness and K-trivial. They are all equal. This survey serves as an introduction to these coincidence results, obtained in [25] and [11]. The focus is on providing backdoor access to the proofs.

1. Outline of the results

All sets will be subsets of $\mathbb{N}$ unless otherwise stated. $K(x)$ denotes the prefix free complexity of a string x. A set A is K-trivial if, within a constant, each initial segment of A has minimal prefix free complexity. That is, there is $c \in \mathbb{N}$ such that

$$\forall n \; K(A \upharpoonright n) \leq K(0^n) + c.$$

This class was introduced by Chaitin [5] and further studied by Solovay (unpublished). Note that the particular effective epresentation of a number n by a string (unary here) is irrelevant, since up to a constant $K(n)$ is independent from the representation.

A is low for Martin-Löf randomness if each Martin-Löf random set is already Martin-Löf random relative to A. This class was defined in Zambella [30], and studied by Kučera and Terwijn [18].

In this survey we will see that the two classes are equivalent [25]. Further concepts have been introduced: to be a basis for ML-randomness

*The author is supported by the Marsden fund of New Zealand, UOA 319.

(Kučera [17]), and to be low for K (Muchnik jr, in a seminar at Moscow State, 1999). They will also be eliminated, by showing equivalence with K-triviality. All the equivalent definitions show different aspects of the same notion. In particular, while low for K, low for random and basis for ML-randomness are forms of computational weakness, K-trivial intuitively means being far from random.

Solovay (1975) proved the existence of a non-computable K-trivial. Kučera and Muchnik each showed the existence of a non-computable set in the class introduced. For the class of low for ML-random sets, existence was only shown in 1997 [18]. All examples were c.e., except for Solovay's example of a K-trivial, which was only Δ_2^0. Later this was improved to a c.e. example by Kummer (unpublished), and Calude & Coles [3].

The main purpose of this paper is to survey the coincidence results obtained in [25] and [11] and to present the proof ideas in an accessible way. However, in Subsection 3.2 we provide some new facts about the cost function construction of a K-trivial set. We also include a sketch of a proof that each K-trivial is low via the golden run method, which is simplest application of this method. Facts quoted without reference can be found in [9], or in my forthcoming book [21]. However, for the ease of the reader we recall some facts here. Throughout, "Martin-Löf" will be abbreviated by "ML". Schnorr's Theorem states that Z is ML-random iff for some c, $\forall n \ K(Z \upharpoonright n) \geq n - c$. Thus Z is ML-random if for each n, $K(Z \upharpoonright n)$ is near its maximal value $n + K(0^n)$. To say that A is K-trivial means that A is far from ML-random, because $K(A \upharpoonright n)$ is minimal (all up to constants).

An example of a ML-random set is Chaitin's halting probability,

$$\Omega = \sum_{U(\sigma)\downarrow} 2^{-|\sigma|},$$

where U is the reference universal prefix free machine.

If x, y are expressions, then $x \leq^+ y$ denotes that $x \leq y + c$ for a constant c independent of the values of x and y.

2. Computational weakness

2.1. *Three classes*

First we will discuss the low for K sets, the low for ML-randomness sets, and the bases for ML-randomness.

In general, adding an oracle A to the computational power of the universal machine decreases $K(y)$. A is low for K if this is not so. In other words,

$$\forall y \ K(y) \leq^+ K^A(y).$$

Let $\mathcal{M}$ denote this class, introduced by Andrej A. Muchnik in 1999, who proved that there is a c.e. noncomputable $A \in \mathcal{M}$. We defer a proof of existence till later. Here is a useful observation.

Proposition 2.1. *If A is low for K, then A is GL_1, namely, $A' \leq_T A \oplus \emptyset'$.*

Proof. If t is least such that $e \in A'_t$, then $K^A(t) \leq^+ 2\log e$. To see this, recall that the number e has the prefix free code $0^{|\sigma|}1\sigma$ where σ is the string representing e in binary. Consider the prefix free machine with oracle A that, on input a prefix free code for e searches for t such that $J^A(e)$ converges at stage t. This machine shows that $K^A(t) \leq^+ 2\log e$. Also $K(t) \leq^+ K^A(t)$ by hypothesis on A, so $K(t) \leq 2\log e + c$ for some constant c. Now $\emptyset'$ can compute $s = \max\{U(\sigma) : |\sigma| \leq 2\log e + c\}$, where U is the reference universal prefix free machine. Then $e \in A' \Leftrightarrow e \in A'_s$. $\square$

Let MLR denote the class of Martin-Löf-random sets. Because an oracle A increases the power of tests, $\mathsf{MLR}^A \subseteq \mathsf{MLR}$. In general one would expect this inclusion to be proper. Zambella [30] defined A to be low for ML-randomness if

$$\mathsf{MLR}^A = \mathsf{MLR}. \tag{2.1}$$

In 1997, Kucera and Terwijn proved that there is a non-computable c.e. set that is low for ML-randomness [18]. To see that low for K implies low for ML-randomness, first note that Schnorr's Theorem relativizes: Z is Martin-Löf random relative to A iff for some c, $\forall n \ K^A(Z \upharpoonright n) \geq n - c$. Now, since MLR can be defined in terms of K, and MLR^A in terms of K^A, low for K implies low for ML-randomness. Thus, the existence of a non-computable c.e. set that is low for ML-randomness also follows from Muchnik's result.

Kučera [17] introduced a further concept expressing computational weakness. He studied sets A such that

$$A \leq_T Z \text{ for some } Z \in \mathsf{MLR}^A.$$

That is, A can be computed from a set that is ML-random relative to A. While Kučera used the term "basis for 1-RRA", we will call such a set a basis for ML-randomness. There is no connection to basis theorems.

If A is low for ML-randomness then A is a basis for ML-randomness. For, by the Kučera-Gács Theorem there is a ML-random Z such that $A \leq_T Z$. Then Z is ML-random relative to A.

2.2. The existence and equivalence theorems

In the following We will discuss two theorems:

Theorem 2.2: There is a c.e. non-computable basis for ML-randomness [17].
Theorem 2.5: Each basis for ML-randomness is low for K [11].

Now two concepts are gone. For, we have already obtained the easy inclusions

low for $K \Rightarrow$ low for ML-randomness $\Rightarrow$ basis for ML-randomness.

Then, by the second Theorem, all three classes are the same. In particular, the Theorems together imply the result of Muchnik that there is a non-computable c.e. low for K set.

How about the fourth concept, K-triviality? While the implication "low for $K \Rightarrow K$-trivial" is immediate, the converse, "K-trivial $\Rightarrow$ low for K", is hard. The proof is carried out separately from all of the above and will be discussed in Section 3.

Chaitin [5] proved that each K-trivial set is Δ_2^0, by an elegant short argument involving the coding theorem. Thus a set that is low for ML-randomness is Δ_2^0. This answers an open question of Kučera and Terwijn [18]. I first gave a direct proof of this [22], introducing techniques which I later extended in order to prove Theorem 2.6 below.

We now proceed to the first Theorem.

Theorem 2.2. Kučera [17]. *There is a c.e. non-computable set A that is a basis for ML-randomness.*

The proof sketched here differs a bit from Kučera's original one. One combines the following two results. The first comes from Kučera's priority free solution to Post's problem. A function $f : \mathbb{N} \mapsto \mathbb{N}$ is called *diagonally non-computable* if $\forall e \neg f(e) = \Phi_e(e)$.

Theorem 2.3. Kučera [16]. *Let Z be Δ_2^0 and diagonally non-computable. Then there is a simple set $A \leq_T Z$.*

Each ML-random set is diagonally non-computable. So we may apply Theorem 2.3 to a low ML-random set Z (say), and then use the following lemma of Hirschfeldt, Nies and Stephan in order to obtain a simple basis for ML-randomness.

Lemma 2.4. [11]. *If $Z <_T \emptyset'$ is ML-random and $A \leq_T Z$ is c.e., then Z is already ML-random relative to A.*

To prove the Lemma, one argues that one can turn a ML-test relative to A which Z fails into a plain ML-test. This uses the incompleteness of Z.

Theorem 2.3 is easier to prove under the stronger hypothesis that Z is ML-random, and this is the only case we need here. Under this stronger hypothesis, it can be proved without using the Recursion Theorem. In fact, Kučera had first thought of this special case, and only later he generalized it to diagonally non-computable Z, where the Recursion Theorem is needed. We sketch the proof of Theorem 2.3 for a ML-random set Z.

Proof. A *Solovay test* $\mathcal{G}$ is given by an effective enumeration of strings $\sigma_0, \sigma_1, \ldots$, such that $\sum_i 2^{-|\sigma_i|} < \infty$. It is not hard to see that Z is ML-random iff for each Solovay test $\mathcal{G} = \sigma_0, \sigma_1, \ldots$, for almost all i, $\sigma_i \npreceq Z$.

We will enumerate A and a Solovay test $\mathcal{G}$. To make A simple, we meet the requirements

$$S_e : \ |W_e| = \infty \Rightarrow A \cap W_e \neq \emptyset.$$

Construction. At stage $s > 0$, if S_e is not satisfied yet, see if there is an x, $2e \leq x < s$, such that

$$x \in W_{e,s} - W_{e,s-1} \ \ \& \ \ \forall t_{x < t < s} Z_t \restriction e = Z_s \restriction e.$$

If so, put x into A. Put the string $\sigma = Z_s \restriction e$ into $\mathcal{G}$. Declare S_e satisfied.

Clearly A is simple (in fact, A can even be made promptly simple). Also, $\mathcal{G}$ is a Solovay test since the requirement S_e contributes at most 2^{-e} to $\mathcal{G}$.

To see $A \leq_T Z$, choose s_0 such that $\sigma \npreceq Z$ for any σ enumerated into $\mathcal{G}$ after stage s_0. Given an input $x \geq s_0$, using Z, compute $t > x$ such that $Z_t \restriction x = Z \restriction x$. Then $x \in A \ \Leftrightarrow \ x \in A_t$, for if we put x into A at a stage $s > t$ for the sake of S_e, then $e < x$, so we also put σ into $\mathcal{G}$ where $\sigma = Z_s \restriction e = Z \restriction e$. This contradicts the fact that $\sigma \npreceq Z$. $\qquad \square$

Note that the enumeration into A is heavily restrained. If $Z \restriction e$ changes another time at stage s, then no $x < s$ can be enumerated after s for the sake of S_e. Z can restrict S_e in this way as late and as often as it wants.

Now we discuss the second result, which we call the hungry sets theorem.

Theorem 2.5. *Hirschfeldt, Nies, Stephan* [11]. *If A is a basis for ML-randomness, then A is low for K.*

We actually obtained the conclusion that A is K-trivial, by a very similar proof. A full proof of the present version is in [21].

The Kraft-Chaitin Theorem. We use the KC-Theorem as a tool. A c.e. set $W \subseteq \mathbb{N} \times 2^{<\omega}$ is a Kraft-Chaitin set (KC set) if

$$\sum_{\langle r,y \rangle \in W} 2^{-r} \leq 1.$$

(Note that some values of r can occur several times in this sum.) The KC-Theorem states that, from a Kraft-Chaitin set L, one can effectively obtain a prefix free machine M such that

$$\forall r, y [\langle r, y \rangle \in L \iff \exists w \, (|w| = r \; \& \; M(w) = y)].$$

Thus, we enumerate requests $\langle r, y \rangle$ ("give a description of y that has length r"). The weight of this request is 2^{-r}. If their total weight is at most 1, then each request will be fulfilled: there actually is an M-description of length r. A critical point in proofs applying the KC-theorem is to verify that the set of requests is in fact KC, namely the total weight is at most 1. We will now outline proof of the hungry sets theorem, and reveal the reason for its culinary name. To ensure the sets enumerated are KC-sets, we use the method of accounting.

Proof. Given a Turing functional Φ, we define a ML-test $(V_d^X)_{d \in \mathbb{N}^+}$ relative to oracle X (later, we use this test for $X = A$). Suppose $A = \Phi^Z$. The goal is this: if $Z \notin V_d^A$ then A is low for K, with constant $d + \mathcal{O}(1)$.

To realise this goal, we also build a uniformly c.e. sequence $(L_d)_{d \in \mathbb{N}^+}$ of KC sets. For each computation of the universal prefix free machine $\mathbf{U}$,

$$\mathbf{U}^\eta(\sigma) = y \text{ where } \eta \preceq A,$$

(that is, whenever y has a description σ with oracle A), we want to ensure there is a description without an oracle that is only by a constant longer. Thus we want to put a request

$$\langle |\sigma| + d + 1, y \rangle$$

into L_d. The problem is that we don't know A, so we don't know which η's to take. So L_d could fail to be KC. To avoid this, the description $\mathbf{U}^\eta(\sigma) = y$ first has to prove itself worthy.

Recall that we are building an auxiliary ML-test (V_d) relative to A. If $Z \notin V_d$ then L_d works. We effectively enumerate open sets $C_{d,\sigma}^\eta$ and let

$$V_d^A = \bigcup_{\eta \prec A} C_{d,\sigma}^\eta.$$

While $\mu(C_{d,\sigma}^\eta) < 2^{-|\sigma|-d}$, $C_{d,\sigma}^\eta$ is hungry. We feed it with fresh oracle strings α, where $\eta \prec \Phi^\alpha$.

All the open sets $[C_{d,\sigma}^{\eta}]^{\preceq}$ are disjoint. When $\mu(C_{d,\sigma}^{\eta})$ exceeds $2^{-|\sigma|-d-1}$, we put the request $\langle |\sigma| + d + 1, y \rangle$ into L_d. We can account the weight of those requests against the measure of the sets $C_{d,\sigma}^{\eta}$, since the measure of $C_{d,\sigma}^{\eta}$ is greater than the weight of the request. This shows that each L_d is a KC set. Because Z is ML-random relative to A, there is some d such that whenever $\mathbf{U}^{\eta}(\sigma) = y$ in the relevant case that $\eta \preceq A$, the request $\langle |\sigma| + d + 1, y \rangle$ we are after is enumerated into L_d. Thus the following fact (named to honor the Brazilian president Lula) can be verified.

Fome Zero Lemma. *Suppose $Z \notin V_d$. Then for each description $\mathbf{U}^{\eta}(\sigma) = y$, where $\eta \prec A$, $\mu(C_{d,\sigma}^{\eta}) = 2^{-|\sigma|-d}$. In other words, $\mu(C_{d,\sigma}^{\eta})$ is hungry no more at the end of time.* $\square$

Discussion. For each partial computable functional Φ, let $S_A^{\Phi} = \{Z : A = \Phi^Z\}$. For each $n > 0$, let $S_{A,n}^{\Phi} = [\{\sigma : A \upharpoonright n = \Phi^{\sigma}\}]^{\preceq}$. Then $S_{A,n}^{\Phi}$ is open and c.e. relative to A, uniformly in n. Moreover, $S_A^{\Phi} = \bigcap_n S_{A,n}^{\Phi}$. Thus S_A^{Φ} is a Π_2^0 class relative to A, and $\mu S_A^{\Phi} = 0$ is equivalent to $\lim_n \mu S_{A,n}^{\Phi} = 0$.

Let us compare a few facts related to this.

- If A is non-computable, then $\mu S_A = 0$ [28].
- If A is ML-random, then (after leaving out the first few components), $(S_{A,n}^{\Phi})_{n \in \mathbb{N}}$ is a ML-test relative to A, a fact from [20]. In other words, there is c such that $\forall n \, \mu S_{A,n} \leq 2^{-n+c}$.
- By the proof of the hungry sets theorem, based on Φ, one can build an oracle ML-test (V_d) such that, whenever A is not low for K, then $S_A \subseteq \bigcap_d V_d^A$. Then, since there is a universal ML test, the whole class $\{Z : A \leq_T Z\}$ is ML-null relative to A. (The converse holds as well: if A is low for K then $\Omega \geq_T A$ is ML-random relative to A, so the class is not ML-null relative to A.)

These facts suggest that, the more random A is, the fewer sets compute it, where "fewer" is taken in the sense of how effective the null set S_A^{Φ} is. The only case where S_A^{Φ} is merely a Π_2^0 null set is when A is low for K, or equivalently, K-trivial.

2.3. *Lowness for other randomness notions*

Next, we digress a bit in order to study lowness for randomness notions implied by ML-randomness (for more details, see [9]). The definition of those classes is the exact analog of (2.1). Each computable set is low for the randomness notion in question. The question is whether there are others, and if so, how to characterize the class. Unexpected things happen.

For my purposes, a *martingale* is a function $M : \{0,1\}^* \mapsto \mathbb{Q}_0^+$ such that

$$M(x0) + M(x1) = 2M(x)$$

(here $\mathbb{Q}_0^+$ is the set of non-negative rationals). M *succeeds* on Z if $\limsup_n M(Z \upharpoonright n) = \infty$, and the class of the sets where M succeeds is denoted Success(M). Z is *computably random* if no computable martingale M succeeds on Z. That is, $M(Z \upharpoonright n)$ is bounded.

While a martingale always bets on the next position, a *non-monotonic betting strategy* can choose some position that has not been visited yet. Z is *Kolmogorov-Loveland random* (KLRand) if not even a non-monotonic betting strategy can succeed on Z. It is easy to verify that

$$\mathsf{MLR} \subseteq \mathsf{KLR} \subset \mathsf{CR},$$

where MLR, KLR and CR denote the classes of Martin-Löf-random, KL-random and computably random sets, respectively. Whether the first inclusion is proper is a major open problem [19, Question 6.2].

Let us discuss the associated lowness notions. Recall that each low for ML-randomness set is a basis for ML-randomness, hence low for ML-randomness implies low for K by Theorem 2.5. First we consider a strengthening of this result, which actually is the version in which it first appeared.

Theorem 2.6. [25]. *If* $\mathsf{MLR} \subseteq \mathsf{CR}^A$ *then A is low for K.*

The converse implication is easy, since low for K implies low for ML-randomness by Schnorr's Theorem, as discussed above.

We sketch the proof of Theorem 2.6.

Proof. Let R be any c.e. open set such that $\mu R < 1$ and Non-MLRand $\subseteq R$, for instance, $R = \{Z : \exists n K(Z \upharpoonright n) \leq n - 1\}$. We will define a Turing functional L such that L^A is a martingale. If $\mathsf{MLR} \subseteq \mathsf{CRand}^A$ then Success(L^A) $\subseteq$ Non-MLRand, and the following lemma applies to $N = L^A$. It says that there is a basic open cylinder $[v] \not\subseteq R$ such that, for each x extending v, if $N(x)$ is large, then x is not too random as a string because x is in R.

Lemma 2.7. *Let N be any martingale such that* Success(N) $\subseteq$ *Non-MLR. Then there are $v \in 2^{<\omega}$ and $m \in \mathbb{N}$ such that $[v] \not\subseteq R$, and*

$$\forall x \succeq v \, [N(x) \geq 2^m \Rightarrow x \in R]. \tag{2.2}$$

To prove the Theorem, let us at first assume we know the witnesses v, m in the Lemma. Thus

$$\forall x \succeq v \, [L^A(x) \geq 2^m \Rightarrow x \in R].$$

The proof parallels the proof of the hungry sets theorem. (The present argument actually was given first, in [25].) Once again, when we see a description $\mathbf{U}^\eta(\sigma) = y$ where $\eta \preceq A$, we want a corresponding set C_σ^η such that $\mu C_\sigma^\eta \geq 2^{-|\sigma|-c}$, c some constant (there is no analog of the parameter d here). While $\mu C_\sigma^\eta < 2^{-|\sigma|-c}$, the set is hungry. The sets for different descriptions have to be disjoint. We feed a set C_σ^η in small servings, as follows: at stage s, pick a clopen set D, $\mu D = \epsilon$ of long strings $x \succeq v$, $D \cap R_s = \emptyset$. Here ϵ is an appropriate small quantity. Define $L^\eta(x) \geq 2^m$ for each $x \in D$ and put D into C_σ^η. If $\eta \prec A$, then D will go into R eventually. Once this happens, repeat with a new set, but again of measure ϵ, as long as C_σ^η is hungry. The fact that the old set D has to enter R before we pick a new one enables us to stuff the right sets up to the desired measure, while the ones where $\eta \not\prec A$ only get a small serving outside of R. To make the sets C_σ^η disjoint, we simply ensure the different portions outside R they are fed with are disjoint.

Actually, we do not know the right witnesses v, m. But it is enough to let L be an infinite weighted sum, over all possible witnesses, of the martingales obtained for those witnesses. See [25] for the rather tricky details. $\square$

Next we consider the class of low for KL-random sets If A is low for KLRand, then

$$\mathsf{MLR} \subseteq \mathsf{KLR} = \mathsf{KLR}^A \subseteq \mathsf{CR}^A.$$

Therefore, the following is a consequence of Theorem 2.6.

Corollary 2.8. *Each low for KL-random set is low for K.*

As one would expect, it is an open problem whether the two classes are the same.

Next we show that the only low for computably random sets are the computable ones [25]. An earlier result in this direction was obtained in joint work with Benjamin Bedregal, then at UFRN, Natal, Brazil.

Theorem 2.9. [2]. *Each low for computably random set A is of hyperimmune free degree.*

But also, by Theorem 2.6 each low for computably random set is K-trivial, and hence Δ_2^0. Since the only hyper-immune free Δ_2^0 sets are the computable ones, we have

Theorem 2.10. [25]. *Each low for computably random set A is computable.*

This answers Question 4.8 in Ambos-Spies and Kucera [1] in the negative. It was conjectured this way by R. Downey.

An *order function* is a non-decreasing unbounded computable function. A notion weaker still than computable randomness is the following. Z is *Schnorr random* (SRand) if no computable martingale M succeeds fast on Z, in the sense that there is an order function h (for instance, $h(n) = \lfloor \log n \rfloor$) such that $M(Z \upharpoonright n) \geq h(n)$ for infinitely many n. Equivalently, Z passes each Schnorr test, namely each ML-test $(V_n)_{n \in \mathbb{N}}$ such that $\mu(V_n) = 2^{-n}$. Just as in the case of lowness for ML-randomness, the associated lowness notion can be characterized in a combinatorial way. A is *computably traceable* if the value $f(x)$ of each $f \leq_T A$ is in a small effectively given set $D_{g(x)}$: g is a computable function depending on f, and $|D_{g(x)}| \leq h(x)$ for an order function h not depending on g. Each computably traceable set is hyper-immune free.

A is low for Schnorr tests if for each Schnorr test $(V_n)_{n \in \mathbb{N}}$ relative to A, there is a Schnorr test $(S_n)_{n \in \mathbb{N}}$ such that $\bigcap_n V_n \subseteq \bigcap_m S_m$. Clearly each set that is low for Schnorr tests is low for Schnorr randomness. Terwijn and Zambella [29] proved that A is low for Schnorr *tests* iff A is computably traceable. They asked if this is also the same as being low for Schnorr randomness. In fact a stronger result holds.

Theorem 2.11. [15]. *The following are equivalent.*

  (i) Each computably random set is Schnorr random relative to A
  (ii) A is computably traceable.

One key ingredient is Theorem 2.9, which persists when one weakens the hypothesis to: each computably random set is Schnorr random relative to A.

3. Far from random

3.1. *Brief introduction to K-triviality.*

For a string y, up to constants, $K(|y|) \leq K(y)$, since one can compute $|y|$ from y (where $|y|$ is represented in binary). A set A is K-trivial if, for some $b \in \mathbb{N}$

$$\forall n \, K(A \restriction n) \leq K(0^n) + b,$$

namely, the K complexity of all initial segments is minimal up to a constant. This notion is opposite to ML-randomness: Schnorr's Theorem (see [9]) says that Z is ML-random iff $\exists b \forall n \, K(Z \restriction n) \geq n - b$. Thus Z is ML-random if all the complexities $K(Z \restriction n)$ are near the upper bound $n + K(n)$, while Z is K-trivial if they have the minimal possible value $K(n)$ (all within constants). If one defines K-triviality using the plain Kolmogorov complexity C instead of K, then one obtains nothing beyond the computable sets [4]. However, Chaitin still managed to prove that the K-trivial sets are Δ_2^0 [5]. As mentioned in the introduction, Solovay (unpublished, 1975) constructed a non-computable K-trivial set A , which was Δ_2^0, as expected, but not c.e.

3.2. *Constructions*

In [10] a short "definition" of a promptly simple K-trivial set is given, which had been anticipated by various researchers (for instance Kummer and Zambella) and is similar to the earlier construction of a non-computable c.e. low for ML-randomness set [18]. We meet the prompt simplicity requirements

$$S_e\colon \ |W_e| = \infty \Rightarrow \exists s \exists x \, [x \in W_{e,s} - W_{e,s-1} \ \& \ x \in A_s].$$

The key ingredient is the "cost function"

$$c(x,s) = \sum_{x < y \leq s} 2^{-K_s(y)}.$$

The c.e. set A is given by letting $A_0 = \emptyset$ and, for $s > 0$,

$A_s = A_{s-1} \cup \{x : \exists e$	
$W_{e,s} \cap A_{s-1} = \emptyset$	we haven't met e-th prompt simplicity requirement
$x \in W_{e,s} - W_{e,s-1}$	we can meet it, via x
$x \geq 2e$	to make A co-infinite
$c(x,s) \leq 2^{-e}\}$	to ensure A is K-trivial.

To see that each S_e is met, note that

$$\forall e \exists y \forall s > y \, [c(y,s) < 2^{-e}]. \tag{3.1}$$

So if $x \geq y$ enters W_e at a stage $s > y$ then x can be enumerated into A.

The K-triviality of A is shown by enumerating a KC-set L such that $\langle K(n) + 2, A \upharpoonright n \rangle \in L$ for each n. By convention, let $K_0(y) = \infty$ for each y. When $r = K_s(y) < K_{s-1}(y)$ we enumerate a request $\langle r + 2, A_s \upharpoonright y \rangle$ into L. The total weight of those requests is $\leq \Omega/4$. When x enters A to meet S_e, then all the initial segments of A from $x + 1$ on change. So for each y such that $x < y < s$, we enumerate a request $\langle K_s(y) + 2, A_s \upharpoonright y \rangle$. The weight added to L is $c(x, s)/4$. Since each S_e is active at most once, the total weight added in this way is at most $(\sum_e 2^{-e})/4 = 1/2$.

Reverse computability theory. Recall the fragments of Peano arithmetic $I\Sigma_1$ (induction over Σ_1 formulas) and $B\Sigma_1$ (for each Σ_1 function f and each x, $f([0,x])$ is bounded). For each $\mathcal{M} \models I\Sigma_1$, there is a promptly simple K-trivial set in $\mathcal{M}$ (Hirschfeldt and Nies). It suffices to verify (3.1) in $\mathcal{M}$. Suppose it fails for $e \in \mathcal{M}$. Consider the Σ_1 formula $\phi(m, e)$ given by

$$\exists u\, [|u| = m\ \&\ \forall i\,(0 \leq i < m \Rightarrow c(u_i, u_{i+1}) > 2^{-e})].$$

By $I\Sigma_1$ and the failure of (3.1) for e, $\mathcal{M} \models \forall m\, \phi(m, e)$. Now let $m = 2^e + 1$ and $u \in M$ be the witness for m. Then, in $\mathcal{M}$,

$$\Omega \geq \sum_{0 \leq i \leq 2^e} c(u_i, u_{i+1}) \geq (2^e + 1)2^{-e} > 1,$$

contradiction.

On the other hand, $B\Sigma_1$ is not sufficient to verify the construction, because of work of Chong and Slaman. Let $\mathcal{M} \models I\Delta_1$. $A \subseteq \mathcal{M}$ is *regular* if for each $n \in \mathcal{M}$, $A \upharpoonright n$ is a string of $\mathcal{M}$ (i.e., encoded by an element of $\mathcal{M}$). Each K-trivial set $A \subseteq \mathcal{M}$ is regular, since $A \upharpoonright n$ has a prefix free description in $\mathcal{M}$, for each n. There is a saturated $\mathcal{M} \models B\Sigma_1$ with a Σ_1 cofinal f whose domain is the standard part. In such an $\mathcal{M}$, each regular c.e. set A is computable.

Necessity of the cost function method, c.e. case. Suppose the c.e. set A is K-trivial via a constant b. Then one can think of A as being built by the cost function construction, when restricting to an appropriate computable set of stages $\{s_i : i \in \mathbb{N}\}$. For each s, one can effectively determine an $f(s) > s$ such that $\forall n < s\, K(A \upharpoonright n) \leq K(n) + b\, [f(s)]$. Let $s(0) = 0$ and

$$s(i + 1) = f(s(i)). \tag{3.2}$$

Proposition 3.1. *Let A be c.e. and K-trivial via b. Then*

$$\sum \{c(x, s(i)) : x < s(i) \text{ is minimal s.t. } A_{s(i)}(x) \neq A_{s(i+1)}(x)\} \leq 2^b.$$

Proof. Let $x_i < s(i)$ be minimal such that $A_{s(i)}(x) \neq A_{s(i+1)}(x)$. For each y, $x_i < y \leq s(i)$, by definition there is at stage $s(i+1)$ a prefix free description of $A_{s(i+1)} \upharpoonright y$ of length $\leq K_{s(i+1)}(y) + b$. Let D_i be the open set generated by such descriptions of a $A_{s(i+1)} \upharpoonright y$, $x_i < y \leq s_i$. Since A is c.e., the strings $A \upharpoonright y$ described at different stages $s(i+1)$ are distinct, so that $D_i \cap D_j = \emptyset$ for $i \neq j$. Hence $\sum_i \mu D_i \leq 1$.

Since

$$c(x, s_i) = \sum\nolimits_{x_i < y \leq s(i)} 2^{-K_{s(i)}(y)} \leq \sum\nolimits_{x_i < y \leq s(i)} 2^{-K_{s(i+1)}(y)} \leq 2^b \mu D_i,$$

this shows $\sum_i c(x_i, s(i)) \leq 2^b$, as required. $\qquad\qquad\qquad\square$

Using deeper methods, Proposition 3.1 can be extended to all K-trivial sets. See Subsection 3.4.

We have seen two constructions of a non-computable c.e. K trivial set A. Both are injury free.

(i) Take a low ML-random set Z. For instance, let Z be the bits of Ω in the even positions, then Z is low by [26]. Now build $A \leq_T Z$ using Kučera's method in Theorem 2.3. Then A is a basis for ML-randomness, hence low for K, and hence K-trivial.

(ii) The cost function construction.

By the extended form of Proposition 3.1, each K-trivial set can be thought of as being obtained via a cost function construction. It is an open question whether each K-trivial set can be obtained via (i):

Question 3.2. *If A is K-trivial, is there a ML-random set $Z <_T \emptyset'$ such that Z is Turing above A?*

In Subsection 3.4, we will see that each K trivial set is Turing below a c.e. K-trivial set. So there is no need to require that the given K-trivial set A is c.e. See [19, Question 4.6] for more details.

3.3. *A is K-trivial iff A is low for K*

We will get there in small steps. First we consider the fact that each K-trivial is wtt-incomplete. Showing the downward closure of the K-trivials under $\leq_{wtt}$ is easy: Suppose $B = \Gamma^A$, where Γ is a wtt reduction procedure with a computable bound f on the use. Then, for each n, within constants,

$$K(A \upharpoonright n) \leq K(A \upharpoonright f(n)) \leq K(f(n)) \leq K(n).$$

Now, since the *wtt*-complete set Ω is ML-random and hence not K–trivial, no K–trivial set A satisfies $\emptyset' \leq_{wtt} A$. To introduce some new techniques, we give a direct proof of wtt-incompleteness.

Suppose that $\emptyset' \leq_{wtt} A$ for a K-trivial A. We build an c.e. set B, and by the Recursion Theorem we can assume we are given a total *wtt*-reduction Γ such that $B = \Gamma^A$, whose use is bounded by a computable function g.

We also build a KC-set L. Thus we enumerate requests $\langle r, n \rangle$ and have to ensure the total weight is at most 1. By the Recursion Theorem, we may assume the coding constant d for L is given in advance. Then, putting $\langle r, n \rangle$ into L causes $K(n) \leq r + d$ and hence $K(A \restriction n) \leq r + b + d$, where b is the triviality constant. (In fact we apply the Double Recursion Theorem.) Let

$$\mathbf{k} = 2^{b+d+1}.$$

Let $n = g(\mathbf{k})$ (the use bound). We wait till $\Gamma^A(\mathbf{k})$ converges, and put the single request $\langle r, n \rangle$ into L, where $r = 1$. Our total cost is $1/2$.

Each time the opponent (named Otto here) has a prefix free description of $A \restriction n$ of length $\leq r + b + d$, we force $A \restriction n$ to change, by putting into B the largest number $\leq \mathbf{k}$ which is not yet in B. If we reach $\mathbf{k} + 1$ such changes, then his total cost is

$$(\mathbf{k} + 1)2^{-(b+d+1)} > 1,$$

contradiction.

Turing-incompleteness. Consider the more general result that each K-trivial set is T-incomplete [10]. There is no recursive bound on the use of $\Gamma^A(\mathbf{k})$. The problem now is that Otto might, before giving a description of $A_s \restriction n$, move this use beyond n, thereby depriving us of the possibility to cause further changes of $A \restriction n$. The solution is to carry out many attempts in parallel, based on computations $\Gamma^A(m)$ for different m. Each time the use of such a computation changes, the attempt is cancelled. What we placed in L for this attempt now becomes garbage. We have to ensure that the weight of the garbage does not build up too much, otherwise L is not a KC set.

More details: j-sets The following is a way to keep track of the number of times Otto had to give new descriptions of strings $A_s \restriction n$. We only consider the stages $s(0) < s(1) < s(2) < \ldots$ where A looks K-trivial with constant b, defined as in 3.2. We write *stage* (in italics) when we mean a stage of this type.

At *stage* t, a finite set E is a j-set if for each $n \in E$

- first we put a request $\langle r_n, n \rangle$ into L
- and then j times at *stages* $s < t$ Otto had to give new descriptions of $A_s \upharpoonright n$ of length $r_n + b + d$.

A c.e. set with an enumeration $E = \bigcup E_t$ is a j-set if E_t is a j-set at each *stage* t.

For $E \subseteq \mathbb{N}$, the weight is defined by $wt(E) = \sum\{2^{-r_n} : n \in E\}$. The weight of a $\mathbf{k}$-set is at most $1/2$.

Lemma 3.3. *If the c.e. set E is a $\mathbf{k}$-set, $\mathbf{k} = 2^{b+d+1}$ as defined above, then* $wt(E) \leq 1/2$.

This is so because $\mathbf{k}$ times Otto has to match our description of n, which has length r_n, by a description of a string $A_s \upharpoonright n$ that is at most $b + d$ longer.

Procedures. Assume A is K-trivial and Turing complete. As in the case of wtt-incompleteness, we attempt to build a $\mathbf{k}$-set $F_{\mathbf{k}}$ of weight $> 1/2$ and reach a contradiction.

The procedure P_j ($2 \leq j \leq \mathbf{k}$) enumerates a j-set F_j. The construction begins calling $P_{\mathbf{k}}$, which calls $P_{\mathbf{k}-1}$ many times, and so on down to P_2, which enumerates L (and F_2).

Each procedure P_j is called with rational parameters $q, \beta \in [0, 1]$. The *goal q* is the weight it wants F_j to reach. When the procedure reaches its goal it returns. The *garbage quota β* is how much garbage it is allowed to produce.

Decanter model. We visualize this construction by a machine similar to Lerman's pinball machine. However, since we enumerate rational quantities instead of single objects, we replace the balls there by amounts of a precious liquid, 1955 Biondi-Santi Brunello wine.

Our machine consists of decanters $F_k, F_{k-1}, \ldots, F_0$. At any stage F_j is a j set. F_{j-1} can be emptied into F_j.

The procedure $P_j(q, \beta)$, $2 \leq j \leq \mathbf{k}$, wants to add a weight of q to F_j. In the beginning it picks a new number m targeted for B. It fills F_{j-1} up to q and then returns, by emptying it into F_j. All numbers put into F_{j-1} are above $\gamma^A(m)$. The emptying is done by enumerating m into B and hence causing another A-change.

The emptying device for F_{j-1} is the $\gamma^A(m)$-marker. It is depicted as a hook, which besides being used once on purpose may go off finitely often by itself (this is the visualization of a premature A-change). When F_{j-1} is emptied into F_j then $F_{j-2}, \ldots, F_0$ are spilled on the floor.

Though the recursion starts by calling P_k with goal 1, wine is first poured into the highest decanter F_0, and thereby into the left domain of L. We want to ensure that at least half the wine we put into F_0 reaches F_k. Recall that the parameter β is the amount of garbage $P_j(q, \beta)$ allows. If v is 1+the number of times the emptying device $\gamma^A(m)$ has gone off by itself, then P_j lets P_{j-1} fill F_{j-1} in portions of $2^{-v}\beta$ (ie it calls P_{j-1} with goal $2^{-v}\beta$ as often as necessary). Then, when F_{j-1} is emptied into F_j, at most $2^{-v}\beta$ can be lost because of being in higher decanters $F_{j-2}, \ldots, F_0$. Altogether the garbage due to $P_j(q, \beta)$ is at most $\beta \sum_{v \geq 1} 2^{-v} = \beta$.

Let us stress this key idea: when we have to cancel a run $P_j(q, \beta)$ because of a premature A-change, what becomes garbage is *not* F_{j-1}, but rather what the sub-procedures called by this run were working on. The set F_{j-1} already is a $j-1$-set, so all we need is another A-change, which is provided here by the cancellation itself, as opposed to being caused actively once the run reaches its goal.

Who enumerates L? The bottom procedures $P_2(q, \beta)$, which is where the recursion reaches ground. It puts requests $\langle r_n, n \rangle$ into L and the top decanter F_0, where n is large and $2^{-r_n} = 2^{-v}\beta$ for v as above. Once it sees the corresponding $A \upharpoonright n$ description, it empties F_0 into F_1. However, if the hook $\gamma^A(m)$ belonging to P_2 moves before that, then F_0 is spilled on the floor, while F_1 is emptied into F_2.

So much for the discussion of Turing incompleteness. Next, we improve this to lowness.

Theorem 3.4. *Each K-trivial set is low.*

Proof. Let $J^A(e)$ denote $\Phi_e^A(e)$. A procedure $P_j(q, \beta)$ is started when $J^A(e)$ newly converges. The goal q is $\alpha 2^{-e}$, where α is the garbage quota of the procedure of type P_{j+1} that called it (assuming $j < k$).

For different e they run in parallel, so we now have a tree of decanters. We cannot change A actively any more (and we are happy if it doesn't). However, this creates a new type of garbage, where $P_j(q, \beta)$ reaches its goal, but no A change happens after that would allow us to empty F_{j-1} into F_j. In this case no more procedures are started because of a new convergence of $J^A(e)$. So the total weight of garbage of this type is $\leq \sum_e 2^{-e}\alpha$, which can be tolerated.

The man with the golden run. The initial procedure P_k never returns, since it has goal 1, while a k-set has weight at most $1/2$ by Lemma 3.3. So there must be a golden run of a procedure $P_{j+1}(p, \alpha)$: it doesn't reach its goal, but all the subprocedures it calls either reach their goals or are cancelled by premature A changes. The golden run shows that A is low: When the run of the subprocedure P_j based on a computation $J^A(e)$ returns, then we guess that $J^A(e)$ converges. If A changes below the use of $J^A(e)$, then P_{j+1} receives the fixed quantity $2^{-e}\alpha$. In this case we change the guess back to "divergent". This can only happen r times where $r = 2^e p/\alpha$, else P_{j+1} reaches its goal. (Note that α is chosen of the form 2^{-l}.) □

This proof actually shows that A is super-low: the number of changes in the approximation of A' is computably bounded. The lowness index is not obtained uniformly: we needed to know which run is golden. This non-uniformity is necessary [10,24].

We are now ready for the full result.

Theorem 3.5. *A is K-trivial iff A is low for K.*

This was obtained joint with Hirschfeldt, via a modification of my result that the K-trivial sets are closed downward under $\leq_T$. It implies lowness, as we have seen an easy proof that each low for K set is GL_1, and each K-trivial set is Δ_2^0.

Proof. A procedure $P_j(q, \beta)$ is started when $U^A(\sigma) = y$ newly converges. The goal q is $\alpha 2^{-|\sigma|}$. Thus the construction is similar to the one in the proof of Theorem 3.4. Again, it is necessary to call procedures based on different inputs σ in parallel. So we have a tree of decanters. At the golden run node $P_{j+1}(p, \alpha)$, we can show that A is low for K, by emulating the cost function construction of a low for K set. When $P_j(q, \beta)$ associated with $U^A(\sigma) = y$ returns, we have the right to put a request $\langle |\sigma| + c, y \rangle$ into a set W (where $c = 1 + \log_2(p/\alpha)$). An A change has a cost, since we put a request for a wrong computation. The fact that P_{j+1} does not reach its goal implies that the cost is bounded. Hence W is a KC-set. □

3.4. *Further applications of the golden run method*

Here are two further applications.

1. The cost function construction is necessary even for K-trivial Δ_2^0 sets A. This can be used to show that there is a c.e. K-trivial set Turing above A [25].

2. A real number r is left-c.e. if $\{q \in \mathbb{Q} : q < r\}$ is c.e. For each K-trivial set A, the relativized Chaitin probability Ω^A is left-c.e. [8]. (The converse also holds here, in case that A is Δ_2^0: if Ω^A is left-c.e. then it is Turing complete, so A is a basis for ML-randomness, hence K-trivial.)

4. Effective descriptive set theory

Π_1^1 sets of numbers are a high-level analog of the c.e. sets, where the steps of an effective enumeration are recursive ordinals. For details see [27]. Hjorth and Nies [13] have studied the analogs of K and of ML-randomness based on Π_1^1-sets. The analog of K in the Π_1^1 setting is denoted $\widetilde{K}$. The analogs of the KC-theorem and Schnorr's Theorem hold, but the proofs take considerable extra effort due to the extra complication of limit stages. There is a Π_1^1-set of numbers which is $\widetilde{K}$-trivial and not hyperarithmetical. In contrast,

Theorem 4.1. *If A is low for Π_1^1-ML-randomness, then A is hyperarithmetical.*

So K-trivial and low for ML-randomness differ in the Π_1^1-setting.

Proof. First we show that $\omega_1^A = \omega_1^{CK}$. This is used to prove that A is in fact $\widetilde{K}$-trivial at some $\eta < \omega_1^{CK}$, namely

$$\forall n \ \widetilde{K}_\eta(A \restriction n) \leq \widetilde{K}_\eta(n) + b.$$

Then A is hyperarithmetical, by the same argument Chaitin used to show that K-trivial sets are Δ_2^0: The collection of Z which are $\widetilde{K}$-trivial at η form a hyperarithmetical tree of width $O(2^b)$. $\square$

5. Subclasses of the K-trivials

Next we look at subclasses of the K-trivial sets, downward closed under Turing reducibility, which may be proper. We will mostly restrict ourselves to the c.e. K-trivial sets, which is a minor restriction here since each K-trivial is Turing below a c.e. one, see subsection 3.4.

5.1. *ML-coverable and ML-noncuppable sets*

We have already seen a subclass of the c.e. K-trivials that is downward closed: the c.e. sets A such that there a ML-random set $Z <_T \emptyset'$ Turing above A. Let us call a c.e. set of that kind *ML-coverable*. In Question 3.2 we ask if each (c.e.) K-trivial set is ML-coverable.

A Δ_2^0 set A is *ML-cuppable* if

$$A \oplus Z \equiv_T \emptyset' \text{ for some ML-random } Z <_T \emptyset'.$$

Here is a further subclass: the c.e. sets that are not ML-cuppable (this recent development was initiated by Kučera in 2004).

Many sets *are* ML-cuppable: If A is not K-trivial, then $A \not\leq_T \Omega^A$ by the hungry sets theorem 2.5, and $A' \equiv_T \Omega^A \oplus A \geq_T \emptyset'$. If A is also low, then $Z = \Omega^A <_T \emptyset'$, so A is ML-cuppable. This shows for instance that each c.e. non-K-trivial set B is ML-cuppable, since one can split it into low c.e. sets, $B = A_0 \cup A_1$, and one of them is also not K-trivial. So the ML-noncuppable c.e. sets are K-trivial.

Theorem 5.1. [23]. *There is a promptly simple set which is not ML-cuppable.*

The proof combines cost functions with the priority method.

Question 5.2. *Can a K-trivial set be ML-cuppable?*

The cost functions in the proof Theorem 5.1 are much more restricting then the one used to characterize the K-trivial sets. This gives some weak evidence that the question has an affirmative answer. The same applies to the Kučera construction of a simple A below a Δ_2^0 ML-random $Z \upharpoonright e$ related to Question 3.2. Recall here that, if $Z \upharpoonright e$ changes another time at s, then $[0, s)$ becomes taboo for S_e. This can happen as often and as late as the computable approximation of Z determines.

5.2. *A common subclass*

By recent work of Hirschfeldt and Nies, and later Miller, there is natural class $\mathcal{L}$ which is a subclass of both the ML-coverable and the ML-noncuppable sets. $\mathcal{L}$ determines an ideal in the c.e. Turing degrees. The following notion is the key. B is $\emptyset'$-trivializing (also called *almost complete*) if $\emptyset'$ is K-trivial relative to B. That is, there is $c \in \mathbb{N}$ such that

$$\forall n \; K^B(\emptyset' \upharpoonright n) \leq K^B(0^n) + c$$

Such a set is high, in fact $\emptyset'' \leq_{tt} B'$. One can obtain Turing incomplete $\emptyset'$-trivializing c.e. sets via the Jockusch-Shore pseudojump inversion.

Theorem 5.3. [14]. *For each c.e. operator W there is a c.e. set C such that*

$$W^C \oplus C \equiv_T \emptyset'.$$

I observed in [25] that pseudojump inversion, applied to the c.e. operator W given by the cost function construction, yields an incomplete but

almost complete c.e. set. A recent result shows that pseudojumps can also
be inverted via ML-random sets.

Theorem 5.4. *Nies 2006, see* [21]. *The conclusion of Theorem 5.3 also
holds for "C ML-random".*

The proof, which was simplified with the help of J. Miller, combines the
techniques to prove the Low Basis Theorem with the methods used in the
proof of Theorem 5.3. While this proof uses full approximation, Kučera has
given an alternative proof purely based on forcing with Π_1^0-classes.

Corollary 5.5. *There is a ML-random almost complete Δ_2^0-set.*

Now let

$$\mathcal{L} = \{A : \ A \text{ is c.e. } \& \ \forall Z$$
$$Z \text{ ML-random, almost complete } \Rightarrow \ A \leq_T Z\}.$$

Clearly, $\mathcal{L}$ determines an ideal in the Turing degrees. Hirschfeldt proved
that there is a promptly simple set in $\mathcal{L}$. By the previous corollary, each
$A \in \mathcal{L}$ is ML-coverable, and hence K-trivial.

Surprisingly, $\mathcal{L}$ also is a subclass of the ML-noncuppable sets. The reason
for this inclusion is that each potential Δ_2^0 ML-random cupping partner Z
of a K-trivial A is almost complete. The proof, due to Hirschfeldt (Dec.
2005), involves a relativization of the van Lambalgen Theorem: for any sets
Z, B and A such that $Z \in \mathsf{MLR}^A$, we have $B \oplus Z \in \mathsf{MLR}^A$ iff $B \in \mathsf{MLR}^{Z \oplus A}$.
Now argue as follows. For each set B,

$$B \in \mathsf{MLR}^Z \Rightarrow B \oplus Z \in \mathsf{MLR}$$
$$\Rightarrow B \oplus Z \in \mathsf{MLR}^A$$
$$\Rightarrow B \in \mathsf{MLR}^{Z \oplus A}$$
$$\Rightarrow B \in \mathsf{MLR}^{\emptyset'}.$$

Thus $\emptyset'$ is low for ML-randomness relative to Z, and hence $\emptyset'$ is K-trivial
relative to Z (using that $Z \leq_T \emptyset'$).

Unless $\mathcal{L}$ coincides with the class of c.e. K-trivial sets, $\mathcal{L}$ is an ideal of
the type we were looking for. The proof that there is a promptly simple set
in $\mathcal{L}$ later was both simplified and generalized.

Theorem 5.6. *Hirschfeldt, Miller 2006. Let $\mathcal{C}$ be a Σ_3^0 null class. Then
there is a promptly simple A such that $A \leq_T Z$ for each ML-random $Z \in \mathcal{C}$.*

Proof. The proof uses a cost function argument. First suppose that $\mathcal{C}$ is a Π_2^0-class. Then $\mathcal{C} = \bigcap_x V_x$ for an effective sequence $(V_x)_{x \in \mathbb{N}}$ of Σ_1^0-classes such that $V_x \supseteq V_{x+1}$ for each x. Let $(V_{x,s})_{s \in \mathbb{N}}$ be an effective ascending sequence of clopen sets approximating V_x, and let $c(x, s) = \mu V_{x,s}$. Then (3.1) holds because $\lim_x \mu V_x = 0$. Now run the cost function construction from subsection 3.2, using this new definition of $c(x, s)$, and obtain a promptly simple set A. When x enters A at stage s, enumerate $V_{x,s}$ into a Solovay test $\mathcal{G}$ (that is, put σ into $\mathcal{G}$, for all σ of length s such that $[\sigma] \subseteq V_{x,s}$). The construction ensures that $\mathcal{G}$ is indeed a Solovay test, by the definition of $c(x, s)$. To see that $A \leq_T Z$ for any ML-random $Z \in \mathcal{C}$, choose s_0 such that $\sigma \not\preceq Z$ for any σ enumerated into $\mathcal{G}$ after stage s_0. Given an input $x \geq s_0$, using Z, compute $t > x$ such that $Z \in V_{x,t}$. Then $x \in A \Leftrightarrow x \in A_t$, for if we put x into A at a later stage, this would throw Z out of $V_{x,t}$.

If $\mathcal{C}$ is a Σ_3^0-class, we extend the argument slightly: we have $\mathcal{C} = \bigcup_i \bigcap_x (V_x^i)_{i,x \in \mathbb{N}}$, for an effective double sequence $(V_x^i)_{i,x \in \mathbb{N}}$ of Σ_1^0-classes such that $V_x^i \supseteq V_{x+1}^i$ for each i, x. Run the cost function construction from subsection 3.2, now based on the function

$$c(x, s) = \sum_i 2^{-i} \mu V_{x,s}^i.$$

To show (3.1) for this new cost function, note that, given k, there is x_0 such that

$$\forall i \leq k + 1 \; \forall x \geq x_0 \; \mu V_x^i \leq 2^{-k-1}.$$

Then $c(x, s) \leq 2^{-k}$ for all $x \geq x_0$ and all s, since the total contribution of terms $2^{-i} \mu V_{x,s}^i$ for $i \geq k + 2$ to $c(x, s)$ is bounded by 2^{-k-1}.

For each i, build a Solovay test $\mathcal{G}_i$, by enumerating $V_{x,s}^i$ into $\mathcal{G}_i$ when x enters A at stage s. The sum of all $2^{-|\sigma|}$, for strings σ enumerated into $\mathcal{G}_i$, is bounded by 2^{i+1}. If $Z \in \mathcal{C}$ is ML-random, then choose i such that $Z \in \bigcap_x V_x^i$, and argue as before that $A \leq_T Z$ using $\mathcal{G}_i$. $\qquad\square$

To obtain a promptly simple set in $\mathcal{L}$, it is now sufficient to observe that the Σ_3^0-class of almost complete sets is a null class (for instance, because the larger class of high degrees has measure 0, but one can also give a direct proof).

If $\mathcal{C} = \{Z\}$, for a ML-random Δ_2^0 set Z. Then $\mathcal{C}$ is a Π_2^0 class: let Z_x be the string of length x approximating Z at stage x, and let $V_{x,s} = \bigcup_{x < y < s} [Z_y \upharpoonright m_y]$, where m_y is least such that $Z_y(m_y) \neq Z_{y-1}(m_y)$, so that $\mathcal{C} = \bigcap_x V_x$. Now the proof turns into the proof of Kučera's Theorem 2.3, for the special case that Z is a ML-random set, discussed in subsection 2.2.

5.3. Is there a characterization of K-triviality independent of randomness and K?

Figueira, Nies and Stephan [12] have tried ultra-lowness, a strengthening of super-lowness: For each unbounded nondescending computable function h, A' has an approximation that changes at most $h(x)$ times at x. This property only depends on the Turing degree of A. They build a c.e. non-computable such set, via a construction that resembles the cost function construction. Recently Cholak, Downey and Greenberg [7] have proved that the class of c.e. ultra-low sets is a proper subclass of the c.e. K-trivials.

References

1. K. Ambos-Spies and A. Kučera. Randomness in computability theory. In Peter Cholak, Steffen Lempp, Manny Lerman, and Richard Shore, editors, *Computability Theory and Its Applications: Current Trends and Open Problems*. American Mathematical Society, 2000.
2. B. Bedregal and A. Nies. Lowness properties of reals and hyper-immunity. In *Wollic 2003*, volume 84 of *Electronic Notes in Theoretical Computer Science*. Elsevier, 2003. Available at
   http://www.elsevier.nl/locate/entcs/volume84.html.
3. Cristian S. Calude and Richard J. Coles. Program-size complexity of initial segments and domination reducibility. In *Jewels are forever*, pages 225–237. Springer, Berlin, 1999.
4. G. Chaitin. A theory of program size formally identical to information theory. *J. Assoc. Comput. Mach.*, 22:329–340, 1975.
5. G. Chaitin. Information-theoretical characterizations of recursive infinite strings. *Theoretical Computer Science*, 2:45–48, 1976.
6. P. Cholak, N Greenberg, and J. Miller. Uniform almost everwhere domination. *J. Symbolic Logic*, to appear.
7. P. Cholak, R. Downey and N. Greenberg. Strong jump traceability: the computably enumerable case. *To appear.*
8. R. Downey, D. Hirschfeldt, J. Miller, and A. Nies. Relativizing Chaitin's halting probability. *J. Math. Logic* 5, No. 2 (2005) 167-192. .
9. R. Downey, D. Hirschfeldt, A. Nies, and S. Terwijn. Calibrating randomness. *Bull. Symb. Logic*, to appear
10. Rod G. Downey, Denis R. Hirschfeldt, André Nies, and Frank Stephan. Trivial reals. In *Proceedings of the 7th and 8th Asian Logic Conferences*, pages 103–131, Singapore, 2003. Singapore Univ. Press.
11. D. Hirschfeldt, A. Nies, and F. Stephan. Using random sets as oracles. To appear.
12. S. Figueira, A. Nies and F. Stephan. Lowness properties and approximations of the jump. *Annals of pure and applied logic*, to appear.
13. G. Hjorth and A. Nies. Randomness in effective descriptive set theory. To appear.

14. C. G. Jockusch, Jr. and R. A. Shore. Pseudo-jump operators I: the r.e. case. *Trans. Amer. Math. Soc.*, 275: 599–609, 1983.
15. B. Kjos-Hanssen, A. Nies, and F. Stephan. Lowness for the class of Schnorr random sets. *SIAM J. Computing*, to appear.
16. A. Kučera. Measure, Π_1^0-classes and complete extensions of PA. In *Recursion theory week (Oberwolfach, 1984)*, volume 1141 of *Lecture Notes in Math.*, pages 245–259. Springer, Berlin, 1985.
17. A. Kučera. On relative randomness. *Ann. Pure Appl. Logic*, 63:61–67, 1993.
18. A. Kucera and S. Terwijn. Lowness for the class of random sets. *J. Symbolic Logic*, 64:1396–1402, 1999.
19. J. Miller and A. Nies. Randomness and computability: open questions. To appear.
20. J. Miller and L. Yu. On initial segment complexity and degrees of randomness. To appear in Trans. Amer. Math. Soc.
21. A. Nies. Computability and Randomness. *Oxford University Press*. To appear in the series Oxford Logic Guides.
22. A. Nies. Low for random sets: the story. Preprint; available at http://www.cs.auckland.ac.nz/nies/papers/.
23. A. Nies. Non-cupping and randomness. *Proc. AMS*, to appear.
24. A. Nies. Reals which compute little. In *Logic Colloquium '02*, Lecture Notes in Logic, pages 260–274. Springer–Verlag, 2002.
25. A. Nies. Lowness properties and randomness. *Adv. in Math.*, 197:274–305, 2005.
26. A. Nies, F. Stephan, and S. A. Terwijn, *Randomness, relativization, and Turing degrees*, J. Symbolic Logic 70 (2005), 515–535.
27. G. Sacks. *Higher Recursion Theory.* Perspectives in Mathematical Logic. Springer–Verlag, Heidelberg, 1990.
28. G. Sacks. Degrees of Unsolvability, volume 55 of *Annals of Mathematical Studies*. Princeton University Press, 1963.
29. S. Terwijn and D. Zambella. Algorithmic randomness and lowness. *J. Symbolic Logic*, 66:1199–1205, 2001.
30. D. Zambella. On sequences with simple initial segments. ILLC technical report ML 1990 -05, Univ. Amsterdam, 1990.

A LOWER CONE IN THE wtt DEGREES OF NON-INTEGRAL EFFECTIVE DIMENSION

André Nies

Department of Computer Science, University of Auckland
Private Bag 92019, Auckland, New Zealand
E-mail: andre@cs.auckland.ac.nz

Jan Reimann

Institut für Informatik, Ruprecht-Karls-Universität Heidelberg
Im Neuenheimer Feld 294, 69120, Heidelberg, Germany
E-mail: reimann@math.uni-heidelberg.de

For any rational number r, we show that there exists a set A (weak truth-table reducible to the halting problem) such that any set B weak truth-table reducible to it has effective Hausdorff dimension at most r, where A itself has dimension at least r. This implies, for any rational r, the existence of a wtt-lower cone of effective dimension r.

1. Introduction

Since the introduction of effective dimension concepts by Lutz [8, 9], considerable effort has been put into studying the effective or resource-bounded dimension of objects occurring in computability or complexity theory. However, up to now there are basically only three types of examples known for individual sets of non-integral effective dimension: The first consists of sets obtained by 'diluting' a Martin-Löf random set with zeroes (or any other computable set). The second example comprises all sets which are random with respect to a Bernoulli distribution on Cantor space. Here Lutz transferred a classic result by Eggleston [1] to show that if μ is a (generalized) Bernoulli measure, then the effective dimension of a Martin-Löf

The first author is supported by the Marsden fund of New Zealand, UOA 319.

μ-random set coincides with the entropy H(μ) of the measure μ. Finally, the third example is a parameterized version of Chaitin's Ω introduced by Tadaki [23].

An obvious question is whether there exist examples of non-integral effective dimension among classes of central interest to computability theory, such as cones or degrees. It is interesting to note that all the examples mentioned above actually produce sets which are Turing equivalent to a Martin-Löf random set. Therefore, one cannot use them to obtain Turing cones of non-integral dimension.

However, when restricted to many-one reducibility, Reimann and Terwijn [15] showed that the lower cone of a Bernoulli random set cannot contain a set of higher dimension than the random set it reduces to, thereby obtaining many-one lower cones of non-integral effective dimension. But the proof does not transfer to weaker reducibilities. Using a different approach, Stephan [22] was able to construct an oracle relative to which there exists a wtt-lower cone of positive effective dimension at most $1/2$.

In this paper we construct, for an arbitrary rational number r, a wtt-lower cone of effective Hausdorff dimension r. This result was independently announced by Hirschfeldt and Miller [3]. The case of Turing reducibility seems much more difficult and remains a major open problem in the field (see also [12]).

Notation. Our notation is fairly standard. 2^{ω} denotes Cantor space, the set of all infinite binary sequences. We identify elements of 2^{ω} with subsets of the natural numbers $\mathbb{N}$ by means of the characteristic function, thus elements of 2^{ω} are generally called *sets*, whereas subsets of 2^{ω} are called *classes*. Sets will be denoted by upper case letters like A, B, C, or X, Y, Z, classes by calligraphic upper case letters $\mathcal{A}, \mathcal{B}, \dots$.

Strings are finite initial segments of sets will be denoted by lower case Latin or Greek letters such as u, v, w, x, y, z or σ, τ. $2^{<\omega}$ will denote the set of all strings. The *initial segment of length n*, $A \restriction n$, of a set A is the string of length n corresponding to the first n bits of A.

Given two strings v, w, v is called a *prefix* of w, written $v \preceq w$, if there exists a string x such that $vx = w$, where vx is the concatenation of v and x. If w is strictly longer than v, we write $v \prec w$. This extends in a natural way to hold between strings and sets. A set of strings is called *prefix free* if no element has a prefix (other than itself) in the set.

Initial segments induce a standard topology on 2^{ω}. The basis of the topology is formed by the *basic open cylinders* (or just *cylinders*, for short).

Given a string $w = w_0 \ldots w_{n-1}$ of length n, these are defined as

$$[w] = \{A \in 2^\omega : A \restriction n = w\}.$$

Imposing this topology turns 2^ω into a totally disconnected Polish space. A class is clopen in 2^ω if and only it is the union of finitely many cylinders.

Finally, λ denotes Lebesgue measure on 2^ω, generated by setting $\lambda[\sigma] = 2^{-|\sigma|}$ for every string σ. For each measurable $\mathcal{C} \subseteq 2^\omega$, recall that the conditional probability is

$$\lambda(\mathcal{C} \mid \sigma) = \lambda(\mathcal{C} \cap [\sigma])2^{|\sigma|}.$$

For all unexplained notions from computability theory we refer to any standard textbook such as [14] or [19], for details on Kolmogorov complexity, the reader may consult [7]; [13] will provide background on the use of measure theory, especially martingales, in the theory of algorithmic randomness.

In the proof of our main result we will use so-called *Kraft-Chaitin sets*. A Kraft-Chaitin set L is a c.e. set of pairs $\langle l, x \rangle$ (called *requests*), where l is a natural number and x is a string, and $\sum_L 2^{-l} \leq 1$. It is a fundamental result in algorithmic randomness that if L is a Kraft-Chaitin set, then $K(x) \leq^+ l$ if $\langle l, x \rangle \in L$.

2. Effective Dimension

In this section we briefly introduce the concept of effective Hausdorff dimension. As we deal exclusively with Hausdorff dimension, we shall in the following often suppress "Hausdorff" and speak simply of *effective dimension*. For a more detailed account of effective dimension notions we refer to [15].

Hausdorff dimension is based on *Hausdorff measures*, which can be effectivized in the same way Martin-Löf tests effectivize Lebesgue measure on Cantor space.

Definition 2.1. Let $0 \leq s \leq 1$ be a rational number. A class $\mathcal{X} \subseteq 2^\omega$ has *effective s-dimensional Hausdorff measure* 0 (or simply is *effectively $\mathcal{H}^s$-null*) if there is a uniformly computably enumerable sequence $\{C_n\}_{n \in \mathbb{N}}$ of sets of strings such that for every $n \in \mathbb{N}$,

$$\mathcal{X} \subseteq \bigcup_{\sigma \in C_n} [\sigma] \quad \text{and} \quad \sum_{w \in C_n} 2^{-s|w|} \leq 2^{-n}.$$

It is obvious that if $\mathcal{X}$ is effectively $\mathcal{H}^s$-null for some rational $s \geq 0$, then it is also effectively $\mathcal{H}^t$-null for any rational $t > s$. This justifies the following definition.

Definition 2.2. The *effective Hausdorff dimension* $\dim_{\mathrm{H}}^1 \mathcal{X}$ of a class $\mathcal{X} \subseteq 2^\omega$ is defined as

$$\dim_{\mathrm{H}}^1 \mathcal{X} = \inf\{s \in \mathbb{Q}^+ : \mathcal{X} \text{ is effectively } \mathcal{H}^s\text{-null}\}$$

The classical (i.e. non-effective) notion of Hausdorff dimension can be interpreted as the right "scaling factor" of $\mathcal{X}$ with respect to Lebesgue measure. The effective theory, however, allows for an interpretation in terms of *algorithmic randomness*. There exist singleton classes, i.e. sets, of positive dimension (whereas in the classical setting every countable class is of Hausdorff dimension zero). In fact, effective dimension has a strong stability property [9]: For any class $\mathcal{X}$ it holds that

$$\dim_{\mathrm{H}}^1 \mathcal{X} = \sup\{\dim_{\mathrm{H}}^1\{A\} : A \in \mathcal{X}\}. \tag{2.1}$$

That is, the effective dimension of a class is completely determined by the dimension of its members (viewed as singleton classes). We simplify notation by writing $\dim_{\mathrm{H}}^1 A$ in place of $\dim_{\mathrm{H}}^1\{A\}$. The effective dimension of a set can be regarded as an indicator of its *degree of randomness*. This is reflected in the following theorem.

Theorem 2.3. *For any set $A \in 2^\omega$ it holds that*

$$\dim_{\mathrm{H}}^1 A = \liminf_{n \to \infty} \frac{K(A \upharpoonright n)}{n}.$$

In other words, the effective dimension of an individual set equals its lower asymptotic entropy. In the following, we will use $\underline{K}(A)$ to denote $\liminf K(A \upharpoonright n)/n$. Theorem 2.3 was first explicitly proved in [11], but much of it is already present in earlier works on Kolmogorov complexity and Hausdorff dimension, such as [17] or [20]. The result can be derived quite easily from the existence of a universal semimeasure (discrete or continuous) by using the *coding theorem*, as observed by Reimann [15] and Staiger [21].

Examples for effective dimension. As mentioned in the introduction, there are mainly three types of examples of sets of non-integral effective dimension.

(1) If $0 < r < 1$ is rational, let $Z_r = \{\lfloor n/r \rfloor : n \in \mathbb{N}\}$. Given a Martin-Löf random set X, define X_r by

$$X_r(m) = \begin{cases} X(n) & \text{if } m = \lfloor n/r \rfloor, \\ 0 & \text{otherwise.} \end{cases}$$

Then, using Theorem 2.3, it is easy to see that

$$\dim_{\mathrm{H}}^1 X_r = r.$$

This technique can be refined to obtain sets of effective dimension s, where $0 \le s \le 1$ is any Δ_2^0-computable real number (see e.g. [10]).

(2) Given a Bernoulli measure μ_p with bias $p \in \mathbb{Q} \cap [0,1]$, the effective dimension of any set that is Martin-Löf random with respect to μ_p equals the entropy of the measure $\mathrm{H}(\mu_p) = -[p \log p + (1-p) \log(1-p)]$ [8]. This is an effective version of a classical theorem due to Eggleston [1].

(3) Let U be a universal, prefix-free machine. Given a computable real number $0 < s \le 1$, the binary expansion of the real number

$$\Omega^{(s)} = \sum_{\sigma \in \mathrm{dom}(U)} 2^{-\frac{|\sigma|}{s}}$$

has effective dimension s. This was shown by Tadaki [23]. Note that $\Omega^{(1)}$ is just Chaitin's Ω.

2.1. *Effective Dimension of cones and degrees*

Fundamental results by Gacs [2] and Kučera [6] showed that every set is Turing reducible to a Martin-Löf random one. Since a Martin-Löf random set has effective dimension 1, it follows from (2.1) that every Turing upper cone is of effective dimension 1. Even more, Reimann [15] was able to show that every many-one upper cone has classical Hausdorff dimension 1 (and hence effective dimension 1, too). This contrasts a classical result by Sacks [18] which shows that the Turing upper cone of a set has Lebesgue measure zero unless the set is recursive.

As regards lower cones and degrees, the situation is different. First, using coding at very sparse locations along with symmetry of algorithmic information, one can show that effective dimension is closed upwards in the weak truth-table degrees, that is, for any sets $A \le_{wtt} B$, the weak truth table-degree of B contains a set C of dimension $\dim_{\mathrm{H}}^1 A$. It is sufficient to choose a computable set R of density $\lim_n |R \cap \{0, \ldots, n-1\}|/n = 1$, and let C equal A on R and B on the complement of R. It follows that the dimension of the weak truth-table degree and the weak truth-table lower cone of a set coincide. The same holds for Turing reducibility.

All three types of examples mentioned above compute a Martin-Löf random set, albeit for different reasons.

It is obvious that any diluted set X_r computes a Martin-Löf random sequence. Furthermore, Levin [25] and independently Kautz [4] showed that any sequence which is random with respect to a computable probability measure on 2^ω (which includes the Bernoulli measures μ_p with rational bias) computes a Martin-Löf random set. Finally, for every rational s, the set Ω_s is a left-c.e. real number. Furthermore, it is not hard to see that it computes a fixed-point free (fpf) function. Hence it follows from the Arslanov completeness criterion that Ω_s is Turing complete and therefore computes a Martin-Löf random set as well.

Regarding stronger reducibilities, Reimann and Terwijn [15] showed that a many-one reduction cannot increase the entropy of a set random with respect to a Bernoulli measure μ_p, p rational. It follows that the many-one lower cone of such a set has effective dimension $H(\mu_p)$.

However, this result does not extend to weaker reducibilities such as truth-table reducibility, since for such measures the Levin-Kautz result holds for a total Turing reduction.

Recently, using a different approach, Stephan [22] was able to construct an oracle relative to which there exists a wtt-lower cone of positive effective dimension at most $1/2$. In the next section we improve this by showing that, for an arbitrary rational number r, there exists an (unrelativized) weak truth-table lower cone of effective Hausdorff dimension at most r.

3. The Main Result

Theorem 3.1. *For each rational α, $0 \leq \alpha \leq 1$, there is a set $A \leq_{\text{wtt}} \emptyset'$ such that $\underline{K}(A) = \alpha$ and $\underline{K}(Z) \leq \alpha$ for each $Z \leq_{\text{wtt}} A$.*

Proof. Let $\mathcal{P}$ be the Π^0_1-class given by

$$\mathcal{P} = \{Z : \forall n \geq n_0 \, K(Z \restriction n) \geq \lfloor \alpha n \rfloor\},$$

where n_0 is chosen so that $\lambda \mathcal{P} \geq 1/2$. Recall that each Π^0_1-class comes with an effective clopen approximation, so we assume that there exists an effective sequence (P_s) of finite sets of strings such that $\mathcal{P} = \bigcap_s P_s$. To facilitate readability we mostly identify finite sets of strings with the clopen class they induce. (If we want to explicitly denote the clopen class induced by some finite set S of strings, we write $[S]^{\preceq}$.) As usual, it is useful to imagine $\mathcal{P}$ and the P_s as sets of infinite paths through trees.

Lemma 3.2. *Let $\mathcal{C}$ be a clopen class such that $\mathcal{C} \subseteq P_s$ and $\mathcal{C} \cap P_t = \emptyset$ for stages $s < t$. Then $\Omega_t - \Omega_s \geq (\lambda \mathcal{C})^\alpha$.*

Proof. Each minimal string in $\mathcal{C}$ has a substring x that receives a description of length at most $\alpha|x|$ between s and t. Thus there is a prefix free set

$\{x_1, \dots, x_m\}$ such that all $[x_i] \cap P_s \neq \emptyset$, $\mathcal{C} \subseteq \bigcup_i [x_i]$, and $K_t(x_i) \leq \alpha |x_i|$. Then, since the function $y \mapsto y^\alpha$ is concave,

$$\Omega_t - \Omega_s \geq \sum_i 2^{-\alpha |x_i|} \geq (\sum_i 2^{-|x_i|})^\alpha \geq (\lambda \mathcal{C})^\alpha.$$

This proves the lemma.

Now we build A on P, thus $\underline{K}(A) \geq \alpha$. To ensure $\underline{K}(Z) \leq \alpha$ for each $Z \leq_{\text{wtt}} A$, we meet the requirements R_j, for each $j = \langle e, b \rangle > 0$.

$$R_j : Z = \Psi_e(A) \ \Rightarrow \ \exists k \geq j \, K(Z \upharpoonright k) \leq^+ \beta k,$$

where $\beta = \alpha + 2^{-b} < 1$, and $(\Psi_e)_{e \in \mathbb{N}}$ is a uniform listing of wtt reduction procedures, with partial computable use bound g_e, such that

$$\forall k \, (m = g_e(k) \downarrow \Rightarrow m \geq \beta k / 2).$$

Thus we only consider reductions which do not turn a short oracle string into a long output string. This is sufficient because a short oracle string would be enough to compress an initial segment of Z. More precisely, consider the plain machine S given by $S(0^e 1 \sigma) \simeq \Phi_e^\sigma$ (where that $(\Phi_e)_{e \in \mathbb{N}}$ is a uniform listing of Turing reduction procedures). Using S, we see that $\Phi_e^\sigma = x$ implies $C(x) \leq^+ |\sigma| + e + 1$. Hence $|\sigma| < \beta k / 2$ implies $K(x) \leq^+ \beta |x|$.

We let $A = \bigcup_j \sigma_j$ where σ_j is a string of length m_j. Both m_j and σ_j are controlled by R_j for $j > 0$. At any stage s, we have $\sigma_{j-1,s} \prec \sigma_{j,s}$ and

$$\lambda(\mathcal{P} \mid \sigma_j) \geq 2^{-2j-1}. \tag{3.1}$$

We let $m_0 = 0$ and hence $\sigma_0 = \emptyset$, so (3.1) is also true for $j = 0$.

We construct a Kraft-Chaitin set L. Each R_j may enumerate into L in order to ensure $K(Z \upharpoonright k) \leq^+ \beta k$.

The idea behind the construction is as follows. We are playing the following R_j strategy. We define a length k_j where we intend to compress Z, and let m_j be the use bound of Ψ_e, $g_e(k_j)$. We define σ_j of length m_j in a way that, if $x = \Psi_e^{\sigma_j}$ is defined then we compress it down to βk_j, by putting an appropriate request into L. The opponent's answer could be to remove σ_j from $\mathcal{P}$. But in that case, the measure he spent for this removal exceeds what we spent for our request, so we can account ours against his. Of course, usually σ_j is much longer than x. So we will only compress x when the measure of oracle strings computing it is large, and use Lemma 3.2.

For each $j, m \in \mathbb{N}$ and each stage t, let

$$G_{j,m,t} = \{\sigma : |\sigma| = m \ \& \ \lambda(P_t \mid \sigma) \geq 2^{-2j}\}.$$

Informally, let us call a string σ of length m_j *good for R_j at stage t* if $\sigma \succ \sigma_{j-1,t}$ and $\sigma \in G_{j,m_j,t}$. These are the only oracle strings R_j looks at. The reason to allow the conditional measure to drop from 2^{-2j+1} at σ_{j-1} down to 2^{-2j} is that we want a sufficiently large measure of them.

Lemma 3.3. *There is an effective sequence (u_j) of natural numbers such that the following holds. Whenever ρ is a string such that $\lambda(P_t \cap [\rho]) \geq 2^{-(2j-1)-|\rho|}$, then for each $m > |\rho|$,*

$$\lambda(G_{j,m,t} \cap [\rho]) \geq 2^{-u_j-|\rho|}.$$

Proof. For each measurable $\mathcal{C}$, one obtains a martingale by letting

$$M_{\mathcal{C}}(\sigma) = \lambda(\mathcal{C} \mid \sigma) = \lambda(\mathcal{C} \cap [\sigma])2^{|\sigma|}.$$

Now let $\mathcal{C} = 2^\omega - \mathcal{P}$. Let $d = 2j - 1$. By hypothesis $M_{\mathcal{C}}(\rho) \leq 1 - 2^{-d}$, so we may apply the so-called Kolmogorov inequality, which bounds the measure of strings $\sigma \succ \rho$ where M can reach $1 - 2^{-(d+1)}$:

$$\lambda(\{\sigma : |\sigma| = m \ \& \ M_{\mathcal{C}}(\sigma) \geq 1 - 2^{-(d+1)}\} \mid \rho) \leq \frac{1 - 2^{-d}}{1 - 2^{-(d+1)}}.$$

Now it suffices to determine u_j such that $1 - 2^{-u_j} \geq (1 - 2^{-d})/(1 - 2^{-(d+1)})$, since then $\lambda(G_{j,m,t} \mid \rho) \leq 2^{-u_j}$. This proves the lemma. ∎

R_j compresses x when the measure of good strings computing it is large (4. below). If each good string σ computing x later becomes very bad, in the sense that the conditional measure $\lambda(\mathcal{P} \mid \sigma)$ dropped down to half, then we can carry out the accounting argument mentioned above and therefore choose a new x (5. below).

The *construction* at stage $s > 0$ consists in letting the requirements R_j, for $j = 0 \ldots s$, carry out one step of their strategy. Each time σ_j is newly defined, all the strategies R_l, $l > j$, are initialized. Suppose $j > 0$, that $\rho = \sigma_{j-1}$ is defined already, (3.1) holds for $j - 1$ and $m_{j-1} = |\rho|$. Let

$$\text{init}(j) = j + 2 + \text{the number of times } R_j \text{ has been initialized.}$$

1. *Let k_j be so large that $\beta k_j \geq \max(\text{init}(j), 2m_{j-1})$, and, where $r_j = u_j + m_{j-1} + k_j$,*

$$\alpha(r_j + 2j + 1) \leq \beta k_j - \text{init}(j). \tag{3.2}$$

2. *While $g_e(k_j)$ is undefined, let $m_j = m_{j-1}+1$, let $\sigma_{j,s}$ be the leftmost extension of σ_{j-1} of length m_j such that $\lambda(P_s \mid \sigma_{j,s}) \geq 2^{-2j}$, and stay at 2. Else let $m_j = g_e(k_j)$ and go to 3.*

3. Let $G_s = G_{j,m_j,s} \cap [\rho]$. While there exists, let $\sigma_{j,s}$ be the leftmost string $\sigma \in G_s$ such that $\Psi_e^\sigma \upharpoonright k_j \uparrow$, and stay at 3. Else, for each string y of length k_j, let

$$S_y = \{\sigma \succ \rho : |\sigma| = m_j \ \& \ \Psi_{e,s}^\sigma = y\},$$

and go to 4.

4. Let x be the leftmost string of length k_j such that $\lambda(G_s \cap S_x) \geq 2^{-r_j}$ where $r_j = u_j + m_{j-1} + k_j$ as above. Put a request

$$\langle \lceil \beta k_j \rceil, x \rangle$$

into L. Note that x exists since $\lambda G_s \geq 2^{-u_j - m_{j-1}}$ by Lemma 3.3, and $(G_s \cap S_y)_{|y|=k_j}$ is a partition of G_s into at most 2^{k_j} sets (because we passed 3). Of course, $G_s \cap S_x$ may shrink later, in which case we have to try a new x. Eventually we will find the right one.

5. From now on, let $\sigma_{j,s}$ be the leftmost string σ in S_x that satisfies (3.1), namely, $\lambda(P_s \mid \sigma) \geq 2^{-2j-1}$. If there is no such σ, then we need to pick a new x, so go back to 4. (We had picked the wrong x. As indicated earlier, we will have to verify that this change is allowed, i.e., the contribution to L does no become too large. The reason is that the opponent had to add at least a measure $2^{-\alpha(r_j+2j+1)}$ of new descriptions to the universal prefix free machine in order to make $G_s \cap S_x$ shrink sufficiently, and we will account the cost of our request against the measure of his descriptions.)

Verification.

Claim 3.1. For each j, $\sigma_j = \lim_s \sigma_{j,s}$ exists.

This is trivial for $j = 0$. Suppose $j > 0$ and inductively the claim holds for $j - 1$. Once σ_{j-1} is stable at stage s_0, we define a final value k_j in 1. If $m_j = g_e(k_j)$ is undefined then $|\sigma_j| = |\sigma_{j-1}| + 1$ and σ_j can change at most once after s_0. Otherwise σ_j can change at most 2^{m_j} times till we reach 4. As remarked above, we always can choose some x in 4. If we cannot find σ in 5. any longer, then $\lambda(G_s \cap S_x) < 2^{-r_j}$, so we will discard this x and not pick it in 4. any more. Since there is an x such that $\lambda(G_t \cap S_x) \geq 2^{-r_j}$ for all $t \geq s_0$, eventually we will stay at 5. Since all strings σ_j we try here have length m_j, eventually σ_j stabilizes. This proves the claim.

Note that $|\sigma_j| > |\sigma_{j-1}|$, so $A = \bigcup_j \sigma_j$ defines a set. Let $A_s = \bigcup_j \sigma_{j,s}$. It is easy to verify that for each x, the number of changes of $A_s(x)$ is computably bounded in x, since the values $m_{j,s}$ are nondecreasing in s, and in 3. and 5., we only move to the right. Thus $A \leq_{\text{wtt}} \emptyset'$.

Claim 3.2. L is Kraft-Chaitin set.

By the definition of $\mathrm{init}(j)$, it suffices to verify that for each value $v = \mathrm{init}(j)$, the weight of the contributions of R_j to L is at most 2^{-v+1}. When a request $\langle r, x \rangle$ is enumerated by R_j at stage s, we distinguish two cases.

Case 1. The strategy stays at 5. after s or R_j is initialized. Then this was the last contribution, and it weighs at most 2^{-v} since we chose k_j in a way that $\beta k_j \geq v$.

Case 2. Otherwise, that is, the strategy gets back to 4. at a stage $t > s$. Then, for each $\sigma \in G_s \cap S_x$, $\lambda(P_t \mid \sigma) < 2^{-2j-1}$ (while, by the definition of G_s, we had $\lambda(P_s \mid \sigma) \geq 2^{-2j}$). Now consider the clopen class

$$\mathcal{C} = P_s \cap [G_s \cap S_x]^{\preceq} - P_t.$$

Since $\lambda(G_s \cap S_x) \geq 2^{-r_j}$,

$$\lambda\mathcal{C} \geq \sum_{\sigma \in G_s \cap S_x} 2^{-|\sigma|}(\lambda(P_s|\sigma) - \lambda(P_t|\sigma))$$

$$\geq \sum_{\sigma \in G_s \cap S_x} 2^{-|\sigma|}(2^{-2j} - 2^{-2j-1}) = \sum_{\sigma \in G_s \cap S_x} 2^{-|\sigma|}2^{-2j-1} \geq 2^{-r_j-2j-1}.$$

Clearly $\mathcal{C} \subseteq P_s$ and $\mathcal{C} \cap P_t = \emptyset$, so by Lemma 3.2 and (3.2),

$$\Omega_t - \Omega_s \geq 2^{-\alpha(r_j+2j+1)} \geq 2^{-\beta k_j + v},$$

hence the total contributions in Case 2 weigh at most $2^{-v}\Omega$. Together, the contribution is at most 2^{-v+1}.

Claim 3.3. If $Z \leq_{\mathrm{wtt}} A$ then $\underline{K}(Z) \leq \alpha$.

It suffices to show that each requirement R_j is met. For then, if $Z \leq_{\mathrm{wtt}} A$ either, the reduction has small use (see remarks at the beginning of the proof) or it is included in the list (Ψ_e). In the latter case, meeting $R_{\langle e,b \rangle}$ for each b ensures $\underline{K}(Z) \leq \alpha$.

By the first claim, σ_j and k_j reach a final value at some stage s_0. If $\Psi_e^A \upharpoonright k_j \uparrow$ then R_j is met. Otherwise, the strategy for R_j gets to 4. after s_0, and enumerates a request into L which ensures $K(Z \upharpoonright k_j) \leq^+ \beta k_j$.

This concludes the proof of Theorem 3.1.

4. Concluding Remarks

It remains an open problem whether there exists a Turing lower cone of non-integral effective dimension (see [12]). This case appears to be much harder. It is, for instance, not even known whether there exists a set of non-integral dimension which does not compute a Martin-Löf random set.

The best known result in this direction is that there exists a computable, non-decreasing, unbounded function f and a set A such that $K(A \upharpoonright n) \geq f(n)$ and A does not compute a Martin-Löf random set. This has been independently proved by Reimann and Slaman [16] and Kjos-Hanssen, Merkle, and Stephan [5].

References

1. H. G. Eggleston. The fractional dimension of a set defined by decimal properties. *Quart. J. Math., Oxford Ser.*, 20:31–36, 1949.
2. P. Gács. Every sequence is reducible to a random one. *Inform. and Control*, 70(2-3):186–192, 1986.
3. D. R. Hirschfeldt and J. S. Miller. personal communication, 2005.
4. S. M. Kautz. *Degrees of Random sequences*. PhD thesis, Cornell University, 1991.
5. B. Kjos-Hanssen, W. Merkle, and F. Stephan. Kolmogorov complexity and the Recursion Theorem. Manuscript, submitted for publication, 2005.
6. A. Kučera. Measure, Π_1^0-classes and complete extensions of PA. In *Recursion theory week (Oberwolfach, 1984)*, volume 1141 of *Lecture Notes in Math.*, pages 245–259. Springer, Berlin, 1985.
7. M. Li and P. Vitányi. *An introduction to Kolmogorov complexity and its applications*. Graduate Texts in Computer Science. Springer-Verlag, New York, 1997.
8. J. H. Lutz. Dimension in complexity classes. In *Proceedings of the Fifteenth Annual IEEE Conference on Computational Complexity*, pp. 158–169. IEEE Computer Society, 2000.
9. J. H. Lutz. Gales and the constructive dimension of individual sequences. In *Automata, languages and programming (Geneva, 2000)*, pp. 902–913. Springer, Berlin, 2000.
10. J. H. Lutz. The dimensions of individual strings and sequences. *Inform. and Comput.*, 187(1):49–79, 2003.
11. E. Mayordomo. A Kolmogorov complexity characterization of constructive Hausdorff dimension. *Inform. Process. Lett.*, 84(1):1–3, 2002.
12. J. S. Miller and A. Nies. Randomness and computability: Open questions. Submitted for publication, 2005.
13. A. Nies, *Computability and Randomness*, to appear.
14. P. Odifreddi. *Classical recursion theory*, volume 125 of *Studies in Logic and the Foundations of Mathematics*. North-Holland Publishing Co., Amsterdam, 1989.
15. J. Reimann. Computability and fractal dimension. Doctoral dissertation, Universität Heidelberg, 2004.
16. J. Reimann and T. A. Slaman. Randomness, entropy, and reducibility. Manuscript, in preparation, 2005.

17. B. Y. Ryabko. Noise-free coding of combinatorial sources, Hausdorff dimension and Kolmogorov complexity. *Problemy Peredachi Informatsii*, 22(3):16–26, 1986.
18. G. E. Sacks. *Degrees of unsolvability*. Princeton University Press, Princeton, N.J., 1963.
19. R. I. Soare. *Recursively enumerable sets and degrees*. Perspectives in Mathematical Logic. Springer-Verlag, Berlin, 1987.
20. L. Staiger. Kolmogorov complexity and Hausdorff dimension. *Inform. and Comput.*, 103(2):159–194, 1993.
21. L. Staiger. Constructive dimension equals Kolmogorov complexity. *Information Process. Lett.*, 93:149–153, 2005.
22. F. Stephan. Hausdorff dimension and weak truth-trable reducibility. Submitted for publication, 2005.
23. K. Tadaki. A generalization of Chaitin's halting probability Ω and halting self-similar sets. *Hokkaido Math. J.*, 31(1):219–253, 2002.
24. S. A. Terwijn. Complexity and randomness. Course Notes, 2003.
25. A. K. Zvonkin and L. A. Levin. The complexity of finite objects and the basing of the concepts of information and randomness on the theory of algorithms. *Uspehi Mat. Nauk*, 25(6(156)):85–127, 1970.

A MINIMAL rK-DEGREE

Alexander Raichev

Department of Computer Science, University of Auckland
Private Bag 92019, Auckland 1001, New Zealand
E-mail: raichev@cs.auckland.ac.nz

Frank Stephan

School of Computing and Department of Mathematics
National University of Singapore, 3 Science Drive 2, Singapore 117543
E-mail: fstephan@comp.nus.edu.sg

We construct a minimal rK-degree, continuum many, in fact. We also show that every minimal sequence, that is, a sequence with minimal rK-degree, must have very low descriptional complexity, that every minimal sequence is rK-reducible to a random sequence and that there is a random sequence with no minimal sequence rK-reducible to it.

1. Introduction

This article continues the study of relative randomness via rK-reducibility initiated in [3] and pursued in [9].

One of the most popular definitions of absolute algorithmic randomness states that an infinite binary sequence R is random if it is incompressible, that is, if

$$\exists d\, \forall n\,.\, K(R[n]) \geq n - d,$$

where $K(\sigma)$ is the prefix-free descriptional complexity of the string σ. Under this same paradigm of incompressibility, one can define relative algorithmic randomness as follows. An infinite binary sequence A is less random than an infinite binary sequence B if A is completely compressible given B, that is, if

$$\exists d\, \forall n\,.\, K(A[n]\,|\,B[n]) < d,$$

where $K(\sigma|\tau)$ is the conditional prefix-free descriptional complexity of σ given τ. In this case, we write $A \leq_{rK} B$ for short and say "A is rK-reducible to B".[a]

The rK-reducibility, which is fairly easily seen to be reflexive and transitive, enjoys the following properties, all of which we will use throughout.

Theorem 1.1 (Downey, Hirschfeldt and LaForte [3]). *For infinite binary sequences A and B, $A \leq_{rK} B$ is equivalent to both of*

- $\exists d\, \forall n\,.\, C(A[n] \mid B[n]) < d,$
- *There exists a partial computable function φ such that*

$$\exists d\, \forall n\, \exists i < d\,.\, \varphi(i, B[n]) = A[n],$$

and implies all three of

- $\exists d\, \forall n\,.\, K(A[n]) \leq K(B[n]) + d,$
- $\exists d\, \forall n\,.\, C(A[n]) \leq C(B[n]) + d,$
- $A \leq_T B.$

Notice that the fifth bullet says that $\leq_{rK}$ really is a reducibility in the sense of computability theory as there is a way to compute A from B. Moreover, a sequence is computable iff it is rK-reducible to any other sequence, that is, the computable sequences are those of least relative randomness, as they should be.

In what follows we answer a basic question: is there a sequence of minimal relative randomness, that is, a sequence with only the computable sequences strictly less random ($<_{rK}$) than it? Indeed, as our title indicates, there is such a degree and this is our main result. In fact, there are even continuum many such minimal degrees. We prove these claims in Section 2 and follow them up in Section 3 with three notes on such minimal sequences.

Before beginning, let us set some notation and conventions. $\mathbb{N}$ will denote the set of natural numbers $\{0, 1, 2, \ldots\}$, $\{0, 1\}^*$ the set of binary strings and $\{0, 1\}^\infty$ the set of infinite binary sequences. 'String' and 'sequence' without further qualification will mean 'binary string' and 'infinite binary sequence', respectively. For strings σ and τ, $|\sigma|$ will denote the length of σ, and $\sigma\tau$ or, when that might cause confusion, $\sigma \,\widehat{}\, \tau$ the concatenation of σ and τ. Also $\sigma \subseteq \tau$ and $\sigma \subset \tau$ will mean σ is an initial segment of τ and σ is a proper

---

[a]The 'rK' stands for 'relative Kolmogorov' complexity, where 'relative' refers to the fact that one uses 'conditional complexity' and 'Kolmogorov' is just used in place of 'plain Kolmogorov' or 'prefix-free Kolmogorov' since by Theorem 1.1 it does not matter which variant is considered.

initial segment of τ, respectively. For a sequence A and a positive natural n, $A[n]$ will denote the length n initial segment of A, that is, the string $\langle A(0), A(1), \ldots, A(n-1) \rangle$. Trees are subsets of $\{0,1\}^*$ closed under initial segments. A path of a tree T is a sequence such that every initial segment of it is a member of T. The set of all paths of T will be denoted by $[T]$. A Π_1^0-class is a set of paths through a Π_1^0-tree.

Lastly, our notation for other notions from the theory of computability follows the books of Odifreddi [8] and Soare [10].

2. The main result

Theorem 2.1. *There is a minimal rK-degree.*

Proof. We construct a special binary tree, suitable paths of which will have minimal rK-degree. Roughly speaking, we make the set of splitting nodes of our tree very sparse so that any incomputable path of hyperimmune-free Turing degree can be recovered in two guesses from its image under an rK-reduction. More precisely, we build a Π_1^0 tree T (a tree whose complement is recursively enumerable) such that

(1) T has no computable paths;
(2) for every computable function $\Phi : \mathbb{N} \to \mathbb{N}$ (thought of as a functional) and for every path X of T there is a string $\star \subset X$ such that
  (a) either Φ^X and Φ^Y are compatible for every path Y of T extending $\star$
  (b) or Φ^Y and Φ^Z are incompatible for all distinct paths Y, Z of T extending $\star$,
(3) the set S of splitting nodes of T is very sparse, to wit, for all computable functions $g : \mathbb{N} \to \mathbb{N}$ we have

$$\forall^\infty \sigma \in S \, \forall \tau \in S . \sigma \subset \tau \to g(|\sigma|) < |\tau|.$$

Constructing T. We build T in stages, beginning with the full binary tree and pruning it stage by stage. To describe this pruning we use moving markers in the style of [11]. For notational niceness stage subscripts are suppressed whenever possible.

Let $\{m_\sigma \mid \sigma \in \{0,1\}^*\} \subseteq \{0,1\}^*$ denote the set of markers of T. These are on the splitting nodes of T. At stage zero, $T = \{0,1\}^*$ and each $m_\sigma = \sigma$. At later stages when necessary T is pruned via the CUT procedure. If $\sigma \subset \tau$ then $\text{CUT}(m_\sigma, m_\tau)$ cuts off all paths of T that extend m_σ but not m_τ and then updates the positions of all the markers, preserving their order, as follows: m_σ moves to m_τ, each $m_{\sigma\epsilon}$ moves to $m_{\tau\epsilon}$ and all other markers stay where they are. Since CUT is the only action ever taken, T will be a perfect tree without leaves at every stage.

At stage $s > 0$ the construction runs as follows, where each check is performed only when the markers involved have indices of length $\leq s$; also, the computations involved are only up to stage s.

- If there exist σ, $i < 2$ and $e \leq |\sigma|$ such that for all $x \leq |\sigma|$, $\Phi_e(x) = m_{\sigma i}(x)$, then CUT$(m_\sigma, m_{\sigma(1-i)})$.
- If there exist σ, δ, ϵ and $e \leq |\sigma|$ such that $\Phi_e^{m_{\sigma 0}}$ and $\Phi_e^{m_{\sigma 1}}$ are compatible for all arguments up to $|\sigma|$, but $\Phi_e^{m_{\sigma 0 \delta}}$ and $\Phi_e^{m_{\sigma 1 \epsilon}}$ are incompatible at some argument up to $|\sigma|$, then CUT$(m_{\sigma 0}, m_{\sigma 0 \delta})$ and CUT$(m_{\sigma 1}, m_{\sigma 1 \epsilon})$.
- If there exist σ, τ, υ and $e \leq |\sigma|$ such that $\sigma \subset \tau \subset \upsilon$ and $|m_\tau| \leq \Phi_e(|m_\sigma|) < |m_\upsilon|$, then CUT$(m_\tau, m_\upsilon)$.

It is not difficult to check that each marker eventually settles and that, in the limit, T satisfies properties (1)-(3).

A suitable path of T. Let A be a path of T of hyperimmune-free Turing degree.[b] Such a path exists by the Hyperimmune-Free Basis Theorem [5] since $[T]$ is a nonempty Π_1^0 class (since T is a Π_1^0 tree). We show that A has minimal rK-degree. By (1), A is incomputable. Let $B \leq_{\mathrm{rK}} A$ be an incomputable set. We need to show that $A \leq_{\mathrm{rK}} B$. To this end, observe that $B \leq_{\mathrm{T}} A$ and, in fact, $B \leq_{tt} A$ since A has hyperimmune-free Turing degree; see page 589 in Odifreddi's book [8]. Let Φ be a computable functional (total on all oracles) that witnesses the truth-table reduction.

We come now to the heart of the argument: building an rK-reduction from B to A. Let $\star$ be the magic string of (2) for Φ and A. Given $B[n]$ for n sufficiently large, run through the computable approximation (that thins) to T until a stage t is reached such that T_t (the stage t approximation of T) has at most two extensions of $\star$ of length n with extensions in T_t that map to $B[n]$ under Φ. The key here is that such a stage is guaranteed to exist by Lemma 2.2 below. To find these extensions effectively from $B[n]$ we use the fact that Φ is total on all oracles and has a computable use function. Output the (at most) two strings of length n found; one will be $A[n]$. Except for finitely many short lengths, this procedure describes an rK-reduction from B to A. Extending it to all lengths gives the final reduction.   □

---

[b]That is, for every total function $f \leq_{\mathrm{T}} A$, there exists a computable function g such that for all x, $g(x) \geq f(x)$. Put more concisely, every total function computable from A is majorized by a computable one.

Lemma 2.2. *Let $\star$ be the magic string of (2) for Φ and A. For almost all lengths n and almost all stages t, T_t has at most two extensions of $\star$ of length n with extensions in T_t that map to $B[n]$ under Φ.*

Proof. Let φ be the computable use function for the tt-reduction Φ. Let f be the function defined for $m \geq |\star|$ by $f(m)$ equals the first stage s such that for all strings $\nu \supset A[m] \,\widehat{}\, (1 - A(m))$ of length $\varphi(s)$ on T_s, there exists $x \leq s$ such that $\Phi^\nu(x)\downarrow \neq \Phi^A(x)$. (Notice that all ν extend $\star$.) For $m < |\star|$, define $f(m)$ to be 0, say. It is unimportant. Note that f is total, for if not, then for all s there exists a string $\nu_s \supset A[m] \,\widehat{}\, (1 - A(m))$ of length $\varphi(s)$ on T_s such that for all $x \leq s$, $\Phi^{\nu_s}(x) = \Phi^A(x)$. (Remember that Φ is total on all oracles.) Then the sequence Y defined by $Y(n) = \liminf_s \nu_s(n)$ is a path of T different from A such that $\Phi^Y = \Phi^A$. But this is a contradiction, because (2b) holds for $X = A$ since B is incomputable. Also, f is A-computable by definition. Thus, since A has hyperimmune-free Turing degree, there is a computable function g majorizing f.

Now, fix n bigger than the length of $\star$, the length that (3) takes effect for g and the length of the first splitting node of A on T. Let τ be the last splitting node of T on $A[n]$ and let $\sigma \subset \tau$ be any other splitting node of T extending $\star$. Then by (3) we have for $s = f(|\sigma|)$ that

$$s \leq g(|\sigma|) < |\tau| \leq n.$$

So by stage s, every $\nu \in T_s$ extending $A[|\sigma|] \,\widehat{}\, (1 - A(|\sigma|)) = \sigma \,\widehat{}\, (1 - A(|\sigma|))$ will have some number $x \leq s$ with $\Phi^\nu(x)\downarrow \neq \Phi^A(x) = B(x)$, so that ν cannot map to $B[n]$ under Φ. Since σ was an arbitrary splitting node of T below the last splitting node of $A[n]$, we see that only the strings extending the last splitting node of $A[n]$ can map to $B[n]$ under Φ. So the result holds. □

In fact, by a generalized Hyperimmune-Free Basis Theorem below, the tree of the proof of Theorem 2.1 has continuum many paths of hyperimmune-free Turing degree. Thus, since every rK-degree is countable, there are continuum many minimal rK-degrees.

Theorem 2.3. *Every nonempty Π^0_1 class with no computable members has $2^{\aleph_0}$ paths of hyperimmune-free Turing degree.*

Proof. By basic facts from the theory of Π^0_1 classes, we can assume without loss of generality that our Π^0_1 class is the set of paths through a binary tree T_0 that is infinite, computable and has no computable paths. We modify

slightly the proof of the Hyperimmune-Free Basis Theorem in [5] by way
of an extra parameter sequence X. For each sequence X we construct (uni-
formly relative to $X \oplus \emptyset''$) computable subtrees $S_1 \supset T_1 \supseteq S_2 \supset T_2 \supseteq \ldots$
of T_0 such that their only common path Y has hyperimmune-free Turing
degree. We then show that the map $X \mapsto Y$ is one-to-one.

To this end, fix X and, starting from T_0, let S_e and T_e be defined
recursively as follows. Let $U_{e,x}$ be the computable tree $\{\tau \mid \Phi^\tau_{e,|\tau|}(x)\!\uparrow\}$.

(1) If for all x, $T_e \cap U_{e,x}$ is finite, then let $S_e = T_e$. Otherwise, choose x
    least such that $U_{e,x}$ is infinite and let $S_e = T_e \cap U_{e,x}$.
(2) Since S_e is an infinite tree with no computable paths, it has at least
    two paths. Let σ be the length-lexicographic least node of S_e such that
    $\sigma 0$ and $\sigma 1$ have paths in S_e through them.
(3) Let $T_{e+1} = \{\tau \in S_e \mid \tau \subseteq \sigma \hat{\ } X(e) \vee \tau \supset \sigma \hat{\ } X(e)\}$.

By induction each $[T_e]$ and $[S_e]$ is nonempty, so that $\bigcap_e [T_e] \cap [S_e]$
is nonempty, being the intersection of a decreasing sequence of closed
nonempty sets in the compact space $\{0,1\}^\infty$. Choose (the unique) sequence
$Y \in \bigcap_e [T_e] \cap [S_e]$. It will have hyperimmune-free Turing degree, for fix a
natural e and consider the function Φ^Y_e. If for every x, $T_e \cap U_{e,x}$ is finite,
then there is a total and computable function g majorizing Φ^Y_e where g is
chosen as

$$g(x) = \max\{\Phi^\tau_{e,|\tau|}(x) \mid \tau \in T_e \wedge |\tau| = l_x\}$$

and l_x is the least value such that $\Phi^\tau_{e,|\tau|}(x)$ is defined for each $\tau \in T_e$ of
length l_x. If there exists some x such that $T_e \cap U_{e,x}$ is infinite, then $\Phi^\tau_{e,|\tau|}(x)$
is undefined for infinitely many $\tau \in T_e$ and S_e is the set of all these τ. Since
all prefixes of Y are in S_e, this means $\Phi^Y_e(x)$ is undefined, so that Φ^Y_e is
not total.

Also, the map $X \mapsto Y$ is one-to-one, for if two sequences X_1 and X_2
differ and e is the first place at which this happens, then the corresponding
trees $S_e(X_1)$ and $S_e(X_2)$ are the same, but the intersection of $T_{e+1}(X_1)$
and $T_{e+1}(X_2)$ is finite since one contains the nodes above $\sigma 0$ and the other
the ones above $\sigma 1$. Thus $Y(X_1)(|\sigma|) \neq Y(X_2)(|\sigma|)$. $\square$

3. Three notes

From now on let us call a sequence with minimal rK-degree a 'minimal
sequence'. As one might expect, minimal sequences have low initial segment
complexity. Indeed, so low that they are close to being computable in the

sense of Chaitin's characterization (see [1]): a sequence X is computable iff $\exists d \, \forall n \, . \, C(X[n]) \leq C(n) + d$.

Proposition 3.1. *If A is a minimal sequence, then for any computable unbounded increasing function $g : \mathbb{N} \to \mathbb{N}$,*

$$\exists d \, \forall n \, . \, C(A[n]) \leq C(n) + g(n) + d \quad and$$
$$\exists d \, \forall n \, . \, K(A[n]) \leq K(n) + g(n) + d.$$

In particular, A cannot be random.

For the proof, we need the notion of dilutions.

Definition 3.2. For $X \in \{0,1\}^\infty$ and $f : \mathbb{N} \to \mathbb{N}$ strictly increasing, the *f-dilution of X* is the sequence defined by

$$X_f(n) = \begin{cases} X(m) & \text{if } n = f(m) \text{ for some (unique) } m \\ 0 & \text{else.} \end{cases}$$

Notice that for any sequence X and any strictly increasing computable function f, $X_f \leq_{\mathrm{rK}} X$ and $X_f \equiv_{\mathrm{T}} X$. Now we prove Proposition 3.1.

Proof. Fix A and g as in the hypothesis. The idea is that since A is a minimal sequence, it is rK-reducible to every one of its computable dilutions. Picking a dilution appropriate to g will give the desired complexity bound.

We prove the bound for K. The argument for C is identical. Define the function $f : \mathbb{N} \to \mathbb{N}$ recursively by

$$f(0) = 0;$$
$$f(x) = \text{the least } n \text{ such that } n > f(x-1) \text{ and } g(n) \geq 4x.$$

Since g is unbounded and increasing, f is well-defined. Also, by construction f is computable, total and strictly increasing. Furthermore, for any given n, if x is the greatest number such that $f(x) \leq n$, then $g(n) \geq 4x$.

Since A is minimal, $A \leq_{\mathrm{rK}} A_f$ via some $[\varphi, e]$. Now fix n and choose x greatest such that $f(x) \leq n$. Observe that inserting zeros into $A[x]$ in the appropriate computable places produces $A_f[n]$. So to describe $A[n]$, besides a few computable partial functions given ahead of time, one only needs the correct $i < e$ such that $\varphi(i, A_f[n]) = A[n]$, the value n and $A[x]$. This information can be coded, up to a uniform constant, by a string of length $K(n) + 2K(A[x])$. The factor of 2 comes from concatenating strings in a prefix-free way. So, up to a uniform additive constant, for all n,

$$K(A[n]) \leq K(n) + 2K(A[x]) \leq K(n) + 4x \leq K(n) + g(n),$$

as desired. Now fixing g as, say, $g(n) = \lfloor \lg(n+1) \rfloor$, we see that A cannot be random. $\qquad\square$

Using dilutions again, we also get the following.

Proposition 3.3. *Every minimal sequence is rK-reducible to a random sequence.*

Proof. Fix a minimal sequence A and choose a random sequence $R \geq_{\text{wtt}} A$ with the use being majorized by $f(n) = 2n$. This is possible since every sequence has such a random; see [4, 6] and also [7] for a more recent proof using martingales. Then $R \geq_{\text{rK}} A_f \geq_{\text{rK}} A$, by the minimality of A, as desired. $\qquad\square$

Do all sequences have randoms rK-above them? That question is still open and seemingly difficult. We end with one last note, a contrast to Proposition 3.3.

Proposition 3.4. *There is a random sequence with no minimal sequence rK-reducible to it.*

Proof. Let R be a random sequence of hyperimmune-free Turing degree. Such a sequence exists by the Hyperimmune-Free Basis Theorem applied to the complement of any member of a universal Martin-Löf test. Then R has no minimal sequence reducible to it.

To see this, assume (toward a contradiction) there is some minimal sequence A such that $A \leq_{\text{rK}} R$. Since R has hyperimmune-free Turing degree, so does A and $A \leq_{\text{tt}} R$. Since A is incomputable and truth-table reducible to a random, A is Turing equivalent to some random S (see [2]). Since A has hyperimmune-free Turing degree, $S \leq_{\text{tt}} A$ via some computable function with computable use function f. Thus, disregarding floor functions and uniform constants for ease of reading, we have that for all n,

$$
\begin{array}{rll}
n & \leq & K(S[n]) \qquad\qquad\qquad\text{(since S is random)} \\
& \leq & K(A[f(n)]) + K(n) \qquad\text{(using the tt-reduction)} \\
& \leq & K(f(n)) + \lg n + K(n) \quad\text{(by Proposition 3.1)} \\
& \leq & 2K(n) + \lg n \qquad\qquad\text{(since f is computable)} \\
& \leq & 4 \lg n,
\end{array}
$$

a contradiction. $\qquad\square$

Remark 3.5. Do maximal rK-degrees exist? That basic question is still open.

Acknowledgements

We would like to thank the Institute for Mathematical Sciences at the National University of Singapore for organizing the outstanding Computational Prospects of Infinity workshop which made this work possible. Also, thanks to Steffen Lempp and Sasha Rubin for their helpful comments.

References

1. Gregory J. Chaitin. Information-theoretic characterizations of recursive infinite strings. *Theoretical Computer Science*, 2(1):45–48, 1976.
2. Osvald Demuth. Remarks on the structure of tt-degrees based on constructive measure theory. *Commentationes Mathematicae Universitatis Carolinae*, 29(2):233–247, 1988.
3. Rod G. Downey, Denis R. Hirschfeldt and Geoff LaForte. Randomness and reducibility. *Journal of Computer and System Sciences*, 68(1):96–114, 2004.
4. Péter Gács. Every sequence is reducible to a random one. *Information and Control*, 70(2-3):186–192, 1986.
5. Carl G. Jockusch, Jr. and Robert I. Soare. Π_1^0 classes and degrees of theories. *Transactions of the American Mathematical Society*, 173:33–56, 1972.
6. Antonín Kučera. Measure, Π_1^0-classes and complete extensions of PA. In *Recursion Theory Week (Oberwolfach, 1984)*, volume 1141 of *Lecture Notes in Mathematics*, pages 245–259. Springer, Berlin, 1985.
7. Wolfgang Merkle and Nenad Mihailović. On the construction of effective random sets. *The Journal of Symbolic Logic*, 69(3):862–878, 2004.
8. Piergiorgio Odifreddi. *Classical recursion theory*, volume 125 of *Studies in Logic and the Foundations of Mathematics*. North-Holland Publishing Co., Amsterdam, 1989.
9. Alexander Raichev. Relative randomness and real closed fields. *The Journal of Symbolic Logic*, 70(1):319–330, 2005.
10. Robert I. Soare. *Recursively enumerable sets and degrees*. Perspectives in Mathematical Logic. Springer-Verlag, Berlin, 1987.
11. Frank Stephan. On the structures inside truth-table degrees. *The Journal of Symbolic Logic*, 66(2):731–770, 2001.

DIAMONDS ON $\mathcal{P}_\kappa\lambda$

Masahiro Shioya

Institute of Mathematics, University of Tsukuba
Tsukuba, 305-8571, Japan
E-mail: shioya@math.tsukuba.ac.jp

Dedicated to Professor Katsuya Eda on the occasion of his 60th birthday

Assuming $2^\omega = 2^{<\kappa}$ we construct a diamond on $\mathcal{P}_\kappa\lambda$ for every cardinal $\lambda > \kappa$. We also construct a model in which $2^\omega < 2^{\omega_1}$ holds and $\mathcal{P}_{\omega_2}\lambda$ carries a diamond for every $\lambda > \omega_2$.

1. Introduction

Let κ be a regular uncountable cardinal. The combinatorial principle $\Diamond_\kappa$ was introduced by Jensen [9] to construct Suslin trees in the constructible universe L. In [8] Jech extended the notion to the context of $\mathcal{P}_\kappa\lambda$: Suppose that λ is a cardinal $\geq \kappa$ and S is a stationary subset of $\mathcal{P}_\kappa\lambda$.

A diamond on S is a map of the form $g : S \to \mathcal{P}_\kappa\lambda$ such that if $A \subset \lambda$, then $\{x \in S : g(x) = A \cap x\}$ is stationary.

It is easy to see that $\Diamond_\kappa$ holds iff $\mathcal{P}_\kappa\kappa$ carries a diamond. In this paper we are concerned with diamonds on $\mathcal{P}_\kappa\lambda$ in the case $\kappa < \lambda$.

Building on work of Shelah [14], Donder and Matet [3] proved

Theorem 1. *$\mathcal{P}_\kappa\lambda$ carries a diamond for every $\lambda > 2^{<\kappa}$.*

See [17] for corrections to their proof. More recently Shelah [16] established

Theorem 2. *$\mathcal{P}_{\omega_1}\lambda$ carries a diamond for every $\lambda > \omega_1$.*

Corollary 3. *The club filter on $\mathcal{P}_{\omega_1}\lambda$ is not 2^λ-saturated for any $\lambda > \omega_1$.*

Partially supported by JSPS Grant-in-Aid for Scientific Research No. 16540094. The author wishes to thank Professor Shelah for sending a draft of his unpublished book in 2000. He also acknowledges the referee's suggestions for improving the presentation.

What about diamonds on $\mathcal{P}_\kappa\lambda$ in the case $\omega_1 < \kappa < \lambda$? In [4] Foreman and Magidor proved that the club filter on $\mathcal{P}_\kappa\lambda$ can be 2^{ω_1}-saturated. In particular it is consistent that there is no diamond on $\mathcal{P}_\kappa\lambda$. Their model is the forcing extension of L by $\mathrm{Add}(\omega_1, \lambda^{++})^L$ if $\mathrm{cf}^L \lambda \geq \kappa$, and by $\mathrm{Add}(\omega_1, \lambda^{+++})^L$ otherwise. Here $\mathrm{Add}(\omega_1, \rho)$ denotes the poset for adjoining ρ Cohen subsets of ω_1.

Through an attempt to understand the proof of Theorem 2, we get

Theorem 4. *Assume* $2^\omega = 2^{<\kappa}$. *Then* $\mathcal{P}_\kappa\lambda$ *carries a diamond for every* $\lambda > \kappa$.

We prove Theorem 4 in §3. Our proof invokes Theorem 1 and presumably differs from Shelah's proof of Theorem 2. Unfortunately we could follow his argument only in the case $\lambda = \omega_2$. In §6 we present Shelah's proof (as we understand it) in somewhat greater generality. Subsequently König and Todorčević [11] found their own proof of Theorem 2, which works uniformly for every $\lambda > \omega_1$.

Is it consistent that $\mathcal{P}_\kappa\lambda$ carries a diamond in the case $2^\omega < 2^{<\kappa}$? In view of the Foreman–Magidor model the case $2^\omega < 2^{\omega_1}$ is particularly interesting. In §5 we give a positive answer to the question:

Theorem 5. *Suppose that* GCH *holds,* $\omega_1 < \kappa < \nu < \theta$, ν *is* θ-*supercompact and* ρ *is a regular cardinal* $\leq \theta$. *Then* $\mathcal{P}_\kappa\lambda$ *carries a diamond for every* $\lambda > \kappa$ *after forcing with* $\mathrm{Col}(\kappa, \nu) \times \mathrm{Add}(\omega_1, \rho)$.

Here $\mathrm{Col}(\kappa, \nu)$ denotes the Levy collapse of ν to the successor of κ.

In the model of Theorem 5, $\mathcal{P}_\kappa\kappa^+$ can be partitioned into $(\kappa^+)^{<\kappa}$ disjoint stationary sets, and hence every club set in $\mathcal{P}_\kappa\kappa^+$ has size $(\kappa^+)^{<\kappa}$. In particular Theorem 5 with $\kappa = \omega_2$ and $\rho = \theta = \nu^+$ improves a result of Baumgartner [1] that every club set in $\mathcal{P}_{\omega_2}\omega_3$ has size $\omega_3^{\omega_1} > \omega_3^\omega$ in the model. So by work of Magidor [12] some large cardinal hypothesis is necessary in Theorem 5. We also remark that $(\kappa^+)^{<\kappa} = (\kappa^+)^{\omega_1}$ holds whenever $\mathcal{P}_\kappa\kappa^+$ carries a diamond and $\kappa > \omega_1$. This follows from another result of Baumgartner [1] that $\mathcal{P}_\kappa\kappa^+$ has a club subset of size at most $(\kappa^+)^{\omega_1}$.

Our proof of Theorem 5 involves a reflection principle for stationary sets in $\mathcal{P}_{\omega_1}\theta$. Recall from [5].

Stationary Reflection in $\mathcal{P}_{\omega_1}\theta$ holds iff for every stationary $S \subset \mathcal{P}_{\omega_1}\theta$ there is $\omega_1 \subset X \subset \theta$ of size ω_1 such that $S \cap \mathcal{P}_{\omega_1}X$ is stationary in $\mathcal{P}_{\omega_1}X$.

Our principle κ-SR (for Stationary κ-Reflection) in $\mathcal{P}_{\omega_1}\theta$ asserts that the reflecting set X has size κ as well as extends κ. See §4 for the precise statement. The original Stationary Reflection is ω_1-SR and κ-SR in $\mathcal{P}_{\omega_1}\theta$ holds in the model of Theorem 5. Applications of κ-SR can be found in [18] as well.

Here is the main result of this paper, of which Theorem 5 is a corollary:

Theorem 6. *Assume* $\omega_1 < \kappa = \kappa^\omega \leq 2^{\omega_1} = 2^{<\kappa}$ *and* κ-SR *in* $\mathcal{P}_{\omega_1}2^{\kappa^+}$. *Then* $\mathcal{P}_\kappa \lambda$ *carries a diamond for every* $\lambda > \kappa$.

We prove Theorem 6 in §5. Presumably the large cardinal hypothesis of Theorem 5 is not optimal because of our approach via Theorem 6. Nonetheless it establishes the connection between Theorems 4 and 5.

2. Preliminaries

For background material we refer the reader to [10]. Throughout the paper κ denotes a regular uncountable cardinal.

For the rest of this section we fix a cardinal $\lambda > \kappa$. Let $f : [\lambda]^{<\omega} \to \mathcal{P}_\kappa \lambda$. Define
$$C(f) = \{z \subset \lambda : f``[z]^{<\omega} \subset \mathcal{P}(z)\}.$$
For $x \subset \lambda$ let $\mathrm{cl}_f\, x$ be the closure of x under f, i.e. the smallest set $z \in C(f)$ with $x \subset z$. It is well-known that the club filter on $\mathcal{P}_\kappa \lambda$ is generated by the sets of the form $\mathcal{P}_\kappa \lambda \cap C(f)$. A set of the form $\mathcal{P}_\kappa \lambda \cap C(f)$ is called σ-club. It is also known that the club filter on $\mathcal{P}_\kappa \lambda$ is generated by the σ-club sets together with the set $\{x \in \mathcal{P}_\kappa \lambda : x \cap \kappa \in \kappa\}$.

In addition to κ, λ we fix regular cardinals μ, ν such that
$$\omega \leq \mu < \kappa < \nu \leq \lambda.$$
Define
$$S^\mu_\nu = \{\gamma < \nu : \mathrm{cf}\,\gamma = \mu\}.$$
Recall from [15] that a club guessing sequence on S^μ_ν is a map $\langle c_\gamma : \gamma \in S^\mu_\nu \rangle$ such that

- c_γ is an unbounded subset of γ of order-type μ and
- if $D \subset \nu$ is club, then $\{\gamma \in S^\mu_\nu : c_\gamma \subset D\}$ is stationary in ν.

Gitik and Shelah [7] constructed such a sequence even with an additional property:

Lemma 7. *Suppose that* $\mu < \kappa < \nu$ *are all regular. Then there is a club guessing sequence* $\langle c_\gamma : \gamma \in S^\mu_\nu \rangle$ *such that if* $\gamma \in S^\mu_\nu \cap \lim\{\alpha < \nu : \mathrm{cf}\,\alpha \geq \kappa\}$, *then* $c_\gamma \subset \{\alpha < \gamma : \mathrm{cf}\,\alpha \geq \kappa\}$.

Here $\lim X$ denotes the set of limit points of X. See [17] for a simplified proof of Lemma 7.

Fix a club guessing sequence

$$\langle c_\gamma : \gamma \in S_\nu^\omega \rangle.$$

For $x \in \mathcal{P}_\kappa \lambda$ with cf sup$(x \cap \nu) = \omega$ define $r(x) \in [\omega]^\omega$ as follows: Let $\gamma = \sup(x \cap \nu)$ and list c_γ in increasing order as $\{\gamma_n : n < \omega\}$. Set

$$r(x) = \{n < \omega : x \cap (\gamma_{n+1} - \gamma_n) \neq \emptyset\}.$$

Proposition 8 is a special case of Theorem 3 of [6]:

Proposition 8. *If* $r \in [\omega]^\omega$, *then*

$$\{x \in \mathcal{P}_\kappa \lambda : \text{cf} \sup(x \cap \nu) = \omega \wedge r(x) = r\}$$

is stationary.

Toward our own proof of Proposition 8 we fix some notation. First we define a tree ordering $\lhd$ on $[\nu]^{<\omega}$ by proper end-extension:

$$a \lhd b \text{ iff } a = b \cap \beta \text{ for some } \beta \in b.$$

Suppose that T is a subtree of $[\nu]^{<\omega}$, i.e. a subset of $[\nu]^{<\omega}$ closed under initial segments. Let $a \in T$. Define

$$T^a = \{b \in [\nu]^{<\omega} : a \unlhd a \cup b \in T\}.$$

Note that T^a is a subtree of $[\nu]^{<\omega}$. Next define

$$\text{suc}_T(a) = \{\alpha < \nu : a \lhd a \cup \{\alpha\} \in T\}.$$

We say that T is stationary if $\text{suc}_T(a)$ is stationary in ν for every $a \in T$. A map $h : T \to \nu$ is called regressive if $h(a) < \min a$ or $h(a) = 0$ for every $a \in T$. Finally $[T]$ denotes the set of infinite branches through T:

$$[T] = \{B \in [\nu]^\omega : \forall \beta \in B(B \cap \beta \in T)\}.$$

Let us recall from [17] Fodor's lemma for trees together with a proof:

Lemma 9. *Suppose that* T *is a stationary subtree of* $[S_\nu^\kappa]^{<\omega}$ *and* $h : T \to \nu$ *is regressive. Then there is a stationary subtree* T' *of* T *such that* $h``T'$ *is bounded in* ν.

Proof. For $\gamma < \nu$ let

$$T_\gamma = \{b \in T : \forall a \unlhd b(h(a) < \gamma)\}.$$

It is easy to see that T_γ is a subtree of T.

Claim. *There is* $\gamma < \nu$ *such that if* $C \subset \nu$ *is club, then* $[T_\gamma] \cap [C]^\omega \neq \emptyset$.

Proof. Assume to the contrary for each $\gamma < \nu$ there is a club $C_\gamma \subset \nu$ such that $[T_\gamma] \cap [C_\gamma]^\omega = \emptyset$. Then

$$C = \Delta_{\gamma < \nu} C_\gamma$$

is club in ν. Since T is a stationary subtree of $[\nu]^{<\omega}$, there is $B \in [T] \cap [C]^\omega$. Set

$$\alpha = \min B.$$

Note that cf $\alpha = \kappa > \omega$ by $\{\alpha\} \in T \subset [S_\nu^\kappa]^{<\omega}$. Since $h(B \cap \beta) < \min B = \alpha$ for every $\beta \in B$, there is $\gamma < \alpha$ such that $h(B \cap \beta) < \gamma$ for every $\beta \in B$. Hence $B \in [T_\gamma]$. Since $B \subset C = \Delta_{\gamma < \nu} C_\gamma$ and $\gamma < \alpha = \min B$, we have $B \subset C_\gamma$. Thus $B \in [T_\gamma] \cap [C_\gamma]^\omega$. This contradicts that $[T_\gamma] \cap [C_\gamma]^\omega = \emptyset$, as desired. □

Fix γ as in the Claim. Define

$$T' = \{b \in T_\gamma : \forall a \trianglelefteq b \forall C \subset \nu \text{ club } ([T_\gamma^a] \cap [C]^\omega \neq \emptyset)\}.$$

We claim that T' is as desired. It is easy to see that T' is a subtree of T_γ. Hence $h\text{``}T' \subset \gamma$. Note that $\emptyset \in T'$ by the Claim. It remains to prove

Claim. T' *is a stationary subtree of* $[\nu]^{<\omega}$.

Proof. Suppose $b \in T'$. We need to prove that $\text{suc}_{T'}(b)$ is stationary in ν. Assume to the contrary

$$D \cap \text{suc}_{T'}(b) = \emptyset$$

for some club $D \subset \nu$. For $\alpha < \nu$ we choose a club $E_\alpha \subset \nu$ as follows: If $\alpha \in D$ and $b \trianglelefteq b \cup \{\alpha\} \in T_\gamma$, then there is a club $E_\alpha \subset \nu$ such that $[T_\gamma^{b \cup \{\alpha\}}] \cap [E_\alpha]^\omega = \emptyset$ by $b \in T'$ and $b \cup \{\alpha\} \notin T'$. Otherwise let $E_\alpha = \nu$. Then

$$E = D \cap \Delta_{\alpha < \nu} E_\alpha$$

is club in ν. Since $b \in T'$, there is $B \in [T_\gamma^b] \cap [E]^\omega$. Set

$$\alpha = \min B.$$

Then $b \cup \{\alpha\} \in T_\gamma$ and $B - \{\alpha\} \in [T_\gamma^{b \cup \{\alpha\}}]$. Note that $\alpha \in D$ by $B \subset E \subset D$. Since $B - \{\alpha\} \subset E \subset \Delta_{\alpha < \nu} E_\alpha$, we have $B - \{\alpha\} \subset E_\alpha$. Thus $B - \{\alpha\} \in [T_\gamma^{b \cup \{\alpha\}}] \cap [E_\alpha]^\omega$. This contradicts that $[T_\gamma^{b \cup \{\alpha\}}] \cap [E_\alpha]^\omega = \emptyset$, as desired. □

This completes the proof. □

Lemma 10 is included in the proof of Main Lemma 1 of [17].

Lemma 10. *Suppose T is a stationary subtree of $[S_\nu^\kappa]^{<\omega}$ and $F : T \to \mathcal{P}_\kappa \lambda$. Then there are a stationary subtree T^* of T and $h : T^* \to \nu$ such that if $a \lhd a \cup b \in T^*$, then $F(a \cup b) \cap \min b \subset h(a)$.*

Proof. By recursion on $n < \omega$ we define a stationary subtree $T_n \subset T$ and a map $h_n : T_n \cap [\nu]^n \to \nu$ as follows: First set $T_0 = T$. Suppose we have defined T_n. Suppose $a \in T_n \cap [\nu]^n$. Then T_n^a is a stationary subtree of $[S_\nu^\kappa]^{<\omega}$. Define a map on T_n^a by

$$b \mapsto \sup(F(a \cup b) \cap \min b).$$

Since $|F(a \cup b)| < \kappa = \operatorname{cf} \min b$, the map is regressive on T_n^a. By Lemma 9 there are a stationary subtree $(T_n^a)'$ of T_n^a and $h_n(a) < \nu$ such that if $b \in (T_n^a)'$, then $F(a \cup b) \cap \min b \subset h_n(a)$. Let T_{n+1} be the subtree of T_n such that

- $T_{n+1} \cap [\nu]^n = T_n \cap [\nu]^n$ and
- $T_{n+1}^a = (T_n^a)'$ for every $a \in T_{n+1} \cap [\nu]^n$.

Then T_{n+1} is stationary.

It is easy to see that $T^* = \bigcup_{n<\omega} T_n \cap [\nu]^n$ and $h = \bigcup_{n<\omega} h_n$ are as desired. $\qquad\square$

We are now ready for

Proof of Proposition 8. Suppose

$$r = \{n(k) : k < \omega\} \in [\omega]^\omega$$

and $f : [\lambda]^{<\omega} \to \mathcal{P}_\kappa \lambda$. We need to find $x \in \mathcal{P}_\kappa \lambda \cap C(f)$ such that

- $\operatorname{cf} \sup(x \cap \nu) = \omega$ and
- $r(x) = \{n(k) : k < \omega\}$.

By Lemma 10 there are a stationary subtree T^* of $[S_\nu^\kappa]^{<\omega}$ and $h : T^* \to \nu$ such that if $a \lhd a \cup b \in T^*$, then $\operatorname{cl}_f(a \cup b) \cap \min b \subset h(a)$. Define

$$D = \{\gamma < \nu : (\operatorname{cl}_f \gamma) \cap \nu = \gamma \wedge \forall a \in T^* \cap [\gamma]^{<\omega}(h(a) < \gamma \in \lim \operatorname{suc}_{T^*}(a))\}.$$

Then D is club. Take $\gamma \in S_\nu^\omega \cap D$ with $c_\gamma \subset D$. Recall that $\{\gamma_n : n < \omega\}$ lists c_γ in increasing order.

By recursion on $k < \omega$ we choose $\alpha_k < \nu$ so that

- $\gamma_{n(k)} < \alpha_k < \gamma_{n(k)+1}$ and
- $\{\alpha_i : i \le k\} \in T^*$

as follows: Suppose we have defined α_i for $i < k$. Since $\gamma_{n(k)+1} \in D$ and $\{\alpha_i : i < k\} \in T^* \cap [\gamma_{n(k)+1}]^{<\omega}$, there is $\alpha_k < \gamma_{n(k)+1}$ such that $\gamma_{n(k)} < \alpha_k \in \operatorname{suc}_{T^*}\{\alpha_i : i < k\}$. Then $\{\alpha_i : i \le k\} \in T^*$.

Now define

$$x = \mathrm{cl}_f\{\alpha_k : k < \omega\} \in \mathcal{P}_\kappa \lambda \cap C(f).$$

We claim that x is as desired. First we prove $\sup(x \cap \nu) = \gamma$. It suffices to show

$$\cdot\gamma = \sup_{k<\omega} \gamma_{n(k)} = \sup_{k<\omega} \alpha_k \leq \sup(x \cap \nu) \leq \sup((\mathrm{cl}_f \gamma) \cap \nu) = \gamma.$$

The second equality follows from $\gamma_{n(k)} < \alpha_k < \gamma_{n(k)+1}$. The first inequality follows from $\{\alpha_k : k < \omega\} \subset x \cap \nu$. The second inequality follows from $x = \mathrm{cl}_f\{\alpha_k : k < \omega\} \subset \mathrm{cl}_f \gamma$. The last equality follows from $\gamma \in D$.

It remains to prove

Claim. $r(x) = \{n(k) : k < \omega\}$.

Proof. One direction follows from $\alpha_k \in x \cap (\gamma_{n(k)+1} - \gamma_{n(k)})$ for every $k < \omega$.

For the converse we prove that $x \cap \gamma_{n(k+1)} \subset \gamma_{n(k)+1}$ for every $k < \omega$. It suffices to show that if $k < \bar{k} < \omega$, then

$$(\mathrm{cl}_f\{\alpha_i : i \leq \bar{k}\}) \cap \gamma_{n(k+1)} \subset (\mathrm{cl}_f\{\alpha_i : i \leq \bar{k}\}) \cap \alpha_{k+1}$$

$$\subset h(\{\alpha_i : i \leq k\}) < \gamma_{n(k)+1}.$$

The first inclusion follows from $\gamma_{n(k+1)} < \alpha_{k+1}$. The second inclusion follows from the choice of h and $\{\alpha_i : i \leq k\} \lhd \{\alpha_i : i \leq \bar{k}\} \in T^*$. The inequality follows from $\gamma_{n(k)+1} \in D$ and $\{\alpha_i : i \leq k\} \in T^* \cap [\gamma_{n(k)+1}]^{<\omega}$. $\square$

This completes the proof. $\square$

3. Diamonds on sets of countable cofinality

This section is devoted to

Proof of Theorem 4. If $\lambda > 2^{<\kappa}$, we are done by Theorem 1. So let us assume $\lambda \leq 2^{<\kappa}$. Then $\lambda^{<\kappa} = 2^{<\kappa} = 2^\omega$.

Fix a disjoint family

$$\{I_n : n < \omega\} \subset [\omega]^\omega$$

such that $n < \min I_n$ for every $n < \omega$. Since $2^\omega = \lambda^{<\kappa}$, there is a map

$$\varphi : \mathcal{P}(\omega) \to \mathcal{P}_\kappa \lambda$$

such that $\varphi``[I_n]^\omega = \mathcal{P}_\kappa \lambda$ for every $n < \omega$. Fix a regular cardinal ν with $\kappa < \nu \leq \lambda$. Then there is a club guessing sequence

$$\langle c_\gamma : \gamma \in S_\nu^\omega \rangle.$$

For $x \in \mathcal{P}_\kappa \lambda$ with $\operatorname{cf} \sup(x \cap \nu) = \omega$ define $g(x) \in \mathcal{P}_\kappa \lambda$ as follows: Let $\gamma = \sup(x \cap \nu)$ and list c_γ in increasing order as $\{\gamma_n : n < \omega\}$. Set

$$r(x) = \{n < \omega : x \cap (\gamma_{n+1} - \gamma_n) \neq \emptyset\},$$
$$g(x) = \bigcup\nolimits_{n \in r(x)} \varphi(I_n \cap r(x)).$$

Claim. g *is a diamond on* $\{x \in \mathcal{P}_\kappa \lambda : \operatorname{cf} \sup(x \cap \nu) = \omega\}$.

Proof. Suppose that $A \subset \lambda$ and $f : [\lambda]^{<\omega} \to \mathcal{P}_\kappa \lambda$. We need to find $x \in \mathcal{P}_\kappa \lambda \cap C(f)$ such that

- $\operatorname{cf} \sup(x \cap \nu) = \omega$ and
- $g(x) = A \cap x$.

By Lemma 10 there are a stationary subtree T^* of $[S_\nu^\kappa]^{<\omega}$ and $h : T^* \to \nu$ such that if $a \lhd a \cup b \in T^*$, then $\operatorname{cl}_f(a \cup b) \cap \min b \subset h(a)$. Define

$$D = \{\gamma < \nu : (\operatorname{cl}_f \gamma) \cap \nu = \gamma \wedge \forall a \in T^* \cap [\gamma]^{<\omega} (h(a) < \gamma \in \lim \operatorname{suc}_{T^*}(a))\}.$$

Then D is club. Take $\gamma \in S_\nu^\omega \cap D$ with $c_\gamma \subset D$.

By recursion on $k < \omega$ we choose $n(k) < \omega$, $\alpha_k < \nu$ and $J_k \in [I_{n(k)}]^\omega$ so that

- $n(k) < n(k+1)$,
- $\gamma_{n(k)} < \alpha_k < \gamma_{n(k)+1}$ and
- $\{\alpha_i : i \leq k\} \in T^*$

as follows: First set $n(0) = 0$. Suppose we have defined α_i, $n(i)$ and J_i for $i < k$. If $k > 0$, let

$$n(k) = \min(\bigcup\nolimits_{i < k} J_i - (n(k-1) + 1)).$$

Since $\gamma_{n(k)+1} \in D$ and $\{\alpha_i : i < k\} \in T^* \cap [\gamma_{n(k)+1}]^{<\omega}$, there is $\alpha_k < \gamma_{n(k)+1}$ such that $\gamma_{n(k)} < \alpha_k \in \operatorname{suc}_{T^*}\{\alpha_i : i < k\}$. Then $\{\alpha_i : i \leq k\} \in T^*$. Since $\varphi``[I_{n(k)}]^\omega = \mathcal{P}_\kappa \lambda$, there is $J_k \in [I_{n(k)}]^\omega$ such that

$$\varphi(J_k) = A \cap \operatorname{cl}_f\{\alpha_i : i \leq k\}.$$

Note that if $k \leq \bar{k} < \omega$, then $n(k) \leq n(\bar{k}) < \min I_{n(\bar{k})} \leq \min J_{\bar{k}}$. Hence

$$n(k) = \min(\bigcup\nolimits_{i < \omega} J_i - (n(k-1) + 1))$$

for every $k > 0$. Thus

$$\{n(k) : k < \omega\} = \{0\} \cup \bigcup\nolimits_{i < \omega} J_i.$$

Now define

$$x = \operatorname{cl}_f\{\alpha_k : k < \omega\} \in \mathcal{P}_\kappa \lambda \cap C(f).$$

We claim that x is as desired. By the proof of Proposition 8 we have $\sup(x \cap \nu) = \gamma$. It remains to prove $g(x) = A \cap x$. By the proof of Proposition 8 we have $r(x) = \{n(k) : k < \omega\}$. Hence

$$r(x) = \{0\} \cup \bigcup_{i < \omega} J_i.$$

Let $k < \omega$. Since $J_k \subset I_{n(k)} \subset \omega - \{0\}$ and $I_{n(k)} \cap J_i = \emptyset$ for every $i \neq k$, we have

$$I_{n(k)} \cap r(x) = I_{n(k)} \cap \bigcup_{i < \omega} J_i = J_k.$$

Hence by the choice of J_k

$$\varphi(I_{n(k)} \cap r(x)) = \varphi(J_k) = A \cap \mathrm{cl}_f\{\alpha_i : i \leq k\}.$$

Therefore

$$g(x) = \bigcup_{n \in r(x)} \varphi(I_n \cap r(x))$$
$$= \bigcup_{k < \omega} \varphi(I_{n(k)} \cap r(x))$$
$$= \bigcup_{k < \omega} A \cap \mathrm{cl}_f\{\alpha_i : i \leq k\}$$
$$= A \cap x.$$

□

This completes the proof. □

4. The principle of Stationary κ-Reflection

Recall our convention from §2 that κ denotes a regular uncountable cardinal. In addition let θ be a cardinal $> \kappa$ in this section. Define

κ-SR in $\mathcal{P}_{\omega_1}\theta$ holds iff for every stationary $S \subset \mathcal{P}_{\omega_1}\theta$ there is $\kappa \subset X \subset \theta$ of size κ such that $S \cap \mathcal{P}_{\omega_1} X$ is stationary in $\mathcal{P}_{\omega_1} X$.

Let us begin by recalling from [13]

Lemma 11. *Every stationary set in* $\mathcal{P}_{\omega_1}\theta$ *remains stationary after forcing with a countably closed poset.*

Proof. Suppose that a condition p forces $\dot{f} : [\theta]^{<\omega} \to \mathcal{P}_{\omega_1}\theta$. It suffices to give a club subset of $\{x \in \mathcal{P}_{\omega_1}\theta : \exists q \leq p(q \Vdash x \in C(\dot{f}))\}$.

Since the poset is countably closed, we can choose by recursion on the length of $s \in {}^{<\omega}\theta$ a condition $p_s \leq p$ and $f_s : [\mathrm{ran}\, s]^{<\omega} \to \mathcal{P}_{\omega_1}\theta$ so that

- p_s forces $\dot{f}|[\mathrm{ran}\, s]^{<\omega} = f_s$ and
- if $s \subset \bar{s}$, then $p_{\bar{s}} \leq p_s$.

Define

$$D = \{x \in \mathcal{P}_{\omega_1}\theta : \forall s \in {}^{<\omega}x(f_s\text{``}[\operatorname{ran} s]^{<\omega} \subset \mathcal{P}(x))\}.$$

Then D is club. We claim that D is as desired. Suppose $x \in D$. We need to find $q \leq p$ that forces $\dot{f}\text{``}[x]^{<\omega} \subset \mathcal{P}(x)$. Fix a bijection

$$\tau : \omega \to x.$$

Since the poset is countably closed, the descending sequence $\langle p_{\tau|n} : n < \omega \rangle$ has a lower bound $q \leq p$. We claim that q is as desired. It suffices to prove that for every $n < \omega$, $p_{\tau|n}$ forces

$$\dot{f}\text{``}[\tau\text{``}n]^{<\omega} = f_{\tau|n}\text{``}[\tau\text{``}n]^{<\omega} \subset \mathcal{P}(x).$$

The first equality follows from the choice of $p_{\tau|n}$ and $f_{\tau|n}$. The second equality follows from $x \in D$ and $\tau|n \in {}^{<\omega}x$. $\qquad\square$

Proposition 12 is proved in effect in [2] and [5].

Proposition 12. *Suppose that $\kappa < \nu \leq \theta$, ν is θ-supercompact and ρ is a cardinal. Then κ-SR in $\mathcal{P}_{\omega_1}\theta$ holds after forcing with $\operatorname{Col}(\kappa, \nu) \times \operatorname{Add}(\omega_1, \rho)$.*

Proof. Let $j : V \to M$ witness that ν is θ-supercompact. Set

$$P = \operatorname{Col}(\kappa, \nu) \times \operatorname{Add}(\omega_1, \rho),$$
$$Q = \operatorname{Col}(\kappa, j(\nu) - \nu) \times \operatorname{Add}(\omega_1, j(\rho) - j\text{``}\rho).$$

Then $j(P) = (j\text{``}P) \times Q$.

Fix a V-generic $G \subset P$ and a $V[G]$-generic $H \subset Q$. Set

$$j(G) = (j\text{``}G) \times H$$

and extend j to $j : V[G] \to M[j(G)]$.

In $V[G]$ let $S \subset \mathcal{P}_{\omega_1}\theta$ be stationary. Note that $j\text{``}\theta \in M$ and $|j\text{``}\theta| = \kappa$ in $M[j(G)]$. By elementarity of j it suffices to show that $j(S) \cap \mathcal{P}_{\omega_1}(j\text{``}\theta)$ is stationary in $\mathcal{P}_{\omega_1}(j\text{``}\theta)$ in $M[j(G)]$. Unless otherwise stated, we work in $V[j(G)]$.

Since Q is countably closed in $V[G]$, S remains stationary in $\mathcal{P}_{\omega_1}\theta$ by Lemma 11. Hence $\{j\text{``}x : x \in S\}$ is stationary in $\mathcal{P}_{\omega_1}(j\text{``}\theta)$. Since each $x \in S$ is countable in $V[G]$, we have $\{j\text{``}x : x \in S\} = \{j(x) : x \in S\} \subset j(S)$. Hence $j(S) \cap \mathcal{P}_{\omega_1}(j\text{``}\theta)$ is stationary in $\mathcal{P}_{\omega_1}(j\text{``}\theta)$. Note that $\mathcal{P}_{\omega_1}(j\text{``}\theta)$ is absolute between $M[j(G)]$ and $V[j(G)]$ by ${}^{\omega}M \subset M$ and the countable closure of $j(P)$ in V. Since stationariness is downward absolute, we get the desired conclusion. $\qquad\square$

Lemma 13 can be found in Shelah's proof [13] of Chang's Conjecture in the Levy collapse of a measurable cardinal to ω_2.

Lemma 13. *Suppose that $\theta = 2^{\kappa^+}$ and D is σ-club in $\mathcal{P}_\kappa 2^\theta$. Then there is a map $d : [2^\theta]^{<\omega} \to \mathcal{P}_{\omega_1} 2^\theta$ such that*

- $\mathcal{P}_\kappa 2^\theta \cap C(d) \subset D$ and
- *if $z \in \mathcal{P}_\kappa 2^\theta \cap C(d)$ and $u \in \mathcal{P}_\kappa \kappa^+$, then*

$$\mathrm{cl}_d(z \cup u) \cap \theta = \mathrm{cl}_d((z \cap \theta) \cup u) \cap \theta.$$

Proof. List all functions of the form $e : [\kappa^+]^{<\omega} \to \mathcal{P}_{\omega_1} \theta$ as $\{e_\zeta : \zeta < \theta\}$. By recursion on $n < \omega$ we define $d_n : [2^\theta]^{<\omega} \to \mathcal{P}_{\omega_1} 2^\theta$ and $\zeta_n : [2^\theta]^{<\omega} \to \theta$ as follows: Since $\{z \in D : \forall \zeta \in z \cap \theta(e_\zeta\text{``}[z \cap \kappa^+]^{<\omega} \subset \mathcal{P}(z))\}$ is σ-club in $\mathcal{P}_\kappa 2^\theta$, there is $d_0 : [2^\theta]^{<\omega} \to \mathcal{P}_{\omega_1} 2^\theta$ such that

$$\mathcal{P}_\kappa 2^\theta \cap C(d_0) \subset \{z \in D : \forall \zeta \in z \cap \theta(e_\zeta\text{``}[z \cap \kappa^+]^{<\omega} \subset \mathcal{P}(z))\}.$$

Suppose we have defined d_n. Define ζ_n and d_{n+1} by

$$e_{\zeta_n(a)} = \langle \mathrm{cl}_{d_n}(a \cup b) \cap \theta : b \in [\kappa^+]^{<\omega} \rangle,$$
$$d_{n+1}(a) = d_n(a) \cup \{\zeta_n(a)\}.$$

Define $d : [2^\theta]^{<\omega} \to \mathcal{P}_{\omega_1} 2^\theta$ by

$$d(a) = \bigcup_{n<\omega} d_n(a).$$

We claim that d is as desired. First note that

$$\mathcal{P}_\kappa 2^\theta \cap C(d) \subset \mathcal{P}_\kappa 2^\theta \cap C(d_0) \subset D.$$

To see the second item, let $z \in \mathcal{P}_\kappa 2^\theta \cap C(d)$ and $u \in \mathcal{P}_\kappa \kappa^+$. It suffices to show that

$$\begin{aligned}
\mathrm{cl}_d(z \cup u) \cap \theta &= \bigcup \{\mathrm{cl}_{d_n}(z \cup u) \cap \theta : n < \omega\} \\
&= \bigcup \{\mathrm{cl}_{d_n}(a \cup b) \cap \theta : n < \omega \wedge a \in [z]^{<\omega} \wedge b \in [u]^{<\omega}\} \\
&= \bigcup \{e_{\zeta_n(a)}(b) : n < \omega \wedge a \in [z]^{<\omega} \wedge b \in [u]^{<\omega}\} \\
&\subset \mathrm{cl}_d((z \cap \theta) \cup u).
\end{aligned}$$

The first equality follows from the definition of d. To see the inclusion, let $n < \omega$, $a \in [z]^{<\omega}$ and $b \in [u]^{<\omega}$. Since $z \in C(d)$, we have

$$\zeta_n(a) \in d_{n+1}(a) \cap \theta \subset d(a) \cap \theta \subset z \cap \theta.$$

Note that $\mathrm{cl}_d((z \cap \theta) \cup u) \in \mathcal{P}_\kappa 2^\theta \cap C(d) \subset \mathcal{P}_\kappa 2^\theta \cap C(d_0)$. Hence by the choice of d_0 we have $e_{\zeta_n(a)}(b) \subset \mathrm{cl}_d((z \cap \theta) \cup u)$, as desired. $\qquad\square$

Recall that

$(\kappa^+, \kappa) \twoheadrightarrow (\omega_1, \omega)$ holds iff for every $f : [\kappa^+]^{<\omega} \to \mathcal{P}_{\omega_1}\kappa^+$ there is $y \in C(f)$ of size ω_1 such that $y \cap \kappa$ is countable.

In particular Chang's Conjecture is $(\omega_2, \omega_1) \twoheadrightarrow (\omega_1, \omega)$. The idea of the proof of Theorem 6 in the next section dates back to that of Chang's Conjecture from Stationary Reflection (see [2]). Let us present the proof in somewhat greater generality:

Proposition 14. *Assume κ-SR in $\mathcal{P}_{\omega_1}2^{\kappa^+}$. Then $(\kappa^+, \kappa) \twoheadrightarrow (\omega_1, \omega)$ holds.*

Proof. Suppose $f : [\kappa^+]^{<\omega} \to \mathcal{P}_{\omega_1}\kappa^+$. We need to find $y \in C(f)$ of size ω_1 such that $|y \cap \kappa| = \omega$.

Set

$$\theta = 2^{\kappa^+}.$$

List all functions of the form $e : [\theta]^{<\omega} \to \mathcal{P}_{\omega_1}\theta$ as $\{e_\zeta : \zeta < 2^\theta\}$. Define

$$D = \{z \in \mathcal{P}_\kappa 2^\theta : z \cap \kappa^+ \in C(f) \wedge \forall \zeta \in z(z \cap \theta \in C(e_\zeta))\}.$$

Then D is σ-club. By Lemma 13 there is a map $d : [2^\theta]^{<\omega} \to \mathcal{P}_{\omega_1}2^\theta$ such that

- $\mathcal{P}_\kappa 2^\theta \cap C(d) \subset D$ and
- if $z \in \mathcal{P}_\kappa 2^\theta \cap C(d)$ and $\alpha < \kappa^+$, then

$$\mathrm{cl}_d(z \cup \{\alpha\}) \cap \theta = \mathrm{cl}_d((z \cap \theta) \cup \{\alpha\}) \cap \theta.$$

Define

$$C = \{x \in \mathcal{P}_{\omega_1}\theta : \exists \alpha < \kappa^+(\alpha \notin x \wedge \mathrm{cl}_d(x \cup \{\alpha\}) \cap \kappa = x \cap \kappa)\}.$$

Claim. *C has a club subset.*

Proof. Suppose that S is stationary in $\mathcal{P}_{\omega_1}\theta$. We need to prove $S \cap C \neq \emptyset$.

By κ-SR in $\mathcal{P}_{\omega_1}\theta$ there is $\kappa \subset X \subset \theta$ of size κ such that $S \cap \mathcal{P}_{\omega_1}X$ is stationary in $\mathcal{P}_{\omega_1}X$. Take $\alpha \in \kappa^+ - X$. Since $\{z \in \mathcal{P}_{\omega_1}2^\theta \cap C(d) : \alpha \in z\}$ is club in $\mathcal{P}_{\omega_1}2^\theta$, there is $z \in \mathcal{P}_{\omega_1}2^\theta \cap C(d)$ such that $\alpha \in z$ and $z \cap X \in S$.

Set

$$x = z \cap X.$$

Then $x \in S \cap \mathcal{P}_{\omega_1}X$. We claim that $x \in C$. First note that $\alpha \notin x$ by $\alpha \notin X$. It remains to prove

$$x \cap \kappa \subset \mathrm{cl}_d(x \cup \{\alpha\}) \cap \kappa \subset z \cap \kappa = z \cap X \cap \kappa = x \cap \kappa.$$

The second inclusion follows from $x \cup \{\alpha\} \subset z \in C(d)$. The first equality follows from $\kappa \subset X$.   □

Take $\zeta < 2^\theta$ with

$$\mathcal{P}_{\omega_1}\theta \cap C(e_\zeta) \subset C.$$

By recursion on $\xi < \omega_1$ we choose $z_\xi \in \mathcal{P}_{\omega_1}2^\theta \cap C(d)$ and $\alpha_\xi < \kappa^+$ so that if $\iota < \xi$, then $z_\iota \subset z_\xi$: First take $z_0 \in \mathcal{P}_{\omega_1}2^\theta \cap C(d)$ with $\zeta \in z_0$. Suppose we have defined z_ξ. Since $\zeta \in z_0 \subset z_\xi \in \mathcal{P}_{\omega_1}2^\theta \cap C(d) \subset D$, we have

$$z_\xi \cap \theta \in \mathcal{P}_{\omega_1}\theta \cap C(e_\zeta) \subset C.$$

Hence there is $\alpha_\xi \in \kappa^+ - z_\xi$ such that

$$\mathrm{cl}_d((z_\xi \cap \theta) \cup \{\alpha_\xi\}) \cap \kappa = z_\xi \cap \kappa.$$

Set

$$z_{\xi+1} = \mathrm{cl}_d(z_\xi \cup \{\alpha_\xi\}).$$

Then $z_\xi \subset z_{\xi+1} \in \mathcal{P}_{\omega_1}2^\theta \cap C(d)$ and $\alpha_\xi \in z_{\xi+1} - z_\xi$. Suppose ξ is limit and we have defined z_ι for $\iota < \xi$. Set

$$z_\xi = \bigcup_{\iota < \xi} z_\iota.$$

Then $z_\xi \in \mathcal{P}_{\omega_1}2^\theta \cap C(d)$.

Now define

$$y = \bigcup_{\xi < \omega_1} z_\xi \cap \kappa^+.$$

We claim that y is as desired. Note that $z_\xi \cap \kappa^+ \in \mathcal{P}_{\omega_1}\kappa^+ \cap C(f)$ by $z_\xi \in \mathcal{P}_{\omega_1}2^\theta \cap C(d) \subset D$. Since $\{z_\xi \cap \kappa^+ : \xi < \omega_1\}$ is increasing, we have $y \in \mathcal{P}_{\omega_2}\kappa^+ \cap C(f)$. Hence $|y| = \omega_1$ by $\{\alpha_\xi : \xi < \omega_1\} \subset y$. It remains to prove $|y \cap \kappa| = \omega$. It suffices to show that for every $\xi < \omega_1$

$$z_{\xi+1} \cap \kappa = \mathrm{cl}_d(z_\xi \cup \{\alpha_\xi\}) \cap \kappa = \mathrm{cl}_d((z_\xi \cap \theta) \cup \{\alpha_\xi\}) \cap \kappa = z_\xi \cap \kappa.$$

The second equality follows from the choice of d. The last equality follows from the choice of α_ξ. $\qquad\square$

For later purposes we fix some notation. We define a tree ordering $\lhd$ on $[\kappa^+]^{<\omega_1}$ by proper end-extension:

$$a \lhd b \text{ iff } a = b \cap \beta \text{ for some } \beta \in b.$$

We say that T is a subtree of $[\kappa^+]^{<\omega_1}$ if

- T is a subset of $[\kappa^+]^{<\omega_1}$ closed under initial segments and
- if $b \in [\kappa^+]^{<\omega_1}$ has a limit order-type and $a \in T$ for every $a \lhd b$, then $b \in T$.

Suppose that $T \subset [\kappa^+]^{<\omega_1}$ is a subtree. For $a \in T$ define

$$\mathrm{suc}_T(a) = \{\alpha < \kappa^+ : a \lhd a \cup \{\alpha\} \in T\}.$$

We say that T is unbounded (resp. stationary) if $\mathrm{suc}_T(a)$ is unbounded (resp. stationary) in κ^+ for every $a \in T$. Finally define

$$[T] = \{B \in [\kappa^+]^{\omega_1} : \forall \beta \in B(B \cap \beta \in T)\}.$$

Proposition 15 is the direct antecedent of Theorem 6. This, together with Proposition 12, gives another proof of Baumgartner's result from [1] mentioned in §1.

Proposition 15. *Assume* $\kappa > \omega_1$ *and* κ-*SR in* $\mathcal{P}_{\omega_1} 2^{\kappa^+}$. *Then every club subset of* $\mathcal{P}_\kappa \kappa^+$ *has size at least* 2^{ω_1}.

Proof. Suppose $f : [\kappa^+]^{<\omega} \to \mathcal{P}_{\omega_1}\kappa^+$. It suffices to prove

$$|\{x \in \mathcal{P}_\kappa \kappa^+ \cap C(f) : x \cap \kappa \in \kappa\}| \geq 2^{\omega_1}.$$

Set

$$\theta = 2^{\kappa^+}.$$

List all functions of the form $e : [\theta]^{<\omega} \to \mathcal{P}_{\omega_1}\theta$ as $\{e_\zeta : \zeta < 2^\theta\}$. For $\gamma \in \kappa^+ - \kappa$ fix a bijection

$$\pi_\gamma : \kappa \to \gamma.$$

Define

$$D = \{z \in \mathcal{P}_\kappa 2^\theta : z \cap \kappa^+ \in C(f) \wedge \forall \zeta \in z(z \cap \theta \in C(e_\zeta)) \wedge$$
$$\forall \gamma \in z \cap (\kappa^+ - \kappa)(\pi_\gamma``(z \cap \kappa) = z \cap \gamma)\}.$$

Then D is σ-club. By Lemma 13 there is a map $d : [2^\theta]^{<\omega} \to \mathcal{P}_{\omega_1} 2^\theta$ such that

- $\mathcal{P}_\kappa 2^\theta \cap C(d) \subset D$ and
- if $z \in \mathcal{P}_\kappa 2^\theta \cap C(d)$ and $u \in \mathcal{P}_\kappa \kappa^+$, then

$$\mathrm{cl}_d(z \cup u) \cap \theta = \mathrm{cl}_d((z \cap \theta) \cup u) \cap \theta.$$

Define

$$S(x) = \{\alpha < \kappa^+ : \mathrm{cl}_d(x \cup \{\alpha\} \cup \sup(x \cap \kappa)) \cap \kappa = \sup(x \cap \kappa)\},$$
$$C = \{x \in \mathcal{P}_{\omega_1}\theta : S(x) \text{ is unbounded in } \kappa^+\}.$$

Claim. *C has a club subset.*

Proof. Suppose that S is stationary in $\mathcal{P}_{\omega_1}\theta$. We need to prove $S \cap C \neq \emptyset$.

By κ-SR in $\mathcal{P}_{\omega_1}\theta$ there is $\kappa \subset X \subset \theta$ of size κ such that $S \cap \mathcal{P}_{\omega_1} X$ is stationary in $\mathcal{P}_{\omega_1} X$. Fix a bijection

$$\pi : \kappa \to X.$$

Since $\{x \in \mathcal{P}_{\omega_1} X : \pi``(x \cap \kappa) = x\}$ is club,

$$\{x \in S \cap \mathcal{P}_{\omega_1} X : \pi``(x \cap \kappa) = x\}$$

is stationary in $\mathcal{P}_{\omega_1} X$. Hence by $\kappa \subset X$

$$\{x \cap \kappa : \pi``(x \cap \kappa) = x \in S \cap \mathcal{P}_{\omega_1} X\}$$

is stationary in $\mathcal{P}_{\omega_1}\kappa$. Thus

$$S' = \{\sup(x \cap \kappa) : \pi``(x \cap \kappa) = x \in S \cap \mathcal{P}_{\omega_1} X\}$$

is stationary in κ.

Since $\{z \in \mathcal{P}_\kappa 2^\theta \cap C(d) : \pi``(z \cap \kappa) \subset z \wedge z \cap \kappa \in \kappa\}$ is club in $\mathcal{P}_\kappa 2^\theta$,

$$S'' = \{z \in \mathcal{P}_\kappa 2^\theta \cap C(d) : \pi``(z \cap \kappa) \subset z \wedge z \cap \kappa \in S'\}$$

is stationary in $\mathcal{P}_\kappa 2^\theta$. Hence $\bigcup S'' = 2^\theta$. Since $\{z \cap \kappa : z \in S''\} \subset S' \subset \kappa$, there is $\delta \in S'$ such that

$$S^* = \kappa^+ \cap \bigcup\{z \in S'' : z \cap \kappa = \delta\}$$

is unbounded in κ^+. Since $\delta \in S'$, there is $x \in S$ such that

- $\pi``(x \cap \kappa) = x$ and
- $\sup(x \cap \kappa) = \delta$.

We claim that $x \in C$. It suffices to show $S^* \subset S(x)$. To see this, let $\alpha \in S^*$. Then there is $z \in S''$ such that

- $\alpha \in z$ and
- $z \cap \kappa = \delta$.

Since $\sup(x \cap \kappa) = \delta$, it suffices to show

$$\delta \subset \mathrm{cl}_d(x \cup \{\alpha\} \cup \delta) \cap \kappa \subset z \cap \kappa = \delta.$$

To see the second inclusion, note that

$$x = \pi``(x \cap \kappa) \subset \pi`` \sup(x \cap \kappa) = \pi``\delta = \pi``(z \cap \kappa) \subset z,$$

hence $x \cup \{\alpha\} \cup \delta \subset z \in C(d)$. $\qquad\square$

Take $\zeta < 2^\theta$ with

$$\mathcal{P}_{\omega_1}\theta \cap C(e_\zeta) \subset C.$$

Since $\{z \in \mathcal{P}_\kappa 2^\theta \cap C(d) : \zeta \in z \wedge z \cap \kappa \in \kappa\}$ is club in $\mathcal{P}_\kappa 2^\theta$, there is $z \in \mathcal{P}_\kappa 2^\theta \cap C(d)$ such that

- $\zeta \in z$ and
- $\delta = z \cap \kappa \in \kappa$ has cofinality ω.

Since $d : [2^\theta]^{<\omega} \to \mathcal{P}_{\omega_1} 2^\theta$, there is a countable $y \subset z$ such that

- $\zeta \in y \in C(d)$ and
- $\sup(y \cap \kappa) = \delta$.

Since $y \cup \delta \subset z \in C(d)$, we have $\delta \subset \mathrm{cl}_d(y \cup \delta) \cap \kappa \subset z \cap \kappa = \delta$, and hence $\mathrm{cl}_d(y \cup \delta) \cap \kappa = \delta$.

Define

$$T = \{a \in [\kappa^+]^{<\omega_1} : \mathrm{cl}_d(y \cup a \cup \delta) \cap \kappa = \delta\}.$$

Claim. T *is an unbounded subtree of* $[\kappa^+]^{<\omega_1}$.

Proof. It is easy to see that T is a subtree of $[\kappa^+]^{<\omega_1}$. Note that $\emptyset \in T$ by the previous paragraph. Suppose $a \in T$. It remains to give an unbounded subset of

$$\{\alpha < \kappa^+ : \mathrm{cl}_d(y \cup a \cup \{\alpha\} \cup \delta) \cap \kappa = \delta\}.$$

Set

$$\bar{y} = \mathrm{cl}_d(y \cup a).$$

Then $y \subset \bar{y} \subset \mathrm{cl}_d(y \cup a \cup \delta)$. Hence by $a \in T$ we have

$$\delta = \sup(y \cap \kappa) \subset \sup(\bar{y} \cap \kappa) \subset \sup(\mathrm{cl}_d(y \cup a \cup \delta) \cap \kappa) = \delta.$$

Thus $\sup(\bar{y} \cap \kappa) = \delta$. Since $\zeta \in y \subset \bar{y} \in \mathcal{P}_{\omega_1} 2^\theta \cap C(d) \subset D$, we have

$$\bar{y} \cap \theta \in \mathcal{P}_{\omega_1}\theta \cap C(e_\zeta) \subset C.$$

Hence

$$S(\bar{y} \cap \theta) = \{\alpha < \kappa^+ : \mathrm{cl}_d((\bar{y} \cap \theta) \cup \{\alpha\} \cup \delta) \cap \kappa = \delta\}$$

is unbounded. We claim that $S(\bar{y} \cap \theta)$ is as desired. Suppose $\alpha \in S(\bar{y} \cap \theta)$. It suffices to show

$$\delta \subset \mathrm{cl}_d(y \cup a \cup \{\alpha\} \cup \delta) \cap \kappa \subset \mathrm{cl}_d(\bar{y} \cup \{\alpha\} \cup \delta) \cap \kappa$$
$$= \mathrm{cl}_d((\bar{y} \cap \theta) \cup \{\alpha\} \cup \delta) \cap \kappa = \delta.$$

The second inclusion follows from $y \cup a \subset \bar{y}$. The first equality follows from the choice of d. The last equality follows from $\alpha \in S(\bar{y} \cap \theta)$. $\qquad\square$

Claim. *There is a subtree T^* of T such that*

- $|\operatorname{suc}_{T^*}(a)| = 2$ *for every* $a \in T^*$ *and*
- *if* $a \lhd b \in T^*$, $a \lhd a \cup \{\gamma\} \in T^*$ *and* $\gamma \notin b$, *then* $\gamma \notin \operatorname{cl}_d(y \cup b \cup \delta)$.

Proof. Suppose $a \in T$. Since $\operatorname{suc}_T(a)$ is unbounded in κ^+, there is $\beta_a \in \operatorname{suc}_T(a)$ such that $|\operatorname{suc}_T(a) \cap \beta_a - \kappa| = \kappa$. Since $|\operatorname{cl}_d(y \cup a \cup \{\beta_a\} \cup \delta)| < \kappa$, there is $\alpha_a \in \operatorname{suc}_T(a) \cap \beta_a - \kappa$ such that $\alpha_a \notin \operatorname{cl}_d(y \cup a \cup \{\beta_a\} \cup \delta)$.

Let T^* be the subtree of T such that if $a \in T^*$, then $\operatorname{suc}_{T^*}(a) = \{\alpha_a, \beta_a\}$. We claim that T^* is as desired. To see the second item, let $a \lhd b \in T^*$. Set

$$z_b = \operatorname{cl}_d(y \cup b \cup \delta).$$

It suffices to prove $\{\alpha_a, \beta_a\} \not\subset z_b$. Assume $\beta_a \in z_b$. We prove $\alpha_a \notin z_b$. Set

$$z_\beta = \operatorname{cl}_d(y \cup a \cup \{\beta_a\} \cup \delta).$$

Since $\alpha_a < \beta_a$, it suffices to show

$$\alpha_a \notin z_\beta \cap \beta_a = \pi_{\beta_a} \text{``}(z_\beta \cap \kappa) = \pi_{\beta_a} \text{``} \delta = \pi_{\beta_a} \text{``}(z_b \cap \kappa) = z_b \cap \beta_a.$$

The non-membership follows from the choice of α_a. The first equality follows from $\beta_a \in z_\beta \in \mathcal{P}_\kappa 2^\theta \cap C(d) \subset D$. The second and third equalities follow from $a \cup \{\beta_a\} \in T^* \subset T$ and $b \in T^* \subset T$ respectively. The last equality follows from $\beta_a \in z_b \in \mathcal{P}_\kappa 2^\theta \cap C(d) \subset D$. $\quad\square$

For $B \in [T^*]$ set

$$x_B = \operatorname{cl}_d(y \cup B \cup \delta) \cap \kappa^+.$$

We claim that $\{x_B : B \in [T^*]\}$ is a subset of $\{x \in \mathcal{P}_\kappa \kappa^+ \cap C(f) : x \cap \kappa \in \kappa\}$ of size 2^{ω_1}. First note that $x_B \in \mathcal{P}_\kappa \kappa^+$ by $\kappa > \omega_1$. Next we have $x_B \in C(f)$ by $\operatorname{cl}_d(y \cup B \cup \delta) \in \mathcal{P}_\kappa 2^\theta \cap C(d) \subset D$. Since $B \cap \beta \in T^* \subset T$ for $\beta \in B$, we have

$$x_B \cap \kappa = \operatorname{cl}_d(y \cup B \cup \delta) \cap \kappa = \bigcup_{\beta \in B} \operatorname{cl}_d(y \cup (B \cap \beta) \cup \delta) \cap \kappa = \delta \in \kappa.$$

Since $|[T^*]| = 2^{\omega_1}$, it remains to prove that the map $B \mapsto x_B$ is injective. Suppose that $A \neq B$ are in $[T^*]$. Then there are a common initial segment a of A and B, and $\gamma \in A - B$ such that $a \lhd a \cup \{\gamma\} \in T^*$. Then $\gamma \in x_A$. On the other hand, since $B \cap \beta \in T^*$ for $\beta \in B$, we have

$$\gamma \notin \bigcup_{\beta \in B} \operatorname{cl}_d(y \cup (B \cap \beta) \cup \delta) \cap \kappa^+ = \operatorname{cl}_d(y \cup B \cup \delta) \cap \kappa^+ = x_B$$

by the Claim, as desired. $\quad\square$

5. Diamonds on sets of cofinality ω_1

In this section we prove Theorem 6 and deduce Theorem 5 as a corollary.

Proof of Theorem 6. If $\lambda > 2^{<\kappa}$, we are done by Theorem 1. So let us assume $\lambda \le 2^{<\kappa}$. Then $\lambda^{<\kappa} = 2^{<\kappa} = 2^{\omega_1}$.

Fix a disjoint family

$$\{I_\eta : \eta < \omega_1\} \subset [\omega_1]^{\omega_1}$$

such that $\eta < \min I_\eta$ for every $\eta < \omega_1$. Since $2^{\omega_1} = \lambda^{<\kappa}$, there is a map

$$\varphi : \mathcal{P}(\omega_1) \to \mathcal{P}_\kappa \lambda$$

such that $\varphi``[I_\eta]^{\omega_1} = \mathcal{P}_\kappa \lambda$ for every $\eta < \omega_1$. By lemma 7 there is a club guessing sequence

$$\langle c_\gamma : \gamma \in S^{\omega_1}_{\kappa^+} \rangle$$

such that if $\gamma \in S^{\omega_1}_{\kappa^+} \cap \lim S^\kappa_{\kappa^+}$, then $c_\gamma \subset \{\alpha < \gamma : \mathrm{cf}\,\alpha = \kappa\}$.

For $x \in \mathcal{P}_\kappa \lambda$ with $\mathrm{cf}\,\sup(x \cap \kappa^+) = \omega_1$ define $g(x) \in \mathcal{P}_\kappa \lambda$ as follows: Let $\gamma = \sup(x \cap \kappa^+)$ and list c_γ in increasing order as $\{\gamma_\eta : \eta < \omega_1\}$. Set

$$r(x) = \{\eta < \omega_1 : x \cap (\gamma_{\eta+1} - \gamma_\eta) \ne \emptyset\},$$

$$g(x) = \bigcup_{\eta \in r(x)} \varphi(I_\eta \cap r(x)).$$

Claim. *g is a diamond on $\{x \in \mathcal{P}_\kappa \lambda : \mathrm{cf}\,\sup(x \cap \kappa^+) = \omega_1\}$.*

Proof. Suppose that $A \subset \lambda$ and $f : [\lambda]^{<\omega} \to \mathcal{P}_{\omega_1} \lambda$. We need to find $x \in \mathcal{P}_\kappa \lambda \cap C(f)$ such that

- $\mathrm{cf}\,\sup(x \cap \kappa^+) = \omega_1$,
- $x \cap \kappa \in \kappa$ and
- $g(x) = A \cap x$.

(Moreover we have $\mathrm{cf}(x \cap \kappa) = \omega$.)

Set

$$\theta = 2^{\kappa^+}.$$

Then $\theta \ge 2^{<\kappa} \ge \lambda$. List all functions of the form $e : [\theta]^{<\omega} \to \mathcal{P}_{\omega_1} \theta$ as $\{e_\zeta : \zeta < 2^\theta\}$. For $\gamma \in \kappa^+ - \kappa$ fix a bijection

$$\pi_\gamma : \kappa \to \gamma.$$

Define

$$D = \{z \in \mathcal{P}_\kappa 2^\theta : z \cap \lambda \in C(f) \wedge \forall \zeta \in z(z \cap \theta \in C(e_\zeta)) \wedge$$
$$\forall \gamma \in z \cap (\kappa^+ - \kappa)(\pi_\gamma``(z \cap \kappa) = z \cap \gamma)\}.$$

Then D is σ-club. By Lemma 13 there is a map $d : [2^\theta]^{<\omega} \to \mathcal{P}_{\omega_1} 2^\theta$ such that

- $\mathcal{P}_\kappa 2^\theta \cap C(d) \subset D$ and
- if $z \in \mathcal{P}_\kappa 2^\theta \cap C(d)$ and $u \in \mathcal{P}_\kappa \kappa^+$, then

$$\mathrm{cl}_d(z \cup u) \cap \theta = \mathrm{cl}_d((z \cap \theta) \cup u) \cap \theta.$$

Define

$$S(x) = \{\alpha \in S^\kappa_{\kappa^+} : \mathrm{cl}_d(x \cup \{\alpha\} \cup \sup(x \cap \kappa)) \cap \kappa = \sup(x \cap \kappa)\},$$

$$C = \{x \in \mathcal{P}_{\omega_1}\theta : S(x) \text{ is stationary in } \kappa^+\}.$$

Subclaim. *C has a club subset.*

Proof. Suppose that S is stationary in $\mathcal{P}_{\omega_1}\theta$. We need to prove $S \cap C \neq \emptyset$.

By κ-SR in $\mathcal{P}_{\omega_1}\theta$ there is $\kappa \subset X \subset \theta$ of size κ such that $S \cap \mathcal{P}_{\omega_1} X$ is stationary in $\mathcal{P}_{\omega_1} X$. Fix a bijection

$$\pi : \kappa \to X.$$

Since $\{x \in \mathcal{P}_{\omega_1} X : \pi``(x \cap \kappa) = x\}$ is club,

$$\{x \in S \cap \mathcal{P}_{\omega_1} X : \pi``(x \cap \kappa) = x\}$$

is stationary in $\mathcal{P}_{\omega_1} X$. Hence by $\kappa \subset X$

$$\{x \cap \kappa : \pi``(x \cap \kappa) = x \in S \cap \mathcal{P}_{\omega_1} X\}$$

is stationary in $\mathcal{P}_{\omega_1}\kappa$. Thus

$$S' = \{\sup(x \cap \kappa) : \pi``(x \cap \kappa) = x \in S \cap \mathcal{P}_{\omega_1} X\}$$

is stationary in κ.

Since $\{z \in \mathcal{P}_\kappa 2^\theta \cap C(d) : \pi``(z \cap \kappa) \subset z \wedge z \cap \kappa \in \kappa\}$ is club in $\mathcal{P}_\kappa 2^\theta$,

$$S'' = \{z \in \mathcal{P}_\kappa 2^\theta \cap C(d) : \pi``(z \cap \kappa) \subset z \wedge z \cap \kappa \in S'\}$$

is stationary in $\mathcal{P}_\kappa 2^\theta$. Hence $\bigcup S'' = 2^\theta$. Since $\{z \cap \kappa : z \in S''\} \subset S' \subset \kappa$, there is $\delta \in S'$ such that

$$S^* = S^\kappa_{\kappa^+} \cap \bigcup \{z \in S'' : z \cap \kappa = \delta\}$$

is stationary in κ^+. Since $\delta \in S'$, there is $x \in S$ such that

- $\pi``(x \cap \kappa) = x$ and
- $\sup(x \cap \kappa) = \delta$.

We claim that $x \in C$. It suffices to show $S^* \subset S(x)$. To see this, let $\alpha \in S^*$. Then there is $z \in S''$ such that

- $\alpha \in z$ and
- $z \cap \kappa = \delta$.

Since $\sup(x \cap \kappa) = \delta$, it suffices to show

$$\delta \subset \mathrm{cl}_d(x \cup \{\alpha\} \cup \delta) \cap \kappa \subset z \cap \kappa = \delta.$$

To see the second inclusion, note that

$$x = \pi``(x \cap \kappa) \subset \pi`` \sup(x \cap \kappa) = \pi``\delta = \pi``(z \cap \kappa) \subset z,$$

hence $x \cup \{\alpha\} \cup \delta \subset z \in C(d)$. $\qquad\square$

Take $\zeta < 2^\theta$ with

$$\mathcal{P}_{\omega_1}\theta \cap C(e_\zeta) \subset C.$$

Since

$$\{z \in \mathcal{P}_\kappa 2^\theta \cap C(d) : \zeta \in z \wedge z \cap \kappa \in \kappa\}$$

is club in $\mathcal{P}_\kappa 2^\theta$, there is $z \in \mathcal{P}_\kappa 2^\theta \cap C(d)$ such that

- $\zeta \in z$ and
- $\delta = z \cap \kappa \in \kappa$ has cofinality ω.

Since $d : [2^\theta]^{<\omega} \to \mathcal{P}_{\omega_1} 2^\theta$, there is a countable $y \subset z$ such that

- $\zeta \in y \in C(d)$ and
- $\sup(y \cap \kappa) = \delta$.

Since $y \cup \delta \subset z \in C(d)$, we have $\delta \subset \mathrm{cl}_d(y \cup \delta) \cap \kappa \subset z \cap \kappa = \delta$. Hence $\mathrm{cl}_d(y \cup \delta) \cap \kappa = \delta$.

Define

$$T = \{a \in [S_{\kappa^+}^\kappa]^{<\omega_1} : \mathrm{cl}_d(y \cup a \cup \delta) \cap \kappa = \delta\}.$$

Subclaim. *T is a stationary subtree of $[\kappa^+]^{<\omega_1}$.*

Proof. It is easy to see that T is a subtree of $[\kappa^+]^{<\omega_1}$. Note that $\emptyset \in T$ by the previous paragraph. Fix $a \in T$. It suffices to give a stationary subset of $\{\alpha \in S_{\kappa^+}^\kappa : \mathrm{cl}_d(y \cup a \cup \{\alpha\} \cup \delta) \cap \kappa = \delta\}$.

Set

$$\bar{y} = \mathrm{cl}_d(y \cup a).$$

Then $y \subset \bar{y} \subset \mathrm{cl}_d(y \cup a \cup \delta)$. Hence by $a \in T$ we have

$$\delta = \sup(y \cap \kappa) \subset \sup(\bar{y} \cap \kappa) \subset \sup(\mathrm{cl}_d(y \cup a \cup \delta) \cap \kappa) = \delta.$$

Thus $\sup(\bar{y} \cap \kappa) = \delta$. Since $\zeta \in y \subset \bar{y} \in \mathcal{P}_{\omega_1} 2^\theta \cap C(d) \subset D$, we have

$$\bar{y} \cap \theta \in \mathcal{P}_{\omega_1}\theta \cap C(e_\zeta) \subset C.$$

Hence

$$S(\bar{y} \cap \theta) = \{\alpha \in S_{\kappa^+}^\kappa : \mathrm{cl}_d((\bar{y} \cap \theta) \cup \{\alpha\} \cup \delta) \cap \kappa = \delta\}$$

is stationary.

We claim that $S(\bar{y} \cap \theta)$ is as desired. Fix $\alpha \in S(\bar{y} \cap \theta)$. It suffices to show

$$\delta \subset \mathrm{cl}_d(y \cup a \cup \{\alpha\} \cup \delta) \cap \kappa \subset \mathrm{cl}_d(\bar{y} \cup \{\alpha\} \cup \delta) \cap \kappa$$
$$= \mathrm{cl}_d((\bar{y} \cap \theta) \cup \{\alpha\} \cup \delta) \cap \kappa = \delta.$$

The second inclusion follows from $y \cup a \subset \bar{y}$. The first equality follows from $\bar{y} \in \mathcal{P}_{\omega_1} 2^\theta \cap C(d)$ and $\{\alpha\} \cup \delta \in \mathcal{P}_\kappa \kappa^+$. The last equality follows from $\alpha \in S(\bar{y} \cap \theta)$. $\qquad\square$

Subclaim. *There are a stationary subtree* $T^* \subset T$ *and* $h^* : T^* \to \kappa^+$ *such that if* $a \vartriangleleft a \cup b \in T^*$, *then* $\mathrm{cl}_d(y \cup a \cup b \cup \delta) \cap \min b \subset h^*(a)$.

Proof. Let $a \in T$. Define a map on $\mathrm{suc}_T(a)$ by

$$\alpha \mapsto \sup(\mathrm{cl}_d(y \cup a \cup \{\alpha\} \cup \delta) \cap \alpha).$$

Since $|\mathrm{cl}_d(y \cup a \cup \{\alpha\} \cup \delta)| < \kappa = \mathrm{cf}\,\alpha$, the map is regressive on $\mathrm{suc}_T(a)$. Hence there are a stationary $S_a \subset \mathrm{suc}_T(a)$ and $h(a) < \kappa^+$ such that

$$\mathrm{cl}_d(y \cup a \cup \{\alpha\} \cup \delta) \cap \alpha \subset h(a)$$

for every $\alpha \in S_a$.

Thus there is a stationary subtree $T^* \subset T$ such that $\mathrm{suc}_{T^*}(a) = S_a$ for every $a \in T^*$. We claim that T^* and $h^* = h|T^*$ are as desired. Suppose $a \vartriangleleft a \cup b \in T^*$. Set

$$\alpha = \min b.$$

Since $a \cup \{\alpha\}$ and $a \cup b$ are in T^*, we have

$$\mathrm{cl}_d(y \cup a \cup \{\alpha\} \cup \delta) \cap \kappa = \mathrm{cl}_d(y \cup a \cup b \cup \delta) \cap \kappa = \delta.$$

Since $\mathrm{cl}_d(y \cup a \cup \{\alpha\} \cup \delta)$ and $\mathrm{cl}_d(y \cup a \cup b \cup \delta)$ are in $\mathcal{P}_\kappa 2^\theta \cap C(d) \subset D$, we have

$$\mathrm{cl}_d(y \cup a \cup b \cup \delta) \cap \alpha = \pi_\alpha ``\delta = \mathrm{cl}_d(y \cup a \cup \{\alpha\} \cup \delta) \cap \alpha \subset h^*(a),$$

as desired. $\qquad\square$

Define

$$E = \{\gamma < \kappa^+ : \mathrm{cl}_d(y \cup \gamma) \cap \kappa^+ = \gamma \wedge$$
$$\forall a \in T^* \cap [\gamma]^{<\omega_1} (h^*(a) < \gamma \in \lim \mathrm{suc}_{T^*}(a))\}.$$

Since $\kappa^\omega = \kappa$, E is unbounded and hence $\lim E$ is club. Take

$$\gamma \in S^{\omega_1}_{\kappa^+} \cap \lim S^\kappa_{\kappa^+} \cap \lim E$$

with $c_\gamma \subset \lim E$. Since $\gamma \in S^{\omega_1}_{\kappa^+} \cap \lim S^\kappa_{\kappa^+}$, we have

$$c_\gamma \subset \{\alpha < \gamma : \mathrm{cf}\,\alpha = \kappa\}.$$

Since $\{\gamma < \kappa^+ : \operatorname{cf} \gamma > \omega\} \cap \lim E \subset E$ by definition of E, we have

$$c_\gamma \cup \{\gamma\} \subset (S^\kappa_{\kappa^+} \cup S^{\omega_1}_{\kappa^+}) \cap \lim E \subset E.$$

Recall that $\{\gamma_\eta : \eta < \omega_1\}$ lists c_γ in increasing order.

By recursion on $\xi < \omega_1$ we choose $\eta(\xi) < \omega_1$, $\alpha_\xi < \kappa^+$ and $J_\xi \in [I_{\eta(\xi)}]^{\omega_1}$ so that

- if $\iota < \xi$, then $\eta(\iota) < \eta(\xi)$,
- $\gamma_{\eta(\xi)} < \alpha_\xi < \gamma_{\eta(\xi)+1}$ and
- $\{\alpha_\iota : \iota \leq \xi\} \in T^*$

as follows: First set $\eta(0) = 0$. Suppose we have defined $\eta(\iota)$, α_ι and J_ι for $\iota < \xi$. If $\xi > 0$, let

$$\eta(\xi) = \min(\bigcup_{\iota < \xi} J_\iota - \sup_{\iota < \xi}(\eta(\iota) + 1)).$$

Since $\gamma_{\eta(\xi)+1} \in E$ and $\{\alpha_\iota : \iota < \xi\} \in T^* \cap [\gamma_{\eta(\xi)+1}]^{<\omega_1}$, there is $\alpha_\xi < \gamma_{\eta(\xi)+1}$ such that $\gamma_{\eta(\xi)} < \alpha_\xi \in \operatorname{suc}_{T^*}\{\alpha_\iota : \iota < \xi\}$. Then $\{\alpha_\iota : \iota \leq \xi\} \in T^*$. Since $\varphi``[I_{\eta(\xi)}]^{\omega_1} = \mathcal{P}_\kappa \lambda$, there is $J_\xi \in [I_{\eta(\xi)}]^{\omega_1}$ such that

$$\varphi(J_\xi) = A \cap \operatorname{cl}_d(y \cup \{\alpha_\iota : \iota \leq \xi\} \cup \delta).$$

Note that if $\xi \leq \bar{\xi} < \omega_1$, then $\eta(\xi) \leq \eta(\bar{\xi}) < \min I_{\eta(\bar{\xi})} \leq \min J_{\bar{\xi}}$. Hence

$$\eta(\xi) = \min(\bigcup_{\iota < \omega_1} J_\iota - \sup_{\iota < \xi}(\eta(\iota) + 1))$$

for every $\xi > 0$. Thus

$$\{\eta(\xi) : \xi < \omega_1\} = \{0\} \cup \bigcup_{\iota < \omega_1} J_\iota.$$

Now define

$$x = \operatorname{cl}_d(y \cup \{\alpha_\xi : \xi < \omega_1\} \cup \delta) \cap \lambda.$$

We claim that x is as desired. First note that $x \in \mathcal{P}_\kappa \lambda$ by $\kappa > \omega_1$. Since $\operatorname{cl}_d(y \cup \{\alpha_\xi : \xi < \omega_1\} \cup \delta) \in \mathcal{P}_\kappa 2^\theta \cap C(d) \subset D$, we have

$$x = \operatorname{cl}_d(y \cup \{\alpha_\xi : \xi < \omega_1\} \cup \delta) \cap \lambda \in C(f).$$

Next we prove $\sup(x \cap \kappa^+) = \gamma \in S^{\omega_1}_{\kappa^+}$. It suffices to show

$$\gamma = \sup_{\xi < \omega_1} \gamma_{\eta(\xi)} = \sup_{\xi < \omega_1} \alpha_\xi \leq \sup(x \cap \kappa^+) \leq \sup(\operatorname{cl}_d(y \cup \gamma) \cap \kappa^+) = \gamma.$$

The second equality follows from $\gamma_{\eta(\xi)} < \alpha_\xi < \gamma_{\eta(\xi)+1}$. The first inequality follows from $\{\alpha_\xi : \xi < \omega_1\} \subset x \cap \kappa^+$. The second inequality follows from $x = \operatorname{cl}_d(y \cup \{\alpha_\xi : \xi < \omega_1\} \cup \delta) \cap \lambda \subset \operatorname{cl}_d(y \cup \gamma) \cap \lambda$. The last equality follows from $\gamma \in E$.

To see that $x \cap \kappa \in \kappa$, it suffices to show

$$x \cap \kappa = \mathrm{cl}_d(y \cup \{\alpha_\xi : \xi < \omega_1\} \cup \delta) \cap \kappa$$

$$= \bigcup\nolimits_{\xi < \omega_1} \mathrm{cl}_d(y \cup \{\alpha_\iota : \iota \leq \xi\} \cup \delta) \cap \kappa = \delta.$$

The last equality follows from $\{\alpha_\iota : \iota \leq \xi\} \in T^* \subset T$.

It remains to prove $g(x) = A \cap x$.

Subclaim. $r(x) = \{\eta(\xi) : \xi < \omega_1\}$.

Proof. One direction follows from $\alpha_\xi \in x \cap (\gamma_{\eta(\xi)+1} - \gamma_{\eta(\xi)})$ for every $\xi < \omega_1$.

For the converse we prove that $x \cap \gamma_{\eta(\xi)+1} \subset \gamma_{\eta(\xi)+1}$ for every $\xi < \omega_1$. It suffices to show that if $\xi < \bar\xi < \omega_1$, then

$$\mathrm{cl}_d(y \cup \{\alpha_\iota : \iota \leq \bar\xi\} \cup \delta) \cap \gamma_{\eta(\xi+1)} \subset \mathrm{cl}_d(y \cup \{\alpha_\iota : \iota \leq \bar\xi\} \cup \delta) \cap \alpha_{\xi+1}$$

$$\subset h^*(\{\alpha_\iota : \iota \leq \xi\}) < \gamma_{\eta(\xi)+1}.$$

The first inclusion follows from $\gamma_{\eta(\xi+1)} < \alpha_{\xi+1}$. The second inclusion follows from the choice of h^* and $\{\alpha_\iota : \iota \leq \xi\} \lhd \{\alpha_\iota : \iota \leq \bar\xi\} \in T^*$. The inequality follows from $\gamma_{\eta(\xi)+1} \in E$ and $\{\alpha_\iota : \iota \leq \xi\} \in T^* \cap [\gamma_{\eta(\xi)+1}]^{<\omega_1}$. $\square$

Hence

$$r(x) = \{0\} \cup \bigcup\nolimits_{\iota < \omega_1} J_\iota.$$

Let $\xi < \omega_1$. Since $J_\xi \subset I_{\eta(\xi)} \subset \omega_1 - \{0\}$ and $I_{\eta(\xi)} \cap J_\iota = \emptyset$ for every $\iota \neq \xi$, we have

$$I_{\eta(\xi)} \cap r(x) = I_{\eta(\xi)} \cap \bigcup\nolimits_{\iota < \omega_1} J_\iota = J_\xi.$$

Hence by the choice of J_ξ

$$\varphi(I_{\eta(\xi)} \cap r(x)) = \varphi(J_\xi) = A \cap \mathrm{cl}_d(y \cup \{\alpha_\iota : \iota \leq \xi\} \cup \delta).$$

Therefore

$$g(x) = \bigcup\nolimits_{\eta \in r(x)} \varphi(I_\eta \cap r(x))$$

$$= \bigcup\nolimits_{\xi < \omega_1} \varphi(I_{\eta(\xi)} \cap r(x))$$

$$= \bigcup\nolimits_{\xi < \omega_1} A \cap \mathrm{cl}_d(y \cup \{\alpha_\iota : \iota \leq \xi\} \cup \delta)$$

$$= A \cap x.$$

$\square$

This completes the proof. $\square$

Proof of Theorem 5. Assume first $\rho < \nu$. Then $2^{<\kappa} = \kappa$ holds in the model. So we are done by Theorem 1.

Assume next $\rho \geq \nu$. Then the model satisfies $\omega_1 < \kappa = \kappa^\omega < \nu = \kappa^+ \leq \rho = 2^{\omega_1} = 2^{<\kappa} \leq 2^{\kappa^+} \leq \theta$. Moreover κ-SR in $\mathcal{P}_{\omega_1}\theta$ holds in the model by Proposition 12. Hence we are done by Theorem 6. $\square$

6. Appendix

This section presents Shelah's construction of a diamond on $\mathcal{P}_{\omega_1}\omega_2$ (as we understand it) in somewhat greater generality. This gives another proof of Theorem 4 in the case $\lambda = \kappa^+$. The point here is that the proof works uniformly regardless of the size of $2^{<\kappa}$.

Proposition 16. *Assume $2^\omega = 2^{<\kappa}$. Then $\mathcal{P}_\kappa\kappa^+$ carries a diamond.*

Proof. Fix a disjoint family

$$\{I_n : n < \omega\} \subset [\omega]^\omega$$

such that $n < \min I_n$ for every $n < \omega$. Since $2^\omega = 2^{<\kappa}$, for each $z \in \mathcal{P}_\kappa\kappa^+$ there is a map

$$\psi(z) : \mathcal{P}(\omega) \to \mathcal{P}(z)$$

such that $\psi(z)``[I_n]^\omega = \mathcal{P}(z)$ for every $n < \omega$. Fix a club guessing sequence

$$\langle c_\gamma : \gamma \in S^\omega_{\kappa^+}\rangle.$$

For $x \in \mathcal{P}_\kappa\kappa^+$ with $\operatorname{cf}\sup x = \omega$ define $g(x) \in \mathcal{P}_\kappa\kappa^+$ as follows: Let $\gamma = \sup x$ and list c_γ in increasing order as $\{\gamma_n : n < \omega\}$. Set

$$r(x) = \{n < \omega : x \cap (\gamma_{n+1} - \gamma_n) \neq \emptyset\},$$
$$g(x) = \bigcup_{n \in r(x)} \psi(x \cap \gamma_n)(I_n \cap r(x)).$$

Claim. *g is a diamond on $\{x \in \mathcal{P}_\kappa\kappa^+ : \operatorname{cf}\sup x = \omega\}$.*

Proof. Suppose that $A \subset \kappa^+$ and $f : [\kappa^+]^{<\omega} \to \mathcal{P}_\kappa\kappa^+$. We need to find $x \in \mathcal{P}_\kappa\kappa^+ \cap C(f)$ such that

- $\operatorname{cf}\sup x = \omega$ and
- $g(x) = A \cap x$.

For $\gamma \in \kappa^+ - \kappa$ fix a bijection

$$\pi_\gamma : \kappa \to \gamma.$$

We may assume that if $x \in C(f)$, then $\pi_\gamma``(x \cap \kappa) = x \cap \gamma$ for every $\gamma \in x - \kappa$.

For $\delta < \kappa$ let

$$T_\delta = \{a \in [\kappa^+]^{<\omega} : \operatorname{cl}_f(a \cup \delta) \cap \kappa = \delta\}.$$

It is easy to see that T_δ is a (possibly empty) subtree of $[\kappa^+]^{<\omega}$.

Subclaim. *There are* $\delta < \kappa$, *a stationary subtree* T^* *of* T_δ *and* $h : T^* \to \kappa^+$ *such that if* $a \lhd a \cup b \in T^*$, *then* $\mathrm{cl}_f(a \cup b \cup \delta) \cap \min b \subset h(a)$.

Proof. First we give $\delta < \kappa$ such that $[T_\delta] \cap [C]^\omega \neq \emptyset$ for every club $C \subset \kappa^+$. Assume to the contrary for each $\delta < \kappa$ there is a club $C_\delta \subset \kappa^+$ such that $[T_\delta] \cap [C_\delta]^\omega = \emptyset$. Then

$$C = \bigcap\nolimits_{\delta < \kappa} C_\delta$$

is club in κ^+. Hence there is $B \subset S_{\kappa^+}^\kappa \cap C$ of order-type ω. Take $\delta < \kappa$ so that $\sup(\mathrm{cl}_f(B \cup \gamma) \cap \kappa) < \delta$ for every $\gamma \lessdot \delta$. Then $\mathrm{cl}_f(B \cup \delta) \cap \kappa = \delta$. Hence $\mathrm{cl}_f((B \cap \beta) \cup \delta) \cap \kappa = \delta$ for every $\beta \in B$. Thus $B \in [T_\delta] \cap [C]^\omega \subset [T_\delta] \cap [C_\delta]^\omega$. This contradicts that $[T_\delta] \cap [C_\delta]^\omega = \emptyset$, as desired.

Define

$$T' = \{b \in T_\delta : \forall a \unlhd b \forall C \subset \kappa^+ \text{ club } ([T_\delta^a] \cap [C]^\omega \neq \emptyset)\}.$$

Then $\emptyset \in T'$ by the previous paragraph. T' is a stationary subtree of T_δ by the proof of Lemma 9. By Lemma 10 there are a stationary subtree T^* of T' and $h : T^* \to \kappa^+$ as required above. $\qquad\square$

Define

$$D = \{\gamma \in \kappa^+ - \kappa : \mathrm{cl}_f \gamma = \gamma \wedge \forall a \in T^* \cap [\gamma]^{<\omega} (h(a) < \gamma \in \lim \mathrm{suc}_{T^*}(a))\}.$$

Then D is club. Take $\gamma \in S_{\kappa^+}^\omega \cap D$ with $c_\gamma \subset D$. Recall that $\{\gamma_n : n < \omega\}$ lists c_γ in increasing order.

By recursion on $k < \omega$ we choose $n(k) < \omega$, $\alpha_k < \kappa^+$ and $J_k \in [I_{n(k)}]^\omega$ so that

- $n(k-1) < n(k)$,
- $\gamma_{n(k)} < \alpha_k < \gamma_{n(k)+1}$ and
- $\{\alpha_i : i \leq k\} \in T^*$

as follows: First set $n(0) = 0$. Suppose we have defined α_i, $n(i)$ and J_i for $i < k$. If $k > 0$, let $n(k) = \min(\bigcup_{i<k} J_i - (n(k-1)+1))$. Since $\gamma_{n(k)+1} \in D$ and $\{\alpha_i : i < k\} \in T^* \cap [\gamma_{n(k)+1}]^{<\omega}$, there is $\alpha_k < \gamma_{n(k)+1}$ such that $\gamma_{n(k)} < \alpha_k \in \mathrm{suc}_{T^*}\{\alpha_i : i < k\}$. Then $\{\alpha_i : i \leq k\} \in T^*$. By the choice of ψ there is $J_k \in [I_{n(k)}]^\omega$ such that

$$\psi(\mathrm{cl}_f(\{\alpha_i : i \leq k\} \cup \delta) \cap \gamma_{n(k)})(J_k) = A \cap \mathrm{cl}_f(\{\alpha_i : i \leq k\} \cup \delta) \cap \gamma_{n(k)}.$$

By the proof of Proposition 8 we have

$$\{n(k) : k < \omega\} = \{0\} \cup \bigcup\nolimits_{i<\omega} J_i.$$

Now define

$$x = \mathrm{cl}_f(\{\alpha_k : k < \omega\} \cup \delta) \in \mathcal{P}_\kappa \kappa^+ \cap C(f).$$

We claim that x is as desired. By the proof of Proposition 8 we have $\sup x = \gamma$. It remains to prove $g(x) = A \cap x$.

Subclaim. $r(x) = \{n(k) : k < \omega\}$.

Proof. One direction follows from $\alpha_k \in x \cap (\gamma_{n(k)+1} - \gamma_{n(k)})$ for every $k < \omega$.

For the converse we prove that $x \cap \gamma_{n(k+1)} \subseteq \gamma_{n(k)+1}$ for every $k < \omega$. It suffices to show that if $k < \bar{k} < \omega$, then

$$\mathrm{cl}_f(\{\alpha_i : i \le \bar{k}\} \cup \delta) \cap \gamma_{n(k+1)} \subseteq \mathrm{cl}_f(\{\alpha_i : i \le \bar{k}\} \cup \delta) \cap \alpha_{k+1}$$
$$\subseteq h(\{\alpha_i : i \le k\}) < \gamma_{n(k)+1}.$$

The first inclusion follows from $\gamma_{n(k+1)} < \alpha_{k+1}$. The second inclusion follows from the choice of h and $\{\alpha_i : i \le k\} \lhd \{\alpha_i : i \le \bar{k}\} \in T^*$. The inequality follows from $\gamma_{n(k)+1} \in D$ and $\{\alpha_i : i \le k\} \in T^* \cap [\gamma_{n(k)+1}]^{<\omega}$. □

Hence

$$r(x) = \{0\} \cup \bigcup_{i<\omega} J_i.$$

Let $k < \omega$. Since $J_k \subseteq I_{n(k)} \subseteq \omega - \{0\}$ and $I_{n(k)} \cap J_i = \emptyset$ for every $i \ne k$, we have

$$I_{n(k)} \cap r(x) = I_{n(k)} \cap \bigcup_{i<\omega} J_i = J_k.$$

Since $\{\alpha_i : i \le k\} \in T^* \subset T_\delta$, we have

$$\mathrm{cl}_f(\{\alpha_i : i \le k\} \cup \delta) \cap \kappa = \delta.$$

Hence

$$x \cap \kappa = \bigcup_{k<\omega} \mathrm{cl}_f(\{\alpha_i : i \le k\} \cup \delta) \cap \kappa = \delta.$$

Since $\mathrm{cl}_f(\{\alpha_i : i \le k\} \cup \delta)$ and x are in $C(f)$, we have

$$\mathrm{cl}_f(\{\alpha_i : i \le k\} \cup \delta) \cap \alpha_k = \pi_{\alpha_k} \text{``}\delta = x \cap \alpha_k.$$

Hence by $\gamma_{n(k)} < \alpha_k$

$$\mathrm{cl}_f(\{\alpha_i : i \le k\} \cup \delta) \cap \gamma_{n(k)} = x \cap \gamma_{n(k)}.$$

Thus by the choice of J_k

$$\psi(x \cap \gamma_{n(k)})(I_{n(k)} \cap r(x)) = \psi(x \cap \gamma_{n(k)})(J_k) = A \cap x \cap \gamma_{n(k)}.$$

Therefore

$$
\begin{aligned}
g(x) &= \bigcup\nolimits_{n \in r(x)} \psi(x \cap \gamma_n)(I_n \cap r(x)) \\
&= \bigcup\nolimits_{k < \omega} \psi(x \cap \gamma_{n(k)})(I_{n(k)} \cap r(x)) \\
&= \bigcup\nolimits_{k < \omega} A \cap x \cap \gamma_{n(k)} \\
&= A \cap x.
\end{aligned}
$$

□

This completes the proof. □

The proofs of Theorem 6 and Proposition 16 can be combined to give

Proposition 17. *Assume* $\omega_1 < \kappa = \kappa^\omega \le 2^{\omega_1} = 2^{<\kappa}$ *and* κ-*SR in* $\mathcal{P}_{\omega_1} 2^{\kappa^+}$. *Then there is a diamond on* $\{x \in \mathcal{P}_\kappa \kappa^+ : \operatorname{cf} \sup x = \omega_1\}$.

When $2^{<\kappa} = \kappa$, Proposition 17 gives a new example of a diamond: The diamond of Theorem 6 in this case lies on the set $\{x \in \mathcal{P}_\kappa \kappa^+ : \operatorname{cf} \sup x = \omega\}$.

References

[1] Baumgartner, J., *On the size of closed unbounded sets*, Ann. Pure Appl. Logic 54 (1991) 195–227.

[2] Bekkali, M., Topics in Set Theory, Lecture Notes in Math. 1476, Springer, Berlin, 1991.

[3] Donder, H.-D. and Matet, P., *Two cardinal versions of diamond*, Israel J. Math. 83 (1993) 1–43.

[4] Foreman, M. and Magidor, M., *Mutually stationary sequences of sets and the non-saturation of the non-stationary ideal on* $P_\kappa(\lambda)$, Acta Math. 186 (2001) 271–300.

[5] Foreman, M., Magidor, M. and Shelah, S., *Martin's Maximum, saturated ideals, and non-regular ultrafilters. Part I*, Ann. Math. 127 (1988) 1–47.

[6] Foreman, M. and Todorčević, S., *A new Löwenheim–Skolem theorem*, Trans. Amer. Math. Soc. 357 (2005) 1693–1715.

[7] Gitik, M. and Shelah, S., *Less saturated ideals*, Proc. Amer. Math. Soc. 125 (1997) 1523–1530.

[8] Jech, T., *Some combinatorial problems concerning uncountable cardinals*, Ann. Math. Logic 5 (1973) 165–198.

[9] Jensen, R., *The fine structure of the constructible hierarchy*, Ann. Math. Logic 4 (1972) 229–308.

[10] Kanamori, A., The Higher Infinite, Springer Monogr. in Math., Springer, Berlin, 2003.

[11] König, B. and Todorčević, S., *Diamond indexed by countable sets*, typed notes.

[12] Magidor, M., *Representing sets of ordinals as countable unions of sets in the core model*, Trans. Amer. Math. Soc. 317 (1990) 91–126.

[13] Shelah, S., Proper Forcing, Lecture Notes in Math. 940, Springer, Berlin, 1982.

[14] ———, Around Classification Theory of Models, Lecture Notes in Math. 1182, Springer, Berlin, 1986.

[15] _____, Cardinal Arithmetic, Oxford Logic Guides 29, Oxford Univ. Press, New York, 1994.

[16] _____, Nonstructure Theory, Oxford Univ. Press, New York, to be published.

[17] Shioya, M., *Splitting $\mathcal{P}_\kappa \lambda$ into maximally many stationary sets*, Israel J. Math. 114 (1999) 347–357.

[18] _____, *Stationary reflection and the club filter*, J. Math. Soc. Japan 59 (2007) 1045–1065.

RIGIDITY AND BIINTERPRETABILITY
IN THE HYPERDEGREES

Richard A. Shore

Department of Mathematics, Cornell University
Ithaca NY 14853, USA
E-mail: shore@math.cornell.edu

Slaman and Woodin have developed and used set-theoretic methods to prove some remarkable theorems about automorphisms of, and definability in, the Turing degrees. Their methods apply to other coarser degree structures as well and, as they point out, give even stronger results for some of them. In particular, their methods can be used to show that the hyperarithmetic degrees are rigid and biinterpretable with second order arithmetic. We give a direct proof using only older coding style arguments to prove these results without any appeal to set-theoretic or metamathematical considerations. Our methods also apply to various coarser reducibilities.

1. Introduction

Slaman and Woodin [2009] (see also Slaman [1991] and [2008]) have developed and used set-theoretic and metamathematical techniques to prove some remarkable theorems about the Turing degrees, $\mathcal{D}_T$. These techniques include forcing over models of ZFC to make the set of reals in the ground model countable in the generic extension as well as absoluteness arguments. One key result is that every relation on $\mathcal{D}_T$ invariant under automorphisms and definable in second order arithmetic is actually definable in $\mathcal{D}_T$. They also prove that the double jump is invariant and hence definable. (This result was then used by Shore and Slaman [2000] to prove that the Turing jump itself is definable in $\mathcal{D}_T$.) As other examples, we mention their results that every degree above $0''$ is fixed under every automorphism; there are at most countably many automorphisms of $\mathcal{D}_T$; and in fact every 5-generic $\mathbf{g}$ is an automorphism base (i.e. if π is an automorphism of $\mathcal{D}_T$ and $\pi(\mathbf{g}) = \mathbf{g}$ then π is the identity map).

Slaman [1991] points out that their methods apply to a wide array of degree structures, often giving stronger results based on specific special properties of the reducibility. For example, in the arithmetic degrees, $\mathcal{D}_a$, every automorphism is the identity on the degrees above $0^{(\omega)}$, the first arithmetic jump of 0, while the hyperdegrees $\mathcal{D}_h$ are rigid and biinterpretable with second order arithmetic. Thus every relation on $\mathcal{D}_h$ is definable if and only if it is definable in second order arithmetic.

(We say that X is arithmetic in Y, $X \leq_a Y$, if $X \leq_T Y^{(n)}$ for some $n \in \omega$ and X is hyperarithmetic in Y, $X \leq_h Y$, if $X \leq_T Y^{(\alpha)}$ for some ordinal α recursive in Y where $Y^{(\alpha)}$ is the αth iterate of the Turing jump applied to Y. Kleene showed (see Sacks [1990, II.1-2]) that $X \leq_h Y$ if and only if X is $\Delta_1^1(Y)$. A degree structure $\mathcal{D}$ is biinterpretable with second order arithmetic if there is a definable standard model of arithmetic (or class of structures all isomorphic to $\mathbb{N}$) with definable schemes for both quantification over subsets of the model and a relation matching degrees with codes for sets in the model which are of the specified degrees. Of course, this immediately gives the desired result on definability of relations on $\mathcal{D}$. See Slaman and Woodin [2009] for more details.)

Our goal here is to prove first that $\mathcal{D}_h$ is rigid by a direct coding argument similar to that used in Abraham and Shore [1986] to prove (under mild set theoretic hypotheses such as $\aleph_1^{L[r]}$ being countable for every real r) that the constructibility degrees of reals are rigid. Both arguments are based on lattice embedding construction from the Turing degrees as in Shore [1982a]. The methods needed do the required embeddings are just Cohen-like forcing in the setting of the hyperarithmetic hierarchy. We also use ideas from Shore [1981] and [1982] to make the codings sufficiently effective to make the recovery of the set coded hyperarithmetic in the top degree of the embedded lattice. We then use the coding methods of Slaman and Woodin [1986] (in the hyperarithmetic setting) to translate the rigidity proof to one of biinterpretability. We exploit the specific proof of rigidity to do this translation and so avoid the need to definably deal with automorphisms as is possible using the set-theoretic methods and absoluteness results of Slaman and Woodin.

In Section 2, we present the proof of rigidity in abstract terms based on the existence of a coding scheme satisfying certain properties. In Section 3, we describe the specific lattices we employ that implement these coding requirements. Then, in Section 4, we use Cohen forcing in the hyperarithmetic setting as introduced by Feferman [1965] and presented in Sacks [1990] to show that all countable lattices (with 0) can be embedded into

$\mathcal{D}_h$. Finally, in Section 5, we describe the translation to biinterpretability and comment on the applicability of our methods to other coarser degree structures mentioned in Slaman [1991].

2. Rigidity

Intuitively, our basic idea is to code any given set X of degree $\mathbf{x}$ into $\mathcal{D}_h$ "near $\mathbf{x}$" in such a way as to be able to uniquely pick out the set X in a way that is definably tied to the degree $\mathbf{x}$. We want some structure $\mathcal{L}_X$ from which X can be "easily" recovered and that we can embed into $\mathcal{D}_h$ near $\mathbf{x}$.

First we explain what constitutes "easily" recoverable. In the setting of the Turing degrees, the underlying obstacle to improving the results fixing, for example, the cone above $0''$ under all automorphisms is the complexity of the notion of Turing reducibility. The relation $X \leq_T Y$ is $\Sigma_3(X, Y)$ and, in general, a formula of the form $(\exists X \leq_T Y)\Phi(X, Y)$ is $\Sigma_3(Y)$ even when Φ is recursive. Thus the best one can hope for is that a formula about $\mathcal{D}_T$ in the language with $\leq_T$ is that it will be Σ_3 in the degrees mentioned and even this only for positive existential formulas. (As the sets Σ_3 in X determine the degree of X'' and vice versa, this is the source of the ubiquitous nature of the double jump, and of $0''$ in particular, in the results on $\mathcal{D}_T$.) In the setting of the constructibility degrees of reals, $\mathcal{D}_c$, the relation $X \leq_c Y$ is itself constructible and so any quantifier free relation in the language with $\leq_c$ on degrees is itself constructible. It is that advantage that permits the coding proof of rigidity of Abraham and Shore [1986] to work in $\mathcal{D}_c$ but leaves it a couple of jumps short in $\mathcal{D}_T$. The situation for hyperarithmetic reducibility is intermediate but close enough to that for constructibility to allow a slightly modified proof to work. Not only is $X \leq_h Y$ a $\Pi_1^1(X, Y)$ relation but a formula $\exists X \leq_h Y\Phi(X, Y)$ is equivalent to a Π_1^1 formula when Φ is itself Π_1^1. Thus a positive coding of both X and its complement $\bar{X}$ (i.e. one that can be decoded by considering only positive existential formulas in the language with $\leq_h$) would show both X and $\bar{X}$ to be Π_1^1, and so hyperarithmetic, in the coding degrees. In all of these settings, the addition of $\vee$ to the language does not cost any more as we can go effectively between X, Y and $X \oplus Y$.

To be a bit more specific we work with lattices with 0 and 1 and plan to associate with each set X a lattice $\mathcal{M}_X$ which codes membership in both X and $\bar{X}$ by a recursive list of positive elementary formulas (in $\leq$ and $\vee$). (Note that we must avoid using $\wedge$ in our decoding formulas since it costs another quantifier.) Thus we will have a lattice $\mathcal{M}_X$ such that, for any lattice embedding $f : \mathcal{M}_X \to \mathcal{D}_h$, $X \leq_h f(1_{\mathcal{M}_X})$. Actually, we work with partial lattices. A partial lattice is a partial order in which join and infimum are determined by relations which define partial functions satisfying the usual conditions (in terms of order) for $\vee$ and $\wedge$ when defined. Our constructions are no different for partial lattices than for lattices and realizing this means that when describing a structure we do not have to specify all the infima and suprema but only the ones relevant to our concerns. Thus whenever we say lattice below we include partial lattices as well.

Next, the notion of "near $\mathbf{x}$" might suggest that we want $f(1_M) \leq_h \mathbf{x}$. This is done for some arguments in $\mathcal{D}_T$ once one is above, for example, $0'$ or $0''$ or is considering degrees with some other special property. It was also done in $\mathcal{D}_c$ below Cohen generic reals (Abraham and Shore [1986]) and a similar argument would work in $\mathcal{D}_h$. Instead, we employ one in which "near" means that $f(0_{\mathcal{M}}) = \mathbf{x}$. This procedure avoids some additional argumentation used for $\mathcal{D}_c$ by building more into our structure. We form a new lattice $\mathcal{L}_X$ from two disjoint copies $\mathcal{M}_X$ and $\hat{\mathcal{M}}_X$ of our original one by letting $1_{\mathcal{L}_X}$ be the join of $1_{\mathcal{M}_X}$ and $1_{\hat{\mathcal{M}}_X}$ and $0_{\mathcal{L}_X}$ be their infimum.

Now, suppose we have an embedding $f : \mathcal{L}_X \to \mathcal{D}_h$ that takes $0_{\mathcal{L}_X}$ to $\mathbf{x}$ and we consider any automorphism π of $\mathcal{D}_h$. As it is an automorphism, π carries the image of $\mathcal{L}_X$ under f to another image $\pi f(\mathcal{L}_X)$ in $\mathcal{D}_h$. The coding scheme assumed above insures that $\mathbf{x} \leq_h \pi f(1_{\mathcal{M}_X}), \pi f(1_{\hat{\mathcal{M}}_X})$. On the other hand, as f is a lattice embedding, $\mathbf{x} \equiv_h f(1_{\mathcal{M}_X}) \wedge f(1_{\hat{\mathcal{M}}_X})$ and so applying the automorphisms gives $\pi(\mathbf{x}) \equiv_h \pi f(1_{\mathcal{M}_X}) \wedge \pi f(1_{\hat{\mathcal{M}}_X})$. Thus $\mathbf{x} \leq_\mathbf{h} \pi(\mathbf{x})$. The same argument applied to π^{-1} gives $\mathbf{x} \leq_\mathbf{h} \pi^{-1}(\mathbf{x})$ and so $\pi(\mathbf{x}) \leq_h \mathbf{x}$. Thus $\mathbf{x} \equiv_\mathbf{h} \pi(\mathbf{x})$ for every automorphism π of $\mathcal{D}_h$. To prove rigidity all we have to do now is describe, for each X, a lattice $\mathcal{L}_X$ with the desired properties and prove that it can be embedded in $\mathcal{D}_h$ with $0_{\mathcal{L}_X}$ going to $\mathbf{x}$.

3. Coding

The essential ingredient in making X "easily" recoverable from the coding lattice $\mathcal{L}_X$ is the "effectively generated" model of arithmetic introduced in Shore [1982],[1981]. We here need only the successor function and the coding of the set X. The elements of our lattice that generate a copy of $\mathbb{N}$ are designated by d_0, e_0, e_1, f_0 and f_1. The element of the lattice corresponding to $n \in \mathbb{N}$ is d_n. The generating scheme that implements the successor function on $\mathbb{N}$ is determined by the following requirements:

(1) $(d_{2n} \vee e_0) \wedge f_1 = d_{2n+1}$ and
(2) $(d_{2n+1} \vee e_1) \wedge f_0 = d_{2n+2}$.

These conditions clearly guarantee that we can enumerate the d_n recursively in the lattice structure and write a recursive list of quantifier free formulas in this language which define each of them. We wish to convert this procedure and these formulas into ones that are positive in the language with just $\leq$ and $\vee$ at least to the extent that we can use them to code X and $\bar{X}$ (with the aid of other parameters c and $\bar{c}$). In Shore [1982] and [1981] the lattices were embedded as initial segments with the d_n as minimal elements and so it sufficed to say, for example, that $d_{2n+1} \leq (d_{2n} \vee e_0), f_1$ and $d_{2n+1} \neq 0$. As for being different from 0, we can simply add two other parameters p and q and require in our lattice that $q \neq 0$ and $p \vee d_n \geq q$ for each n. Thus we can say of an x that we view as a candidate for being one of the d_n that $x \vee p \geq q$ in place of saying that $x \neq 0$. In the context of coding X (done by exact pairs outside the basic lattice rather than by internal elements in Shore [1981] for reasons extraneous to our concerns here) we can replace the initial segment features of the structure with additional purely lattice theoretic requirements on the coding parameters. Specifically, we require the lattice to have two additional parameters c_X and $\bar{c}_X$ such that $d_n \leq c_X$ for $n \in X$, $d_n \wedge c_X = 0$ for $n \notin X$, $d_n \leq \bar{c}_X$ for $n \notin X$ and $d_n \wedge \bar{c}_X = 0$ for $n \in X$.

We now show how to recursively generate positive existential formulas $\phi_n(x)$ using just $\leq$ and $\vee$ such that, in any lattice $\mathcal{L}_X$ with elements $d_0, e_0, e_1, f_0, f_1, p$ and q as described, $\phi_n(x)$ holds of x if and only $0 < x \leq d_n$. Given such formulas, our requirements on c and $\bar{c}$ allow us to define X by $n \in X \Leftrightarrow \exists x(\phi_n(x) \,\&\, x \leq c_X)$ and $n \notin X \Leftrightarrow \exists x(\phi_n(x) \,\&\, x \leq \bar{c}_X)$. As we have already noted, when interpreted in an isomorphic copy of $\mathcal{L}_X$ in $\mathcal{D}_h$, such formulas are equivalent to ones Π_1^1 in the relevant parameters.

We begin by setting ϕ_0 to be $x = d_0$. Recursively, we let $\phi_{2n+1}(x)$ be $\exists z(\phi_{2n}(z) \,\&\, x \leq z \vee e_0, f_1 \,\&\, q \leq x \vee p)$ and $\phi_{2n+2}(x)$ be $\exists z(\phi_{2n+1}(z) \,\&\, x \leq z \vee e_1, f_0 \,\&\, q \leq x \vee p)$. Consider any x such that $\phi_{2n+1}(x)$ holds. We then have a z as described such that, by induction, $0 < z \leq d_{2n}$. Thus $z \vee e_0 \leq d_{2n} \vee e_0$ and so $x \leq d_{2n} \vee e_0, f_1$. As $d_{2n+1} = (d_{2n} \vee e_0) \wedge f_1$, $x \leq d_{2n+1}$ as required. Of course, $q \leq x \vee p$ guarantees that $x > 0$ as well. The argument for ϕ_{2n+2} is essentially the same.

We now turn to embedding countable lattices in $\mathcal{D}_h$.

4. Embedding lattices

In this section, we describe how the elementary methods of Shore [1982a] in the Turing degrees can be used to embed any countable (partial) lattice with 0 in $\mathcal{D}_h$ preserving 0. Relativization to any degree $\mathbf{x}$ supplies the desired embeddings of $\mathcal{L}_X$ (the partial lattice generated as specified above by elements $d_0, e_0, e_1, f_0, f_1, p, q, c$ and $\bar{c}$) as it is recursive in X as a partial lattice, i.e. the partial order, the partial functions $\vee$ and $\wedge$ and their domains are recursive in X.

The standard lattice representation arguments (originally from Jonsson [1953] but translated into the language of Lerman [1971] or [1983] and as presented also in Shore [1982a]) give our desired representation theorem. (A simple proof without the requirement for 0 is in Shore [1982a]. Adding the requirement that the value of each function in the representation is 0 at 0 at the beginning presents no difficulties nor does relativization to X.)

Theorem 4.1. *Let $\{p_i\}$ enumerate a recursively presentable partial lattice $\mathcal{P}$ with p_0 its least element. There is a uniformly recursive array of functions $\alpha_n : \omega \to \omega$ such that for all i, j, k, n, m:*

0. $\alpha_n(0) = 0$,

1. $p_i \le p_j \Rightarrow \alpha_n(j) = \alpha_m(j) \to \alpha_n(i) = \alpha_m(i)$ and

      $p_i \not\le p_j \Rightarrow \exists q, r(\alpha_q(j) = \alpha_r(j) \ \& \ \alpha_q(i) \ne \alpha_r(i))$,

2. $p_i \vee p_j = p_k \Rightarrow [\alpha_n(i) = \alpha_m(i) \ \& \ \alpha_n(j) = \alpha_m(j) \ \to \ \alpha_n(k) = \alpha_m(k)]$,

3. $p_i \wedge p_j = p_k \ \& \ \alpha_n(k) = \alpha_m(k) \Rightarrow \exists q_1, q_2, q_3 [\alpha_n(i) = \alpha_{q_1}(i) \ \& \ \alpha_{q_1}(j) = \alpha_{q_2}(j) \ \& \ \alpha_{q_2}(i) = \alpha_{q_3}(i) \ \& \ \alpha_{q_3}(j) = \alpha_m(j)]$.

We can define an embedding of $\mathcal{P}$ into $\mathcal{D}_h$ from any sufficiently generic function $g : \omega \to \omega$ by setting the image of p_i to be the degree of the function h_i defined by $h_i(n) = \alpha_{g(n)}(i)$. Intuitively, if one iterates the construction of trees of $(n+1)$-generics inside ones of n-generics into the transfinite taking appropriate diagonal-like intersections of the trees at limits one gets paths P which are generic at each level γ of the hyperarithmetic hierarchy and make $P^{(\gamma)} \equiv_T P \oplus 0^{(\gamma)}$. If one does this uniformly and properly for all recursive γ one gets a generic P (indeed one recursive in $\mathcal{O}$) such that, in addition, $\omega_1^P = \omega_1^{CK}$ and so everything hyperarithmetic in P is recursive in $P \oplus 0^{(\gamma)}$ for some recursive γ. This reduces the arguments about hyperarithmetic reducibility to ones about Turing reducibility and so the correctness of the embedding in the hyperarithmetic setting can be read off from the proof for the Turing degrees.

Formalizing this idea seems to require a hierarchy of languages in which one can talk about formulas (and terms representing the sets constructed) at each level of the hyperarithmetic hierarchy. The needed facts can probably be extracted from, e.g. MacIntyre [1977] who modifies the basic approach to Cohen forcing in the hyperarithmetic setting as introduced by Feferman [1965] (or from the analysis of a more general setting in Jockusch and Shore [1984]). We describe a somewhat coarser but more readily available analysis based on essentially the presentation in Sacks [1990] of Feferman's results. For convenience we make the purely notational change of using a function symbol $\mathcal{G}$ in our forcing language in place of one $\mathcal{T}$ for a set as in Sacks [1990] and so the conditions consist of consistent finite conjunctions of formulas of the form $\mathcal{G}(n) = m$ thought off as nonempty subbasic open subsets of ω^ω (in place of 2^ω). (If one prefers, one can keep the set version and code our desired function into a set in any standard way.). We also introduce standard terms $\mathcal{H}_i$ for the h_i with the specified interpretations and the obvious forcing relations. By generic we now mean generic for the (obvious extension of the) ramified language $\mathcal{L}(\omega_1^{CK}, \mathcal{G})$ defined in Sacks [1990, III.4], i.e. for every formula $\mathcal{F}$ there is a condition satisfied by the generic that decides $\mathcal{F}$. Also note that there is a term of the language $\hat{x}\Phi(x)$ for each ranked formula Φ that denotes the set of numbers satisfying Φ which define all the elements of the structure $\mathcal{M}(\omega_1^{CK}, g)$. For generic g with $\omega_1^g = \omega_1^{CK}$, the structure consists of all sets hyperarithmetic in g as is shown there as well. Of course, the sets hyperarithmetic in the h_i are represented by terms $t(\mathcal{H}_i)$ built up from $\mathcal{H}_i$, i.e. ones $\hat{x}\Phi(x)$ where Φ (hereditarily) contains $\mathcal{H}_i$ but no other $\mathcal{H}_j$ or $\mathcal{G}$. The existence of functions g generic in this sense with $g \leq_h \mathcal{O}$ and so $\omega_1^g = \omega_1^{CK}$ as well as all the usual facts about generic objects can be found in Sacks [1990, IV.3]. We can now easily argue as in the Turing degrees that we have the desired (partial) lattice embedding.

Theorem 4.2. *Let $\mathcal{P}$ be a recursive partial lattice (with 0); let α_n be a recursive representation of $\mathcal{P}$ as in Theorem 4.1; and g be an $\mathcal{M}(\omega_1^{CK})$ generic. The map taking p_i to the hyperdegree of h_i as defined above preserves the partial lattice structure of $\mathcal{P}$.*

Proof. The argument is standard. As $h_0(n) = \alpha_{g(n)}(0) = 0$ for every n by 4.1.0, 0 is preserved. If $p_i \leq p_j$ then $h_i(n) = \alpha_n(i) = \alpha_m(i)$ for any m such that $\alpha_m(j) = \alpha_n(j) = h_j(n)$ and so h_i is even recursive in h_j as the array α_n is uniformly recursive. Thus the embedding preserves order. Similarly if $p_i \vee p_j = p_k$ we can compute $h_k(n) = \alpha_{g(n)}(k)$ recursively from $h_i(n)$ and $h_j(n)$ by finding any α_m such that $\alpha_m(i) = h_i(n) = \alpha_{g(n)}(i)$ and $\alpha_m(j) = h_j(n) = \alpha_{g(n)}(j)$ as by 4.1.1, $\alpha_m(k) = \alpha_{g(n)}(k) = h_k(n)$ for any such m. Thus the embedding preserves join when defined in $\mathcal{P}$. The arguments for preserving $\not\leq$ and $\wedge$ depend on genericity.

Suppose $p_i \not\leq p_j$ and consider any term $t(\mathcal{H}_j)$ of the language generated by $\mathcal{H}_j$. We wish to show that $t(h_j) \neq h_i$ as this implies that $h_i \not\leq_h h_j$. If it were otherwise, there would be a condition $q \Vdash t(\mathcal{H}_j) = \mathcal{H}_i$. By 4.1.2 there are n and m such that $\alpha_n(j) = \alpha_m(j)$ but $\alpha_n(i) \neq \alpha_m(i)$. Let r be an extension of q containing the formula $\mathcal{G}(z) = n$ which decides a values for $t(\mathcal{H}_j)(z)$ and forces $g(z) = n$ for some z not mentioned in q and let r' be the same condition as r except that it contains $\mathcal{G}(z) = m$ in place of $\mathcal{G}(z) = n$. As the interpretation of $\mathcal{H}_j$ is the same in any two generics extending r and r' which differ only at z both conditions force the same value for $t(\mathcal{H}_j)(z)$. On the other hand, $r \Vdash \mathcal{H}_i(z) = \alpha_n(i)$ while $r' \Vdash \mathcal{H}_i(z) = \alpha_m(i)$. As $\alpha_n(i) \neq \alpha_m(i)$ by our choice of n and m we have the desired contradiction. Thus the embedding preserves $\not\leq$.

Finally, suppose that $p_i \wedge p_j = p_k$. We already know that h_i and h_j are hyperarithmetic (indeed recursive) in h_k by the preservation of order. We wish to show that any $f \leq_h h_i, h_j$ is hyperarithmetic in h_k. We take terms t_0 and t_1 and a condition q satisfied by g that forces $t_0(\mathcal{H}_i) = t_1(\mathcal{H}_j)$ and describe a procedure hyperarithmetic in h_k that defines $f = t_0(h_i) = t_1(h_j)$. The definition of f from h_k as a function on ω is given by the following procedure: to find $f(u)$ find any r extending q which forces some particular (necessarily common) value for $t_0(\mathcal{H}_i)(u)$ and $t_1(\mathcal{H}_j)(u)$ such that $h_k(z)(= \alpha_{g(z)}(k)) = \alpha_{r(z)}(k)$ for every z in the domain of r. We claim that this procedure is hyperarithmetic in h_k and provides the true value of these terms evaluated on h_i and h_j, respectively. It is, of course, recursive in h_k to check that a forcing condition satisfies the second requirement. As forcing for sentences of fixed rank is a hyperarithmetic relation, this procedure produces a hyperarithmetic reduction of f to h_k as long as it always produces the correct value.

To see that the procedure always produces the correct values, suppose there are $u, v \neq w \in \omega$ and some r extending q as described such that $r \Vdash t_0(\mathcal{H}_i)(u) = t_1(\mathcal{H}_j)(u) = v$ but $t_0(h_i) = w = t_1(h_j)$. As g is generic there is an s extending q satisfied by g that forces $t_0(\mathcal{H}_i)(u) = t_1(\mathcal{H}_j)(u) = w$. We now work for a contradiction. Assume without loss of generality that the domains of r and s are the same. For each z in this common domain, $\alpha_{r(z)}(k) = \alpha_{s(z)}(k)$ by our assumption on r. By 4.1.3, we can choose for each such z numbers $q_{z,l}$ for $l = 1, 2, 3$ witnessing the conclusion of 4.1.3 for $\alpha_{r(z)}$ and $\alpha_{s(z)}$ in place of α_n and α_m. We define forcing conditions q_l by $q_l(z) = q_{z,l}$. We extend these conditions to generics g^l (with g^0 extending r) by simply copying g when they are not defined. (Finite changes in a generic keep the function generic.) It is clear from the requirements of 4.1.3

that $h_i = h_i^1$, $h_j^1 = h_j^2$, $h_i^2 = h_i^3$ and $h_j^3 = h_j$. Thus by our choice of q and the first equality, $v = t_0(h_i)(u) = t_0(h_i^1)(u)$. Our choice of q and the next equalities give $t_0(h_i^1)(u) = t_1(h_j^1)(u) = t_1(h_j^2)(u) = t_0(h_i^2)(u) = t_0(h_i^3)(u) = t_1(h_j^3)(u) = t_1(h_j)(u)$ but this last term has value $w \neq v$ by assumption for the desired contradiction. $\square$

This theorem shows that the partial lattices described in the Section 3 can be embedded in $\mathcal{D}_h$ as required to implement the proof of rigidity described there and in Section 2. Thus we have our direct proof of rigidity.

Theorem 4.3. (Slaman and Woodin) *The hyperdegrees are rigid.*

5. Biinterpretability

Slaman and Woodin [2009] prove for $\mathcal{D}_T$ that rigidity implies biinterpretability by showing that one can describe their full analysis of persistence of automorphisms within the degree structure itself. As mentioned in Slaman [1991] their methods apply to many other degree structures. Based on our direct proof of rigidity for $\mathcal{D}_h$ we can derive biinterpretability using only the coding methods of Slaman and Woodin [1986] to code each countable relation (there on $\mathcal{D}_T$, here on $\mathcal{D}_h$) by finitely many parameters (uniformly in the arity of the relation) in the setting of the hyperarithmetic degrees. These methods will also apply to the coarser degree structures mentioned in Slaman [1991].

There is not much needed here beyond pointing out that the coding constructions of Slaman and Woodin [1986] work the same way for $\mathcal{D}_T$ as long as one uses Cohen-like forcing in the hyperarithmetic setting in place of the arithmetic one used in their work. Indeed, the arguments here are even a bit simpler since we do not need the careful calculations done there of how much genericity is needed. We give a brief description of the argument.

The first step is to prove that any antichain of degrees $\mathbf{a}_i$ is definable from three parameters $\mathbf{b}, \mathbf{g}_1$ and $\mathbf{g}_2$. We take $B \in \mathbf{b}$ to be an upper bound on the $\mathbf{a}_i$. The other parameters are defined by a forcing construction. One begins with a sequence of representatives $A_i \in \mathbf{a}_i$ such that A_i is recursive in any of its infinite subsets. The notion of forcing, $\mathcal{P}$, consists of triples $p = \langle p_1, p_2, p_3 \rangle$ where $p_1, p_2 \in 2^{<\omega}$ have the same length (which we also call the length of p) and $p_3 \in \omega$. Extension is defined by $q \leq p \Leftrightarrow q_1 \supseteq p_1$ & $q_2 \supseteq p_2$ & $q_3 \geq p_3$ & $\forall k \leq p_3 \forall a \in A_k(|p_1| < \langle k, a \rangle \leq |q_1| \rightarrow q_1(\langle k, a \rangle) = q_2(\langle k, a \rangle))$. Let G be any generic for $\mathcal{P}$ in the sense of hyperarithmetic forcing analogous to the one in Sacks [1990] as described above relativized to B with basic terms $\mathcal{A}_k$ for the A_k, $\mathcal{B}$ for B and $\mathcal{G}_1, \mathcal{G}_2$ for the unions of first and second coordinates of $\mathcal{G}$ as well. If G_1, G_2 are the union of the first coordinates of G then the set $\{\mathbf{a}_i | i \in \omega\}$ of hyperdegrees is definable as the set of minimal solutions $\mathbf{x}$ in $\mathcal{D}_h$ below $\mathbf{b}$ of the equation $(\mathbf{g}_1 \vee \mathbf{x}) \wedge (\mathbf{g}_2 \vee \mathbf{x}) \neq \mathbf{x}$.

To see that each $\mathbf{a}_k$ satisfies the equation, consider a condition $p \in G$ with $p_3 \geq k$ and the set $C(k) = \{m \in A_k | \; |p_1| < \langle k, m \rangle \in G_1\}$. It is immediate from its definition that $C(k) \leq_T G_1 \oplus A_k$ and it follows from the choice of p and the definition of extension that $C(k) = \{m \in A_k | \; |p_1| < \langle k, m \rangle \in G_2\}$ and so $C(k) \leq_T G_2 \oplus A_k$ as well. To see that $C(k) \not\leq_h A_k$ suppose to the contrary and choose a term $t_1(\mathcal{G}_1)$ for $C(k)$ and consider a term t so that $t(\mathcal{A}_k)$ is a standard name for a given set hyperarithmetic in A_k and a condition $q \in G$ extending p such that $q \Vdash t(\mathcal{A}_k) = t_1(\mathcal{G}_1)$. Let $m \in A_k$ be larger than $|q_1|$. We can clearly find an extension r of q such that $r_1(\langle k, m \rangle) \neq t(A_k)(\langle k, m \rangle)$ as the right hand side is fixed independently of the choice of generic. This gives the desired contradiction.

Next, we suppose we have some $C \leq_h B$ and $D \leq_h G_1 \oplus C, G_2 \oplus C$ such that $D \not\leq_h C$ and prove that $A_k \leq_h C$ for some k. Choose terms $t_C(\mathcal{B}), t_D(\mathcal{B}), t_1(\mathcal{G}_1 \oplus t_c(\mathcal{B})), t_2(\mathcal{G}_2 \oplus t_c(\mathcal{B}))$ representing the relevant sets and a condition $p \in G$ such that $p \Vdash t_D(\mathcal{B}) = t_1(\mathcal{G}_1 \oplus t_c(\mathcal{B})) = t_2(\mathcal{G}_2 \oplus t_c(\mathcal{B}))$. There must be infinitely many n with conditions q^n, r^n of p of common length at least n extending p that force different values for $t_1(\mathcal{G}_1 \oplus t_c(\mathcal{B}))(n)$ as otherwise we could compute D hyperarithmetically in C by finding for each n (other than the finitely many assumed exceptions) any condition q of length at least n extending p that forced a value for $t_1(\mathcal{G}_1 \oplus t_c(\mathcal{B}))(n)$ and know that it is the correct value. As forcing for formulas of fixed rank is hyperarithmetic in the parameter C, this would contradict our assumption that $D \not\leq_h C$. We can thus find such q^n and r^n hyperarithmetically in C. We can interpolate a sequence of conditions $s^{n,1}, \ldots, s^{n,m}$ between q^n and r^n so that the successive $s^{n,i}$ differ at exactly one number. We can then extend the $s^{n,i}$ to $\hat{s}^{n,i}$ which also differ only at that same location and each force a value for $t_1(\mathcal{G}_1 \oplus t_c(\mathcal{B}))(n) = t_2(\mathcal{G}_2 \oplus t_c(\mathcal{B}))(n)$. As the values at the two ends are different there must be an i such that $\hat{s}^{n,i}$ and $\hat{s}^{n,i+1}$ force different values. If the one location $\langle j, m \rangle$ at which they differed were not such that $j < p_3$ and $m \in A_j$ then we could form a single condition $s = \langle \hat{s}_1^{n,i}, \hat{s}_2^{n,i+1}, u \rangle$ for u the maximum of the third coordinates of the two conditions which would extend p and have the same values for G_1 and G_2 as $\hat{s}^{n,i}$ and $\hat{s}^{n,i+1}$, respectively. Thus s would force the same value for $t_1(\mathcal{G}_1 \oplus t_c(\mathcal{B}))(n)$ as $\hat{s}_1^{n,i}$ while it would force the same value for $t_2(\mathcal{G}_2 \oplus t_c(\mathcal{B}))(n)$ as $\hat{s}^{n,i+1}$. As these are different and s extends p this would be a contradiction. Thus there are infinitely many pairs of conditions (extending p) differing only at one point $\langle j, m \rangle$ with $j < p_3$ which force different answers. Again, as the forcing relation for formulas of fixed rank is hyperarithmetic, we can find infinitely many such hyperarithmetically in the parameter C. As all of these must have $m \in A_j$ by our argument above and there must be infinitely many with the same $j < p_3$, we can find infinitely many $m \in A_j$ for this j hyperarithmetically in C. Thus by our choice of A_j, $A_j \leq_h C$ as required.

Finally, Slaman and Woodin [1986] show how to convert an arbitrary countable relation on the degrees into an antichain so that the original relation is definable from the antichain and parameters. Suppose R is an n-ary relation on the degrees less than $\mathbf{b}$, $\langle \mathbf{b}_j | j \in \omega \rangle$ lists the degrees below $\mathbf{b}$ with representatives B_j, and $G_{i,j}$ for $1 \leq i \leq n$ and $j \in \omega$ are mutually Cohen generics over $B \in \mathbf{b}$ (in the sense of hyperarithmetic forcing) of degrees $\mathbf{g}_i$ (for example the appropriate columns of a single generic). Now all of the sets $\{\mathbf{b}_j \vee \mathbf{g}_{i,j} | j \in \omega\} = S_i$ and $\{\mathbf{g}_{i,j} | j \in \omega\} = T_i$ are antichains in $\mathcal{D}_h$ and so definable as above as is $U = \{\mathbf{g}_{1,j_1} \vee \ldots \vee \mathbf{g}_{n,j_n} | R(\mathbf{b}_{j_1}, \ldots, \mathbf{b}_{j_n})\}$. Thus $R(\mathbf{x}_1, \ldots, \mathbf{x}_n)$ if and only if there are $\mathbf{y}_i \in T_i$ and $\mathbf{z}_i \in S_i$ such that the join of the $\mathbf{y}_i$ is in U and for each i, $\mathbf{x}_i \vee \mathbf{y}_i = \mathbf{z}_i$.

Thus we can definably quantify over all countable relations on $\mathcal{D}_h$. In particular, we can definably describe a class of parameters that each define, by some fixed scheme, the standard model of arithmetic with a scheme for coding subsets of the model by degrees as well. This allows us to translate our proof of rigidity into biinterpretability. For any coding scheme for a model of arithmetic we can define a relation between degrees and codes of sets in such a model that associates degrees with sets of that degree. To be specific, we say that a degree $\mathbf{x}$ is associated with a set X coded in the model if for every set Y coded in the model there is a (partial) lattice isomorphic to $\mathcal{L}_Y$ in $\mathcal{D}_h$ with least element $\mathbf{x}$ if and only if (in the model) $Y \leq_h X$. As each $\mathcal{L}_Y$ is recursive in Y, it can be described in the model using the code for X and the apparatus of arithmetic in the model. The required images of $\mathcal{L}_Y$ in $\mathcal{D}_h$ and isomorphism between it and the version coded in the model of arithmetic can then be specified by other countable relations on the degrees of the model and ones above $\mathbf{x}$ and below some $\mathbf{z}$ representing the top of the lattice. Thus we have our direct proof of biinterpretability.

Theorem 5.1. (Slaman and Woodin) *The structure $\mathcal{D}_h$ of the hyperdegrees is biinterpretable with second order arithmetic.*

We close with a comment on another view of hyperarithmetic reducibility and its implications for some other degree structures.

We have mentioned two views of the hyperarithmetic sets. The first sees them as the sets recursive in $0^{(\alpha)}$ for some recursive ordinal α. The second as the Δ_1^1 sets. A third view sees them as the subsets of ω constructed in Gödel's L before the first nonrecursive ordinal, ω_1^{CK}. In the last view, we see X as hyperarithmetic in Y if $X \in L_{\omega_1^Y}[Y]$, i.e. X is constructed before the first ordinal not recursive in Y with the use of a predicate for Y in the language. This view of $\leq_h$ has a natural generalization when one sees ω_1^{CK} as the first Σ_1 admissible ordinal and $L_{\omega_1^{CK}}$ as the least admissible set containing ω. The relativization sees $L_{\omega_1^Y}[Y]$ as the least Σ_1 admissible set containing Y. The suggested reducibility generalizes Σ_1 to Σ_n and we say that $X \leq_{\Sigma_n} Y$ if X is a member of the least Σ_n admissible set containing Y. The associated degrees are called the Σ_n-admissible degrees in Slaman [1991]. As noted there, the methods of Slaman and Woodin carry over to these degrees as well. So do the ones presented here.

The notions of forcing to be considered are the same. The universes are now of the form $L_{\omega_n^X}[X]$ where ω_n^X is the first ordinal α such that $L_{\omega_n^X}[X]$ is Σ_n admissible. Cohen forcing in these settings has been considered in α-recursion theory. The crucial point for the forcing constructions is the preservation of Σ_n admissibility. Once this is established, the arguments for e.g. incomparability and infima requirements are the same. Discussions of the forcing and preservation of Σ_n admissibility can be found, for example, in Chong [1984] for $n = 1$ and for larger n in Shore [1974] albeit mixed in there with a more complicated priority argument. The essential ingredient for our decoding analysis was that the decoding of X (and so also $\bar{X}$) was given by a formula which was Σ_1 over the structure being considered and that X being Δ_1 over the structure for Y guarantees that X is reducible to Y. (Note that if X is Δ_1 over $L_{\omega_n^Y}[Y]$ then $X \in L_{\omega_n^Y}[Y]$ as even Σ_1 admissibility gives a bound β on the witnesses needed to demonstrate that $n \in X$ or $n \notin X$ for each $n \in \omega$. We can then define X over $L_\beta[Y]$ and so see that $X \leq_{\Sigma_n} Y$.) The formulas decoding $n \in X$ (and $n \in \bar{X}$) were all of the form that there are various sets reducible to a fixed set Z with properties described by positive formulas in the orderings. Each set Σ_n-reducible to Z is given simply by an ordinal less than ω_n^Z and the relations of one set being constructible from another by a given ordinal and an ordinal being Σ_n admissible relative to a given set are certainly Δ_1 (over even any Σ_1 admissible) in the sets and ordinals. Thus the decoding of X from an embedding of $\mathcal{M}_X$ produces a set Σ_n-reducible to the image of $1_{\mathcal{M}_X}$ as required. Our forcing arguments then give a direct proof of the analogous results.

Theorem 5.2. (Slaman and Woodin) *The structures of the Σ_n-degrees are for $n > 1$ are rigid and biinterpretable with second order arithmetic.*

Acknowledgments

This work was partially supported by NSF Grant DMS-0100035 and the IMS of the National University of Singapore. The material in this paper was presented in a talk at the IMS workshop Computational Prospects of Infinity in July, 2005. We also want to thank Ted Slaman for some conversations and suggestions.

References

1. Chong, C. T. [1984], *Techniques of Admissible Recursion Theory*, LNM **1106**, Springer-Verlag, Berlin.
2. Feferman, S. [1965], Some applications of the notion of forcing and generic sets, *Fund. Math.* **56**, 325-45.
3. Jockusch, C. G. Jr. and Shore, R. A. [1984], Pseudo-jump operators II: transfinite iterations, hierarchies and minimal covers, *J. Symb. Logic* **49**, 1205-1236.
4. Jónsson, B. [1953], On the representations of lattices, *Math. Scand.* 1, 193-206.
5. Lerman, M. [1971], Initial segments of the degrees of unsolvability, *Ann. Math.* **93**, 365-89.
6. Lerman, M. [1983], *Degrees of Unsolvability*, Perspectives in Mathematical Logic, Springer-Verlag, Berlin.
7. MacIntyre, J. [1977], Transfinite extensions of Friedberg's completeness criterion, *J. Symb. Logic* **42**, 1-10.
8. Sacks, G. E. [1990], *Higher Recursion Theory*, Perspectives in Mathematical Logic, Springer-Verlag, Berlin.
9. Abraham, U. and Shore, R. A. [1986], The degrees of constructibility below a Cohen real, *J. London Math. Soc.* (3) **53** (1986), 193-208.
10. Shore, R. A. [1974], Σ_n sets which are Δ_n-incomparable (uniformly), *J. Symb. Logic* **39** (1974), 295-304
11. Shore, R. A. [1981], The theory of the degrees below $0'$, *J. London Math. Soc.* (3) **24** (1981), 1-14.
12. Shore, R. A. [1982], On homogeneity and definability in the first order theory of the Turing degrees, *J. Symb. Logic* **47** (1982), 8-16.
13. Shore, R. A. [1982a], Finitely generated codings and the degrees r.e. in a degree **d**, *Proc. Am. Math. Soc.* **84**, 256-263.
14. Shore, R. A. and Slaman, T. A. [2000], Defining the Turing jump, *Math. Research Letters* **6** (1999), 711-722.

15. Slaman, T. A. [1991], Degree structures, in *Proc. Int. Cong. Math., Kyoto 1990*, Springer-Verlag, Tokyo, 303-316.

16. Slaman, T. A. [2008], Global properties of the Turing Degrees and Turing Jump, in *Computational Prospects of Infinity, Part I: Tutorial*, IMS Lecture Notes Series **14**, World Scientific, 83–101.

17. Slaman, T. A. and Woodin, H. W. [1986], Definability in the Turing degrees, *Illinois J. Math.* **30**, 320-334.

18. Slaman, T. A. and Woodin, H. W. [2009], Definability in degree structures, *Journal of Mathematical Logic*, to appear.

SOME FUNDAMENTAL ISSUES CONCERNING DEGREES OF UNSOLVABILITY

Stephen G. Simpson

Department of Mathematics, Pennsylvania State University
McAllister Building, University Park, State College, PA 16802, USA
E-mail: simpson@math.psu.edu
URL: http://www.math.psu.edu/simpson/

Recall that $\mathcal{R}_T$ is the upper semilattice of recursively enumerable Turing degrees. We consider two fundamental, classical, unresolved issues concerning $\mathcal{R}_T$. The first issue is to find a specific, natural, recursively enumerable Turing degree $\mathbf{a} \in \mathcal{R}_T$ which is $> \mathbf{0}$ and $< \mathbf{0}'$. The second issue is to find a "smallness property" of an infinite, co-recursively enumerable set $A \subseteq \omega$ which ensures that the Turing degree $\deg_T(A) = \mathbf{a} \in \mathcal{R}_T$ is $> \mathbf{0}$ and $< \mathbf{0}'$. In order to address these issues, we embed $\mathcal{R}_T$ into a slightly larger degree structure, $\mathcal{P}_w$, which is much better behaved. Namely, $\mathcal{P}_w$ is the lattice of weak degrees of mass problems associated with nonempty Π_1^0 subsets of 2^ω. We define a specific, natural embedding of $\mathcal{R}_T$ into $\mathcal{P}_w$, and we present some recent and new research results.

1. Preface

This paper is based on my talks in Evanston, Illinois, October 24, 2004, and in Singapore, August 8, 2005. The Evanston talk was part of a regional meeting of the American Mathematical Society, held at Northwestern University, October 23–24, 2004. The Singapore talk was part of a workshop entitled Computational Prospects of Infinity, held at the Institute for Mathematical Sciences, National University of Singapore, June 20 through August 15, 2005. This paper was submitted on December 15, 2005 and accepted on June 28, 2006 for publication in the Computational Prospects of Infinity proceedings volume.

The research reported in this paper was partially supported by my NSF research grant DMS-0070718, 2000–2004. Preparation of this paper was

313

partially supported by my NSF research grant DMS-0600823, 2006–2009. These grants are from the National Science Foundation of the U.S.A.

2. Motivation

Recall that $\mathcal{D}_T$ is the upper semilattice consisting of all Turing degrees. Recall also that $\mathcal{R}_T$ is the countable sub-semilattice of $\mathcal{D}_T$ consisting of the recursively enumerable Turing degrees. Both of these semilattices have been principal objects of study in recursion theory for many decades. See for instance the monographs of Sacks [39], Rogers [38], Soare [45], and Odifreddi [35, 36].

Two fundamental, classical, unresolved issues concerning $\mathcal{R}_T$ are:

Issue 1: To find a specific, natural, recursively enumerable Turing degree $\mathbf{a} \in \mathcal{R}_T$ which is $> \mathbf{0}$ and $< \mathbf{0}'$.

Issue 2: To find a "smallness property" of an infinite, co-recursively enumerable set $A \subseteq \omega$ which ensures that the Turing degree $\deg_T(A) = \mathbf{a} \in \mathcal{R}_T$ is $> \mathbf{0}$ and $< \mathbf{0}'$.

These unresolved issues go back to Post's classical 1944 paper, *Recursively enumerable sets of positive integers and their decision problems* [37].

My recent interest in Issue 1 began in 1999 at a conference in Boulder, Colorado [11]. There I heard a talk by Shmuel Weinberger, a prominent topologist and geometer. At the time Weinberger was trying to learn something about the recursively enumerable Turing degrees, $\mathcal{R}_T$, with an eye to applying them in the study of moduli spaces in differential geometry [48], using recursion-theoretic methods pioneered by Nabutovsky [33, 34]. Weinberger was visibly frustrated by the fact that $\mathcal{R}_T$ does not appear to contain any specific, natural examples of recursively enumerable Turing degrees, beyond the two standard examples due to Turing, namely $\mathbf{0}' =$ the Turing degree of the Halting Problem, and $\mathbf{0} =$ the Turing degree of solvable problems. Weinberger expressed his frustration by lamenting the fact that there are no recursively enumerable Turing degrees with specific names such as "Bill" or "Fred".

The purpose of this paper is to show how to address Issues 1 and 2 by passing from *decision problems* to *mass problems*. Specifically, we embed $\mathcal{R}_T$ into a slightly larger degree structure, called $\mathcal{P}_w$, which is much better behaved. In the $\mathcal{P}_w$ context, we obtain satisfactory, positive answers to Issues 1 and 2. In particular, we find that $\mathcal{P}_w$ contains not only the degrees $\mathbf{0}'$ and $\mathbf{0}$ but also many other specific, natural, intermediate degrees. These degrees are assigned specific names such as "Carl", "Stanley", "Klaus", "László", "Per", "Wilhelm", and "Bjørn", as explained below.

What is this wonderful structure $\mathcal{P}_w$? Briefly, $\mathcal{P}_w$ is the lattice of weak degrees of mass problems associated with nonempty Π_1^0 subsets of 2^ω. In order to fully explain $\mathcal{P}_w$, we must first explain (1) mass problems, (2) weak degrees, and (3) nonempty Π_1^0 subsets of 2^ω.

3. Mass problems (informal discussion)

A "decision problem" is the problem of deciding whether a given $n \in \omega$ belongs to a fixed set $A \subseteq \omega$ or not. To compare decision problems, we use Turing reducibility. Recall that $A \leq_T B$ means that A is Turing reducible to B, i.e., A can be computed using an oracle for B.

A "mass problem" is a kind of generalized decision problem, whose solution is not necessarily unique. (By contrast, a decision problem has only one solution.) We identify a mass problem with the set of its solutions. Here the solutions are identified with Turing oracles, i.e., elements of ω^ω. The "mass problem" associated with a set $P \subseteq \omega^\omega$ is the problem of "finding" an element of P. The "solutions" of this problem are the elements of P.

A mass problem is said to be "solvable" if it has a computable solution. A mass problem is said to be "reducible" to another mass problem if, given any solution of the second problem, we can use it as a Turing oracle to compute a solution of the first problem. Two mass problems are said to be "equivalent" or "of the same degree of unsolvability", if each is reducible to the other.

4. Mass problems and weak degrees (rigorous definition)

Let P and Q be subsets of $\omega^\omega = \{f \mid f : \omega \to \omega\}$. Viewing P and Q as mass problems, we say that P is *weakly reducible* to Q if

$$(\forall g \in Q)\,(\exists f \in P)\,(f \leq_T g).$$

This is abbreviated $P \leq_w Q$. Thus $P \leq_w Q$ means that, given any solution of the mass problem Q, we can use it as an oracle to compute a solution of the mass problem P.

Definition 4.1. We define $P, Q \subseteq \omega^\omega$ to be *weakly equivalent*, abbreviated $P \equiv_w Q$, if $P \leq_w Q$ and $Q \leq_w P$. The *weak degrees* are the equivalence classes under $\equiv_w$. There is an obvious partial ordering of the weak degrees induced by weak reducibility. Thus $\deg_w(P) \leq \deg_w(Q)$ if and only if $P \leq_w Q$. The partial ordering of all weak degrees is denoted $\mathcal{D}_w$.

It can be shown that $\mathcal{D}_w$ is a complete distributive lattice. The bottom element of $\mathcal{D}_w$ is denoted $\mathbf{0}$. This is the weak degree of solvable mass problems. Thus $\deg_w(P) = \mathbf{0}$ if and only if $P \cap \mathrm{REC} \neq \emptyset$.

5. Digression: Weak vs. strong reducibility

Let P and Q be mass problems, i.e., subsets of ω^ω. We make the following definitions.

Definition 5.1.

(1) As already stated, P is *weakly reducible* to Q, abbreviated $P \leq_w Q$, if for all $g \in Q$ there exists e such that $\{e\}^g \in P$.
(2) P is *strongly reducible* to Q, abbreviated $P \leq_s Q$, if there exists e such that $\{e\}^g \in P$ for all $g \in Q$.

Thus strong reducibility is a uniform variant of weak reducibility.

By a result of Nerode (see Rogers [38], Chapter 9, Theorem XIX), we have an analogy:

$$\frac{\text{weak reducibility}}{\text{Turing reducibility}} = \frac{\text{strong reducibility}}{\text{truth-table reducibility}}.$$

In this paper we shall deal only with weak reducibility.

As a historical note, we mention that weak reducibility goes back to Muchnik 1963 [32], while strong reducibility goes back to Medvedev 1955 [31]. Actually, as mentioned by Terwijn [46], both of these notions ultimately derive from ideas concerning the Brouwer/Heyting/Kolmogorov interpretation of intuitionistic propositional calculus.

6. The lattice $\mathcal{P}_w$ (rigorous definition)

Recall that $\mathcal{R}_T$, the semilattice of recursively enumerable Turing degrees, is a countable sub-semilattice of $\mathcal{D}_T$, the semilattice of all Turing degrees. Analogously we now define $\mathcal{P}_w$, a certain countable sublattice of the lattice $\mathcal{D}_w$ of all weak degrees. In defining $\mathcal{P}_w$, our guiding analogy is:

$$\frac{\mathcal{P}_w}{\mathcal{D}_w} = \frac{\mathcal{R}_T}{\mathcal{D}_T}.$$

The relevant notions are as follows. Let ω^ω be the *Baire space*, i.e., the set of all total functions $f : \omega \to \omega$. Recall that a set $P \subseteq \omega^\omega$ is said to be Π_1^0 if it is of the form $P = \{f \in \omega^\omega \mid \forall n\, R(f, n)\}$ where R is a recursive predicate. Here n ranges over ω, the set of natural numbers. It is well known that $P \subseteq \omega^\omega$ is Π_1^0 if and only if P is the set of all paths through some recursive subtree of $\omega^{<\omega}$. Here $\omega^{<\omega}$ denotes the tree of finite sequences of natural numbers. Recall also that Π_1^0 subsets of ω^ω are sometimes called *effectively closed*, because such a set is just the complement of an *effectively open* set, i.e., the union of a recursive sequence of basic open sets in ω^ω.

Additionally, $P \subseteq \omega^\omega$ is said to be *recursively bounded* if there exists a recursive function h in ω^ω such that $f(n) < h(n)$ for all $f \in P$ and all $n \in \omega$. Recall that recursively bounded Π_1^0 subsets of ω^ω are sometimes called *effectively compact*.

Definition 6.1. $\mathcal{P}_w$ is the set of weak degrees of nonempty, recursively bounded, Π_1^0 subsets of ω^ω. There is an obvious partial ordering of $\mathcal{P}_w$ induced by weak reducibility. Thus $\deg_w(P) \leq \deg_w(Q)$ if and only if $P \leq_w Q$.

Remark 6.2. Many authors including Jockusch/Soare [23] and Groszek/Slaman [20] have studied the Turing degrees of elements of Π_1^0 subsets of ω^ω which are nonempty and recursively bounded. This earlier research is part of the inspiration for our current study of the weak degrees of mass problems associated with such sets, i.e., $\mathcal{P}_w$.

Remark 6.3. It is well known that every recursively bounded Π_1^0 subset of ω^ω is recursively homeomorphic to a recursively bounded Π_1^0 set of a special kind, namely, a Π_1^0 subset of the *Cantor space*,

$$2^\omega = \{0,1\}^\omega = \{X \mid X : \omega \to \{0,1\}\}.$$

See for example Theorem 4.10 of [44]. Here the recursive bounding function is the constant function 2, i.e., $h(n) = 2$ for all $n \in \omega$.

It follows that $\mathcal{P}_w$ may be alternatively defined as the set of weak degrees of nonempty Π_1^0 subsets of 2^ω. Note also that $\mathcal{P}_w$, as a subset of $\mathcal{D}_w$, is partially ordered by weak reducibility. See Figure 1.

Remark 6.4. Some basic facts about $\mathcal{P}_w$ are as follows.

(1) $\mathcal{P}_w$ is a countable sublattice of $\mathcal{D}_w$.
(2) The bottom element of $\mathcal{P}_w$ is $\mathbf{0}$, the same as the bottom element of $\mathcal{D}_w$.
(3) The top element of $\mathcal{P}_w$ is the weak degree of

$$PA = \{\text{completions of Peano Arithmetic}\}.$$

This goes back to Scott/Tennenbaum [40]. See also Jockusch/Soare [23].

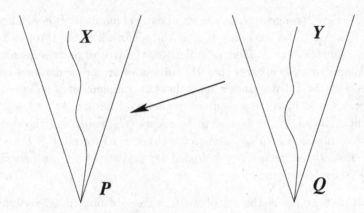

Fig. 1. Weak reducibility of Π^0_1 subsets of 2^ω. In this figure, $P \leq_w Q$ means $(\forall Y \in Q)\,(\exists X \in P)\,(X \leq_T Y)$. Here P and Q are given by infinite recursive subtrees of the full binary tree $\{0,1\}^{<\omega}$ of all finite sequences of 0's and 1's. Also, X and Y are infinite paths through P and Q respectively.

Remark 6.5. $\mathcal{R}_T$ is usually regarded as the smallest or simplest natural sub-semilattice of $\mathcal{D}_T$. Similarly, $\mathcal{P}_w$ may be regarded as the smallest or simplest natural sublattice of $\mathcal{D}_w$. In addition, $\mathcal{P}_w$ resembles $\mathcal{R}_T$ in other respects, as we shall see below. In particular, $\mathcal{P}_w$ includes a copy of $\mathcal{R}_T$, as we now show.

7. Embedding $\mathcal{R}_T$ into $\mathcal{P}_w$

Recall that $\mathcal{R}_T$ = the semilattice of Turing degrees of recursively enumerable subsets of ω, and $\mathcal{P}_w$ = the lattice of weak degrees of nonempty Π^0_1 subsets of 2^ω. The following embedding theorem was obtained by Simpson in 2002 [42].

Theorem 7.1. *There is a specific, natural embedding*

$$\phi : \mathcal{R}_T \hookrightarrow \mathcal{P}_w\,.$$

The embedding ϕ is given by

$$\phi : \deg_T(A) \mapsto \deg_w(\mathrm{PA} \cup \{A\})\,.$$

Here ϕ is one-to-one and preserves the partial ordering $\leq$, the least upper bound operation sup, *and the top and bottom elements.*

Remark 7.2. In Theorem 7.1, the fact that $\deg_w(\mathrm{PA} \cup \{A\})$ belongs to $\mathcal{P}_w$ is not obvious, because $\mathrm{PA} \cup \{A\}$ is usually not a Π^0_1 set. However, it turns out that PA $\cup \{A\}$ is always *of the same weak degree as* a Π_1^0 subset of 2^ω. This fact is a consequence of our Embedding Lemma, Lemma 3.3 of [42], presented as Lemma 12.1 below.

Likewise, it may not be obvious that the embedding $\phi : \mathcal{R}_T \to \mathcal{P}_w$ is one-to-one. However, the one-to-oneness of ϕ can be shown as a consequence of a famous theorem known as the Arslanov Completeness Criterion. This theorem can be found in textbooks, e.g., Soare [45], Theorem V.5.1.

Convention: *Throughout this paper, we shall identify $\mathcal{R}_T$ with its image in $\mathcal{P}_w$ under the embedding ϕ.* We shall also identify each recursively enumerable Turing degree with the weak degree which is its image in $\mathcal{P}_w$ under the embedding ϕ. In particular, we identify $\mathbf{0}', \mathbf{0} \in \mathcal{R}_T$ with the top and bottom elements of $\mathcal{P}_w$. See Figure 2.

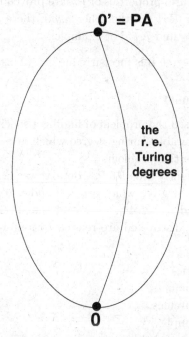

0' = PA

the
r. e.
Turing
degrees

0

Fig. 2.   A picture of $\mathcal{P}_w$, the lattice of weak degrees of nonempty Π_1^0 subsets of 2^ω. Note that $\mathcal{R}_T$, the upper semilattice of recursively enumerable Turing degrees, is embedded in $\mathcal{P}_w$. Moreover, $\mathbf{0}'$ and $\mathbf{0}$ are the top and bottom elements of both $\mathcal{R}_T$ and $\mathcal{P}_w$.

8. Structural properties of $\mathcal{P}_w$

It can be shown that $\mathcal{P}_w$ has many structural features which are similar to those of $\mathcal{R}_T$. Some of the similar features are as follows.

(1) $\mathcal{P}_w$ is a countable distributive lattice. Moreover, every countable distributive lattice is lattice embeddable in every initial segment of $\mathcal{P}_w$. This result is due to Binns/Simpson [4, 9].

(2) The $\mathcal{P}_w$ analog of the Sacks Splittting Theorem holds. In other words, for all $\mathbf{a}, \mathbf{c} > \mathbf{0}$ in $\mathcal{P}_w$ we can find $\mathbf{b}_1, \mathbf{b}_2 \in \mathcal{P}_w$ such that $\mathbf{a} = \sup(\mathbf{b}_1, \mathbf{b}_2)$ and $\mathbf{b}_1 \not\geq \mathbf{c}$ and $\mathbf{b}_2 \not\geq \mathbf{c}$. This result is due to Binns [4, 5].

(3) We conjecture that the $\mathcal{P}_w$ analog of the Sacks Density Theorem holds. This would mean that for all $\mathbf{a}, \mathbf{b} \in \mathcal{P}_w$ with $\mathbf{a} < \mathbf{b}$ there exists $\mathbf{c} \in \mathcal{P}_w$ such that $\mathbf{a} < \mathbf{c} < \mathbf{b}$.

(4) There are some degrees in $\mathcal{P}_w$ with interesting lattice-theoretic properties, such as being meet-reducible or not, and joining to $\mathbf{0}'$ or not. See Theorem 10.2 below. See also Simpson [42, 44].

Note that these structural properties of $\mathcal{P}_w$ are proved by means of priority arguments, just as for $\mathcal{R}_T$. On the other hand, there are some structural differences between $\mathcal{P}_w$ and $\mathcal{R}_T$. For example:

(5) Within $\mathcal{P}_w$, the degree $\mathbf{0}$ is meet-irreducible. (This is trivial.)

9. Response to Issue 1

Recall that Issue 1 posed the problem of finding a specific, natural example of a recursively enumerable Turing degree which is $> \mathbf{0}$ and $< \mathbf{0}'$. We do not know how to solve this problem.

However, in the slightly broader $\mathcal{P}_w$ context, we [42, 44] have discovered many specific, natural degrees which are $> \mathbf{0}$ and $< \mathbf{0}'$. See Figure 3.

Moreover, as noted in [42, 44], several of the specific, natural degrees in $\mathcal{P}_w$ which we have discovered are related to foundationally interesting topics:

- reverse mathematics,
- algorithmic randomness,
- subrecursive hierarchies,
- computational complexity,
- diagonal nonrecursiveness.

See also the additional explanations below.

10. Some specific, natural degrees in $\mathcal{P}_w$

Conside the following specific, natural, weak degrees. Let $\mathbf{r}_n$ be the weak degree of the set of n-random reals [26, 44]. Let $\mathbf{d}$ be the weak degree of

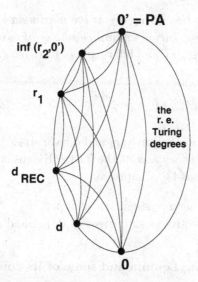

Fig. 3. Some specific, natural degrees in $\mathcal{P}_w$. Note that each of these specific, natural degrees in $\mathcal{P}_w$ is incomparable with all of the recursively enumerable Turing degrees, except $\mathbf{0}'$ and $\mathbf{0}$.

the set of diagonally nonrecursive functions [22]. Let $\mathbf{d}_{\mathrm{REC}}$ be the weak degree of the set of diagonally nonrecursive functions which are recursively bounded.

Not all of these weak degrees belong to $\mathcal{P}_w$. However, we have the following theorem.

Theorem 10.1. *In $\mathcal{P}_w$ we have*

$$\mathbf{0} < \mathbf{d} < \mathbf{d}_{\mathrm{REC}} < \mathbf{r}_1 < \inf(\mathbf{r}_2, \mathbf{0}') < \mathbf{0}'.$$

Moreover, all of these specific, natural degrees in $\mathcal{P}_w$ are incomparable with all of the recursively enumerable Turing degrees, except $\mathbf{0}'$ and $\mathbf{0}$.

Proof. See Simpson [42, 44]. The strict inequalities $\mathbf{d} < \mathbf{d}_{\mathrm{REC}} < \mathbf{r}_1$ follow from Kumabe [28] and Ambos-Spies et al [3]. □

We also have:

Theorem 10.2.

(1) *We may characterize $\mathbf{r}_1$ as the maximum weak degree of a Π_1^0 subset of 2^ω which is of positive measure.*

(2) *We may characterize* $\inf(\mathbf{r}_2, \mathbf{0}')$ *as the maximum weak degree of a* Π_1^0
     subset of 2^ω *whose Turing upward closure is of positive measure.*

(3) *Each of the degrees* $\mathbf{r}_1$ *and* $\inf(\mathbf{r}_2, \mathbf{0}')$ *is meet-irreducible and does not*
     join to $\mathbf{0}'$.

Proof. See Simpson [42, 44].   □

Remark 10.3. The weak degrees $\mathbf{r}_1$ and $\mathbf{d}$ have arisen in the reverse mathematics of measure theory and of the Tietze Extension Theorem, respectively [10, 19]. See also [43], Chapter X.

Remark 10.4. We hereby assign the names "Carl", "Klaus", and "Per" [22, 3, 30] to the respective weak degrees $\mathbf{d}$, $\mathbf{d}_{\mathrm{REC}}$, and $\mathbf{r}_1$ in $\mathcal{P}_w$.

11. The Embedding Lemma and some of its consequences

Several of the results stated above are consequences of the following lemma, due to Simpson [42], which we call the Embedding Lemma:

If $S \subseteq \omega^\omega$ *is* Σ_3^0 *and if* $P \subseteq 2^\omega$ *is nonempty* Π_1^0, *then* $\deg_w(S \cup P) \in \mathcal{P}_w$.

Using the Embedding Lemma we can show that the weak degrees of many specific, natural Σ_3^0 subsets of ω^ω belong to $\mathcal{P}_w$.

Example 11.1.

(1) Let $R_1 = \{X \in 2^\omega \mid X \text{ is 1-random}\}$. Since R_1 is Σ_2^0, it follows by the
     Embedding Lemma that $\mathbf{r}_1 = \deg_w(R_1) \in \mathcal{P}_w$.

(2) Let $R_2 = \{X \in 2^\omega \mid X \text{ is 2-random}\}$. Since R_2 is Σ_3^0, it follows by the
     Embedding Lemma that $\inf(\mathbf{r}_2, \mathbf{0}') = \deg_w(R_2 \cup \mathrm{PA}) \in \mathcal{P}_w$.

(3) Let $D = \{f \in \omega^\omega \mid f \text{ is diagonally nonrecursive}\}$. Since D is Π_1^0, it
     follows by the Embedding Lemma that $\mathbf{d} = \deg_w(D) \in \mathcal{P}_w$.

(4) Let $D_{\mathrm{REC}} = \{f \in D \mid f \text{ is recursively bounded}\}$. Since D_{REC} is Σ_3^0, it
     follows by the Embedding Lemma that $\mathbf{d}_{\mathrm{REC}} = \deg_w(D_{\mathrm{REC}}) \in \mathcal{P}_w$.

(5) Let $A \subseteq \omega$ be recursively enumerable. Since $\{A\}$ is Π_2^0, it follows by
     the Embedding Lemma that $\deg_w(\{A\} \cup \mathrm{PA}) \in \mathcal{P}_w$. This gives our
     embedding of $\mathcal{R}_T$ into $\mathcal{P}_w$.

12. Proof of the Embedding Lemma

We restate the Embedding Lemma as follows.

Lemma 12.1. The Embedding Lemma. *Let $S \subseteq \omega^\omega$ be Σ_3^0. Let $P \subseteq 2^\omega$ be nonempty Π_1^0. Then we can find $Q \subseteq 2^\omega$ nonempty Π_1^0 such that $Q \equiv_w S \cup P$.*

Proof. First use a Skolem function technique to reduce to the case when S is Π_1^0. Namely, letting $S = \{f \in \omega^\omega \mid \exists k \, \forall n \, \exists m \, R(f, k, n, m)\}$ where R is recursive, replace S by the set of all $\langle k \rangle^\frown (f \oplus g) \in \omega^\omega$ such that $\forall n \, R(f, k, n, g(n))$ holds. Clearly the latter set is $\equiv_w S$ and Π_1^0.

Assuming now that S is a Π_1^0 subset of ω^ω, let T_S be a recursive subtree of $\omega^{<\omega}$ such that S is the set of paths through T_S. We may safely assume that, for all $\tau \in T_S$ and all $n <$ the length of τ, $\tau(n) \geq 2$. Let T_P be a recursive subtree of $\{0, 1\}^{<\omega}$ such that P is the set of paths through T_P. Define T_Q to be the set of finite sequences $\rho \in \omega^{<\omega}$ of the form

$$\rho = \sigma_0^\frown \langle m_0 \rangle^\frown \sigma_1^\frown \langle m_1 \rangle^\frown \cdots ^\frown \langle m_{k-1} \rangle^\frown \sigma_k$$

such that $\langle m_0, m_1, \ldots, m_{k-1} \rangle \in T_S$, $\sigma_0, \sigma_1, \ldots, \sigma_k \in T_P$, and $\rho(n) \leq \max(n, 2)$ for all $n <$ the length of ρ. Thus T_Q is a recursive subtree of $\omega^{<\omega}$. Let $Q \subseteq \omega^\omega$ be the set of paths through T_Q.

We claim that $Q \equiv_w S \cup P$. Let $f \in S$ be given. Since T_P is infinite, it contains a recursive sequence of finite sequences τ_n of length n for each n. Setting $g = \tau_{f(0)}^\frown \langle f(0) \rangle^\frown \cdots ^\frown \tau_{f(n)}^\frown \langle f(n) \rangle^\frown \cdots$, we have $g \in Q$ and $g \leq_T f$. This shows that $Q \leq_w S$. Note also that $T_Q \supseteq T_P$, hence $Q \supseteq P$, hence $Q \leq_w P$. We now have $Q \leq_w S \cup P$. Conversely, given $g \in Q$, set

$$I = \{n \mid g(n) \geq 2\} = \{n_0 < n_1 < \cdots < n_k < \cdots\}.$$

If I is infinite, then setting $f(k) = n_k$ for all k, we have $f \in S$ and $f \leq_T g$. If I is finite, say $I = \{n_0 < n_1 < \cdots < n_{k-1}\}$, then setting $n_{-1} = -1$ and $X(i) = g(n_{k-1} + i + 1)$ for all i, we have $X \in P$ and $X \leq_T g$. Thus $Q \geq_w S \cup P$ and our claim is proved.

Note that Q is Π_1^0 and recursively bounded, with bounding function $h(n) = \max(n, 2)$. Therefore, we can find a Π_1^0 set $Q^* \subseteq 2^\omega$ which is recursively homeomorphic to Q, hence weakly equivalent to Q. This proves our lemma. $\square$

13. Some additional, specific degrees in $\mathcal{P}_w$

By the same method as in Theorem 10.1, we can use the Embedding Lemma to identify some additional, specific, natural degrees in $\mathcal{P}_w$. Some of these degrees in $\mathcal{P}_w$ are associated with computational complexity classes, as follows.

Definition 13.1. Let C be a uniformly recursively enumerable class of total recursive functions satisfying some mild closure conditions, as explained in Section 10 of [44]. Let $\mathbf{d}_C$ be the weak degree of the set of diagonally non-recursive functions which are bounded by some $h \in C$. As a consequence of the Embedding Lemma, we have $\mathbf{d}_C \in \mathcal{P}_w$. See Section 10 of [44] for a detailed justification of our claim that these degrees $\mathbf{d}_C$ are recursion-theoretically natural.

Remark 13.2. If $C^* \supseteq C$ is another such class of recursive functions, then we have $\mathbf{d}_{C^*} \leq \mathbf{d}_C$. Moreover, according to Theorem 1.9 of [3], we have strict inequality $\mathbf{d}_{C^*} < \mathbf{d}_C$ provided C^* contains a function which "grows much faster than" all functions in C. There are some interesting problems here.

Example 13.3. Let C be any of the following complexity classes:

(1) PR = the class of primitive recursive functions
(2) ER = the class of elementary recursive functions.
(3) PTIME = the class of polynomial-time computable functions.
(4) PSPACE = the class of polynomial-space computable functions.
(5) EXPTIME = the class of exponential-time computable functions, etc.
(6) C_α = the class of recursive functions at levels $< \omega \cdot (1 + \alpha)$ of the transfinite Ackermann hierarchy due to Wainer [47]. Here α is any ordinal number $\leq \varepsilon_0$. Thus $C_0 = $ PR, $C_1 = $ the class of functions which are primitive recursive in the Ackermann function, etc.

For each of these classes C, we have a specific, natural degree $\mathbf{d}_C$ in $\mathcal{P}_w$. Thus we have

$$\mathbf{r}_1 > \mathbf{d}_{\mathrm{PTIME}} > \mathbf{d}_{\mathrm{PSPACE}} > \mathbf{d}_{\mathrm{EXPTIME}} > \mathbf{d}_{\mathrm{ER}} > \mathbf{d}_{\mathrm{PR}} = \mathbf{d}_0$$

in $\mathcal{P}_w$, corresponding to well-known complexity classes. Also, writing $\mathbf{d}_\alpha = \mathbf{d}_{C_\alpha}$ for each $\alpha \leq \varepsilon_0$, we have

$$\mathbf{d}_0 > \mathbf{d}_1 > \cdots > \mathbf{d}_\alpha > \mathbf{d}_{\alpha+1} > \cdots > \mathbf{d}_{\varepsilon_0} > \mathbf{d}_{\mathrm{REC}}$$

in $\mathcal{P}_w$. Moreover, if α is a limit ordinal, then by Remark 10.12 of [44] we have $\mathbf{d}_\alpha = \inf_{\beta < \alpha} \mathbf{d}_\alpha$.

Remark 13.4. We hereby assign the names "Wilhelm", "László", and "Stanley" [1, 25, 47] to the respective weak degrees $\mathbf{d}_0 = \mathbf{d}_{\mathrm{PR}}$, $\mathbf{d}_{\mathrm{ER}}$, and $\mathbf{d}_{\varepsilon_0}$ in $\mathcal{P}_w$.

In addition, let $\mathbf{d}^2$ be the weak degree of the set of $f \oplus g$ such that f is diagonally nonrecursive, and g is diagonally nonrecursive relative to f. More generally (see Section 4 of [42]), we can define $\mathbf{d}^n$ for all $n \geq 1$, and we can extend this into the transfinite.

Theorem 13.5. *In* $\mathcal{P}_w$ *we have*

$$\mathbf{d} = \mathbf{d}^1 < \mathbf{d}^2 < \cdots < \mathbf{d}^n < \cdots < \mathbf{r}_1 \, .$$

Proof. This is a consequence of Kumabe [28]. See Simpson also Section 4 of [42]. □

Remark 13.6. We conjecture that all of the $\mathbf{d}^n$'s, $n \geq 2$ are incomparable with all of the $\mathbf{d}_\alpha$'s, $\alpha \leq \varepsilon_0$, and with $\mathbf{d}_{\mathrm{REC}}$.

14. Positive-measure domination

Up until this point, all of our examples of specific, natural degrees in $\mathcal{P}_w$ have turned out to be incomparable with all of the recursively enumerable Turing degrees, except $\mathbf{0}'$ and $\mathbf{0}$. We now present an example which behaves differently in this respect.

Starting with Dobrinen/Simpson [15] and continuing with Cholak/-Greenberg/Miller [13], Binns et al [7], and Kjos-Hanssen [27], there has been a recent upsurge of interest in domination properties related to the reverse mathematics of measure theory. We consider one such property.

Definition 14.1. $A \in 2^\omega$ is said to be *positive-measure dominating* if every Π_2^0 subset of 2^ω of positive measure includes a $\Pi_1^{0,A}$ set of positive measure. This notion was implicitly introduced in Conjecture 3.1 of [15] and has been developed in [27].

Remark 14.2. Positive-measure domination is related to the reverse mathematics of the following measure-theoretic regularity statement:

Every G_δ set of positive measure includes a closed set of positive measure.

For more on the reverse mathematics of measure-theoretic regularity, see [15].

Remark 14.3. Kjos-Hanssen [27] has shown that the set

$$\mathrm{PMD} = \{A \in 2^\omega \mid A \text{ is positive-measure dominating}\}$$

is Σ_3^0. Setting $\mathbf{m} = \deg_w(\mathrm{PMD})$, we may apply the Embedding Lemma to conclude that $\inf(\mathbf{m}, \mathbf{0}') \in \mathcal{P}_w$. It follows from the results of [13, 7] that $\inf(\mathbf{m}, \mathbf{0}')$ is incomparable with $\mathbf{d}$ and that there exist recursively enumerable Turing degrees $\mathbf{a}$ such that $\mathbf{0} < \inf(\mathbf{m}, \mathbf{0}') < \mathbf{a} < \mathbf{0}'$ in $\mathcal{P}_w$. See Figure 4.

Remark 14.4. We hereby assign the name "Bjørn" [27] to the specific, natural degree $\inf(\mathbf{m}, \mathbf{0}')$ in $\mathcal{P}_w$.

Note added June 30, 2006: Recently Binns, Kjos-Hanssen, Miller and Solomon [8] have shown that $A \in 2^\omega$ is positive-measure dominating if and only if A is almost everywhere dominating, if and only if A is uniformly almost everywhere dominating. In addition, Simpson [41] has shown that any such A is *superhigh*, i.e., $A' \geq_{tt} 0''$, i.e., $0''$ is truth-table reducible to A'.

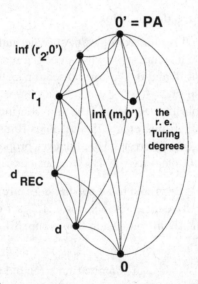

Fig. 4. Another specific, natural degree in $\mathcal{P}_w$. Note that $\inf(\mathbf{m}, \mathbf{0}')$, unlike $\mathbf{d}$, $\mathbf{d}_{\mathrm{REC}}$, $\mathbf{r}_1$, and $\inf(\mathbf{r}_2, \mathbf{0}')$, is bounded above by some recursively enumerable Turing degrees which are strictly less than $\mathbf{0}'$. In addition, $\inf(\mathbf{m}, \mathbf{0}')$ is incomparable with $\mathbf{d}$.

15. Some further specific, natural degrees in $\mathcal{P}_w$

We now mention some further examples of specific, natural, Σ_3^0 subsets of ω^ω which, via the Embedding Lemma, give rise to specific, natural degrees in $\mathcal{P}_w$. These degrees are earmarked for future research.

Example 15.1.

(1) Let $\mathbf{d}^*$ be the weak degree of the set of functions which are diagonally nonrecursive relative to the Halting Problem. This set of functions has arisen in the reverse mathematics of Ramsey's Theorem [21]. The Embedding Lemma tells us that $\inf(\mathbf{d}^*, \mathbf{0}') \in \mathcal{P}_w$.

(2) Let $\mathbf{d}^*_{\mathrm{REC}}$ = the weak degree of the set of functions which are (a) diagonally nonrecursive relative to the Halting Problem, and (b) recursively bounded. The Embedding Lemma tells us that $\inf(\mathbf{d}^*_{\mathrm{REC}}, \mathbf{0}') \in \mathcal{P}_w$.

(3) For each computational complexity class C as considered in Definition 13.1, let $\mathbf{d}^*_C$ = the weak degree of the set of functions which are (a) diagonally nonrecursive relative to the Halting Problem, and (b) bounded by a function in C. The Embedding Lemma tells us that $\inf(\mathbf{d}^*_C, \mathbf{0}') \in \mathcal{P}_w$.

Remark 15.2. It should be interesting to explore the relationships among these newly identified degrees $\inf(\mathbf{d}^*, \mathbf{0}')$, $\inf(\mathbf{d}^*_{\mathrm{REC}}, \mathbf{0}')$, $\inf(\mathbf{d}^*_C, \mathbf{0}')$ in $\mathcal{P}_w$, as well as their relationships with the previously identified degrees $\mathbf{d}$, $\mathbf{d}_{\mathrm{REC}}$, $\mathbf{d}_C$ in $\mathcal{P}_w$, and with reverse mathematics.

Remark 15.3. Another promising source of examples is as follows. Let $Q \subseteq \omega^\omega$ be Σ^0_2 relative to the Halting Problem. Let $\mathbf{s}$ be the weak degree of S, where S is either

$$\{f \in \omega^\omega \mid \exists g \, (g \in Q \text{ and } g \leq_{tt} f')\}$$

or

$$\{f \in \omega^\omega \mid \exists g \, (g \in Q \text{ and } g \leq_{tt} f \oplus 0')\}.$$

Here $\leq_{tt}$ denotes truth-table reducibility, and f' denotes the Turing jump of f. Since S is Σ^0_3, the Embedding Lemma applies. Moreover, if Q is specific and natural, then so is $\mathbf{s}$, hence so is $\inf(\mathbf{s}, \mathbf{0}') \in \mathcal{P}_w$. It should be interesting to explore the relationships among these degrees and others in $\mathcal{P}_w$. The ideas of Jockusch/Stephan [24] and Kjos-Hanssen [7, 27] concerning cohesiveness and superhighness may be relevant.

16. Response to Issue 2

Issue 2 was the problem of finding a "smallness property" of infinite Π^0_1 (i.e., co-recursively enumerable) sets $A \subseteq \omega$ which ensures that the Turing degree of A is $> \mathbf{0}$ and $< \mathbf{0}'$. We do not know how to do this.

However, in the $\mathcal{P}_w$ context, we have identified several "smallness properties" of Π_1^0 sets $P \subseteq 2^\omega$ which ensure that the weak degree of P is $> \mathbf{0}$ and $< \mathbf{0}'$.

Here is one result of this type.

Definition 16.1. A Π_1^0 set $P \subseteq 2^\omega$ is said to be *thin* if, for all Π_1^0 sets $Q \subseteq P$, $P \setminus Q$ is Π_1^0. Equivalently, all Π_1^0 sets $Q \subseteq P$ are of the form $P \cap D$ where D is clopen. A set $P \subseteq 2^\omega$ is said to be *perfect* if it has no isolated points.

Remark 16.2. Nonempty Π_1^0 subsets of 2^ω which are thin and perfect have been constructed by means of priority arguments. Much is known about them. For example, any two such sets are automorphic in the lattice of Π_1^0 subsets of 2^ω under inclusion. See Martin/Pour-El [29], Downey/Jockusch/Stob [16, 17], and Cholak et al [12].

Theorem 16.3. *Let* $\mathbf{p} = \deg_w(P)$ *where* $P \subseteq 2^\omega$ *is* Π_1^0, *nonempty, thin, and perfect. Then* $\mathbf{p}$ *is incomparable with* $\mathbf{r}_1$. *Hence* $\mathbf{0} < \mathbf{p} < \mathbf{0}'$.

Proof. See Simpson [44]. □

One may also consider other smallness properties. As above, let P be a nonempty Π_1^0 subset of 2^ω. The following definition and theorem are due to Binns [6].

Definition 16.4. P is *small* if there is no recursive function f such that for all n there exist n members of P which differ at level $f(n)$ in the binary tree $\{0,1\}^{<\omega}$. For example, let $A \subseteq \omega$ be hypersimple, and let $A = B_1 \cup B_2$ where B_1, B_2 are recursively enumerable. Then $P = \{X \in 2^\omega \mid X$ separates $B_1, B_2\}$ is small.

Theorem 16.5. *If* P *is small, then the weak degree of* P *is* $< \mathbf{0}'$.

Proof. See Binns [6]. □

The following definition and theorem are due to Simpson, unpublished.

Definition 16.6. P is *h-small* if there is no recursive, canonically indexed sequence of pairwise disjoint clopen sets D_n, $n \in \omega$, such that $P \cap D_n \neq \emptyset$ for all n.

Theorem 16.7. *If* P *is h-small, then the weak degree of* P *is* $< \mathbf{0}'$.

17. Summary

We summarize this paper as follows.

There are basic, unresolved issues concerning $\mathcal{R}_T$, the semilattice of recursively enumerable Turing degrees. One of the issues is the lack of specific, natural, recursively enumerable degrees.

In order to address this issue, we have embedded $\mathcal{R}_T$ into $\mathcal{P}_w$, the lattice of weak degrees of nonempty Π^0_1 subsets of 2^ω. We identify $\mathcal{R}_T$ with its image in $\mathcal{P}_w$. In the $\mathcal{P}_w$ context, some of the unresolved issues can be satisfactorily addressed. In particular, $\mathcal{P}_w$ contains many specific, natural degrees which are related to foundationally interesting topics: algorithmic randomness, reverse mathematics, computational complexity.

Note added proof (February 15, 2008):

Subsequent to the writing of this paper, there has been additional progress. (1) Stephen G. Simpson, Mass problems and almost everywhere domination, Mathematical Logic Quarterly, 53:483–492, 2007. (2) Joshua A. Cole and Stephen G. Simpson, Mass problems and hyperarithmeticity, 20 pages, 2006, submitted for publication. (3) Stephen G. Simpson, Medvedev degrees of 2-dimensional subshifts of finite type, 8 pages, 2007, to appear in Ergodic Theory and Dynamical Systems. (4) Stephen G. Simpson, Mass problems and intuitionism, 10 pages, 2007, to appear in Notre Dame Journal of Formal Logic.

References

1. Wilhelm Ackermann. Zum Hilbertschen Aufbau der reellen Zahlen. *Mathematische Annalen*, 99:118–133, 1928.
2. K. Ambos-Spies, G. H. Müller, and G. E. Sacks, editors. *Recursion Theory Week*. Number 1432 in Lecture Notes in Mathematics. Springer-Verlag, 1990. IX + 393 pages.
3. Klaus Ambos-Spies, Bjørn Kjos-Hanssen, Steffen Lempp, and Theodore A. Slaman. Comparing DNR and WWKL. *Journal of Symbolic Logic*, 69:1089–1104, 2004.
4. Stephen Binns. *The Medvedev and Muchnik Lattices of Π^0_1 Classes*. PhD thesis, Pennsylvania State University, August 2003. VII + 80 pages.
5. Stephen Binns. A splitting theorem for the Medvedev and Muchnik lattices. *Mathematical Logic Quarterly*, 49:327–335, 2003.
6. Stephen Binns. Small Π^0_1 classes. *Archive for Mathematical Logic*, 45:393–410, 2006.
7. Stephen Binns, Bjørn Kjos-Hanssen, Manuel Lerman, and David Reed Solomon. On a question of Dobrinen and Simpson concerning almost everywhere domination. *Journal of Symbolic Logic*, 71:119–136, 2006.

8. Stephen Binns, Bjørn Kjos-Hanssen, Joseph S. Miller, and David Reed Solomon. Lowness notions, measure and domination. Preprint, 3 pages, 2006, in preparation, to appear.

9. Stephen Binns and Stephen G. Simpson. Embeddings into the Medvedev and Muchnik lattices of Π_1^0 classes. *Archive for Mathematical Logic*, 43:399–414, 2004.

10. Douglas K. Brown, Mariagnese Giusto, and Stephen G. Simpson. Vitali's theorem and WWKL. *Archive for Mathematical Logic*, 41:191–206, 2002.

11. P. A. Cholak, S. Lempp, M. Lerman, and R. A. Shore, editors. *Computability Theory and Its Applications: Current Trends and Open Problems*, volume 257 of *Contemporary Mathematics*. American Mathematical Society, 2000. XVI + 320 pages.

12. Peter Cholak, Richard Coles, Rod Downey, and Eberhard Herrmann. Automorphisms of the lattice of Π_1^0 classes; perfect thin classes and ANC degrees. *Transactions of the American Mathematical Society*, 353:4899–4924, 2001.

13. Peter Cholak, Noam Greenberg, and Joseph S. Miller. Uniform almost everywhere domination. *Journal of Symbolic Logic*, 71:1057–1072, 2006.

14. S. B. Cooper, T. A. Slaman, and S. S. Wainer, editors. *Computability, Enumerability, Unsolvability: Directions in Recursion Theory*. Number 224 in London Mathematical Society Lecture Notes. Cambridge University Press, 1996. VII + 347 pages.

15. Natasha L. Dobrinen and Stephen G. Simpson. Almost everywhere domination. *Journal of Symbolic Logic*, 69:914–922, 2004.

16. Rodney G. Downey, Carl G. Jockusch, Jr., and Michael Stob. Array nonrecursive sets and multiple permitting arguments. In [2], pages 141–174, 1990.

17. Rodney G. Downey, Carl G. Jockusch, Jr., and Michael Stob. Array nonrecursive degrees and genericity. In [14], pages 93–105, 1996.

18. J.-E. Fenstad, I. T. Frolov, and R. Hilpinen, editors. *Logic, Methodology and Philosophy of Science VIII*. Studies in Logic and the Foundations of Mathematics. North-Holland, 1989. XVII + 702 pages.

19. Mariagnese Giusto and Stephen G. Simpson. Located sets and reverse mathematics. *Journal of Symbolic Logic*, 65:1451–1480, 2000.

20. Marcia J. Groszek and Theodore A. Slaman. Π_1^0 classes and minimal degrees. *Annals of Pure and Applied Logic*, 87:117–144, 1997.

21. Tamara Hummel (now Tamara Lakins). *Effective Versions of Ramsey's Theorem*. PhD thesis, University of Illinois at Urbana/Champaign, 1993.

22. Carl G. Jockusch, Jr. Degrees of functions with no fixed points. In [18], pages 191–201, 1989.

23. Carl G. Jockusch, Jr. and Robert I. Soare. Π_1^0 classes and degrees of theories. *Transactions of the American Mathematical Society*, 173:35–56, 1972.

24. Carl G. Jockusch, Jr. and Frank Stephan. A cohesive set which is not high. *Mathematical Logic Quarterly*, 39:515–530, 1993.

25. Kalmár László. Egyszerű példa eldönthetetlen aritmetikai problémára. *Matematikai és Fizikai Lapok*, 50:1–23, 1943.

26. Steven M. Kautz. *Degrees of Random Sets*. PhD thesis, Cornell University, 1991. X + 89 pages.

27. Bjørn Kjos-Hanssen. Low for random reals and positive measure domination. *Proceedings of the American Mathematical Society.* Preprint, July 2005, 12 pages, to appear.

28. Masahiro Kumabe. A fixed point free minimal degree. Preprint, 1997, 48 pages.

29. Donald A. Martin and Marian B. Pour-El. Axiomatizable theories with few axiomatizable extensions. *Journal of Symbolic Logic*, 35:205–209, 1970.

30. Per Martin-Löf. The definition of random sequences. *Information and Control*, 9:602–619, 1966.

31. Yuri T. Medvedev. Degrees of difficulty of mass problems. *Doklady Akademii Nauk SSSR, n.s.*, 104:501–504, 1955. In Russian.

32. A. A. Muchnik. On strong and weak reducibilities of algorithmic problems. *Sibirskii Matematicheskii Zhurnal*, 4:1328–1341, 1963. In Russian.

33. Alexander Nabutovsky. Einstein structures: existence versus uniqueness. *Geometric and Functional Analysis*, 5:76–91, 1995.

34. Alexander Nabutovsky. Fundamental group and contractible closed geodesics. *Communications on Pure and Applied Mathematics*, 49:1257–1270, 1996.

35. Piergiorgio Odifreddi. *Classical Recursion Theory.* Number 125 in Studies in Logic and the Foundations of Mathematics. North-Holland, 1989. XVII + 668 pages.

36. Piergiorgio Odifreddi. *Classical Recursion Theory, Volume 2.* Number 143 in Studies in Logic and the Foundations of Mathematics. North-Holland, 1999. XVI + 949 pages.

37. Emil L. Post. Recursively enumerable sets of positive integers and their decision problems. *Bulletin of the American Mathematical Society*, 50:284–316, 1944.

38. Hartley Rogers, Jr. *Theory of Recursive Functions and Effective Computability.* McGraw-Hill, 1967. XIX + 482 pages.

39. Gerald E. Sacks. *Degrees of Unsolvability.* Number 55 in Annals of Mathematics Studies. Princeton University Press, 1963. IX + 174 pages.

40. Dana S. Scott and Stanley Tennenbaum. On the degrees of complete extensions of arithmetic (abstract). *Notices of the American Mathematical Society*, 7:242–243, 1960.

41. Stephen G. Simpson. Almost everywhere domination and superhighness. *Mathematical Logic Quarterly*, 53:462–482, 2007.

42. Stephen G. Simpson. An extension of the recursively enumerable Turing degrees. *Journal of the London Mathematical Society*, 75:287–297, 2007.

43. Stephen G. Simpson. *Subsystems of Second Order Arithmetic.* Perspectives in Mathematical Logic. Springer-Verlag, 1999. XIV + 445 pages.

44. Stephen G. Simpson. Mass problems and randomness. *Bulletin of Symbolic Logic*, 11:1–27, 2005.

45. Robert I. Soare. *Recursively Enumerable Sets and Degrees.* Perspectives in Mathematical Logic. Springer-Verlag, 1987. XVIII + 437 pages.

46. Sebastiaan A. Terwijn. The Medvedev lattice of computably closed sets. *Archive for Mathematical Logic*, 45:179–190, 2005.

47. Stanley S. Wainer. A classification of the ordinal recursive functions. *Archiv für Mathematische Logik und Grundlagenforschung*, 13:136–153, 1970.

48. Shmuel Weinberger. *Computers, Rigidity, and Moduli: The Large-Scale Fractal Geometry of Riemannian Moduli Space*. Princeton University Press, 2005. X + 174 pages.

WEAK DETERMINACY AND ITERATIONS OF INDUCTIVE DEFINITIONS

MedYahya Ould MedSalem* and Kazuyuki Tanaka[†]

Mathematical Institute, Tohoku University
Sendai, 980-8578, Japan
*E-mails: * ouldmohamedsalem@yahoo.com*
[†] *tanaka@math.tohoku.ac.jp*

In [5], we proved that Δ_3^0-Det$_0$ follows from Δ_3^1-CA$_0$ + Σ_3^1-IND and that Σ_3^1-IND cannot be dropped. Nevertheless, the exact strength of Δ_3^0-Det$_0$ was left undetermined. In this paper, we settle this problem by showing that it is equivalent to a new axiom $[\Sigma_1^1]^{\text{TR}}$-ID of transfinite combinations of Σ_1^1-inductive definitions in the presence of Π_3^1-transfinite induction.

Keywords: Weak determinacy, Inductive definition, Reverse mathematics.

1. Introduction

In second order arithmetic, the exact strength of determinacy of infinite game has been investigated up to Σ_2^0 ([10], [11] and [12]). Since Davis' proof of Σ_3^0-determinacy can be carried out over Π_3^1-CA$_0$, we have Π_3^1-CA$_0 \vdash \Sigma_3^0$-Det$_0$ (cf. Welch [13]). Σ_4^0-Det cannot be proved in the full system of the second arithmetic (cf. [4] and [3]). In [5], we compared Δ_3^0-Det$_0$ with many popular axioms, but we were not able to characterize it by combinations of them.

In this paper, we characterize Δ_3^0-determinacy in term of inductive definition of transfinitely many Σ_1^1-operators. More precisely, we prove (over ACA$_0 + \Pi_3^1$-TI) the equivalence between the determinacy of the games whose complexity lies in the α-level of difference hierarchy over Π_2^0, and the axiom which asserts the existence of the sets inductively defined by a combination of α-many Σ_1^1-operators.

We work in the framework of second order arithmetic and use mainly the systems ACA_0 and ATR_0 in our investigations. ACA_0 consists of the ordered semi-ring axioms for $(\mathbb{N}, +, \cdot, 0, 1, <)$, Σ^0_1-CA and Σ^0_1-IND. ATR_0 stands for *arithmetical transfinite recursion*, and it is known to be equivalent to $\mathsf{ACA}_0 + \Sigma^0_1$-$\mathsf{Det}_0$. See [9] for further information.

The paper is structured as follows. In section 2, we show that Σ^0_2-determinacy deduces (non-monotone) Σ^1_1-inductive definition, and combining this with the main result of Tanaka [12], we deduce that monotone and non-monotone Σ^1_1-operators have the same power. In section 3, we introduce the inductive definition of a combination of finitely many Σ^1_1-operators and show its equivalence to the determinacy of Boolean combinations of Σ^0_2-sets. Section 4 generalizes the results of section 3 to the transfinite levels. This gives a characterization of Δ^0_3-determinacy by virtue of the Hausdorff-Kuratowski theorem on Δ^0_3-sets.

2. Σ^0_2-Det and Σ^1_1-ID

Let φ be a formula with a distinct variable f ranging over $\mathbb{N}^{\mathbb{N}}$. A two-person *game* G_φ (or simply φ) is defined as follows: player I and player II alternately choose natural numbers (starting with I) to form an infinite sequence $f \in \mathbb{N}^{\mathbb{N}}$ and I (resp. II) wins iff $\varphi(f)$ (resp. $\neg\varphi(f)$). We say that φ is *determinate* if one of the players has a *winning strategy*, that is, a map $\sigma : \mathbb{N}^{<\mathbb{N}} \to \mathbb{N}$ such that the player is guaranteed to win every play f in which he played $f(n) = \sigma(f[n])$ whenever it was his turn to play.

For a class of formulas $\mathcal{C}$, we use $\mathcal{C}$-Det to denote the axiom which states that every φ in $\mathcal{C}$ is determinate. Sometimes, we write $\mathcal{C}$-Det_0 to emphasize that we are talking about the system based on determinacy, i.e., $\mathsf{ACA}_0 + \mathcal{C}$-Det, rather than the axiom itself.

In [11], Σ^0_2-Det was related to the axiom which guarantees the existence of the sets inductively defined by monotone Σ^1_1-operators. In this section, we extend this result to non-monotone Σ^1_1-operators.

An operator $\Gamma : P(\mathbb{N}) \to P(\mathbb{N})$ is Σ^1_1 iff its graph $\{(x, X) : x \in \Gamma(X)\}$ is Σ^1_1. Γ is said to be *monotone* iff $\Gamma(X) \subset \Gamma(Y)$ whenever $X \subset Y$. We will use mon-$\mathcal{C}$ to denote the class of *monotone operators* in $\mathcal{C}$.

Let R be a given relation. The field of R, denote field(R), is the set $\{x : \exists y((x, y) \in R \vee (y, x) \in R)\}$. We say that R is *connected* iff for any $x, y \in$ field(R), we have $(x, y) \in R$ or $(y, x) \in R$. The relation R is said to be *pre-ordering* iff it is reflexive, connected and transitive. R is a *pre-wellordering* iff it is a pre-ordering and well-founded. For $y \in$ field(R), we define R_y and $R_{<y}$ to be $\{x : (x, y) \in R\}$ and $\{x : (x, y) \in R \wedge (y, x) \notin R\}$ respectively. In ACA_0, if R exists then field(R), R_y and $R_{<y}$ also exist.

An axiom of *inductive definition* asserts the existence of a pre-wellordering constructed by iterative application of a given operator.

Definition 2.1. Let $\mathcal{C}$ be a set of $\mathcal{L}_2$ formulas. $\mathcal{C}$-ID asserts that for any operator $\Gamma \in \mathcal{C}$, there exists $W \subset \mathbb{N} \times \mathbb{N}$ such that

(1) W is a pre-wellordering on its field F,
(2) $\forall x \in F \quad W_x = \Gamma(W_{<x}) \cup W_{<x}$,
(3) $\Gamma(F) \subset F$.

We use $\mathcal{C}$-MI to denote [mon-$\mathcal{C}$]-ID. We also remark that for a monotone operator Γ, the second condition of the above definition can be replaced by

$$\forall x \in F \quad W_x = \Gamma(W_{<x}).$$

Let Γ be a Σ_1^1-operator. By Lemma V.4.1 of [9], $n \in \Gamma(X)$ can be written as $\exists f \varphi(n, f, X)$ where $\varphi(n, f, X)$ is a Π_1^0-formula. We often write $n \in \Gamma^f(X)$ to denote the Π_1^0-formula $\varphi(n, f, X)$.

Theorem 2.2. *The following assertions hold.*

- $\mathsf{ATR}_0 \vdash \Sigma_1^1\text{-ID} \to \Sigma_2^0\text{-Det}$,
- $\mathsf{ACA}_0 \vdash \Sigma_2^0\text{-Det} \to \Sigma_1^1\text{-ID}$.

Proof. The first part of the theorem is straightforward from the equivalence between Σ_1^1-MI and Σ_2^0-Det, shown in [12]. For the second part, we shall also adopt the argument of [12] with suitable modification for non-monotone operators.

Let Γ be a Σ_1^1-operator. We construct a Σ_2^0-game G such that player I has no winning strategy, and also that the set W inductively defined by Γ can be easily obtained from a winning strategy of player II. More precisely, Player I starts by playing y^* to ask whether y^* is in the field of W, then II has to answer Yes or No. If II says Yes, he is requested to construct an initial segment V of W with $y^* \in$ field(V). The other player wins if he points out a mistake in the $(\leq y^*)$-segment of V. If II says No, the roles are reversed. We call the player constructing a segment of W by Pro, and the other player Con. Roughly speaking, Pro wins the game iff Con cannot prove that Pro makes an erroneous assertion. The main difficulty in this approach is to keep the complexity of the game as Σ_2^0, but this can be overcome by some unfolding arguments and by avoiding direct talk about well-foundedness.

More formally, the players play (after their roles (Pro and Con) are decided) as follows.

$$\begin{array}{cc} \text{Pro} & \text{Con} \\ \langle v(0), f(0) \rangle & \langle c(0), u(0), g(0) \rangle \\ \vdots & \vdots \\ \langle v(n), f(n) \rangle & \langle c(n), u(n), g(n) \rangle \\ \vdots & \vdots \end{array}$$

where $v, u \in 2^{\mathbb{N}}, f, g \in \mathbb{N}^{\mathbb{N}}$ and $c \in (\{-1\} \cup \mathbb{N})^{\mathbb{N}}$.

For $m, n \in \mathbb{N}$, we identify the pair (m, n) with its code. We write n_0, n_1 for the unique x, y such that $n = (x, y)$. Pro builds a pre-ordering $V = \{(n_0, n_1) : v(n) = 1\}$ with y^* in its field and for each $x \in V_y - V_{<y}$ he gives a witness f for $x \in \Gamma(V_{<y})$. Con's move $c(n) \neq -1$ is used to point out Pro's mistakes. For instance, if there exist m, y such that $m \in \Gamma(V_{<y}) \wedge m \notin V_y$, then Con sets $c(n) = (m, y)$ and starts constricting U and g such that $U = \{(n_0, n_1) : u(n) = 1\} \subset V_{<y}$, $m \in \Gamma^g(U)$ and some extra conditions on U hold. These extra conditions on U are an essential tool for treating non-monotone operators, which is not included in the previous paper [12].

Now, we write down the winning conditions of the game precisely as follows:

(1) We ask Con to make challenges along the order already constructed by Pro in decreasing way and below y^*. That is, at stage n Con may challenge Pro's assertion $v(m) = 0$ if:

- it has been stated by Pro before the stage n that $m_1 \leq_V y^*$ (i.e., $v(m_1, y^*) = 1$), and

- for all previous challenges $c(n') = m' = (m_0', m_1'), n' < n$, it has been stated by Pro before stage n, that $m_1 <_V m_1'$ (i.e., $v(m_1, m_1') = 1 \wedge v(m_1', m_1) = 0$).

If Con disobeys this role, he loses. Supposing that Condition (1) has been obeyed, we give the further winning conditions.

(2) If Con makes no challenge, then Pro wins iff:

- V is a pre-ordering on its field F and $y^* \in F$, and
- for each $y \in F$, $y \in \Gamma^{f_y}(W_{<y})$, where $f_y(n) = f(y, n)$.

(3) If Con makes finite number of challenges. Let n be the last stage such that $c(n) = m \wedge m \neq -1$. Con wins iff:

- $V_{\leq m_1}$ is not pre-ordering, or
- $\neg(\forall y \in F, \ y \in \Gamma^{f_y}(V_{<y}))$, or
- he constructs $U \subset V_{<m_1}$ such that:

$$m_0 \in \Gamma^g(U), \text{ and}$$

$$\forall y \in U, \forall x \ (V_x = V_y \to x \in U), \text{ and}$$

$$\forall y \in U, \ y \in \Gamma^g(U_{<y}),$$

where $U_{<y} = U \cap V_{<y}$.

(4) If Con makes infinitely many challenges, then player II wins.

The above game, due to the forth condition, is Σ_2^0. Hence, it is determinate. On the other hand, player I cannot win in both the roles Pro and Con, and hence II has a winning strategy.

Let τ be II's winning strategy. The above game gives us only an initial segment of the set constructed by Γ. So, we need to glue these segments in the same way as in [12] (page 189) to define $\tilde{W}$. Let F denote the field of the maximal well-founded segment of $\tilde{W}$. The proof of Sublemmas 4.1.2, 4.1.3 and 4.1.5 of [12] doesn't use the monotonicity of Γ. Sublemmas 4.1.4, 4.1.6 and 4.1.7 can be reproved if we take into account the following simple fact: Let p be a τ-consistent play, V^p be the set constructed by Pro in p and $y \in \text{field}(V^p)$. If $F \subset V_{<y}^p$, then we have

- either $F = V_{<y}^p$,
- or $F = V_{<a}^p$ for some $a \in V_{<y}^p$. $\qquad \square$

Finally, we have the following corollary:

Corollary 2.3. $\mathsf{ATR}_0 \vdash \Sigma^0_2\text{-Det} \leftrightarrow \Sigma^1_1\text{-MI} \leftrightarrow \Sigma^1_1\text{-ID}$.

3. Finite levels

In this section, we characterize the determinacy of Boolean combinations of Σ^0_2-formulas in term of inductive definition of a combination of finitely many Σ^1_1-operators. First, let's explain how to iterate a pair of operators Γ_0 and Γ_1. We assume Γ_0 has a distinct parameter X, and so it is also denoted by Γ_0^X. Let Y be any set. We start by the empty set, and iterate applying Γ_0^Y until we get a fixed point, say F_0. Then re-start iterating $\Gamma_0^{\Gamma_1(F_0)}$, get a fixed point F_1. Again, iterate $\Gamma_0^{\Gamma_1(F_0) \cup \Gamma_1(F_1)}$ and so on. The procedure stops when we get a fixed point F such that if F' is the least fixed point of Γ_0^F then $\Gamma_1(F') \subset F$. This can be depicted as follows:

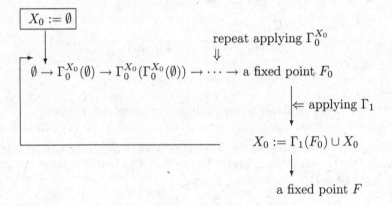

The above procedure can be viewed as an inductive definition by a single operator $[\Gamma_0^X, \Gamma_1]^Y$.

Generally, the iteration of $\Gamma_0^X, \cdots, \Gamma_{k-1}, \Gamma_k$ can be described as follows. We iterate $[\Gamma_0^X, \cdots, \Gamma_{k-1}]^\phi$ until we get a fixed point, say F_0. Then restart iterating $[\Gamma_0^X, \cdots, \Gamma_{k-1}]^{\Gamma_k(F_0)}$, get another fixed point F_1, again iterate $[\Gamma_0^X, \cdots, \Gamma_{k-1}]^{\Gamma_k(F_0) \cup \Gamma_k(F_1)}$ and so on. Eventually, we stop when we get a fixed point F such that if F' is the least fixed point of $[\Gamma_0^X, \cdots, \Gamma_{k-1}]^F$ then $\Gamma_k(F') \subset F$.

The following diagram represents the iteration of four operators Γ_0, Γ_1, Γ_2 and Γ_3.

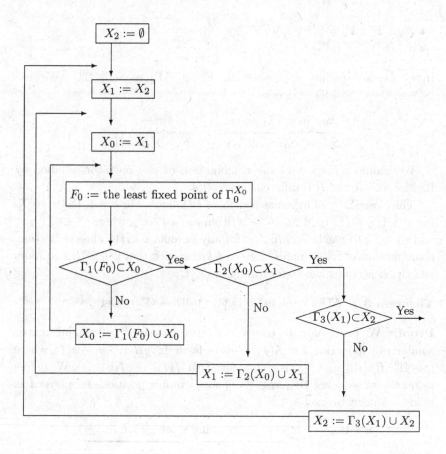

Now, let $S_0, \cdots, S_k$ be $k+1$ collections of operators. $[S_0, S_1, \cdots, S_k]$-ID is the axiom which guarantees the existence of the set inductively defined by $[\Gamma_0^X, \cdots, \Gamma_k]$ for any $\Gamma_0^X \in S_0, \cdots, \Gamma_k \in S_k$. We only give a formal definition of $[S_0, S_1]$-ID.

Definition 3.1. $[S_0, S_1]$-ID asserts that for any $\Gamma_0 \in S_0, \Gamma_1 \in S_1$, there exist a pre-wellordering $W \subset \mathbb{N} \times \mathbb{N}$ on its field F, and V', $\langle V^m \subset \mathbb{N} \times \mathbb{N} : m \in F \rangle$, such that for all $m \in F$

- V^m is a pre-wellordering (on its field F^m),
- $\forall y \in F^m, \ V_y^m = \Gamma_0^{W<m}(V_{<y}^m) \cup V_{<y}^m$,
- $W_m = \Gamma_1(F^m) \cup W_{<m}$,
- $\Gamma_0^{W<m}(F^m) \subset F^m$,
- V' is a pre-wellordering (on its field F'),

- $\forall y \in F',\ V'_y = \Gamma_0^F(V'_{<y}) \cup V'_{<y},$
- $\Gamma_0^F(F') \subset F \wedge \Gamma_1(F') \subset F.$

Here, we are mainly interested in $S_i \in \{\Sigma_1^1, \text{mon-}\Sigma_1^1, \Sigma_1^0\}$. We use $[S_0^{k_0}, S_1^{k_1}, \cdots, S_i^{k_i}]$-ID to denote

$$\overbrace{[S_0, \ldots, S_0,}^{k_0 \text{ times}} \overbrace{S_1, \ldots, S_1,}^{k_1 \text{ times}} \cdots, \overbrace{S_i, \ldots, S_i]}^{k_i \text{ times}}\text{-ID}.$$

We should remark that the combination of two operators studied by Richter and Aczel [7] is different from ours.

The class $(\Sigma_2^0)_k$ of formulas is defined as follows. For $k = 1$, $(\Sigma_2^0)_1 = \Sigma_2^0$. For $k > 1$, $\psi \in (\Sigma_2^0)_k$ iff it can be written as $\neg\psi_1 \wedge \psi_2$, where $\psi_1 \in (\Sigma_2^0)_{k-1}$ and $\psi_2 \in \Sigma_2^0$. It can be shown that for any formula ψ in the class of Boolean combinations of Σ_2^0-formulas, there is a $k \in \omega$ such that $\psi \in (\Sigma_2^0)_k$, or more strictly, ψ is equivalent to a formula in $(\Sigma_2^0)_k$.

Theorem 3.2. $\mathsf{ATR_0} \vdash [\text{mon-}\Sigma_1^1, (\Sigma_1^0)^{k-1}]\text{-ID} \to (\Sigma_2^0)_k\text{-Det}.$

Proof. We prove only the case $k = 2$, since the case $k > 2$ can be treated similarly by induction. Let $A(f)$ be of the form $\exists x \forall y R(x, f[y]) \wedge \psi(f)$, where ψ is Π_2^0, R is Π_0^0 and $f[y]$ is a code for $\langle f(0), f(1), \cdots, f(y-1)\rangle$. We define a transfinite sequence $\langle W_\alpha, \alpha \in Y\rangle$ of sure winning positions for player I as follows: for any ordinal $\alpha \in Y$,

$$u \in W_\alpha \leftrightarrow \underbrace{\exists x}_{(1)} \underbrace{(\text{I has a winning strategy for } A_{u,\alpha,x})}_{(2)},$$

where $A_{u,\alpha,x}(f) \equiv \forall y(R(x, (u * f)[y]) \vee (u * f)[y] \in W_{<\alpha}) \wedge \psi(u * f)$, $u * f$ denotes the concatenation of u and f, and $W_{<\alpha} = \cup_{\beta<\alpha} W_\beta$.

Here, part(1) of the right hand side of this definition is a Σ_1^0-operator. Part(2) can be regarded as the complement of the fixed point of Σ_1^1-monotone operator, see [12]. Thus, W_α is defined by a combination of a Σ_1^0-operator and a Σ_1^1-monotone operator. Hence $W_\infty = \bigcup_{\alpha \in Y} W_\alpha$ exists by $[\text{mon-}\Sigma_1^1, \Sigma_1^0]$-ID.

Claim. For any $u \in \mathbb{N}^{<\mathbb{N}}$, we have

(1) $u \in W_\infty \to$ player I has a winning strategy for A^u,
(2) $u \notin W_\infty \to$ player II has a winning strategy for A^u,

where A^u is defined by $f \in A^u \leftrightarrow u * f \in A$.

The assertion (1) is easy, and so we only show (2). Suppose $u \in \mathbb{N}^{<\mathbb{N}}$ and $u \notin W_\infty$. By Π_2^0-**Det**, for any x, II has a winning strategy $\tau_{u,x}$ in $A_{u,x}$, where

$$A_{u,x}(f) \equiv \forall y(R(x, (u*f)[y]) \vee (u*f)[y] \in W_\infty) \wedge \psi(u*f).$$

If II plays the original game from the position u following $\tau_{u,x}$, then

- either II gets $\neg\psi(u*f)$, which implies that II wins A^u,
- or II reaches a position $u' \supset u$ such that $\neg R(x, u') \wedge u' \notin W_\infty$.

In the later case, II can switch to a new strategy $\tau_{u',x'}$ for any x'. Playing this way, II can construct an infinite sequence: $u \subset u_0 \subset u_1 \subset u_2 \subset \cdots \subset u_n \subset \cdots$ such that $\forall i \neg R(i, u_i)$. If we take f such that $\forall i(u_i \subset f)$, then we get $\forall i \exists y \neg R(i, f[y])$, which implies $f \notin A^u$. Though we use Π_1^1-**AC** to build II's winning strategy from the infinitely many strategies $\tau_{u,x}$, it can be avoided by using the following game: I starts by playing (x, u) such that $x \in \mathbb{N}$ and $u \notin W_\infty$. Then, II wins if $\exists y(\neg R(x, u*f[y]) \wedge u*f[y] \notin W_\infty) \vee \neg\psi(u*f)$. Obviously, I cannot win, and hence II has a winning strategy $\tilde{\tau}$ by Π_2^0-**Det**. Using $\tilde{\tau}$, we can construct a winning strategy of II in the game A^u. This completes the proof of the claim.

Now, we choose $u = \langle\rangle$. From the claim, one of the players has a winning strategy for A, which completes the proof of the main theorem. $\square$

Next, we turn to prove the converse of the previous theorem.

Theorem 3.3. $\mathsf{ACA}_0 \vdash (\Sigma_2^0)_k\text{-}\mathsf{Det} \to [\Sigma_1^1]^k\text{-}\mathsf{ID}$.

Proof. As in the previous theorem, we prove only the case $k = 2$. Let Γ_0, Γ_1 be Σ_1^1-operators. We recall from Definition 3.1 that the structure constructed by $[\Gamma_0, \Gamma_1]$ consists mainly of an outer structure W and some related substructures V^m for every m in the field of W. We construct a $(\Sigma_2^0)_2$-game G in which player I starts with playing some y^* and asks wether it belongs to W or not? Then, according to player II's answer, one of the players is called Pro and requested to build the $(\leq y^*)$-segment of the outer structure W and the related substructures $\langle V^m, m \leq_W y^* \rangle$. The opponent of Pro is called Con. His role is to watch what Pro is constructing and point out if there is any mistake. If he does so, he wins.

Here, we notice that, by $(\Sigma_2^0)_2$-**Det**, we can discuss the entire structure of V^m, for each m in W_{y^*}, rather than an initial segment of it. As for W, doing in the same way as the proof of Theorem 2.2, we will construct

the $\leq y$-segment of W for any y, then glue these initial segments, over
ACA$_0$, to get the entire W. We also mention that Definition 3.1 involves
another structure V'. But this structure can be constructed from W by
Σ_1^1-ID. Finally, Pro and Con play as follows:

$$
\begin{array}{cc}
\text{Pro} & \text{Con} \\
\langle w(0), v(0), f(0) \rangle & \langle c(0), u(0), v'(0), g(0) \rangle \\
\vdots & \vdots \\
\langle w(n), v(n), f(n) \rangle & \langle c(n), u(0), v'(0), g(n) \rangle \\
\vdots & \vdots
\end{array}
$$

where $w, v, u, v' \in 2^{\mathbb{N}}, f, g \in \mathbb{N}^{\mathbb{N}}, c \in (\{-1, 0, 1, 2\} \times \mathbb{N})^{\mathbb{N}}$.

To simplify the notation, we shall not distinguish between the triple
(x, y, z) and its code. For $n \in \mathbb{N}$, n_0, n_1, n_2 will denote the unique x, y, z
such that $n = (x, y, z)$. Pro builds pre-orderings $W = \{(n_0, n_1) : w(n) = 1\}$
and $V^{n_0} = \{(n_1, n_2) : v(n) = 1\}$ for all n_0 in the field of W, together
with witnesses f_m^i, where $f_m^i(n) = f((i, m, n))$ for $i = 0, 1$. We often write
f instead of f_n^i, when i and y are obvious from the context, e.g., $m_2 \in$
$\Gamma_0^{W_{<m_0}, f_m^0}(V_{<m_1}^m)$ and $m_1 \in \Gamma_1^{f_m^1}(F^{m_0})$.

At stage n, Con may challenge Pro's move $w(m) = 0$ by setting $c(n) =$
$(0, m)$, where $m \leq n$. This challenge must be in decreasing way in the sense
of Section 2. Con may also challenge $v(m) = 0$ by setting $c(n) = (1, m)$. In
this case, Con must obey the following conditions:

- it has been established by Pro that $m_0 \leq_W y^*$,
- for all the previous challenges $c(n') = (i, m'), n' < n$, it has been stated
  by Pro that $m_0 \in W_{<m'_1}$ if $i = 0$, and $m_2 \in V^{m_0}_{<m'_2}$ if $i = 1$.

Now, assuming that these conditions have been obeyed, the further winning conditions are given as follows:

(1) If Con makes no challenge, Pro wins if:

- W is a pre-ordering on its field (F_W) with $y^* \in F_W$, and
- for any x in F_W, then V^x is a pre-ordering on its field F^x, and
  $x \in \Gamma_1^f(F^x)$ and $\forall y \in F^x$, $y \in \Gamma_0^{W_{<x}, f}(V_{<y}^x)$.

(2) If Con makes finitely many challenges. Let n be the last stage such that $c(n) \notin \{-1\} \times \mathbb{N}$. We consider the following three cases:

Case 1: $c(n) = (0, m)$. In this case, Con wins if:

- $W_{\leq m_1}$ is not a pre-ordering, or
- $\neg(\forall x \in F_W, \forall y \in F^x, \ y \in \Gamma_0^{W_{<x}, f}(V_{\leq y}^x) \wedge x \in \Gamma_1^f(F^x))$, or
- $m_0 \in \Gamma_1^g(F^{m_1})$.

Case 2: $c(n) = (1, m)$. In this case, Con is challenging $m_1 \notin V_{m_2}^{m_0}$. Thus, he wins if:

- $V_{<m_2}^{m_0}$ is not a pre-ordering, or
- $\neg(\forall x \in F_W, \forall y \in F^x, \ y \in \Gamma_0^{W_{<x}, f}(V_{\leq y}^x) \wedge x \in \Gamma_1^f(F^x))$, or
- he constructs U such that:
  - $U \subset V_{<m_2}^{m_0}$ and $m_1 \in \Gamma_0^{W_{<m_0}, g}(U)$, and
  - U is closed under $=_{V^{m_0}}$, i.e., $\forall x \in U, \forall y (V_y^{m_0} = V_x^{m_0} \to y \in U)$, and
  - $\forall y \in U, \ y \in \Gamma_0^{W_{<m_0}, g}(U_{<y})$,

  where $U_{<y} = U \cap V_{<y}^{m_0}$.

Case 3: $c(n) = (2, m)$. In this case, Con has challenged $w(m) = 0$ and he claims that $W_{<m_1}$ and $\langle F^y, \ y \in W_{m_1} \rangle$ are not suitable. He wins if

- $\neg(\forall x \in F_W, (\forall y \in F^x, \ y \in \Gamma_0^{W_{<x}, f}(V_{\leq y}^x)) \wedge (x \in \Gamma_1^f(F^x)))$ or,
- Con constructs U and $\langle V'^y, \ y \in U \rangle$ such that:
  - $U \subset W_{<m_1} \wedge \forall y (y \in U \to V'^y \subset F^y)$, and
  - $m_0 \in \Gamma_1^g(U)$, and
  - $\forall y \in U \ \forall x \in V'^y, \ (y \in \Gamma_0^{U_{<y}, g}(V_{\leq x}'^y)) \wedge (y \in \Gamma_1^g(V'^y))$, and
  - $\neg(\exists x \exists y \in U (x \in \Gamma_0^{U_{<y}, f}(V^y) \wedge x \notin V'^y))$.

(3) If Con makes infinitely many challenges, he wins.

Since Conditions 1, 2 and 3 can be written (for Con) as a Π_2^0-statement, the game is a $(\Sigma_2^0)_2$-game, and hence it is determinate. On the other hand, it can be shown that player I does not always win. Therefore, player II has a winning strategy τ by $(\Sigma_2^0)_2$-Det. Using τ we can construct the required structure. $\square$

Theorem 3.4. *Assume $0 < k < \omega$. The following assertions are equivalent over* ATR$_0$.

- $(\Sigma_2^0)_k$-Det
- [mon-$\Sigma_1^1, (\Sigma_1^0)^{k-1}$]-ID
- $[\Sigma_1^1]^k$-ID.

Proof. It is straightforward from Theorems 3.2 and 3.3. □

Let $(\Sigma_2^0)_{<\omega}$-Det and $[\Sigma_1^1]^{<\omega}$-ID denote $\bigcup_{k\in\omega}(\Sigma_2^0)_k$-Det and $\bigcup_{k\in\omega}[\Sigma_1^1]^k$-ID respectively. Note that they are infinite sets of sentences, and different from the single sentences $(\Sigma_2^0)_{\mathbb{N}}$-Det and $[\Sigma_1^1]^{\mathbb{N}}$-ID, which will be introduced in the next section. The following assertions are proved in [5].

Lemma 3.5.

- Π_2^1-CA$_0 \vdash \Delta_2^1$-ID$_0$,
- Δ_2^1-MI$_0 \vdash (\Sigma_2^0)_{<\omega}$-Det.

Proof. See Theorem 5.1 and 4.2 of [5]. □

Corollary 3.6. *For any particular* $k \in \omega$, Π_2^1-CA$_0$ *proves* $(\Sigma_2^0)_k$-Det.

Proof. It is straightforward from Lemma 3.5. □

Lemma 3.7. Π_2^1-CA$_0$ *and* $(\Sigma_2^0)_{<\omega}$-Det$_0$ *are proof theoretically equivalent, i.e., they prove the same* Π_1^1-*theorems.*

Proof. See [6]. □

It follows from the previous lemma that Π_2^1-CA$_0$ does not prove the consistency of $(\Sigma_2^0)_{<\omega}$-Det$_0$. This fact will be used in several places throughout this paper.

Lemma 3.8. *For any* k *in* ω, Π_2^1-CA$_0 \vdash \exists\beta_2$-*model of* Δ_2^1-CA$_0 + (\Sigma_2^0)_k$-Det.

Proof. By Theorem VII.7.7 of [9], we have Π_2^1-CA$_0 \vdash \exists \beta_2$-model M such that $M \models \Delta_2^1$-CA$_0$. Since $(\Sigma_2^0)_k$-Det can be stated as a Π_3^1-sentence for every k in ω, and that M satisfies all the true Π_3^1-sentences, we deduce from Corollary 3.6 that $M \models \Delta_2^1$-CA$_0 + (\Sigma_2^0)_k$-Det, which completes the proof. □

Note that $(\Sigma_2^0)_{<\omega}$-Det asserts the determinacy of any Boolean combinations of Σ_2^0-formula, or alternatively, $(\Sigma_2^0)_k$-Det, for any standard natural number k. $(\Sigma_2^0)_{<\omega}$-Det consists of infinitely many sentences and it should not be confused with the axiom $\forall n(\Sigma_2^0)_n$-Det, which can be stated as a single Π_3^1-statement.

Corollary 3.9. *The following assertions are true.*

(1) Δ_2^1-MI$_0 \vdash [\Sigma_1^1]^{<\omega}$-ID,
(2) Δ_2^1-CA$_0 + [\Sigma_1^1]^{<\omega}$-ID $\nvdash \Delta_2^1$-MI$_0$.

Proof. (1) is direct from Theorems 3.4 and Lemma 3.5. For (2), by way of contradiction, suppose that $\Delta_2^1\text{-CA}_0 + [\Sigma_1^1]^{<\omega}\text{-ID} \vdash \Delta_2^1\text{-MI}_0$. Then, for some n_0 in ω, we have

$$\Delta_2^1\text{-CA}_0 + [\Sigma_1^1]^{n_0}\text{-ID} \vdash \Delta_2^1\text{-MI}_0.$$

Thus, by Theorem 3.4 and Lemmas 3.5 and 3.8, we have $\Pi_2^1\text{-CA}_0$ proves the consistency of $(\Sigma_2^0)\text{-Det}_0$, which is a contradiction by Lemma 3.7. $\square$

Theorem 3.10. *For any k in ω, we have*

$$(\Sigma_2^0)_k\text{-Det}_0 \subsetneqq (\Sigma_2^0)_{k+1}\text{-Det}_0 \subsetneqq (\Sigma_2^0)_{<\omega}\text{-Det}_0.$$

Proof. We first show the part $(\Sigma_2^0)_{k+1}\text{-Det}_0 \subsetneqq (\Sigma_2^0)_{<\omega}\text{-Det}_0$. By Lemma 3.8, $\Pi_2^1\text{-CA}_0$ proves the consistency of $(\Sigma_2^0)_k\text{-Det}_0$ for any $k \in \omega$. By Lemma 3.7, $\Pi_2^1\text{-CA}_0$ does not prove the consistency of $(\Sigma_2^0)_{<\omega}\text{-Det}_0$. Thus, we deduce that $(\Sigma_2^0)_{k+1}\text{-Det}_0 \subsetneqq (\Sigma_2^0)_{<\omega}\text{-Det}_0$.

For the part $(\Sigma_2^0)_k\text{-Det}_0 \subsetneqq (\Sigma_2^0)_{k+1}\text{-Det}_0$, suppose there exists $k_0 \in \omega$ such that $(\Sigma_2^0)_{k_0}\text{-Det}_0 \leftrightarrow (\Sigma_2^0)_{k_0+1}\text{-Det}_0$. Then, we can prove by meta-induction that $(\Sigma_2^0)_{k_0}\text{-Det}_0 \leftrightarrow (\Sigma_2^0)_{<\omega}\text{-Det}_0$, which leads to a contradiction by the first part of the proof. $\square$

We recall that Bradfield [1] proved that the sets of the winning positions of $(\Sigma_2^0)_k$-games are exactly the same as the $(k+1)$-level of μ-calculus alternation hierarchy Σ_{k+1}^μ. The following corollary provides a simple proof for a theorem due to Bradfield [2] that the hierarchy $\langle \Sigma_n^\mu, n \in \omega \rangle$ is strict.

Corollary 3.11. *The μ-calculus alternation hierarchy is strict. That is, for any k in ω, we have $\Sigma_k^\mu \subsetneqq \Sigma_{k+1}^\mu$.*

Proof. By way of contradiction, suppose μ-calculus alternation hierarchy were not strict. By Bradfield [1], we would get that $(\Sigma_2^0)_k\text{-Det}_0 \leftrightarrow (\Sigma_2^0)_{k+1}\text{-Det}_0$ for some k, which contradicts with Theorem 3.10. $\square$

4. Transfinite levels

In this section, we prove the equivalence between the determinacy of the transfinite levels of difference hierarchy and the inductive definition of the combination of transfinitely many Σ_1^1-operators.

Let $\prec$ be a recursive well-ordering on $\mathbb{N}$. We define a recursive well-ordering $\prec^*$ on $\mathbb{N} \times \{0, 1\}$ as follows:

$$(x, i) \prec^* (y, j) \quad \text{iff} \quad x \prec y \vee (x = y \wedge i < j).$$

We say that a formula $\varphi(n, i, f)$ is *decreasing along* $\prec^*$ if and only if

$$\forall f \in \mathbb{N}^{\mathbb{N}} \; \forall n \forall i \forall m \forall j \; ((m, j) \prec^* (n, i) \wedge \varphi(n, i, f) \rightarrow \varphi(m, j, f)).$$

We recall that, by Section VIII.3 of [9], recursive ordinals can be regarded, over ATR_0, as first order objects.

Now, let α be a recursive ordinal. We define the class $(\Sigma_2^0)_\alpha$ as follows. $A \in (\Sigma_2^0)_\alpha$ if:

$$\forall f \in \mathbb{N}^{\mathbb{N}}, \;\; A(f) \equiv \exists x (\neg \varphi(x, 1, f) \wedge \varphi(x, 0, f))$$

where $\varphi(x, i, f)$ is a decreasing Π_2^0-formula along some recursive well-ordering relation $\prec^*$ and α is the order type of $\prec^*$. Particularly, A is in $(\Sigma_2^0)_{\mathbb{N}}$ iff $\prec^*$ is the usual order of $\mathbb{N}$.

The inductive definition by infinite sequence of Σ_1^1-operators $\langle \Gamma_0^X, \cdots, \Gamma_n, \cdots, \Gamma_{\mathbb{N}} \rangle$, denoted by $[\Sigma_1^1]^{\mathbb{N}}$-ID, can be described as follows. For every n, we iterate $[\Gamma_0^X, \cdots, \Gamma_n]^\phi$ to construct a structure W^n, then again, for every $n \in \mathbb{N}$, iterate $[\Gamma_0^X, \cdots, \Gamma_n]^{\Gamma_{\mathbb{N}}(\uplus_i W^i)}$ to get a new sequence $\langle W^{0,0}, W^{0,1}, \cdots, W^{0,n}, \cdots \rangle$. Then, for every $n \in \mathbb{N}$, iterate $[\Gamma_0^X, \cdots, \Gamma_n]^{\Gamma_{\mathbb{N}}(\uplus_i W^i) \cup \Gamma_{\mathbb{N}}(\uplus_i W^{0,i})}$ and so on. We stop when we get a fixed point F, such that if F' is the sequence of least fixed points of $[\Gamma_0^X, \cdots, \Gamma_n]^F$, then $\Gamma_{\mathbb{N}}(\uplus F') \subset F$. The following diagram illustrates this iteration.

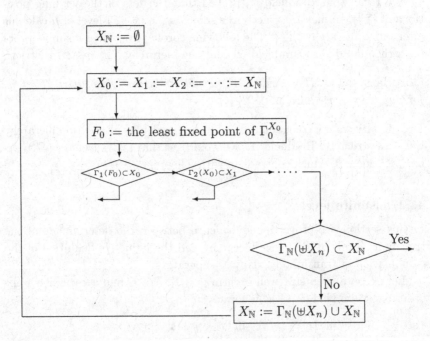

This can be made precise through the following definition.

Definition 4.1.

$[\Sigma_1^1]^{\mathbb{N}}$-ID asserts that, for any sequence $\langle \Gamma_0^X, \cdots, \Gamma_n, \cdots, \Gamma_{\mathbb{N}} \rangle$ of Σ_1^1-operators, there exist $W, V, \langle W^s : s \in \mathbb{N}^{<\mathbb{N}} \rangle$ and $\langle V^t : t \in \mathbb{N}^2 \rangle$ such that

- W, V, W^s and V^t are pre-wellorderings on F, F_V, F^s and F_V^t, respectively.
- $\forall y \in F_V \quad V_y = \Gamma_{\mathbb{N}}^{F_V}(V_{<y}) \cup V_{<y}$
- $\forall m \in F, \forall n, \quad \Gamma_n(F_V^{\langle m,n \rangle}) \subset F^{\langle m,n \rangle}$
- $\forall m \in F \, \forall n \, \forall y \in F_V^{\langle m,n \rangle} \quad V_y^{\langle m,n \rangle} = \Gamma_n^{F^{\langle m,n \rangle}}(V_{<y}^{\langle m,n \rangle}) \cup (V_{<y}^{\langle m,n \rangle})$
- $\forall m \in F \, \forall n \, \forall s \in \{m\} \times \{n\} \times \mathbb{N}^{<n} \, \forall y \in F^s, W_y^s$ is equal to:

$$\begin{cases} \Gamma_{n+1-|s|}(F^{s*\langle y \rangle}) \cup W_{<y}^s & \text{if } |s| < n+1, \\ \Gamma_0^{W^{s[n]}_{<s(n)}}(W_{<y}^s) \cup W_{<y}^s & \text{if } |s| = n+1, \text{ and } s(n) \text{ is not} \\ & (W^{s[n]})\text{-minimal}, \\ \Gamma_0^{W_{<y}}(W_{<y}^s) \cup W_{<y}^s & \text{if } |s| = n+1, \text{ and } s(n) \text{ is } (W^{s[n]})\text{-} \\ & \text{minimal}. \end{cases}$$

- $\forall m \in F \, \forall n \, \forall s \in \{m\} \times \{n\} \times \mathbb{N}^{<n} \, F^s$ is a subset of

$$\begin{cases} \Gamma_0^{F^s}(F_V^{\langle m,n \rangle}) & \text{if } |s| = 2, \\ \Gamma_0^{W^{s[n]}_{<s(n)}}(F^s) & \text{if } |s| = n+1, \text{ and } s(n) \text{ is not } (W^{s[n]})\text{-minimal}, \\ \Gamma_0^{W_{<y}}(F^s) & \text{if } |s| = n+1, \text{ and } s(n) \text{ is } (W^{s[n]})\text{-minimal}. \end{cases}$$

- $W_m = \Gamma_{\mathbb{N}}(\uplus_{n,s} F^{\langle m,n \rangle *s}) \cup W_{<m}$
- $\forall y \in F_V, \quad V_y = \Gamma_0^F(V_{<y}) \cup V_{<y}$
- $\Gamma_{\mathbb{N}}(F_V) \subset F \wedge \Gamma_0^F(F_V) \subset F_V$

where $|s|$ denotes the length of the finite sequence s, and for every $i < |s|$ $s[i+1]$ denotes the sequence $\langle s(0), \cdots, s(i) \rangle$.

We also define $[\text{mon-}\Sigma_1^1, (\Sigma_1^0)^{\mathbb{N}}]$-ID to be the same as $[\Sigma_1^1]^{\mathbb{N}}$-ID except that we take the first operator from mon-Σ_1^1 and all the others from Σ_1^0.

Theorem 4.2. $\text{ATR}_0 + \Pi_3^1\text{-IND} \vdash [\text{mon-}\Sigma_1^1, (\Sigma_1^0)^{\mathbb{N}}]\text{-ID} \to (\Sigma_2^0)_{\mathbb{N}}\text{-Det}.$

Proof. First, we notice that the proofs of Theorems 3.2 and 3.3 can be used to show that:

$$\text{ATR}_0 \vdash \forall n \, (((\Sigma_2^0)_n\text{-Det} \leftrightarrow [\Sigma_1^1]^n\text{-ID}) \to ((\Sigma_2^0)_{n+1}\text{-Det} \leftrightarrow [\Sigma_1^1]^{n+1}\text{-ID})).$$

Thus, $\text{ATR}_0 + \Pi_3^1\text{-IND} \vdash \forall n \, ((\Sigma_2^0)_n\text{-Det} \leftrightarrow [\Sigma_1^1]^n\text{-ID}).$

Let A be a $(\Sigma_2^0)_\mathbb{N}$-set. Then, there exists a decreasing Π_2^0-formula $\varphi(n, i, f)$ such that

$$A(f) \leftrightarrow \exists n(\neg\varphi(n, 1, f) \wedge \varphi(n, 0, f)).$$

Here, $\neg\varphi(n, 1, f)$ is a Σ_2^0-formula, and so it can be written as $\exists x \forall y R(n, 1, x, f[y])$, where R is a Π_0^0-relation. We define a transfinite sequence $\langle W_\alpha, \alpha \in Y \rangle$ of sure winning positions for player I as follows: for any ordinal $\alpha \in Y$,

$$u \in W_\alpha \leftrightarrow \exists n \exists x(\text{I has a winning strategy in } A_{u,\alpha,n,x}),$$

where $A_{u,\alpha,n,x}(f)$ is the following formula:

$$((\forall y R(n, 1, x, u * f[y]) \vee u * f[y] \in W_{<\alpha}) \wedge \varphi(n, 0, u * f)) \vee$$
$$(\exists i < n(\neg\varphi(i, 1, u * f) \wedge \varphi(i, 0, u * f))).$$

By $[\text{mon-}\Sigma_1^1, (\Sigma_1^0)^\mathbb{N}]\text{-ID}$, $W_\infty = \cup_{\alpha \in Y} W_\alpha$ exists. Then by imitating the proof of Theorem 3.2, we can show that:

- $\langle\rangle \in W_\infty \rightarrow$ player I has a winning strategy for A,
- $\langle\rangle \notin W_\infty \rightarrow$ player II has a winning strategy for A.   □

Theorem 4.3. *For any recursive ordinal α, we have*

$$\mathsf{ACA}_0 + \Pi_3^1\text{-TI} \vdash [\text{mon-}\Sigma_1^1, (\Sigma_1^0)^\alpha]\text{-ID} \rightarrow (\Sigma_2^0)_\alpha\text{-Det}.$$

Proof. In the case of $(\Sigma_2^0)_\alpha$, the induction is carried over a recursive well-order instead of the usual order of $\mathbb{N}$. Therefore, we need Π_3^1-TI instead of Π_3^1-IND. We didn't add ATR_0 since it is included in Π_3^1-TI by [8].   □

Next, we turn to prove the reversal of Theorem 4.2.

Theorem 4.4. $\mathsf{ACA}_0 + \Pi_3^1\text{-IND} \vdash (\Sigma_2^0)_\mathbb{N}\text{-Det} \rightarrow [\Sigma_1^1]^\mathbb{N}\text{-ID}$

Proof. Let $\langle \Gamma_0^X, \cdots, \Gamma_l, \cdots, \Gamma_\mathbb{N} \rangle$ be an infinite sequence of Σ_1^1-operators. We construct a $(\Sigma_2^0)_\mathbb{N}$-game in which player I asks about some y^*. Then, one of the players builds the $(\leq y^*)$-segment of the structure constructed by $[\Gamma_0^X, \cdots, \Gamma_N]$. The game goes as follows:

Pro	Con
$\langle w(0), v(0), f(0) \rangle$	$\langle c(0), u(0)g(0) \rangle$
$\vdots$	$\vdots$
$\langle w(n), v(n), f(n) \rangle$	$\langle c(n), u(n)g(n) \rangle$
$\vdots$	$\vdots$

where $w, v, u \in 2^{\mathbb{N}}, f, g \in \mathbb{N}^{\mathbb{N}}, c \in (\{-1, 0, 1, 2\} \times \mathbb{N}^{<\mathbb{N}} \times \mathbb{N})^{\mathbb{N}} \cup (\{-1\} \cup \mathbb{N})^{\mathbb{N}}$.

For $s \in \mathbb{N}^{<\mathbb{N}}$, we define W^s to be $\{(n_0, n_1) : w(\langle n, s \rangle) = 1\}$. The possible mistakes of Pro can be divided in two types:

(1) Mistakes about some ($\leq y$)-segment of W^s, e.g., $\Gamma_{l+1-|s|}(F^{s*\langle y \rangle}) \not\subset W_y^s$.
(2) Mistakes about the field F^s of W^s, e.g., F^s is not the least fixed point of $\Gamma_{l+1-|s|}$.

Type (1) can be treated directly with the help of unfolding tricks. To treat (2), we give Con a chance to win if he constructs a smaller fixed point of $\Gamma_{l+1-|s|}$. The latter case can be seen as a new subgame in which Pro and Con interchange the roles.

Thus, at stage n, Con may challenge Pro's assertion $w(m) = 0$ by setting $c(n) = m$. This kind of challenging should be in decreasing way in the sense of Section 3. Con may as well challenge $(m_0, m_1) \notin W^s$, $m_1 \in F^s$, and use $c(n) = (0, s, m)$ or $c(n) = (1, s, m)$ to point out a mistake of type(1). He can also use $c(n) = (2, s, m)$ to claim that there is a mistake of type (2). In both cases, Con must obey the following rules:

- $s(0) \in W_{y^*}$,

- $\forall n' < n, \begin{cases} c(n') = m' \to (s(0), m_1') \in W \\ c(n') \in \{(0, s', m'), (1, s', m'), (2, s', m')\} \to \\ \quad (\exists i < |s| - 2 \ \ s' = s[i+2]) \vee (s = s' \wedge m_1 \in W_{<m_1'}^s) \end{cases}$

- For any i such that $i < |s| - 2$, $s(i+2) \in F_{s(i+2)}^{s[i+2]}$

The further winning conditions are given as follows:

(1) If Con makes no challenge, then Pro wins if:

- W is pre-ordering on its field F with $y^* \in F$, and
- Pro obeys a standard condition (S), that is, for any $k \in F$, we have $k \in \Gamma_{\mathbb{N}}^f(\uplus_s F^s)$ and $\forall l$, $\forall s \in \{k\} \times \{l\} \times \mathbb{N}^{<l}$

  - W^s is a pre-ordering on its field F^s, and
  - for any y in F^s, we have

$$\begin{cases} y \in \Gamma_{l+1-|s|}^f(F^{s*\langle y \rangle}) & \text{if } |s| < l+1, \\ y \in \Gamma_0^{W_{<y}^s,f}(W_{<y}^s) & \text{if } |s| = l+1, \text{ and } s(l) \text{ is } (W^{s[l]})- \\ & \qquad \text{minimal,} \\ y \in \Gamma_0^{W_{<s(l)}^{s[l]},f}(W_{<y}^s) & \text{if } |s| = l+1, \text{ and } s(l) \text{ is not } (W^{s[l]})- \\ & \qquad \text{minimal.} \end{cases}$$

(2) If Con makes finitely many challenges. Let n be the last stage such that $c(n) \notin \{-1\} \cup (\{-1\} \times \mathbb{N}^{<\mathbb{N}} \times \mathbb{N})$. We need to consider the following four cases:

Case(1): $c(n) = m$. In this case, Con wins iff one of the following conditions holds

- $W_{\leq m_1}$ is not a pre-ordering,
- Pro does not obey condition (S),
- $m_0 \in \Gamma^{W_{<\mathbb{N}}^{m_1},g}(\uplus_s F^s)$.

Case(2): $c(n) = (0, s, m)$. Con wins iff one of the following conditions holds

- $W_{\leq m_1}^s$ is not a pre-ordering,
- Pro does not obey condition (S),
- $m_0 \in \Gamma_{l+1-|s|}^g(F^{s*\langle m_1 \rangle})$ if $|s| < l+2$, and $m_0 \in \Gamma_0^{W_{<y},g}(W_{<y}^s)$ if $|s| = l+2$.

Case(3): $c(n) = (1, s, m)$. In this case, Con is requested to refine $W_{<m_1}^s$ and $\langle F^{s*\langle y \rangle}, \ y \in W_{<m_1}^s \rangle$, and Pro is given a chance to win if he points out a mistake in Con's construction, provided he (i.e., Pro) keeps doing well on what he already started. This case can happen only finitely many times.

Case (4): $c(n) = (2, s, m)$. In this case, Con wins if one of the following cases happened

- $W_{\leq m_1}^s$ is not a pre-ordering,
- Pro does not obey condition (S),
- $|s| = l$ and he constructs U and $\langle V^y, \ y \in U \rangle$ such that
  - $U \subset W_{<m_1}^s$ and it is closed under $=_{W^s}$, and
  - $\forall y \in U$, $V^y \subset F^{s*\langle y \rangle}$ and V^y is closed under $=_{W^{s*\langle y \rangle}}$, and

$- \forall y \in U, \ (y \in \Gamma_0(V^y) \wedge (\forall x \in V^y, \ x \in \Gamma_0^{U_{<y},g}(V^y_{<x}))$, and
$- m_0 \in \Gamma_1^g(U)$

where $U_{<y}$ and $V^y_{<x}$ denote $U \cap W^s_{<m_1}$ and $V^y \cap W^{s*\langle y \rangle}_{<x}$ respectively.

- $|s| = l + 1$ and he constructs $U \subset W^s_{<m_1}$ such that

  $- U$ is closed under $=_{W^s}$, and
  $- \forall y \in U y \in \Gamma_0^{W^{s[l]}_{s(l)}}(U_{<y})$, and
  $- m_0 \in \Gamma_0^{W^{s[l],g}_{s(l)}}(U)$,

  where $U_{<y} = U \cap W^s_{<m_1}$.

(3) If Con makes infinitely many challenges, he wins.

In this game, Case (1) and Case (2) can be written as Σ_2^0-statements. Case (3) leads either to Case(2) or Case(4), which are $(\Sigma_2^0)_2$-statements. Therefore, the game is $(\Sigma_2^0)_{\mathbb{N}}$ and hence it is determinate. On the other hand, it can be easily seen that player II has a winning strategy τ. Using τ, we can build the required structure.

Theorems 4.2 and 4.3 can be generalized easily to any recursive ordinal.

Theorem 4.5. *For any recursive ordinal α, we have*

$$\mathsf{ACA}_0 + \Pi_3^1\text{-TI} \vdash (\Sigma_2^0)_\alpha\text{-Det} \leftrightarrow [\text{mon-}\Sigma_1^1, (\Sigma_1^0)^\alpha]\text{-ID} \leftrightarrow [\Sigma_1^1]^\alpha\text{-ID}.$$

Proof. It is straightforward. $\qquad\square$

Let $[\Sigma_1^1]^{\text{TR}}\text{-ID}$ be the axiom which asserts that for any ordinal α, we have $[\Sigma_1^1]^\alpha\text{-ID}$. By Theorem 3.1 of [5], ψ is a Δ_3^0-formula iff there is an ordinal α such that $\psi \in (\Sigma_2^0)_\alpha$. Therefore, from Corollary 4.2, we have

Corollary 4.6. *The following assertion holds over $\mathsf{ACA}_0 + \Pi_3^1\text{-TI}$:*

$$\Delta_3^0\text{-Det} \leftrightarrow [\text{mon-}\Sigma_1^1, \Sigma_1^0]^{\text{TR}}\text{-ID} \leftrightarrow [\Sigma_1^1]^{\text{TR}}\text{-ID}.$$

where $[\text{mon-}\Sigma_1^1, \Sigma_1^0]^{\text{TR}}\text{-ID}$ is a particular case of $[\Sigma_1^1]^{\text{TR}}\text{-ID}$ in which the first operator is taken to be Σ_1^1-monotone and all the others to be Σ_1^0-operators.

Combining the results of this section and those of [5], we get the following corollary:

Corollary 4.7. *The following two assertions hold.*

(1) $\Pi_2^1\text{-CA}_0 + \Pi_3^1\text{-TI}$ *proves* $[\Sigma_1^1]^{\text{TR}}\text{-ID}_0$,
(2) $\Delta_3^1\text{-CA}_0$ *does not prove* $[\Sigma_1^1]^{\mathbb{N}}\text{-ID}_0$.

Proof. (1) is straightforward from Theorem 5.6 of [5] and Theorem 4.5. For (2), suppose Δ_3^1-$\mathsf{CA}_0 \vdash [\Sigma_1^1]^{\mathbb{N}}$-$\mathsf{ID}_0$. Then, Δ_3^1-CA_0 would prove the existence of an ω-model M of $[\Sigma_1^1]^{\mathbb{N}}$-$\mathsf{ID}_0$. Since M satisfies Σ_∞^1-IND, we get $M \models (\Sigma_2^0)_{\mathbb{N}}$-$\mathsf{Det}_0$ by Theorem 4.2. and hence M is a model of $(\Sigma_2^0)_{<\omega}$-Det_0. Thus, Δ_3^1-CA_0 proves the consistency of $(\Sigma_2^0)_{<\omega}$-Det_0. By Corollary IX.4.12 of [9] Δ_3^1-CA_0 is Π_4^1-conservative over Π_2^1-CA_0 and then particularly, Π_2^1-CA_0 proves the consistency of $(\Sigma_2^0)_{<\omega}$-Det_0, which is a contradiction by Lemma 3.7. □

We recall that the determinacy on the finite levels of the difference hierarchy over Π_1^0-formulas (with set parameters) collapses after the first stage. That is, $(\Sigma_1^0)_2$-$\mathsf{Det}_0 = (\Sigma_1^0)_n$-$\mathsf{Det}_0 = \Pi_1^1$-$\mathsf{CA}_0$ [11]. By contrast, Corollary 3.5 together with the fact that Π_2^1-$\mathsf{CA}_0 \nvdash (\Sigma_2^0)_{<\omega}$-$\mathsf{Det}_0$, shows that the determinacy on the difference hierarchy over Σ_2^0-formulas is strict at least up to the $\mathbb{N}$-level. We do not know whether $(\Sigma_2^0)_{\mathbb{N}}$-$\mathsf{Det}_0$ is strictly weaker than Δ_3^0-Det_0 or not. This is relevant to the question whether we can reduce Π_3^1-TI in the first assertion of Corollary 4.7 to Π_3^1-IND or not.

We also mention that the equivalence between Δ_3^0-Det and $[\Sigma_1^1]^{\mathrm{TR}}$-$\mathsf{ID}$ has been established in the presence of Π_3^1-TI. The questions whether Π_3^1-TI can be reduced to Π_1^1-IND or even whether it can be eliminated remain open.

References

1. J.C. Bradfield, *Fixpoints, games and the difference hierarchy*, Theor. Inform. Appl. 37 (2003), pp. 1-15.
2. J.C. Bradfield, *The modal μ-calculus alternation hierarchy is strict*, Theor. Comput. Sci. 195 (1998), pp. 133-153.
3. H.M. Friedman, *Higher set theory and mathematical practice*, Annals of Mathematical Logic 2 (1971), pp. 325-357.
4. L.A. Harrington and A.S. Kechris, *A basis result for Σ_3^0 sets of reals with an application to minimal covers*, Proc. of A.M.S 53 (1975), pp. 445-448.
5. M.O. Medsalem and K. Tanaka, *Δ_3^0-determinacy, comprehension and induction*, Journal of Symbolic Logic 72 (2007), pp. 452-462.
6. C. Heinatsch and M. Möllerfeld, *The determinacy strength of Π_2^1-comprehension*, preprint.
7. W. Richter and P. Aczel, *Inductive definitions and reflecting properties of admissible ordinals*, J.E. Fenstad and P.G. Hinman, eds., General Recursion Theory (1974), pp. 301-381.
8. S.G. Simpson, *Σ_1^1 and Π_1^1 transfinite induction*, Logic Colloquium 80 (1980), pp. 239-253.
9. S.G. Simpson, *Subsystems of Second Order Arithmetic*, Springer (1999).

10. J.R. Steel, *Determinateness and subsystems of analysis*, Ph.D. thesis, University of California, Berkeley (1977).

11. K. Tanaka, *Weak axioms of determinacy and subsystems of analysis I:* Δ_2^0-*games,* Zeitschrift für mathematische Logik und Grundlagen der Mathematik 36 (1990), pp. 481-491.

12. K. Tanaka, *Weak axioms of determinacy and subsystems of analysis II:* Σ_2^0-*games,* Annals of Pure and Applied Logic 52 (1991), pp. 181-193.

13. P. Welch, *Weak systems of determinacy and arithmetical quasi-inductive definitions,* preprint.

A tt VERSION OF THE POSNER-ROBINSON THEOREM

W. Hugh Woodin

Department of Mathematics
University of California, Berkeley
Berkeley, CA 94720, USA
E-mail: woodin@math.berkeley.edu

1. Introduction

For subsets, X and Y, of ω,

$$X \leq_T Y$$

if X is recursive in Y, and $X =_T Y$ is $X \leq_T Y$ and $Y \leq_T X$.

Similarly if X, Y, and Z are subsets of ω then $X \leq_T Y \vee Z$ if X is recursive in the pair (Y, Z). Thus if

$$\pi : \omega \to \omega \times \omega$$

a recursive bijection, then $X =_T Y \vee Z$ if $X =_T \pi^{-1}[Y \times Z]$.

For each set $X \subseteq \omega$, X' denotes the Turing jump of X. The Posner-Robinson Theorem states that if $X \subset \omega$ and X is not recursive then there exists a set $Y \subset \omega$ such that

$$Y' =_T X \vee Y.$$

Suppose one requires that Y' be more effectively computed from the pair, (X, Y)? Given a set $Y \subset \omega$, define the reducibility, $\leq_{tt_Y}$ as follows where $[\omega]^{<\omega}$ denotes the set of all finite subsets of ω. Suppose that $A \subseteq \omega$, $B \subseteq \omega$ and $C \subseteq \omega$. Then

$$A \leq_{tt_Y} B \vee C$$

355

if there exist a primitive recursive function,

$$f_0 : [\omega]^{<\omega} \times [\omega]^{<\omega} \to [\omega]^{<\omega}$$

and a function

$$f : \omega \to \omega$$

such that

(1) f is recursive in Y,
(2) for all $n < \omega$, $A \cap n = f_0 (B \cap f(n), C \cap f(n))$.

This also defines $A \leq_{tt_Y} B$ (take $C = B$). Finally

$$A =_{tt_Y} B \vee C$$

if $A \leq_{tt_Y} B \vee C$, $B \leq_{tt_Y} A$, and $C \leq_{tt_Y} A$.

For each set $X \subseteq \omega$, $X \in \text{HYP}$ if

$$X \in L_{\omega_1^{CK}}$$

where ω_1^{CK} is the least ordinal, α, such that L_α is admissible. With this notation our main theorem is the following.

Theorem 1.1. *Suppose $X \subset \omega$ and $X \notin \text{HYP}$. Then there exists a set $Y \subseteq \omega$ such that*

$$Y' =_{tt_Y} X \vee Y.$$

In contrast, Riemann and Slaman have proved the following theorem as a corollary of a theorem on relative randomness. Taken together these two theorems give an interesting characterization of HYP.

Theorem 1.2. (Reimann-Slaman) *Suppose that $X_0 \subseteq \omega$ and $X_0 \in \text{HYP}$. Then there exists a set $X \subseteq \omega$ such that*

(1) $X \in \text{HYP}$,
(2) $X_0 \leq_{tt_0} X$,
(3) for all $Y \subseteq \omega$, $Y' \neq_{tt_Y} X \vee Y$.

In the final section of this paper we prove the generalization of Theorem 1.1 to the case of transfinite iterates of the Turing jump up to the level of the hyperjump; cf. Theorem 4.1 and Theorem 4.2. One can easily generalize further but the basic method for these cases seems be all that is needed. On the other hand whether these theorems are optimal in the sense that Theorem 1.1 is optimal, is open–what is missing is the corresponding generalization of Theorem 1.2.

2. Preliminaries

We review a technical device from [3].

Lemma 2.1. *There is a primitive recursive function,*

$$\rho_0 : [\omega]^{<\omega} \to [\omega]^{<\omega},$$

such that

(1) for all $s \subset t$ if $s = t \cap i$ for some $i \in t$ then $\max(\rho_0(s)) < \min(\rho_0(t))$,
(2) for all finite sequences,

$$\langle X_i : i < N \rangle$$

of distinct elements of $\mathcal{P}(\omega)$, for all $k < \omega$, for all $\sigma \subseteq N$, if

$$X_i \cap k \neq X_j \cap k$$

for all $i < j < N$, then there exist $i > k$ such that

$$i \in \left(\cap \{ \rho_0(X_i \cap k) | i \in \sigma \} \right) \cap \left(\cap \{ \omega \backslash \rho_0(X_i \cap k) | i \notin \sigma \} \right).$$

Proof. This is a completely straightforward construction. □

We fix some more notation. For each $X \subset \omega$, let $\sigma_X = \cup \{ \rho_0(X \cap i) | i < \omega \}$. The point of the definition of ρ_0 is that the set

$$\{ \sigma_X | X \subset \omega \}$$

is an *independent set*; if $\mathcal{A} \subset \mathcal{P}(\omega)$ and $\mathcal{B} \subset \mathcal{P}(\omega)$ are pairwise disjoint nonempty finite sets then,

$$\left(\cap \{ \sigma_X | X \in \mathcal{A} \} \right) \cap \left(\cap \{ \omega \backslash \sigma_X | X \in \mathcal{B} \} \right)$$

is infinite. This claim follows easily from the definition of ρ_0.

Suppose that $\sigma \subset \omega$. A (nonempty) set $\mathcal{X} \subset \mathcal{P}(\omega)$ is *independent on σ* if for all disjoint nonempty finite sets, $\mathcal{A}, \mathcal{B} \subset \mathcal{X}$, the set

$$\sigma \cap (\cap \mathcal{A}) \cap (\cap \{ \omega \backslash \tau | \tau \in \mathcal{B} \})$$

is infinite (so σ must be infinite).

Definition 2.2. Let $\mathbb{I}_{\rho_0} \subseteq \mathcal{P}(\omega)$ be the set all sets $\sigma \subseteq \omega$ such that for any cofinite set $\mathcal{A} \subset \mathcal{P}(\omega)$,

$$\{ \sigma_X | X \in \mathcal{A} \}$$

is not independent on σ.

Note that $\sigma \in \mathbb{I}_{\rho_0}$ if and only if there exists a sequence,

$$\langle \mathcal{A}_i : i < \omega \rangle,$$

of pairwise disjoint nonempty finite subsets of $\mathcal{P}(\omega)$ such that for each i,

$$\{\sigma_X | X \in \mathcal{A}_i\},$$

is not independent on σ.

Lemma 2.3. $\mathbb{I}_{\rho_0}$ *is a proper ideal and* $\mathbb{I}_{\rho_0}$ *is* Δ_1^1.

Proof. Clearly $\mathbb{I}_{\rho_0}$ is closed under subsets and clearly $\omega \notin \mathbb{I}_{\rho_0}$.
   We must show that $\mathbb{I}_{\rho_0}$ is closed under finite unions. Suppose

$$\sigma_0, \ldots, \sigma_n \in \mathbb{I}_{\rho_0}$$

and let

$$\sigma = \sigma_0 \cup \cdots \cup \sigma_n.$$

Suppose that $\mathcal{C} \subset \mathcal{P}(\omega)$ is finite. Thus there exist pairwise disjoint finite sets $\mathcal{A}_0, \ldots, \mathcal{A}_n \subset \mathcal{P}(\omega) \backslash \mathcal{C}$ such that for each $i \leq n$,

$$\{\sigma_X | X \in \mathcal{A}_i\}$$

is not independent on σ_i. But $\sigma = \sigma_0 \cup \cdots \cup \sigma_n$, and so it follows that

$$\{\sigma_X | X \in \mathcal{P}(\omega) \backslash \mathcal{C}\}$$

is not independent on σ.
   Thus for each cofinite set $\mathcal{A} \subset \mathcal{P}(\omega)$, the set, $\{\sigma_X | X \in \mathcal{A}\}$, is not independent on σ and so $\sigma \in \mathbb{I}_{\rho_0}$.
   To show that $\mathbb{I}_{\rho_0}$ is Δ_1^1 it suffices to note that if (M, E) is an ω-model of ZFC\Powerset then

$$(\mathbb{I}_{\rho_0})^{(M,E)} = \mathbb{I}_{\rho_0} \cap M.$$

Clearly,

$$(\mathbb{I}_{\rho_0})^{(M,E)} \subseteq \mathbb{I}_{\rho_0},$$

and so it suffices to show that

$$\mathbb{I}_{\rho_0} \cap M \subseteq (\mathbb{I}_{\rho_0})^{(M,E)}.$$

This follows by compactness from the definition of $\mathbb{I}_{\rho_0}$. To see this suppose that $\sigma \in \mathbb{I}_{\rho_0} \cap M$. Suppose $\mathcal{A} \subset \mathcal{P}(\omega) \cap M$ and $\mathcal{A}$ is finite. Since $\sigma \in \mathbb{I}_{\rho_0}$, there exists a finite set $\mathcal{B} \subset \mathcal{P}(\omega) \backslash \mathcal{A}$ such that $\{\sigma_X | X \in \mathcal{B}\}$ is not independent on σ. Let $\mathcal{B} = \mathcal{B}_0 \cup \mathcal{B}_1$ be a decomposition into disjoint sets such that

$$(\cap \{\sigma_X | X \in \mathcal{B}_0\}) \cap (\cap \{\omega \backslash \sigma_X | X \in \mathcal{B}_1\}) \cap \sigma$$

is finite. Let $k < \omega$ be sufficiently large such that

(1) for all $X, Y \in \mathcal{A} \cup \mathcal{B}$, if $X \neq Y$ then $X \cap k \neq Y \cap k$,
(2) $(\cap \{\sigma_X | X \in \mathcal{B}_0\}) \cap (\cap \{\omega \backslash \sigma_X | X \in \mathcal{B}_1\}) \cap \sigma \subset k$.

By compactness there must exist finite disjoint sets $\mathcal{C}_0, \mathcal{C}_1 \subset \mathcal{P}(\omega) \cap M \backslash \mathcal{A}$ such that

(1) $|\mathcal{B}_0| = |\mathcal{C}_0|$ and $\{X \cap k | X \in \mathcal{C}_0\} = \{X \cap k | X \in \mathcal{B}_0\}$,
(2) $|\mathcal{B}_1| = |\mathcal{C}_1|$ and $\{X \cap k | X \in \mathcal{C}_1\} = \{X \cap k | X \in \mathcal{B}_1\}$,
(3) $(\cap \{\sigma_X | X \in \mathcal{C}_0\}) \cap (\cap \{\omega \backslash \sigma_X | X \in \mathcal{C}_1\}) \cap \sigma \subset k$.

Thus there exists an infinite sequence, $\langle \mathcal{A}_i : i < \omega \rangle$, of pairwise disjoint finite subsets of $M \cap \mathcal{P}(\omega)$ such that for each $i < \omega$, $\{\sigma_X | X \in \mathcal{A}_i\}$ is not independent on σ. Therefore $\sigma \in (\mathbb{I}_{\rho_0})^{(M,E)}$. $\qquad\square$

Suppose U is a nonprincipal ultrafilter on ω. Associated to U is a forcing notion, $\mathbb{P}_U$, which is analogous to Prikry forcing.

Definition 2.4. Suppose that U is an ultrafilter on ω. $\mathbb{P}_U$ denotes the partial order defined as follows.

(1) $\mathbb{P}_U$ is the set of all pairs (s, h) such that $s \in [\omega]^{<\omega}$ and

$$h : [\omega]^{<\omega} \to U.$$

(2) The order on $\mathbb{P}_U$ is defined as follows.

$$(s_1, h_1) \leq (s_0, h_0)$$

if

(a) $s_0 = s_1 \cap k$ where $k = \max(s_0) + 1$,
(b) for all $i \in s_1 \backslash s_0$, $i \in h_0(s_1 \cap i)$,
(c) for all $s \in [\omega]^{<\omega}$ if $s_1 = s \cap k$ where $k = \max(s_1) + 1$ then $h_1(s) \subseteq h_0(s)$.

A generic filter $g \subset \mathbb{P}_U$ can be identified with the subset of ω,

$$\cup \{s | (s, h) \in g \text{ for some } h\},$$

it defines.

A good reference for these kinds of generalizations of Prikry forcing is [1] to which we also refer the reader for historical remarks.

Lemma 2.5. *Suppose that U is a nonprincipal ultrafilter on ω, $(s, h) \in \mathbb{P}_U$, and $b \in \mathrm{RO}(\mathbb{P}_U)$. Then there exists $h^* : [\omega]^{<\omega} \to U$ such that either,*

$$(s, h^*) \Vdash \text{``}b \in G\text{''};$$

or, $(s, h^*) \Vdash \text{``}b \notin G\text{''}.$

Proof.

Define sets $R_\alpha \subseteq [\omega]^{<\omega}$ by induction on α.

Let R_0 be the set of all $a \in [\omega]^{<\omega}$ such that

(1) $s \subseteq a$ and $s = a \cap (\max(s) + 1)$,
(2) there exists $h^* : [\omega]^{<\omega} \to U$ such that

$$(a, h^*) \Vdash \text{``}b \in G\text{''}.$$

If α is a limit ordinal then

$$R_\alpha = \cup \{R_\beta | \beta < \alpha\}$$

and for all α,

$$R_{\alpha+1} = \{a \in [\omega]^\omega \,|\, \{k | a \cup \{k\} \in R_\alpha\} \in U\}.$$

Define for each $a \in [\omega]^\omega$,

(1) $\rho_b(a) = \min \{\alpha | a \in R_{\alpha+1}\}$ if $a \in \cup \{R_\alpha | \alpha \in \mathrm{Ord}\}$,
(2) $\rho_b(a) = \infty$.

Define

$$h^* : [\omega]^{<\omega} \to U$$

as follows. Suppose $a \in [\omega]^{<\omega}$.

(1) if $\rho_b(a) = \infty$ then $h^*(a) = \{k | a \subset k$ and $\rho_b(a \cup \{k\}) = \infty\}$.
(2) if $\rho_b(a) = 0$ then $h^*(a) = \{k | a \subset k$ and $\rho_b(a \cup \{k\}) = 0\}$.
(3) if $0 < \rho_b(a) < \infty$ then $h^*(a) = \{k | a \subset k$ and $\rho_b(a \cup \{k\}) < \rho_b(a)\}$.

There are two cases. First suppose that $\rho_b(s) \neq \infty$. Then we claim that

$$(s, h^*) \Vdash \text{``}b \in G\text{''}.$$

Suppose toward a contradiction that there is a condition $(s', h') \leq (s, h^*)$ such that

$$(s', h') \Vdash \text{``}b \notin G\text{''}.$$

Since $(s', h') \leq (s, h^*)$, for all $i \in s' \backslash s$, $i \in h^*(s' \cap i)$, and so either $\rho_b(s) = 0$ or $\rho_b(s') < \rho_b(s)$. In either case there exists $(s'', h'') \leq (s', h')$ such that $\rho_b(s'') = 0$. But then there exists $f : [\omega]^{<\omega} \to U$ such that

$$(s'', f) \Vdash \text{``} b \in G\text{''}.$$

But $(s'', h') \Vdash \text{``} b \notin G\text{''}$ since $(s'', h'') \leq (s', h')$. This is a contradiction and so

$$(s, h^*) \Vdash \text{``} b \in G\text{''}.$$

The second case is that $\rho_b(s) = \infty$. We claim that

$$(s, h^*) \Vdash \text{``} b \notin G\text{''}.$$

If not then there exists $(s', h') \leq (s, h^*)$ such that

$$(s', h') \Vdash \text{``} b \in G\text{''}.$$

But then $\rho_b(s') = 0$. Since $(s', h') \leq (s, h^*)$, for all $i \in s' \backslash s$, $i \in h^*(s' \cap i)$. Therefore it follows easily by induction and the definition of h^*, that $\rho_b(s') = \infty$. This is again a contradiction and so

$$(s, h^*) \Vdash \text{``} b \notin G\text{''}$$

as claimed. $\square$

We fix some notation. Suppose that U is a nonprincipal ultrafilter on ω and that $s \in [\omega]^{<\omega}$. Let $\mathbb{P}_U | s$ denote the set of all $(t, h) \in \mathbb{P}_U$ such that $s \subseteq t$ and $s = t \cap (\max(s) + 1)$.

Suppose that $O \subseteq \mathbb{P}_U$ is open in $\mathbb{P}_U$. Define (as in the proof of Lemma 2.5) sets R_α by induction of α.

Let R_0 be the set of all $a \in [\omega]^{<\omega}$ such that

(1) $s \subseteq a$ and $s = a \cap (\max(s) + 1)$,
(2) there exists $h^* : [\omega]^{<\omega} \to U$ and $q \in O$ such that

$$(s, h^*) \leq q.$$

If α is a limit ordinal then

$$R_\alpha = \cup \{R_\beta | \beta < \alpha\}$$

and for all α,

$$R_{\alpha+1} = \{a \in [\omega]^\omega | \{k | a \cup \{k\} \in R_\alpha\} \in U\}.$$

Suppose there exists an ordinal α such that $s \in R_{\alpha+1}$. Let $\|O, \mathbb{P}_U|s\|$ denote the least such ordinal. Otherwise $\|O, \mathbb{P}_U|s\| = \infty$.

Clearly,

$$\|O, \mathbb{P}_U|s\| = \|O \cap \mathbb{P}_U|s, \mathbb{P}_U|s\|$$

and $\|O, \mathbb{P}_U|s\| < \infty$ if and only if there exists

$$h : [\omega]^{<\omega} \to U$$

such that

$$O \cap \{q \in \mathbb{P}_U | q \leq (s, h)\}$$

is dense below (s, h).

The following lemma records some elementary properties of $\|O, \mathbb{P}_U|s\|$ and the proof of the lemma is immediate from the definition.

Lemma 2.6. *Suppose that O is open in $\mathbb{P}_U$. For each $a \in [\omega]^{<\omega}$ let*

$$\rho(a) = \|O, \mathbb{P}_U|a\|.$$

Then for all $a \in [\omega]^{<\omega}$,

(1) $\rho(a) = \infty$ if and only if $\{k \in \omega | \rho(a \cup \{k\}) = \infty\} \in U$,
(2) if $0 < \alpha < \rho(a) < \infty$ then $\{k \in \omega | \alpha \leq \rho(a \cup \{k\}) < \rho(a)\} \in U$,
(3) if $\rho(a) = 0$ then $\{k \in \omega | \rho(a \cup \{k\}) = 0\} \in U$.

Using Lemma 2.6, Lemma 2.5 can be reformulated as follows.

Lemma 2.7. *Suppose that U is a nonprincipal ultrafilter on ω and that $s \in [\omega]^{<\omega}$.*

Suppose $O \subseteq \mathbb{P}_U$ is open and O^* is a maximal open subset of $\mathbb{P}_U$ such that $O \cap O^* = \emptyset$.

Then;

(1) $\|O, \mathbb{P}_U|s\| < \infty$ or $\|O^*, \mathbb{P}_U|s\| < \infty$;
(2) $\|O, \mathbb{P}_U|s\| = \infty$ or $\|O^*, \mathbb{P}_U|s\| = \infty$.

Lemma 2.8. *Suppose that U is a nonprincipal ultrafilter on ω and that $s \in [\omega]^{<\omega}$.*

For each countable ordinal $\alpha < \omega_1$ there exists a set

$$O \subset \mathbb{P}_U|s$$

such that

(1) O is open in $\mathbb{P}_U|s$,

(2) $\left\|O, \mathbb{P}_U|s\right\| = \alpha$.

Proof. We prove by induction on α, that the statement of the lemma holds at α for all $t \in [\omega]^{<\omega}$.

If $\alpha = 0$ and $t \in [\omega]^{<\omega}$ then let $O = \mathbb{P}_U|t$. Clearly $\left\|O, \mathbb{P}_U|t\right\| = 0$.

Now suppose $\alpha > 0$ and that for all $\beta < \alpha$, for all $t \in [\omega]^{<\omega}$, there exists a set $O_t^\beta \subseteq \mathbb{P}_U|t$ such that O_t^β is open in $\mathbb{P}_U|t$ and such that

$$\left\|O_t^\beta, \mathbb{P}_U|t\right\| = \beta.$$

Fix $s \in [\omega]^{<\omega}$ and let $\langle \beta_i : i < \omega \rangle$ be such that

$$\alpha = \sup\left\{\beta_i | i < \omega\right\}.$$

Let $\langle n_i : i < \omega \rangle$ be an strictly increasing sequence such that $s \subset n_0$ and such that

$$\{n_i | i < \omega\} \in U$$

and such that $\{n_i | i < \omega\} \subset h(s)$. For each $i < \omega$ let $s_i = s \cup \{n_i\}$. Let

$$O = \cup\left\{O_{s_i}^{\beta_i} | i < \omega\right\}.$$

For each $t \in [\omega]^{<\omega}$ such that and for each $\xi < \omega_1$ let $(R_\xi)^t$ denote R_ξ defined using t and $O \cap \mathbb{P}_U|t$.

Note that for all $i_1 < i_2 < i$,

$$s_{i_1} \cap (\max(s) + 1) = s = s_{i_2} \cap (\max(s) + 1)$$

and $|s_{i_1}| = |s_{i_2}| = n + 1$. Therefore for each $\xi < \omega_1$, for each $i < \omega$,

$$(R_\xi)^s \cap \mathbb{P}_U|s_i = (R_\xi)^{s_i}.$$

Thus for each $i < \omega$, $s_i \in (R_{\beta_i+1})^s$ and moreover β_i is the least such ordinal. But

$$\alpha = \sup\left\{\beta_i | i < \omega\right\}$$

and $\{n_i | i < \omega\} = \{k|s \cup \{k\} \in \{s_i | i < \omega\}\}$.

Therefore $s \in (R_{\alpha+1})^s$ and $s \notin (R_{\beta+1})^s$ for any $\beta < \alpha$. $\square$

We shall (in this section) for the most part be primarily interested in forcing over models (M, E), where (M, E) is an ω-model such that

$$(M, E) \vDash \mathrm{ZC} + \text{``}V = L\text{''}.$$

For such models, (M, E), we identify the standard part of (M, E) with its corresponding transitive collapse and so we regard

$$(V_{\omega+\omega})^{(M,E)} \subset V_{\omega+\omega}.$$

Thus for example if $U \in M$ and

$$(M, E) \vDash \text{``}U \text{ is a nonprincipal ultrafilter on } \omega\text{''},$$

then $U \subset \mathcal{P}(\omega) \cap M$ etc.

We fix some notation. Suppose (M, E) is an ω-model,

$$(M, E) \vDash ZC + \text{``}V = L\text{''} ,$$

$U \in M$ and

$$(M, E) \vDash \text{``}U \text{ is a nonprincipal ultrafilter on } \omega\text{''}.$$

Suppose that $g \subset \omega$ and let $\mathcal{F}_g \subset (\mathbb{P}_U)^{(M,E)}$ be the set of all $(s, h) \in (\mathbb{P}_U)^{(M,E)}$ such that

(1) $s = g \cap i_s$,
(2) for all $i \in g \backslash i_s$, $i \in h(g \cap i)$,

where $i_s = \max(s) + 1$.

We say g is (M, E)-generic for $(\mathbb{P}_U)^{(M,E)}$ if $\mathcal{F}_g$ is (M, E)-generic. Clearly if $\mathcal{F}_g$ is (M, E)-generic then for all functions

$$h : [\omega]^{<\omega} \to U$$

with $h \in M$, there exists $k < \omega$ such that for all $i \in g \backslash k$, $i \in h(g \cap i)$.

Lemma 2.9. *Suppose that*

$$(M, \in) \vDash ZC + \text{``}V = L\text{''} ,$$

$U \in M$ *and*

$$(M, \in) \vDash \text{``}U \text{ is a nonprincipal ultrafilter on } \omega\text{''}.$$

Suppose $g \subset \omega$. Then the following are equivalent.

1) g is $(M, \in)$-generic for $(\mathbb{P}_U)^{(M,\in)}$.
2) for all functions

$$h : [\omega]^{<\omega} \to U$$

with $h \in M$, there exists $k < \omega$ such that for all $i \in g \backslash k$, $i \in h(g \cap i)$.

Proof. Clearly (1) implies (2).

Suppose that $g \subset \omega$ and (2) holds for g. Let $D \subset (\mathbb{P}_U)^M$ be open-dense with $D \in M$. We must prove that $\mathcal{F}_g \cap D \neq \emptyset$.

Since D is open-dense, $\left\| D, \mathbb{P}_U \right\|^M < \infty$. Define $\rho : [\omega]^{<\omega} \to \mathrm{Ord}$ by

$$\rho(a) = \left\| D, \mathbb{P}_U | a \right\|^M$$

for each $a \in [\omega]^{<\omega}$ and let

$$A = \left\{ a \in [\omega]^{<\omega} \Big\| D, \mathbb{P}_U | a \big\|^M = 0 \right\}.$$

Let $\pi \in M$ be a function such that for all $a \in A$, $(a, \pi(a)) \in D$.

Define

$$h : [\omega]^{<\omega} \to U$$

by

$$h(a) = \{ k \in \omega \,|\, \rho(a \cup \{k\}) < \rho(a) \}$$

if $\rho(a) > 0$. If $\rho(a) = 0$ then

$$h(a) = \cap \left\{ (\pi(b))(a) \upharpoonright \text{ for some } i < \omega, \, b = a \cap i \text{ and } b \in A \right\}.$$

Clearly $h \in M$ and by Lemma 2.6, for each $a \in [\omega]^{<\omega}$, $h(a) \in U$.

Fix $k < \omega$ such that $k < \omega$ such that for all $i \in g \backslash k$, $i \in h(g \cap i)$. For all $k_1 < k_2$ if $k < k_1$ then either $\rho(g \cap k_2) = 0$ or $\rho(g \cap k_2) < \rho(g \cap k_1)$. Therefore there must exist $i < \omega$ such that $\rho(g \cap i) = 0$. But then

$$(g \cap i, \pi(g \cap i)) \in \mathcal{F}_g \cap D$$

and in particular $\mathcal{F}_g \cap D \neq \emptyset$. $\qquad\square$

Definition 2.10. Suppose (M, E) is an ω-model,

$$(M, E) \vDash \mathrm{ZC} + \text{``}V = L\text{''} ,$$

$U \in M$ and

$$(M, E) \vDash \text{``}U \text{ is a nonprincipal ultrafilter on } \omega\text{''}.$$

A set $g \subset \omega$ is (M, E)-*weakly generic* for $(\mathbb{P}_U)^{(M,E)}$ if for all functions

$$h : [\omega]^{<\omega} \to U$$

with $h \in M$, there exists $k < \omega$ such that for all $i \in g \backslash k$, $i \in h(g \cap i)$.

Note that if $g \subset \omega$ is (M, E)-*weakly generic* for $(\mathbb{P}_U)^{(M,E)}$ then $\mathcal{F}_g$ is a filter on $(\mathbb{P}_U)^{(M,E)}$.

Lemma 2.9 is in general false for ω-models, (M, E). But the proof of Lemma 2.9 shows that the following version does hold.

Lemma 2.11. *Suppose* (M, E) *is an* ω-*model,*

$$(M, E) \vDash \mathrm{ZC} + \text{``}V = L\text{''},$$

$U \in M$ *and*

$$(M, E) \vDash \text{``}U \text{ is a nonprincipal ultrafilter on } \omega\text{''}.$$

Suppose that $g \subset \omega$ *is* (M, E)-*weakly generic for* $(\mathbb{P}_U)^{(M,E)}$, $D \in M$ *is open-dense in* $(\mathbb{P}_U)^{(M,E)}$ *and that* $\|D, \mathbb{P}_U\|^{(M,E)}$ *is in the standard part of* (M, E).
Then $\mathcal{F}_g \cap D \neq \emptyset$.

The next lemma shows how Σ^0_0 and Σ^0_1 assertions about g (with parameters from M) behave relative to weakly generic filters.

Lemma 2.12. *Suppose* (M, E) *is an* ω-*model,*

$$(M, E) \vDash \mathrm{ZC} + \text{``}V = L\text{''},$$

$U \in M$ *and*

$$(M, E) \vDash \text{``}U \text{ is a nonprincipal ultrafilter on } \omega\text{''}.$$

Suppose that $W \in M \cap \mathcal{P}(\omega)$ *and that* $g \subset \omega$ *is* (M, E)-*weakly generic for* $(\mathbb{P}_U)^{(M,E)}$.
Suppose that $\psi(x, y, z)$ *is a* Σ^0_0-*formula and let*

$$O = \Big\{(s, h) \in (\mathbb{P}_U)^{(M,E)} \,\big|\, \text{For some } i < \max(s) \; \psi[i, s, W \cap (\max(s)+1)] \text{ holds}\Big\}.$$

Then the following are equivalent.

(1) $(\exists x \psi)[g, W]$ *holds.*
(2) $\mathcal{F}_g \cap O \neq \emptyset$.
(3) For some $m < \omega$, $\|O, \mathbb{P}_U|g_m\|^{(M,E)} = 0$.

Proof. This is immediate from the definitions. $\qquad\qquad\qquad\qquad \square$

Lemma 2.13. *Suppose* (M, E) *is an* ω-*model,*

$$(M, E) \vDash \mathrm{ZC} + \text{``}V = L\text{''},$$

$U \in M$ *and*

$$(M, E) \vDash \text{``}U \text{ is a nonprincipal ultrafilter on } \omega\text{''}.$$

Then for each open set $O \subset (\mathbb{P}_U)^{(M,E)}$ *such that*

$$\big\| O, \mathbb{P}_U \big\|^{(M,E)} \notin M \cap \mathrm{Ord}$$

there exists $g \subset \omega$ *such that*

1) g *is* (M, E)*-weakly generic for* $(\mathbb{P}_U)^{(M,E)}$,
2) $\mathcal{F}_g \cap O = \emptyset$.

Proof. Let

$$B = \left\{ a \in [\omega]^{<\omega} \,\big\|\, \big\| O, \mathbb{P}_U | a \big\|^{(M,E)} \notin M \cap \mathrm{Ord} \right\}.$$

By Lemma 2.6, for each $a \in [\omega]^{<\omega}$ the following are equivalent.

(1) $a \in B$.
(2) $\{ k < \omega \,|\, a \cup \{k\} \in B \} \in U$.

The point is that if $a \in B$ and

$$\big\| O, \mathbb{P}_U | a \big\|^{(M,E)} \neq \infty$$

then $\big\| O, \mathbb{P}_U | a \big\|^{(M,E)}$ is in the nonstandard part of (M, E) and so there exists $\alpha \in (\mathrm{Ord})^{(M,E)} \backslash \mathrm{Ord}$ such that

$$(M, E) \vDash \text{``}\alpha < \big\| O, \mathbb{P}_U | a \big\| \text{''}.$$

Since $\big\| O, \mathbb{P}_U | a \big\|^M \notin M \cap \mathrm{Ord}$, $\emptyset \in B$.
Let $\langle h_k : k < \omega \rangle$ enumerate all the functions

$$h : [\omega]^\omega \to U$$

such that $h \in M$. Let $g \subset \omega$ be such that for all $i \in g$,

(1) $i \in h_i(g \cap i)$,
(2) $g \cap i \in B$.

Clearly g is (M, E)-weakly generic for $(\mathbb{P}_U)^{(M,E)}$ and $\mathcal{F}_g \cap O = \emptyset$. $\square$

3. The main theorem

We shall need the following lemma which is relatively standard. This lemma is also a special case of Lemma 4.12 which we prove in the next section.

Lemma 3.1. *Suppose* $X \subset \omega$ *and* $X \notin \mathrm{HYP}$. *Then there exists a set* $Z \subsetneq \omega$ *such that:*

(1) $X \notin \mathrm{HYP}_Z$;
(2) $X \leq_{\mathrm{T}} \mathcal{O}_Z$.

The following lemma illustrates the key property of ρ_0 and the associated ideal $\mathbb{I}_{\rho_0}$, see Lemma 2.1 and Definition 2.2 for the relevant definitions.

Lemma 3.2. *Suppose* (M, E) *is an* ω-*model,*

$$(M, E) \vDash \mathrm{ZC},$$

$U \in M$ *and*

$$(M, E) \vDash \text{``}U \text{ is a nonprincipal ultrafilter on } \omega \text{ and } U \cap \mathbb{I}_{\rho_0} = \emptyset\text{''}.$$

Suppose $X \subset \omega$ *and* $X \notin M$. *Then for each set* $A \in U$, *both* $A \cap \sigma_X$ *and* $A \backslash \sigma_X$ *are infinite.*

Proof. Suppose toward a contradiction that for some $n < \omega$, either $A \cap \sigma_X \subset n$ or $A \backslash \sigma_X \subset n$.

By Lemma 2.3,

$$(\mathbb{I}_{\rho_0})^{(M,E)} = M \cap \mathbb{I}_{\rho_0}.$$

Therefore $U \cap \mathbb{I}_{\rho_0} = \emptyset$ and in particular $A \notin \mathbb{I}_{\rho_0}$. But by the definition of $\mathbb{I}_{\rho_0}$ this implies that for some cofinite set $\mathcal{B} \subset \mathcal{P}(\omega)$, the set,

$$\{\sigma_Y | Y \in \mathcal{B}\}$$

is independent on A. Let

$$\mathcal{C} = \{Y \subset \omega | A \cap \sigma_Y \subset n \text{ or } A \backslash \sigma_Y \subset n\}.$$

Then $\mathcal{C}$ is finite and $X \in \mathcal{C}$. But clearly $\mathcal{C}$ is $\Pi_1^0(A)$ and so $\mathcal{C} \subset \mathrm{HYP}_A \subset M$. This implies $X \in M$ which is a contradiction. $\square$

We next prove a technical strengthening of Lemma 2.13, this lemma is really the key to the proof of the main theorem.

Lemma 3.3. *Suppose* (M, E) *is a countable nonstandard* ω-*model,*

$$(M, E) \vDash \mathrm{ZC} + \text{``}V = L\text{''},$$

$U \in M$ *and*

$(M, E) \vDash$ *"U is a nonprincipal ultrafilter on ω and $U \cap \mathbb{I}_{\rho_0} = \emptyset$".*

Suppose $\langle O_n : n < \omega \rangle$ is a sequence of open subsets of $(\mathbb{P}_U)^{(M,E)}$ such that $O_n \in M$ for each $n < \omega$.

Then for each set $X \subset \omega$ such that $X \notin M$, and for each infinite set $G \subset \omega$, there exists $g \subset \omega$ such that

(1) g is (M, E)-weakly generic for $(\mathbb{P}_U)^{(M,E)}$,
(2) for all $i < \omega$, $|g \cap i| < |G \cap i|$,
(3) for each $n < \omega$, $O_n \cap \mathcal{F}_g \neq \emptyset$ if and only if $\left\| O_n, \mathbb{P}_U|g_n \right\|^{(M,E)} \in M \cap \mathrm{Ord}$,
(4) for all $n < \omega$, $\min(g \backslash g_n) \in \sigma_X$ if and only if $\mathcal{F}_g \cap O_n \neq \emptyset$,

where for each $n < \omega$, $g_n = \{i \in g \,|\, |g \cap i| < n\}$.

Proof.

Let $\langle h_k : k < \omega \rangle$ enumerate all functions

$$h : [\omega]^{<\omega} \to U$$

such that $h \in M$.

We define the initial segments, $g_n = \{i \in g \,|\, |g \cap i| < n\}$, by induction on n such that for all $i < n$ if

$$\left\| O_i, \mathbb{P}_U|g_i \right\|^{(M,E)} \notin \mathrm{Ord}$$

then $\left\| O_i, \mathbb{P}_U|g_n \right\|^{(M,E)} \notin \mathrm{Ord}$.

Suppose that g_n is defined. Choose $\alpha_n \in (\mathrm{Ord})^M \backslash \mathrm{Ord}$ such that for each $i \leq n$, if

$$\left\| O_i, \mathbb{P}_U|g_i \right\|^{(M,E)} \notin \mathrm{Ord}$$

then

$$(M, E) \vDash \text{"} \left\| O_i, \mathbb{P}_U|g_n \right\|^{(M,E)} > \alpha_n \text{"}.$$

Let $A \subset \omega$ be the set of all k such that for all $i \leq n$,

(1) $k \in h_i(g_n)$,
(2) if $\left\| O_i, \mathbb{P}_U|g_n \right\|^{(M,E)} \notin \mathrm{Ord}$ then

$$(M, E) \vDash \text{"} \left\| O_i, \mathbb{P}_U|(g_n \cup \{k\}) \right\|^{(M,E)} > \alpha_n \text{"}.$$

Then A is finite intersection of sets in U and so $A \in U$. Therefore by Lemma 3.2, $A \cap \sigma_X$ and $A \backslash \sigma_X$ are each infinite. Choose $k \in A$ such that

(1) $g_n \subset k$,
(2) $|G \cap k| > n + 1$,
(3) if $\left\| O_i, \mathbb{P}_U | g_n \right\|^{(M,E)} \notin \mathrm{Ord}$ then $k \notin \sigma_X$,
(4) if $\left\| O_i, \mathbb{P}_U | g_n \right\|^{(M,E)} \in \mathrm{Ord}$ then $k \in \sigma_X$.

Define $g_{n+1} = g_n \cup \{k\}$. Clearly g_{n+1} satisfies the inductive requirements. This defines the initial segments of g.   □

Theorem 3.4. *Suppose $X \subset \omega$ and $X \notin \mathrm{HYP}$. Then there exists a set $Y \subseteq \omega$ such that*

$$Y' =_{\mathrm{tt}_Y} X \vee Y.$$

Proof.

Let $e : \omega \to V_\omega$ be the bijection such that for all $i < \omega$,

$$e(i) = \{e(j) | j < i \text{ and } b_i(j) = 1\}$$

where for each $i < \omega$, $b_i : i + 1 \to \{0, 1\}$ and

$$i = \sum_{k=0}^{i} b_i(k) 2^k.$$

Notice that for all (infinite) $g, W \subset \omega$, for all Σ_0^0 formulas, $\phi(x, y)$, the following are equivalent:

(1) $(\exists x \phi)[Y]$ holds;
(2) For some $n < \omega$, there exists $k < g_n$ such that

$$\phi[k, \{i | e(i) \in g_n \times W\} \cap 2^{\max(g_n)}]$$

holds where $g_n = \{i \in g | |g \cap i| < n\}$ is the initial segment of g of length n;

where $Y = \{i < \omega | e(i) \in g \times W\}$.

We fix some notation. For each $t \in \mathcal{P}(\omega)$ and for each $k \in \omega$,

(1) if $k \in \mathcal{O}_t$ then $\left\| k, \mathcal{O}_t \right\| = \alpha$ where α is the associated countable ordinal,
(2) otherwise $\left\| k, \mathcal{O}_t \right\| = \infty$.

Fix a set $X \subset \omega$ such that $X \notin \mathrm{HYP}$. Choose an ω-model, (M, E), and $Z \subset \omega$ such that

(1) $(M, E) \vDash \mathrm{ZFC} + \text{``}V = L\text{''}$;
(2) $Z \in M$ and $X \notin M$;
(3) $X \leq_{\mathrm{T}} \mathcal{O}_Z$.

Fix a set $G \subset \omega$ such that

$$X \leq_{\mathrm{g}} \mathcal{O}_Z$$

for all infinite sets $g \subset \omega$ such that

$$|g \cap i| < |G \cap i|$$

for all $i < \omega$.

Let $U \in M$ satisfy

(1) $(M, E) \vDash \text{``}U$ is an ultrafilter on $\omega\text{''}$;
(2) $(M, E) \vDash \text{``}U \cap \mathbb{I}_{\rho_0} = \emptyset\text{''}$.

Let $\pi \in M$ be such that in (M, E) the following hold.

(1) π is a function with domain $[\omega]^{<\omega}$.
(2) For each $a \in [\omega]^{<\omega}$, $\pi(a) \subset \mathbb{P}_U | a$ and $\pi(a)$ is open in $\mathbb{P}_U | a$.
(3) Suppose $a \in [\omega]^n$ and $e(n) \in \omega$. Then $\big\| \pi(a), \mathbb{P}_U | a \big\|^{(M,E)} = \big\| e(n), \mathcal{O}_Z \big\|^{(M,E)}$.

By Lemma 2.8, π exists. Let

$$Y_\pi = \{(a, b) | (b, h) \in \pi(a) \text{ for some } h \in M\}.$$

Let $W = \{i | e(i) \in Z \times Y_\pi\}$. We define a sequence $\langle O_n : n < \omega \rangle$ of open subsets of $(\mathbb{P}_U)^{(M,E)}$ as follows.

(1) Suppose $e(n)$ is a Σ_0^0 formula, $\phi(x, y)$. Then O_n is the set of all $(s, h) \in (\mathbb{P}_U)^{(M,E)}$ such that for some $k < \max(s)$,

$$\phi[k, \{i | e(i) \in s \times W\} \cap 2^{\max(s)}].$$

(2) Suppose $e(n) < \omega$. Then O_n is the set of all $(a, h) \in (\mathbb{P}_U)^{(M,E)}$ such that $|a| > n$ and $(a, h) \in \pi(a_n)$ where $a_n = \{i \in a | |a \cap i| < n\}$.

Let $g \subset \omega$ be such that

(1) g is (M, E)-weakly generic for $(\mathbb{P}_U)^{(M,E)}$,
(2) for all $i < \omega$, $|g \cap i| < |G \cap i|$,

(3) for each $n < \omega$, $O_n \cap \mathcal{F}_g \neq \emptyset$ if and only if $\|O_n, \mathbb{P}_U|g_n\|^{(M,E)} \in M \cap \mathrm{Ord}$,
(4) for each $n < \omega$, $\min(g \backslash g_n) \in \sigma_X$ if and only if $\mathcal{F}_g \cap O_n \neq \emptyset$,

where for each $n < \omega$, $g_n = \{i \in g \mid \|g \cap i\| < n\}$.

By Lemma 3.3, g exists. Let

$$ Y = \{i < \omega \mid e(i) \in g \times W\}. $$

We show that Y witnesses the theorem holds for X.

We first show that $Y' \leq_{\mathrm{tt}_Y} X \vee Y$.

Suppose $\phi(x, y)$ is a Σ_0^0 formula and $e(n) = \exists x \phi(x, y)$. By Lemma 2.12, the definition of O_n, and since g satisfies 1-4, the following are equivalent.

(1) $(\exists x \phi)[Y]$ holds.
(2) $\mathcal{F}_g \cap O_n \neq \emptyset$.
(3) $\min(g \backslash g_n) \in \sigma_X$.

Therefore $Y' \leq_{\mathrm{tt}_Y} X \vee Y$.

We finish by showing that $X \leq_{\mathrm{tt}_Y} Y'$. For this it suffices to show that $\mathcal{O}_Z \leq_{\mathrm{tt}_Y} Y'$ since by the choice of (Z, G) and 2,

$$ X \leq_{\mathrm{tt}_g} \mathcal{O}_Z, $$

which in turn implies by the definition of Y, that $X \leq_{\mathrm{tt}_Y} \mathcal{O}_Z$.

For each $k < \omega$ the following are equivalent where

$$ g_n = \{i \in g \mid g \cap i < n\} $$

and $n = e^{-1}(k)$.

(1) $k \in \mathcal{O}_Z$.
(2) $\|O_n, \mathbb{P}_U|g_n\|^{(M,E)} \in M \cap \mathrm{Ord}$.
(3) $\mathcal{F}_g \cap O_n \neq \emptyset$.
(4) $\{b \mid (g_n, b) \in Y_\pi\} \cap \{g \cap i \mid i < \omega\} \neq \emptyset$.

Therefore $\mathcal{O}_Z \leq_{\mathrm{tt}_Y} Y'$ since $e^{-1}[Y_\pi] \leq_T Y$ (in fact we have $e^{-1}[Y_\pi] \leq_{\mathrm{tt}_0} Y$). $\qquad \square$

4. Further results

Let ω_1^* be the least ordinal which is admissible and a limit of admissible ordinals and let

$$\mathrm{HYP}^* = \mathcal{P}(\omega) \cap L_{\omega_1^*}.$$

For each set $X \subseteq \omega$ let $\mathcal{O}_X$ be the hyperjump of X; this is the Σ_1-theory of the structure,

$$\langle L_\beta[X], \{X\} \rangle$$

where β is the least ordinal α such that $L_\alpha[X]$ is admissible.

Theorem 4.1. *Suppose $X \subset \omega$ and $X \notin \mathrm{HYP}^*$. Then there exists a set $Y \subseteq \omega$ such that $\mathcal{O}_Y =_{\mathrm{tt}_Y} X \vee Y$.*

Before proving this theorem we will prove the following local version. The statement of this local version requires some additional notation.

Suppose $X \subseteq \omega$ and $\alpha < \omega_1^*$. If $\alpha = \omega \cdot \gamma$ then $X^{(\alpha)}$ is the Σ_1-theory of the structure, $(L_\gamma[X], \{X\})$. Otherwise

$$X^{(\alpha)} = Y^{(k)}$$

where $Y = X^{(\gamma)}$ and $\alpha = \omega \cdot \gamma + k$. This defines $X^{(\alpha)}$, the α-jump of X, for all $\alpha < \omega_1^*$, in fact this definition defines the α-jump of X for all $\alpha < \alpha^*$ where α^* is least such that there exists $\beta < \alpha^*$ such that

$$L_\beta[X] \prec_{\Sigma_1} L_{\alpha^*}[X],$$

and it is easy to see that $\alpha^* > \omega_1^*$.

For each $\alpha < \omega_1^*$ and for each $X \subseteq \omega$ let $\mathcal{O}_X^{(\alpha)}$ denote the α-hyperjump of X. Let

$$\mathrm{HYP}_X^{(\alpha)} = \mathcal{P}(\omega) \cap L_{\gamma_\alpha^X}$$

where γ_α^X is the α-th admissible ordinal relative to X. Note that $\mathcal{O}_X^{(\alpha)}$ is the Σ_1-theory of the structure, $(L_{\gamma_\alpha^X}[X], \{X\})$, and so

$$\mathcal{O}_X^{(\alpha)} = X^{(\gamma_\alpha^X)}.$$

Theorem 4.2. *Suppose $\alpha < \omega_1^*$, $X \subset \omega$ and $X \notin \mathrm{HYP}^{(\alpha)}$. Then there exists a set $Y \subseteq \omega$ such that $Y^{(\alpha+1)} =_{\mathrm{tt}_Y} X \vee Y$.*

The proof of Theorem 4.2 requires a number of additional definitions and preliminary lemmas.

Definition 4.3. (ZC) Suppose $V = L$. Let $\mathbb{U}$ be the ultrafilter U on ω such that such that for all $\gamma < \omega_1$,

(1) if $\gamma < \omega_1^*$ and γ is admissible then $U \cap L_{\gamma+1}$ is the L-least ultrafilter on $L_{\gamma+1} \cap \mathcal{P}(\omega)$ such that

  (a) $U \cap L_\gamma \subset U \cap L_{\gamma+1}$,
  (b) $U \cap L_{\gamma+1} \cap \mathbb{I}_{\rho_0} = \emptyset$,
  (c) suppose there exists an ω-model, (M, E), such that

    (i) $(M, E) \in L_{\gamma+1}$ and $\mathcal{P}(\omega) \cap L_\gamma \subset M$,
    (ii) $(M, E) \vDash \mathrm{ZC} + \text{``}V = L\text{''}$,

    Then if (M, E) is the L-least such ω-model and if $(\mathbb{U})^{(M,E)}$ is ultra-filter disjoint from $\mathbb{I}_{\rho_0}$ and extending $U \cap L_\gamma$ then $U \cap M = (\mathbb{U})^{(M,E)}$.

(2) otherwise $U \cap L_{\gamma+1}$ is the L-least ultrafilter on $L_{\gamma+1} \cap \mathcal{P}(\omega)$ such that

  (a) $U \cap L_\gamma \subset U \cap L_{\gamma+1}$,
  (b) $U \cap L_{\gamma+1} \cap \mathbb{I}_{\rho_0} = \emptyset$.

Suppose that (M, E) is an ω-model and

$$(M, E) \vDash \mathrm{ZC} + \text{``}V = L\text{''}.$$

Then $(\mathbb{U})^{(M,E)}$ denotes $\mathbb{U}$ as defined in (M, E).

We show that $(\mathbb{U})^{(M,E)}$ is uniquely defined. For each Σ_1 formula, $\phi(x)$ let $\mathbb{U}_\phi$ be defined with $(\mathbb{U})^{(M,E)}$ in Definition 4.3(1c) replaced by.

$$\{Y \in M \cap \mathcal{P}(\omega) | (M, E) \vDash \phi[Y]\}.$$

Thus for each Σ_1-formula, $\phi(x)$, $(\mathbb{U}_\phi)^{(M,E)}$ is an ultrafilter on $\mathcal{P}(\omega) \cap M$ which is disjoint from $\mathbb{I}_{\rho_0}$.

By the Recursion Theorem there exists a Σ_1-formula ϕ^* such that

$$(\mathbb{U}_{\phi^*})^{(M,E)} = \{Y \in M \cap \mathcal{P}(\omega) | (M, E) \vDash \phi^*[Y]\}$$

for all ω-models

$$(M, E) \vDash \mathrm{ZC} + \text{``}V = L\text{''}.$$

Assume $\mathrm{ZC} + \text{``}V = L\text{''}$. Suppose $\gamma < \omega_1$ is admissible and let (M, E) be the L-least ω-model such that:

(1) $(M, E) \in L_{\gamma+1}$ and $\mathcal{P}(\omega) \cap L_\gamma \subset M$,
(2) $(M, E) \vDash \mathrm{ZC} + \text{``}V = L\text{''}$.

Then $L_\gamma \cap \mathbb{U}_{\phi^*} = L_\gamma \cap (\mathbb{U}_{\phi^*})^{(M,E)}$ and so $\mathbb{U}_{\phi^*} \cap M = (\mathbb{U}_{\phi^*})^{(M,E)}$.

We could simply define

$$(\mathbb{U})^{(M,E)} = (\mathbb{U}_{\phi^*})^{(M,E)}$$

for a specific choice of (ϕ, ϕ^*). However by the following lemma, $(\mathbb{U}_{\phi^*})^{(M,E)}$ does not depend on the choice of (ϕ, ϕ^*).

Lemma 4.4. *There is no sequence*

$$\langle M_i : i < \omega \rangle$$

of ω-models such that for all $i < \omega$,

(1) $M_i \vDash ZC + \text{``}V = L\text{''}$,
(2) $\langle M_k : i < k < \omega \rangle \in M_i$.

Proof. This is an immediate corollary of the main theorem of Steel [2] which states that for any sequence

$$\langle x_i : i < \omega \rangle$$

of subsets of ω there exists $i < \omega$ such that

$$(\langle x_k : i < k < \omega \rangle)' \not\leq_T x_i. \qquad \square$$

By the remarks above we have proved the following lemma.

Lemma 4.5. *Assume $V = L$, $\gamma < \omega_1^*$ and γ is admissible. Let (M, E) be the L-least ω-model such that,*

(i) $(M, E) \in L_{\gamma+1}$ and $\mathcal{P}(\omega) \cap L_\gamma \subset M$,
(ii) $(M, E) \vDash ZC + \text{``}V = L\text{''}$.

Then $\mathbb{U} \cap M = (\mathbb{U})^{(M,E)}$.

We prove the next two lemmas simultaneously by induction on β.

Lemma 4.6. *Suppose $\beta < \omega_1^*$,*

$$h_0 : [\omega]^{n_0} \to \mathbb{U}$$

$h_0 \in L_{\gamma_\beta^0}$ *and $n_0 < \omega$.*
  Then there exists

$$h : [\omega]^{<\omega} \to \mathbb{U}$$

and a function

$$f : [\omega]^{<\omega} \to \{0, 1\}$$

such that

(1) $(f, h) \in L_{\gamma_\beta^0}$,

(2) $h|[\omega]^{n_0} = h_0$,

(3) for all infinite sets $g \subset \omega$ *if* $i \in h(g \cap i)$ *for all* $i \in g \backslash g_{n_0}$ *then*

$$g^{(\beta)} = \{n < \omega | f(g_{n_0+n}) = 1\}$$

where for each $m < \omega$, $g_m = \{i \in g | |g \cap i| < m\}$.

Lemma 4.7. *Suppose* $\beta < \omega_1^*$, (M, E) *is an* ω-*model*,

$$(M, E) \vDash ZC + \text{``}V = L\text{''},$$

$HYP_0^{(\beta)} \subset M$ *and* $M \in L_{\gamma_{\beta+1}^0}$. *Suppose*

$$h_0 : [\omega]^{n_0} \to \mathbb{U} \cap L_{\gamma_\beta^0},$$

$h_0 \in M$ *and* $n_0 < \omega$.

Then there exists

$$h : [\omega]^{<\omega} \to (\mathbb{U})^{(M,E)},$$

and a function

$$f : [\omega]^{<\omega} \to \{0, 1\}$$

such that

(1) $(f, h) \in L_{\gamma_{\beta+1}^0}$,

(2) $h|[\omega]^{n_0} = h_0$,

(3) for all infinite sets $g \subset \omega$ *if* $i \in h(g \cap i)$ *for all* $i \in g \backslash g_{n_0}$ *then*

$$g^{(\beta+1)} = \{n < \omega | f(g_{n_0+n}) = 1\}$$

where for each $m < \omega$, $g_m = \{i \in g | |g \cap i| < m\}$.

Proof.

We begin by proving that for all $\beta < \omega_1^*$, if Lemma 4.7 holds for β then Lemma 4.6 holds for $\beta+1$. This is immediate. Let h_0 and n_0 be given. Since $\beta < \omega_1^*$, γ_β^0 is countable in $L_{\gamma_\beta^0+1}$ and so it follows there exists an ω-model,

$$(M, E) \vDash ZC + \text{``}V = L\text{''},$$

such that

(1) $(M, E) \in L_{\gamma_\beta^0+1}$,

(2) $\mathcal{P}(\omega) \cap L_{\gamma_\beta^0} \subset M$,

(3) $\mathbb{U} \cap M = (\mathbb{U})^{(M,E)}$.

Since Lemma 4.7 holds for β, there exists

$$h : [\omega]^{<\omega} \to \mathbb{U} \cap M,$$

and a function

$$f : [\omega]^{<\omega} \to \{0, 1\}$$

such that

(1) $(f, h) \in L_{\gamma^0_{\beta+1}}$,
(2) $h|[\omega]^{n_0} = h_0$,
(3) for all infinite sets $g \subset \omega$ if $i \in h(g \cap i)$ for all $i \in g \backslash g_{n_0}$ then

$$g^{(\beta+1)} = \{n < \omega | f(g_{n_0+n}) = 1\}.$$

This proves Lemma 4.6 holds for $\beta + 1$.

We prove both Lemma 4.6 and Lemma 4.7 by simultaneous induction on $\beta < \omega_1^*$.

We first prove Lemma 4.7 for $\beta = 0$, note that Lemma 4.6 is essentially trivial for $\beta = 0$. Let $(M, E) \in L_{\gamma_1^0}$, h_0 and n_0 be given. Let $\langle \phi_n(x) : n < \omega \rangle$ be a recursive enumeration of all Σ_0^0 formulas so that for all $X \subset \omega$,

$$X' = \{n | \phi_n[X \cap k] \text{ holds for some } k < \omega\}.$$

For each $n < \omega$ let O_n be the set of all $(s, h) \in (\mathbb{P}_\mathbb{U})^{(M,E)}$ such that $\phi_n[s]$ holds.

Thus $O_n \in M$ and O_n is open in $(\mathbb{P}_\mathbb{U})^{(M,E)}$.

Choose

$$h : [\omega]^{<\omega} \to (\mathbb{U})^{(M,E)}$$

such that:

(1) $h|[\omega]^{n_0} = h_0$,
(2) for all $a \in [\omega]^{<\omega}$, for all n such that $n_0 + n \leq |a|$, if

$$\left\| O_n, \mathbb{P}_\mathbb{U}|a \right\|^{(M,E)} \notin M \cap \mathrm{Ord}$$

then for all $k \in h(a)$, $a \subset k$ and

$$\left\| O_n, \mathbb{P}_\mathbb{U}|a \cup \{k\} \right\|^{(M,E)} \notin M \cap \mathrm{Ord}.$$

Define $f : [\omega]^{<\omega} \to \{0, 1\}$ such that for all $n < \omega$, for all $a \in [\omega]^{n_0+n}$,

(1) $f(a) = 0$ if $\left\| O_n, \mathbb{P}_\mathbb{U}|a \right\|^{(M,E)} \notin M \cap \mathrm{Ord}$,
(2) $f(a) = 1$ otherwise.

Since $M \in L_{\gamma_1^0}$, $(\gamma_0^0)^{(M,E)}$ is not wellfounded and further $M \cap \mathrm{Ord} < \gamma_1^0$ (which implies that the nonstandard part of (M, E) is in $L_{\gamma_0^0}$). Thus (h, f) can be chosen in $L_{\gamma_0^1}$. This proves Lemma 4.7 for $\beta = 0$.

Now suppose that $\beta > 0$, β is a limit ordinal and that both Lemma 4.6, Lemma 4.7 hold for all $\eta < \beta$. We prove that Lemma 4.6 holds for β.

Let $\langle \beta_i : i < \omega \rangle$ be an increasing sequence cofinal in β such that

$$\langle \beta_i : i < \omega \rangle \in L_{\beta+1}$$

and such that for all $i < \omega$, $\beta_i + 1 < \beta_{i+1}$. Because $\beta < \omega_1^*$ it follows that for all $i < \omega$, $\beta_i < \gamma_{\beta_i}^0$.

Let $\langle (h_i, f_i) : i < \omega \rangle$ be a sequence such that for all $i < \omega$,

(1) if $i > 0$ then $h_i : [\omega]^{<\omega} \to \mathbb{U} \cap L_{\gamma_{\beta_i}^0}$,

(2) if $i > 0$ then $f_i : [\omega]^{<\omega} \to \{0, 1\}$, otherwise f_i is the constant function with value 1,

(3) $(h_{i+1}, f_{i+1}) \in L_{\gamma_{\beta_{i+1}}^0}$,

(4) $h_{i+1}|[\omega]^{n_0+i} = h_i|[\omega]^{n_0+i}$ and if $i > 0$ then for all $a \in [\omega]^{<\omega}$,

$$h_{i+1}(a) \subseteq h_i(a),$$

(5) for all infinite sets $g \subset \omega$ if $k \in h_{i+1}(g \cap k)$ for all $k \in g \backslash g_{n_0+i}$ then

$$g^{(\beta_{i+1})} = \{n < \omega | f_{i+1}(g_{n_0+i+n}) = 1\},$$

(6) $(h_{i+1}, f_{i+1}) \in L_{\gamma_{\beta_{i+1}}^0}$.

By the induction hypothesis, $\langle (h_i, f_i) : i < \omega \rangle$ exists. By making L-least choices minimizing the pair of ordinals associated with 5 and 6, we can further suppose that $\langle (h_i, f_i) : i < \omega \rangle$ is definable in L_γ from β where

$$\gamma = \sup \{ \gamma_{\beta_i} | i < \omega \}.$$

Using $\langle (h_i, f_i) : i < \omega \rangle$ one can define (h, f) such that,

(1) $h : [\omega]^{<\omega} \to \mathbb{U} \cap L_\gamma$ and $h|[\omega]^{n_0} = h_0$,

(2) $f : [\omega]^{<\omega} \to \{0, 1\}$,

(3) for all infinite sets $g \subset \omega$ if $k \in h(g \cap k)$ for all $k \in g \backslash g_{n_0}$ then

$$g^{(\beta)} = \{n < \omega | f(g_n) = 1\}$$

where $g_n = \{i \in g | |g \cap i| < n\}$,

(4) $(h, f) \in L_{\gamma_\beta^0}$.

This proves Lemma 4.6 holds for β.

We next prove that for all $\beta < \omega_1^*$, if Lemma 4.6 holds for β then Lemma 4.7 holds for β.

Let (M, E), h_0 and n_0 be given with $(M, E) \in L_{\gamma_{\beta+1}^0}$ and with $\mathrm{HYP}_0^{(\beta)} \subset M$. Let $\langle \phi_n(x) : n < \omega \rangle$ be a recursive enumeration of all Σ_0^0 formulas so that for all $X \subset \omega$,

$$X^{(\beta+1)} = \left\{ n | \phi_n[X^{(\beta)} \cap k] \text{ holds for some } k < \omega \right\}.$$

Let $(h, f) \in L_{\gamma_\beta^0}$ satisfy the conclusion of Lemma 4.6 so that:

(1) $(f, h) \in L_{\gamma_\beta^0}$,
(2) $h|[\omega]^{n_0} = h_0$,
(3) for all infinite sets $g \subset \omega$ if $i \in h(g \cap i)$ for all $i \in g \backslash g_{n_0}$ then

$$g^{(\beta)} = \{n < \omega | f(g_{n_0+n}) = 1\}.$$

Since $\mathrm{HYP}_0^{(\beta)} \subset M$, $(h, f) \in M$.

For each $s \in [\omega]^{<\omega}$ such that $|s| \geq n_0$ let

$$s^* = \{n < \omega | n_0 + n \leq |s| \text{ and } f(s_{n_0+n}) = 1\}$$

Thus for all $g \subset \omega$ if $i \in h(g \cap i)$ for all $i \in g \backslash g_{n_0}$ then

$$g^{(\beta+1)} = \{n | \phi_n[(g \cap k)^*] \text{ holds for some } k, \ \max(g_{n_0+n}) \leq k < \omega\}.$$

Since $\mathrm{HYP}_0^{(\beta)} \subset M$, $(h, f) \in M$. The remainder of the argument is just like the case $\beta = 0$. For each $n < \omega$ let O_n be the set of all $(s', h') \in (\mathbb{P}_\mathbb{U})^{(M,E)}$ such that

(1) $n_0 + n \leq |s'|$,
(2) $\phi_n[s']$ holds.

Thus $O_n \in M$ and O_n is open in $(\mathbb{P}_\mathbb{U})^{(M,E)}$. Choose

$$h^* : [\omega]^{<\omega} \to (\mathbb{U})^{(M,E)}$$

such that:

(1) $h^*|[\omega]^{n_0} = h_0$,
(2) for all $a \in [\omega]^{<\omega}$, for all n such that $n_0 + n \leq |a|$, if

$$\left\| O_n, \mathbb{P}_\mathbb{U}|a \right\|^{(M,E)} \notin M \cap \mathrm{Ord}$$

then for all $k \in h^*(a)$, $a \subset k$, $k \in h(a)$, and

$$\left\| O_n, \mathbb{P}_\mathbb{U}|a \cup \{k\} \right\|^{(M,E)} \notin M \cap \mathrm{Ord}.$$

Define $f^* : [\omega]^{<\omega} \to \{0,1\}$ such that for all $n < \omega$, for all $a \in [\omega]^{n_0+n}$,

(1) $f^*(a) = 0$ if $\big\|O_n, \mathbb{P}_U|a\big\|^{(M,E)} \notin M \cap \mathrm{Ord}$,
(2) $f^*(a) = 1$ otherwise.

Since γ_β^0 is contained in the standard part of (M, E) and since $\beta < \gamma_\beta^0$, $(\gamma_\beta^0)^{(M,E)}$ is defined. Since $(M, E) \in L_{\gamma_{\beta+1}^0}$, $(\gamma_\beta^0)^{(M,E)}$ is not wellfounded and the nonstandard part of (M, E) is also in $L_{\gamma_{\beta+1}^0}$. Thus (h^*, f^*) can be chosen in $L_{\gamma_{\beta+1}^0}$. This proves Lemma 4.7 for β.

To summarize, we have proved that for all $\beta < \omega_1^*$,

(1) if $\beta = 0$ then Lemma 4.7 holds for β,
(2) if $\beta > 0$, β is a limit ordinal and if both Lemma 4.6, Lemma 4.7 hold for all $\eta < \beta$, then both Lemma 4.6, Lemma 4.7 hold for β,
(3) if Lemma 4.7 holds for β then Lemma 4.6 holds for $\beta + 1$,
(4) if Lemma 4.6 holds for β then Lemma 4.7 holds for β.

This completes the induction. $\square$

We actually need Lemma 4.7 relativized to sets $Z \subset \omega$. The proof is identical.

Definition 4.8. (ZC) Suppose that $Z \subset \omega$ and $V = L[Z]$. Let $\mathbb{U}_Z$ be the ultrafilter U on ω such that such that for all $\gamma < \omega_1$,

(1) if $\gamma < (\omega_1^*)_Z$ and γ is Z-admissible then $U \cap L_{\gamma+1}[Z]$ is the $L[Z]$-least ultrafilter on $L_{\gamma+1}[Z] \cap \mathcal{P}(\omega)$ such that

    (a) $U \cap L_\gamma[Z] \subset U \cap L_{\gamma+1}[Z]$,
    (b) $U \cap L_{\gamma+1}[Z] \cap \mathbb{I}_{\rho_0} = \emptyset$,
    (c) suppose there exists an ω-model, (M, E), such that

       (i) $(M, E) \in L_{\gamma+1}[Z]$ and $L_\gamma[Z] \cap \mathcal{P}(\omega) \subset M$,
      (ii) $(M, E) \vDash \mathrm{ZC} + \text{``}V = L[Z]\text{''}$,

       Then if (M, E) is the $L[Z]$-least such ω-model and if $(\mathbb{U}_Z)^{(M,E)}$ is ultrafilter disjoint from $\mathbb{I}_{\rho_0}$ and extending $U \cap L_\gamma[Z]$ then $U \cap M = (\mathbb{U}_Z)^{(M,E)}$.

(2) otherwise $U \cap L_{\gamma+1}[Z]$ is the $L[Z]$-least ultrafilter on $L_{\gamma+1}[Z] \cap \mathcal{P}(\omega)$ such that

    (a) $U \cap L_\gamma[Z] \subset U \cap L_{\gamma+1}[Z]$,
    (b) $U \cap L_{\gamma+1}[Z] \cap \mathbb{I}_{\rho_0} = \emptyset$.

Lemma 4.9. *Suppose $Z \subset \omega$, $\beta < (\omega_1^*)_Z$, (M, E) is an ω-model, $Z \in M$,*

$$(M, E) \vDash \mathrm{ZC} + \text{``}V = L[Z]\text{''},$$

$\mathrm{HYP}_Z^{(\beta)} \subset M$, *and* $(M, E) \in L_{\gamma_{\beta+1}^Z}[Z]$. *Then there exists*

$$h : [\omega]^{<\omega} \to (\mathbb{U}_Z)^{(M,E)},$$

and a function

$$f : [\omega]^{<\omega} \times [\omega]^{<\omega} \to \{0, 1\}$$

such that

(1) $(h, f) \in L_{\gamma_{\beta+1}^0}[Z]$,
(2) for all infinite sets $g \subset \omega$ if $i \in h(g \cap i)$ for all $i \in g$ then

$$(g, Z)^{(\beta+1)} = \{n < \omega | f(g_n, Z \cap \max(g_n)) = 1\}.$$

Proof. This is a special case of Lemma 4.7 relativized to Z (here $n_0 = 0$).

$\square$

Lemma 4.10. *Suppose $0 < \alpha < \omega_1^*$, $Z \subset \omega$, (M, E) is an ω-model,*

$$(M, E) \vDash \mathrm{ZC} + \text{``}V = L[Z]\text{''},$$

$Z_0 \subset \omega$ *and* $\mathrm{HYP}_{Z_0}^{(\alpha)} \subset M$. *Let*

$$\gamma = \sup \left\{ \gamma_\beta^{Z_0} | \beta < \alpha \right\},$$

and suppose that g is (M, E)-weakly generic for $(\mathbb{P}_{\mathbb{U}_Z})^{(M,E)}$.
Then

$$(Z_0)^{(\gamma)} \leq_{\mathrm{tt}_g} (g, Z_0)^{(\alpha)}.$$

Proof. The key point is that $\gamma < \gamma_\alpha^{Z_0}$ and that either γ is admissible relative to Z_0 or a limit of ordinals admissible relative to Z_0. Therefore

$$L_\gamma[Z_0] \cap \mathcal{P}(\omega) \in M$$

and there is a Σ_1^0-formula, $\phi(x_0, x_1, x_2)$, such that:

(1) for all $X \in \mathcal{P}(\omega) \cap L_\gamma[Z_0]$ there exists $n < \omega$ such that for all $m \in g \backslash n$

$$\mathcal{O}_X = \{k | \phi[k, X, g \backslash m]\}.$$

This implies that for each $X \in \mathcal{P}(\omega) \cap L_\gamma[Z_0]$,

$$\mathcal{O}_X \leq_{tt_g} (g, X)'.$$

Therefore (by induction) we can assume that $\alpha > 0$, α is a limit ordinal and for all $\beta < \alpha$,

$$\mathcal{O}_{Z_0}^{(\beta)} \leq_{tt_g} (g, Z_0)^{(\beta+1)}.$$

Thus $\gamma = \omega \cdot \gamma$ and so $(Z_0)^{(\gamma)}$ is the Σ_1-theory of $L_\gamma[Z_0]$ in parameter Z_0. Since α is a limit ordinal, $\alpha = \omega \cdot \xi$ for some $\xi \leq \alpha$ and so $(g, Z_0)^{(\alpha)}$ is the Σ_1-theory of $L_\xi[g, Z_0]$ in parameter (g, Z_0).

Fix a Σ_1-formula, $\psi_0(x_0)$, such that

$$L_{\gamma+1}[Z_0] \models \psi_0[Z_0]$$

and such that $L_\gamma[Z_0] \nvDash \psi_0[Z_0]$ (since $\gamma < \omega_1^*$, $\psi_0(x_0)$ must exist).

By 1 it follows that for each Σ_1 formula $\psi(x_0)$ the following are equivalent

(1) $L_\gamma[Z_0] \models \psi[Z_0]$.
(2) There exists an ω-model (M_0, E_0) such that

    (a) $(M_0, E_0) \in L_\xi[g, Z_0]$,
    (b) $(M_0, E_0) \models Z \backslash \text{Powerset} + \text{``}V = [Z_0]\text{''}$,
    (c) $(M_0, E_0) \models \psi[Z_0]$,
    (d) $(M_0, E_0) \nvDash \psi_0[Z_0]$,
    (e) for all $X \in \mathcal{P}(\omega) \cap M_0$, there exists $n < \omega$ such that for all $m \in g \backslash m$

$$\{k | \phi[k, X, g \backslash m]\} \in M_0.$$

To see that this equivalence holds suppose that (M_0, E_0) satisfies 2. Then by 1,

$$\mathcal{O}_X \in M_0$$

for all $X \in M_0 \cap L_\gamma[Z_0] \cap \mathcal{P}(\omega)$. In addition since $0 < \alpha < \omega_1^*$ and since α is a limit ordinal, γ is not admissible relative to Z_0. Therefore either

$$(M_0, E_0) \cong L_\eta[Z_0]$$

for some $\eta \leq \gamma$ or $\gamma + 1$ must be in the standard part of (M_0, E_0). But the latter cannot happen since

$$(M_0, E_0) \nvDash \psi_0[Z_0]$$

and since $L_{\gamma+1}[Z_0] \vDash \psi_0[Z_0]$. Therefore

$$(M_0, E_0) \cong L_\eta[Z_0]$$

for some $\eta \leq \gamma$ and this proves the equivalence of 1 and 2.

Finally the equivalence of 1 and 2 immediately implies that

$$(Z_0)^{(\gamma)} \leq_{tt_g} (g, Z_0)^{(\alpha)}$$

and this proves the lemma. □

A set $P \subset \mathcal{P}(\omega)$ is *binary-perfect* if P is closed (in the natural topology) and there exists an infinite set $G_P \subset \omega$ such that

(1) for all $A, B \in P$, $A = B$ if and only if $A \cap G_P = B \cap G_P$,
(2) $\mathcal{P}(G_P) = \{A \cap G_P | A \in P\}$.

Note that

$$G_P = \{k_n | n < \omega\},$$

where for each $n < \omega$,

$$k_n = \min \{k < \omega || P|k| = 2^{n+1}\}.$$

Lemma 4.11. *Suppose $\gamma \leq \omega_1^*$, $\omega \cdot \gamma = \gamma$ and $Z \subset \omega$. Then there exists a binary-perfect set $P \subset \mathcal{P}(\omega)$ such that*

(1) $\langle P|n : n < \omega \rangle \leq_{tt_Z} Z^{(\gamma)}$,
(2) for each $C \in P$, C is Cohen generic over $L_\gamma[Z]$,
(3) for each $C, D \in P$ with $C \neq D$, C and D are mutually Cohen generic over $L_\gamma[Z]$,

where for each $n < \omega$, $P|n = \{X \cap n | X \in P\}$.

Proof. Since $\gamma \leq \omega_1^*$ there is a partial map

$$\pi : \omega \to L_\gamma[Z]$$

such that

(1) π is Σ_1-definable over $L_\gamma[Z]$,
(2) π is a surjection.

The sequence $\langle P|n : n < \omega \rangle$ is easily defined from π, recalling that

$$Z^{(\gamma)} = \{\phi | \phi \text{ is a } \Sigma_1\text{-sentence and } L_\gamma[Z] \vDash \phi\}.$$

□

As a corollary we obtain the following two lemmas which we shall need.

Lemma 4.12. *Suppose* $\alpha < \omega_1^*$, $X \subset \omega$ *and* $X \notin \mathrm{HYP}^{(\alpha)}$. *Then there exists a set* $Z \subseteq \omega$ *such that:*

(1) $X \notin \mathrm{HYP}_Z^{(\alpha)}$;
(2) $X \leq_{\mathrm{T}} \mathcal{O}_Z^{(\alpha)}$.

Proof.
Let $\gamma = \gamma_\alpha^0$. By Lemma 4.11 there exists a binary-perfect set $P \subset \mathcal{P}(\omega)$ such that

(1) $\langle P|n : n < \omega \rangle \leq_{\mathrm{tt}_0} 0^{(\gamma)}$,
(2) for each $C \in P$, C is Cohen generic over L_γ,
(3) for each $C, D \in P$ with $C \neq D$, C and D are mutually Cohen generic over L_γ,

where for each $n < \omega$, $P|n = \{X \cap n | X \in P\}$.
For each $n < \omega$ let

$$k_n = \min \left\{ k < \omega \big| |P|k| = 2^{n+1} \right\},$$

and let $G_P = \{k_n | n < \omega\}$.
Let $C, D \in P$ be such that

(1) $X \leq_{\mathrm{tt}_{G_P}} (C, 0^{(\gamma)})$,
(2) $X \leq_{\mathrm{tt}_{G_P}} (D, 0^{(\gamma)})$,
(3) $C \neq D$.

Since C, D are mutually Cohen generic over L_γ, $X \notin L_\gamma[C]$ or $X \notin L_\gamma[D]$. Let $Z = C$ if $X \notin L_\gamma[C]$, otherwise let $Z = D$.
Thus

(1) $\gamma = \gamma_Z^\alpha$,
(2) $X \notin \mathrm{HYP}_Z^{(\alpha)}$,
(3) $X \leq_{\mathrm{T}} \mathcal{O}_Z^{(\alpha)}$. $\qquad\square$

Lemma 4.13. *Suppose* $\alpha < \omega_1^*$, (M, E) *is an* ω-model,

$$(M, E) \vDash \mathrm{ZC} + \text{``}V = L\text{''},$$

$Z_0 \subset \omega$ *and* $\mathrm{HYP}_{Z_0}^{(\alpha)} \subset M$.
Suppose $A \in M \cap \mathcal{P}(\omega)$. *Then there exists* $Z \in \mathcal{P}(\omega) \cap M$ *such that*

(1) $Z_0 \leq_{\mathrm{tt}_0} Z$ *and* $\mathrm{HYP}_Z^{(\alpha)} \subset M$,

(2) for all $g \subset \omega$, if g is (M, E)-weakly generic for $(\mathbb{P}_{U_Z})^{(M,E)}$ then

$$A \leq_{tt_g} (g, Z)^{(\alpha)}.$$

Proof. If $\alpha = 0$ then we can simply take $Z = (A, Z_0)$ and so we can suppose $\alpha > 0$.

Let $\gamma = \sup \left\{ \gamma_\beta^{Z_0} | \beta < \alpha \right\}$. Thus $\omega \cdot \gamma = \gamma$ and $\gamma_\alpha^{Z_0}$ is the least ordinal above γ which is admissible relative to Z_0.

By Lemma 4.11, there exists a binary-perfect set, $P \subset \mathcal{P}(\omega)$, such that

(1) $\langle P|n : n < \omega \rangle \leq_{tt_{Z_0}} Z_0^{(\gamma)}$,
(2) for each $C \in P$, C is Cohen generic over $L_\gamma[Z_0]$,

where for each $n < \omega$, $P|n = \{X \cap n | X \in P\}$.

For each $n < \omega$ let

$$k_n = \min \left\{ k < \omega \| |P|k| = 2^{n+1} \right\},$$

and let $G_P = \{k_n | n < \omega\}$.

Since $\gamma < \gamma_\alpha^{Z_0}$, $Z_0^{(\gamma)} \in M$ and so

$$\langle P|n : n < \omega \rangle \in M.$$

Let $C \in P \cap M$ be such that

$$A \leq_{tt_g} (C, \langle P_n : n < \omega \rangle)$$

for all infinite sets $g \subset \omega$ such that

$$|g \cap i| < |G_P \cap i|$$

for all $i < \omega$.

Let $Z = (Z_0, C)$. Trivially $Z_0 \leq_{tt_0} Z$. Further, for all $\eta < \alpha$,

$$\text{HYP}_Z^{(\eta)} = \text{HYP}_{Z_0}^{(\eta)}[Z]$$

and so it follows that $\text{HYP}_Z^{(\alpha)} \subset M$.

Suppose g is (M, E)-weakly generic for $(\mathbb{P}_{U_Z})^{(M,E)}$. Then by Lemma 4.10

$$(Z_0)^{(\gamma)} \leq_{tt_g} (g, Z_0)^{(\alpha)}.$$

Therefore

$$\langle P|n : n < \omega \rangle \leq_{tt_g} (g, Z_0)^{(\alpha)}.$$

The enumeration function for g eventually dominates all functions in $M \cap \omega^\omega$ (since g is (M, E)-weakly generic for $(\mathbb{P}_{U_Z})^{(M,E)}$) and $G_P \in M$. Therefore

$$A \leq_{tt_g} (\langle P_n : n < \omega \rangle, Z),$$

which implies that $A \leq_{tt_g} (g, Z)^{(\alpha)}$. $\qquad\qquad\square$

We now prove our first transfinite version of Theorem 3.4.

Theorem 4.14. *Suppose $\alpha < \omega_1^*$, $X \subset \omega$ and $X \notin \mathrm{HYP}^{(\alpha)}$. Then there exists a set $Y \subseteq \omega$ such that $Y^{(\alpha+1)} =_{tt_Y} X \vee Y$.*

Proof.

Fix $\alpha < \omega_1^*$ and for each $t \in \mathcal{P}(\omega)$ and for each $k \in \omega$ define $\left\| k, \mathcal{O}_t^{(\alpha)} \right\|$ as follows.

(1) If $k \in \mathcal{O}_t^{(\alpha)}$ then

$$\left\| k, \mathcal{O}_t^{(\alpha)} \right\| = \left\| k, \mathcal{O}_{t^*} \right\|$$

where $t^* = t^{(\gamma^*)}$ and $\gamma^* = \sup \left\{ \gamma_\eta^t | \eta < \alpha \right\}$,

(2) otherwise $\left\| k, \mathcal{O}_t^{(\alpha)} \right\| = \infty$.

Fix a set $X \subset \omega$ such that $X \notin \mathrm{HYP}^{(\alpha)}$. By Lemma 4.12 there exists a set $Z_0 \subseteq \omega$ such that:

(1) $X \notin \mathrm{HYP}_{Z_0}^{(\alpha)}$;
(2) $X \leq_T \mathcal{O}_{Z_0}^{(\alpha)}$.

There must exist two ω-models, (M_0, E_0) and (M_1, E_1) such that

(1) $Z_0 \in M_0$ and $Z_0 \in M_1$,
(2) $(M_0, E_0) \vDash \mathrm{ZC} + \text{``}V = L[Z_0]\text{''}$,
(3) $(M_1, E_1) \vDash \mathrm{ZC} + \text{``}V = L[Z_0]\text{''}$,
(4) $\{(M_0, E_0), (M_1, E_1)\} \in L_{\gamma_{\alpha+1}^{Z_0}}$,
(5) $\mathrm{HYP}_{Z_0}^{(\alpha)} = \mathcal{P}(\omega) \cap M_0 \cap M_1$,

Thus there exists an ω-model, (M, E), such that

(1) $(M, E) \in L_{\gamma_{\alpha+1}^{Z_0}}$ and $\mathrm{HYP}_{Z_0}^{(\alpha)} \subset M$,
(2) $(M, E) \vDash \mathrm{ZC} + \text{``}V = L[Z_0]\text{''}$,

Fix a set $G \subset \omega$ such that for all infinite sets $g \subset \omega$, if

$$|g \cap i| < |G \cap i|$$

for all $i < \omega$ then $X \leq_{tt_g} \mathcal{O}_{Z_0}^{(\alpha)}$, since

$$X \leq_T \mathcal{O}_{Z_0}^{(\alpha)}$$

the set G necessarily exists.

By Lemma 4.13, there exists $(\pi, Z) \in M$ such that

(1) $Z_0 \leq_{\mathrm{tt}_0} Z$,

(2) $\{\pi\} \cup \mathrm{HYP}_Z^{(\alpha)} \subset M$,

(3) for all $n < \omega$, for all $a \in [\omega]^{2n+1}$, $\pi(a) \subset (\mathbb{P}_{\mathbb{U}_Z \cap M})^{(M,E)}$, $\pi(a)$ is open and

$$\left\| \pi(a), \mathbb{P}_{\mathbb{U}_Z \cap M} | a \right\|^{(M,E)} = \left\| n, \mathcal{O}_{Z_0}^{(\alpha)} \right\|^{(M,E)},$$

(4) for all $g \subset \omega$, if g is (M,E)-weakly generic for $(\mathbb{P}_{\mathbb{U}_Z})^{(M,E)}$ then

$$Y_\pi \leq_{\mathrm{tt}_{(g,Z)}} (g, Z)^{(\alpha)}.$$

where

$$Y_\pi = \{(a,b) | (b,h) \in \pi(a) \text{ for some } h \in M\}.$$

By Lemma 4.9, there exist

$$h : [\omega]^{<\omega} \to (\mathbb{U}_Z)^{(M,E)},$$

and a function

$$f : [\omega]^{<\omega} \times [\omega]^{<\omega} \to \{0,1\}$$

such that for all infinite sets $g \subset \omega$ if $i \in h(g \cap i)$ for all $i \in g$ then

$$(g, Z)^{(\alpha+1)} = \{n < \omega | f(g_n, Z \cap \max(g_n)) = 1\}.$$

For each $a \in [\omega]^{<\omega}$, $h(a) \in (\mathbb{U}_Z)^{(M,E)}$, and so since,

$$(\mathbb{U}_Z)^{(M,E)} \cap \mathbb{I}_{\rho_0} = \emptyset$$

and since $X \notin M$, by Lemma 3.2, both $\sigma_X \cap h(a)$ and $\sigma_X \backslash h(a)$ are infinite. Let $g \subset \omega$ be (M,E)-weakly generic for $(\mathbb{P}_{\mathbb{U}_Z})^{(M,E)}$ such that

(1) for all $i \in g$, $i \in h(g \cap i)$,

(2) for all $i < \omega$, $|g \cap i| < |G \cap i|$,

(3) for each $2n < \omega$, $f(g_n, Z \cap \max(g_n)) = 1$ if and only if $\min(g \backslash g_{2n}) \in \sigma_X$,

(4) for each $2n + 1 < \omega$,

$$(g_{2n+1}, g_k) \in Y_\pi$$

for some $k > 2n + 1$ if and only if $n \in \mathcal{O}_Z^{(\alpha)}$,

where as usual for each $n < \omega$, $g_n = \{i \in g | |g \cap i| < n\}$.

Thus $(g, Z)^{(\alpha+1)} \leq_{\mathrm{tt}_g} X \vee g$. Next we show

$$X \leq_{\mathrm{tt}_{(g,Z)}} (g, Z)^{(\alpha+1)}.$$

By 4, $\mathcal{O}_Z^{(\alpha)} \leq_{\mathrm{tt}_{(g,Z)}} (Y_\pi, g)'$ and by the choice of Z,

$$Y_\pi \leq_{\mathrm{tt}_{(g,Z)}} (g, Z)^{(\alpha)},$$

since g is (M, E)-weakly generic for $(\mathbb{P}_{\mathbb{U}_Z})^{(M,E)}$. This implies that

$$\mathcal{O}_Z^{(\alpha)} \leq_{\mathrm{tt}_{(g,Z)}} (g, Z)^{(\alpha+1)}$$

and so by the choice G and 2,

$$X \leq_{\mathrm{tt}_{(g,Z)}} (g, Z)^{(\alpha+1)}.$$

Finally let $Y =_{\mathrm{tt}_0} g \vee Z$. Then

$$Y^{(\alpha+1)} =_{\mathrm{tt}_Y} X \vee Y. \qquad \qquad \square$$

The remainder of this section is devoted to proving Theorem 4.1.

Lemma 4.15. *Suppose $Z \subset \omega$. Then there is a function*

$$h : [\omega]^{<\omega} \to \mathbb{U}_Z \cap L_{(\omega_1^*)^Z}[Z]$$

such that if $g \subset \omega$ is such that $i \in h(g \cap i)$ for all $i \in g$ then

(1) $(\omega_1^*)^Z$ is admissible relative to (g, Z),
(2) $\mathcal{O}_{(g,Z)} =_{\mathrm{tt}_g} (g, \mathcal{O}_Z^*)$.

Proof. We assume $V = L$ so that $\mathbb{U}_Z$ is an ultrafilter in V.

By induction on $\eta < (\omega_1^*)^Z$ it follows that for each $n < \omega$ and for each $a \in [\omega]^{<\omega}$, if

$$\left\| O, \mathbb{P}_{\mathbb{U}_Z} | a \right\| \geq \gamma_\eta^Z$$

then $\left\| O, \mathbb{P}_{\mathbb{U}_Z} | a \right\| = \infty$ where

$$O = \left\{ p \in \mathbb{P}_{\mathbb{U}_Z} \,|\, p \Vdash n \in (g, Z)^{(\eta)} \right\}.$$

From this it follows that if $\eta < (\omega_1^*)^Z$ and

$$\eta = \sup \left\{ \gamma_\xi^Z \,|\, \xi < \eta \right\}$$

then for each Σ_1 formula $\phi(x_0, x_1, x_2)$,

$$\rho_\phi \in L_{\gamma_{\eta+1}^Z}$$

where ρ_ϕ is the function such that

(1) $\mathrm{dom}(\rho_\phi) = [\omega]^{<\omega} \times \eta$,
(2) for all $a \in [\omega]^{<\omega}$, for all $\alpha < \eta$,

$$\rho_\phi(a, \alpha) = \big\| O, \mathbb{P}_{\mathbb{U}_Z} | a \big\|$$

where

$$O = \{p \in \mathbb{P}_{\mathbb{U}_Z} | p \Vdash L_\eta[Z, g] \vDash \phi[g, \alpha, Z]\}.$$

Thus for all formulas $\phi(x_0, x_1, x_2)$ the set of all (a, α, η) such that

(1) $a \in [\omega]^\omega$,
(2) there exists $f : [\omega]^{<\omega} \to \mathbb{U}_Z$ such that

$$(a, f) \Vdash L_\eta[Z, g] \vDash \phi[\alpha, Z, g]$$

(in V relative to $\mathbb{P}_{\mathbb{U}_Z}$),

is Δ_1-definable over $L_{(\omega_1^*)^Z}[Z]$ from Z.

Further if $X \subset \mathbb{P}_{\mathbb{U}_Z} \cap L_{(\omega_1)^*}[Z]$ and

$$X \in L_{(\omega_1)^*}[Z]$$

then for all $a \in [\omega]^{<\omega}$ either

$$\big\| O_X, \mathbb{P}_{\mathbb{U}_Z} \big\| = \infty$$

or $\big\| O_X, \mathbb{P}_{\mathbb{U}_Z} \big\| < (\omega_1^*)^Z$ where

$$O_X = \{p \in \mathbb{P}_{\mathbb{U}_Z} | p \leq q \text{ for some } q \in X\}.$$

Given an infinite set $g \subset \omega$, $\mathcal{F}_g \subset \mathbb{P}_{\mathbb{U}_Z}$ is the set of $(s, h) \in \mathbb{P}_{\mathbb{U}_Z}$ such that

(1) $s = g_n$ where $n = |s|$,
(2) for all $i \in g \backslash s$, $i \in h(g \cap i)$,

and the set g is $L_{(\omega_1^*)^Z}[Z]$-generic for $\mathbb{P}_{\mathbb{U}_Z} \cap L_{(\omega_1^*)^Z}[Z]$ if for all sets

$$X \subset \mathbb{P}_{\mathbb{U}_Z} \cap L_{(\omega_1^*)^Z}[Z]$$

such that X is predense in $\mathbb{P}_{\mathbb{U}_Z} \cap L_{(\omega_1^*)^Z}[Z]$ and such that $X \in L_{(\omega_1^*)^Z}[Z]$, $\mathcal{F}_g \cap X \neq \emptyset$.

Putting everything together, for any infinite set $g \subset \omega$ the following hold.

(1) g is $L_{(\omega_1^*)^Z}[Z]$-generic for $\mathbb{P}_{\mathbb{U}_Z} \cap L_{(\omega_1^*)^Z}[Z]$ if and only it for all functions

$$f : [\omega]^{<\omega} \to \mathbb{U}_Z$$

with $f \in L_{(\omega_1^*)^Z}[Z]$, there exists $k \in \omega$ such that $i \in f(g \cap i)$ for all $i \in g$.

(2) For all $\eta < (\omega_1^*)^Z$, for all formulas $\phi(x_0, x_1, x_2)$ for all $\alpha < \eta$,

$$L_\eta[Z, g] = L_\eta[Z][g]$$

and

$$L_\eta[Z, g] \vDash \phi[Z, g, \alpha]$$

if and only if there exists $p \in \mathcal{F}_g \cap L_{(\omega_1^*)^Z}[Z]$ such that

$$p \Vdash L_\eta[Z, g] \vDash \phi[Z, g, \alpha]$$

in V relative to $\mathbb{P}_{\mathbb{U}_Z}$.

(3) If g is $L_{(\omega_1^*)^Z}[Z]$-generic for $\mathbb{P}_{\mathbb{U}_Z} \cap L_{(\omega_1^*)^Z}[Z]$ then $(\omega_1^*)^Z$ is admissible relative to (g, Z).

Now one can easily define

$$h : [\omega]^{<\omega} \to \mathbb{U}_Z \cap L_{(\omega_1^*)^Z}[Z]$$

such that for all $a \in [\omega]^{<\omega}$, there exists $(a, f) \in \mathbb{P}_{\mathbb{U}_Z}$ such that

$$(a, f) \Vdash |a| \in \mathcal{O}_g$$

(in V relative to $\mathbb{P}_{\mathbb{U}_Z}$) if and only if

$$(a, h) \Vdash |a| \in \mathcal{O}_g.$$

Further one can require in addition that for all

$$f : [\omega]^{<\omega} \to \mathbb{U}_Z$$

with $f \in L_{(\omega_1^*)^Z}[Z]$, there exists $n < \omega$ such that $h(a) \subset h(a)$ for all $a \in [\omega]^{<\omega}$ with $n < |a|$.

Let Y be the set of all $a \in [\omega]^{<\omega}$ such that

$$(a, h) \Vdash |a| \in \mathcal{O}_g.$$

Then Y is Σ_1-definable over $L_{(\omega_1^*)^Z}[Z]$ from Z and therefore for all infinite sets $g \subset \omega$ if $i \in h(g \cap i)$ for all $i \in g$ then

$$\mathcal{O}_{(g,Z)} =_{\text{tt}_g} (g, \mathcal{O}_Z^*).$$

and so h is as required. $\square$

We finish by proving one last generalization of Theorem 3.4.

Theorem 4.16. *Suppose $X \subset \omega$ and $X \notin \mathrm{HYP}^*$. Then there exists a set $Y \subsetneq \omega$ such that $\mathcal{O}_Y =_{\mathrm{tt}_Y} X \vee Y$.*

Proof. By Lemma 4.11 there exists a binary-perfect set $P \subset \mathcal{P}(\omega)$ such that

(1) $\langle P | n : n < \omega \rangle \leq_{\mathrm{tt}_0} 0^{(\gamma)}$,
(2) for each $C \in P$, C is Cohen generic over L_γ,
(3) for each $C, D \in P$ with $C \neq D$, C and D are mutually Cohen generic over L_γ,

where $\gamma = \omega_1^*$, where for each $n < \omega$, $P | n = \{X \cap n | X \in P\}$.

For each $n < \omega$ let

$$k_n = \min \left\{ k < \omega \,||P|k| = 2^{n+1} \right\},$$

and let $G_P = \{k_n | n < \omega\}$.

Let $C, D \in P$ be such that for all infinite sets $g \subset \omega$ such that

$$|g \cap i| < |G_p \cap i|$$

for all $i < \omega$,

(1) $C \neq D$,
(2) $X \leq_{\mathrm{tt}_g} \mathcal{O}_C^*$,
(3) $X \leq_{\mathrm{tt}_g} \mathcal{O}_D^*$.

Since C, D are mutually Cohen generic over $L_{\omega_1^*}$,

$$\mathrm{HYP}^* = \mathrm{HYP}^*[C] \cap \mathrm{HYP}^*[D].$$

Therefore there exists $Z \subset \omega$ such that for all infinite sets $g \subset \omega$ such that

$$|g \cap i| < |G_P \cap i|$$

for all $i < \omega$,

(1) $X \notin \mathrm{HYP}_Z^*$,
(2) $X \leq_{\mathrm{tt}_g} \mathcal{O}_Z^*$,

since one of C or D must work.

Let

$$h : [\omega]^{<\omega} \to \mathbb{U}_Z \cap L_{(\omega_1^*)^Z}[Z]$$

be such that if $g \subset \omega$ is such that $i \in h(g \cap i)$ for all $i \in g$ then

(1) $(\omega_1^*)^Z$ is admissible relative to (g, Z),

(2) $\mathcal{O}_{(g,Z)} =_{\mathrm{tt}_g} (g, \mathcal{O}_Z^*)$.

The function h exists by Lemma 4.15.

Let $g \subset \omega$ be an infinite set such that $i \in h(g \cap i)$ for all $i \in g$ and such that the enumeration function of g dominates G_P.

Let $Y =_{\mathrm{tt}_0} g \vee Z$. Then

$$\mathcal{O}_Z^* \leq_{\mathrm{tt}_Y} X \vee Y$$

and so by 2 and 2

$$\mathcal{O}_Y =_{\mathrm{tt}_Y} X \vee Y. \qquad\qquad \square$$

References

1. A. Blass. Selective ultrafilters and homogeneity. *Ann. Pure Appl. Logic*, 38:215–255, 1988.

2. John Steel. Descending sequences of degrees. *J. Symbolic Logic*, 40(1):59–61, 1975.

3. W. Hugh Woodin. The cardinals below $|[\omega_1]^{<\omega_1}|$. *Ann. Pure Appl. Logic*, 140:161–232, 2006.

CUPPING COMPUTABLY ENUMERABLE DEGREES IN THE ERSHOV HIERARCHY

Guohua Wu*

Division of Mathematical Sciences
School of Physical and Mathematical Sciences
Nanyang Technological University
Singapore 639798, Singapore
E-mail: guohua@ntu.edu.sg

Keywords: Computably enumerable degrees, cupping, jump classes, Ershov hierarchy.

1. Introduction

The study of the degrees below $0'$ starts from Kleene and Post's seminal paper [37]. Research in this area has led to the deep understanding of Turing degrees and the development of various construction techniques. In many constructions, for example, in the construction of minimal degrees below $0'$, it is necessary to combine the priority argument with the sorts of forcing methods employed in the global theory. The difficulties encountered in attempting to "localize" results from the global theory to degrees below $0'$ have led to a better understanding of the nature of various constructions.

The research of the degrees below $0'$ is closely related to the jump operator, which provides a basic scale by which we measure the complexity of unsolvable problems. These research has shed considerable light on the degree theoretic properties of the jump.

One reason for studying the degrees below $0'$ is that all sets being computable in $0'$ can be approximated effectively. That is, a set A is Turing reducible to $\emptyset'$ if and only if A is Δ_2^0 if and only if there is a computable function f such that for any x, $A(x) = \lim_s f(x, s)$.

*The author is partially supported by a start-up grant from NTU and the International Collaboration Grant No. 60310213 of NSFC of China.

This characterization suggests a possible hierarchy of degrees below $\mathbf{0}'$. Note that if A is computably enumerable (c.e. for short), and $\{A_s\}$ is an effective approximation of A, then for any x, $\{A_s(x)\}$ changes at most once. That is, if we define $f(x, s) = A_s(x)$, then f is an approximating function of A such that for any x, there is at most one s such that $f(x, s) \neq f(x, s+1)$. This led to a natural generalization of the class of c.e. sets. A set is n-c.e. if A has an effective approximation $\{A_s\}$ such that $A_0 = \emptyset$ and for all x,

$$|\{s + 1 \mid f(x, s) \neq f(x, s + 1)\}| \leq n.$$

Obviously, the 1-c.e. sets are just the c.e. sets and $\cup_n \{A : A \text{ is } n\text{-c.e.}\}$ is just the Boolean algebra generated Boolean algebra generated by the c.e. sets.

By an n-c.e. degree, we mean a degree of an n-c.e. set. Since any nonzero n-c.e. degree bounds a nonzero c.e. degree, a fact first noticed by Lachlan, no n-c.e. degree is minimal. As a consequence, there are degrees below $\mathbf{0}'$ not n-c.e. for any n.

Ershov [27], [28] extended the hierarchy of n-c.e. sets to transfinite levels as follows. A set $A \subseteq \omega$ is α-*computably enumerable* (α-c.e., for short) relative to a computable system $\mathcal{S}$ of notations for α if and only if there is a *partial computable* (p.c.) function f such that for all k, $A(k) = f(k, b)$, where b is the $\mathcal{S}$-least notation x such that $f(k, x)$ converges.

When $\alpha = \omega$, then the ω-c.e. sets have the following characterization: $A \subseteq \omega$ is ω-c.e. if and only if there are two computable functions $f(x, s), g(x)$ such that for all $x \in \omega$,

(a) $f(x, 0) = 0$,
(b) $\lim_s f(x, s) \downarrow = A(x)$, and
(c) $|\{s + 1 \mid f(x, s) \neq f(x, s + 1)\}| \leq g(x)$.

Let $\mathbf{D}_n$, $\mathbf{D}_\omega$, and $\mathbf{D}(\leq \mathbf{0}')$ denote the set of n-c.e. degrees, the set of ω-c.e. degrees, and the set of degrees below $\mathbf{0}'$ respectively. It is known know that $\mathbf{D}_n \subset \mathbf{D}_\omega \subset \mathbf{D}(\leq \mathbf{0}')$, and for $m < n$, $\mathbf{D}_m \subset \mathbf{D}_n$. $\mathbf{D}_1$ will be also denoted as $\mathbf{R}$.

In this paper, we survey recent progress of cupping computably enumerable degree in the Ershov hierarchy.

Let $\mathbf{a}, \mathbf{b}$ be any two degrees, and A, B are sets in $\mathbf{a}, \mathbf{b}$ respectively. Define $A \oplus B = \{2x : x \in A\} \cup \{2x + 1 : x \in B\}$, and we define $\mathbf{a} \cup \mathbf{b}$ as the degree of $A \oplus B$. $\mathbf{a} \cup \mathbf{b}$, the least upper bound of $\mathbf{a}$ and $\mathbf{b}$, is well-defined. Therefore, all $\mathbf{D}_n$, $\mathbf{D}_\omega$, and $\mathbf{D}(\leq \mathbf{0}')$ are upper semi-lattices with least element $\mathbf{0}$ and greatest element $\mathbf{0}'$.

We say that **b** cups **a** to **c** if $\mathbf{a} \cup \mathbf{b} = \mathbf{c}$. In case when $\mathbf{c} = \mathbf{0}'$, then we say that **a** is cuppable, i.e., if there there is a c.e. degree **b** such that $\mathbf{a} \cup \mathbf{b} = \mathbf{0}'$.

If **a** and **b** have the greatest lower bound, we use $\mathbf{a} \cap \mathbf{b}$ to denote this greatest lower bound. Given a degree **a**, let $\mathbf{a}'$ be the *Turing jump* of **a**. The *jump classes* (or *high/low hierarchy*) are defined by

$$\mathbf{H}_n = \{\mathbf{a} \in \mathcal{R} : \mathbf{a}^{(n)} = \mathbf{0}^{(n+1)}\}, \quad \mathbf{L}_n = \{\mathbf{a} \in \mathcal{R} : \mathbf{a}^{(n)} = \mathbf{0}^{(n)}\},$$

where $\mathbf{x}^{(0)} = \mathbf{x}$ and $\mathbf{x}^{(n+1)}$ is the Turing jump of $\mathbf{x}^{(n)}$. A degree in $\mathbf{H}_n$ ($\mathbf{L}_n$) is called $high_n$ (low_n). When $n = 1$, degrees in $\mathbf{H}_1, \mathbf{L}_1$, are also called *high* degrees, *low* degrees respectively.

2. Cuppable degrees and plus-cupping

In 1963, Sacks proved that two most important theorems for the c.e. degrees, Sacks splitting theorem and Sacks density theorem.

Theorem 2.1. (Sacks Splitting Theorem) *Any nonzero c.e. degree can be split into two smaller ones. That is, if $\mathbf{a} > \mathbf{0}$, then there are c.e. degrees $0 < \mathbf{a}_0, \mathbf{a}_1 < \mathbf{a}$ such that $\mathbf{a}_0 \cup \mathbf{a}_1 = \mathbf{a}$.*

Theorem 2.2. (Sacks Density Theorem) *The c.e. degrees are dense. That is, for any c.e. degrees $\mathbf{b} < \mathbf{a}$, there is a c.e. degree $\mathbf{c}$ such that $\mathbf{b} < \mathbf{c} < \mathbf{a}$.*

By Theorem 2.1, $\mathbf{0}'$ can be split into two incomplete c.e. degrees, and as a consequence, there are incomplete cuppable degrees. Indeed, it is easy to prove that above any incomplete c.e. degree **a**, there are incomplete cuppable degrees. The proof is obvious. If **a** is cuppable, then any incomplete c.e. degree above **a** is also cuppable. If **a** is noncuppable, then we take **b** as any incomplete cuppable degree, and $\mathbf{a} \cup \mathbf{b}$ is incomplete and cuppable, as wanted.

In [5], Ambos-Spies, Lachlan and Soare proved that there are no minimal cuppable degrees.

Theorem 2.3. (Ambos-Spies, Lachlan and Soare [5]) *For any given incomplete c.e. degrees $\mathbf{a}, \mathbf{b}$, if $\mathbf{a} \cup \mathbf{b} = \mathbf{0}'$, then there is a c.e. degree $\mathbf{c}$ below $\mathbf{a}$ cupping $\mathbf{b}$ to $\mathbf{0}'$.*

The continuity problem originated from Lachlan's major subdegree problem. After seeing that every incomputable c.e. set has a major subset (Lachlan [40]), in 1967, Lachlan proposed a notion of *major subdegrees*, analogous to the notion of *major subsets* as follows: for any c.e. degrees $b < a$, say that b is a *major subdegree* of a, if every c.e. degree cupping a to $0'$ also cups b to $0'$. Lachlan asked whether every c.e. degree a between 0 and $0'$ has a major subdegree?

In his thesis [68], Seetapun proved that any nonzero low_2 c.e. degree has a major subdegree.

Theorem 2.4. (Seetapun [68]) *Every nonzero low_2 c.e. degree has a major subdegree.*

In [69], Seetapun provides a further improvement:

Theorem 2.5. (Seetapun [69]) *Given any c.e. degree a with $0 < a < 0'$, there is a c.e. degree $b \not\geq a$ such that any c.e. degree cupping a to $0'$ also cups b to $0'$.*

Recently, Cooper and Li announced in [20] a full solution of the major subdegree problem.

In Theorem 2.3, we want c below a to be as "deep" (in the high/low hierarchy) as possible. In [52], Li, Wu and Zhang introduced the following notion: a c.e. degree a is low_n-*cuppable*, if there is a low_n c.e. degree b cupping a to $0'$.

Theorem 2.6. (Li, Wu and Zhang [52]) *There is a low_2-cuppable, but not low-cuppable degree.*

Recently, Li claimed that there are cuppable degrees not low_2 cuppable. We believe the answer to the following question is positive.

Question 2.1. Is every cuppable degree low_3-cuppable?

In 1978, Harrington [33] (see Fejer and Soare [29]) proposed a much stronger cupping property.— plus-cupping property.

Definition 2.1. A nonzero c.e. degree a is a plus-cupping degree[a] if every nonzero c.e. degree below a is cuppable, i.e. if $0 < b \leq a$, then there is a c.e. degree c cupping b to $0'$.

---

[a] Harrington's original notion of plus-cupping degrees is even stronger: a is plus-cupping, in the sense of Harrington, if for any c.e. degrees b, c, if $0 < b \leq a \leq c$, there is a c.e. degree e below c cupping b to c. The notion given in this paper was given by Fejer and Soare in [29]. The proof of the existence of Harrington's notion of plus-cupping involves a complicated $0'''$ construction.

The following theorem is due to Harrington, which is called the *plus-cupping theorem*.

Theorem 2.7. (Harrington [33]) *Plus-cupping degrees exist.*

Proof sketch: The construction is a gap-cogap argument. To prove the theorem, we will construct a c.e. set A satisfying the following requirements:

$\mathcal{P}_e$: $A \neq \Phi_e$;
$\mathcal{R}_e$: $W_e = \Phi_e^A \Rightarrow W_e$ is computable or $\exists C_e$ c.e. and a p.c. functional Γ_e such that C_e is incomplete and $K = \Gamma_e^{C_e, W_e}$.

$\mathcal{P}_e$ is a simple Friedberg-Muchnik strategy. To satisfy $\mathcal{R}_e$, we will construct a c.e. set C_e such that if W_e is not computable, then C_e cups W_e to K via Γ_e.

Let α be an $\mathcal{R}_e$ strategy. First we define the length agreement function as follows:

Definition 2.2. (1) $\ell(\alpha, s) = \max\{x : \text{for all } y < x, W_{e,s}(y)[s] = \Phi_e^A(y)[s]\}$.
(2) $m(\alpha, s) = \max\{\ell(\alpha, t) : t < s \text{ and } t \text{ is an } \alpha\text{-stage}\}$.

Say that s is α-expansionary if $s = 0$ or $\ell(\alpha, s) > m(\alpha, s)$ and s is an α-stage. Suppose that there are infinitely many α expansionary stages. Then to satisfy $\mathcal{R}_e$, α will construct Γ_e and a c.e. set E such that $K = \Gamma_e^{C_e, W_e}$, $E \neq \Phi_i^{C_e}$ for each i (this ensures that C_e is incomplete) or α will show that W_e is computable. The construction of E is separated into infinitely many substrategies:

$\mathcal{S}_{e,i}$: $E \neq \Phi_i^{C_e}$.

Let β be an $\mathcal{S}_{e,i}$ strategy. β first chooses k as a big number. Wait for a stage s such that $\Phi_i^{C_e}(x)$ converges to 0. However, we cannot put x into E immediately since later, to rectify Γ_e, we need to enumerate γ_e-markers into C_e, and such enumerations can change the computation $\Phi_i^{C_e}(x)$, and our attack by putting x into E becomes useless.

To get around this, we use the gap-cogap method to force $\Phi_i^{C_e}(x)$ to be clear of the γ_e-markers (and hence this computation cannot be injured by this $\mathcal{R}$-strategy), or to show that W_e is computable. That is, when we see $\Phi_i^{C_e}(x)$ converges to 0, at stage s say, we create a link between α and β (we open a gap for A to change), and at the next α expansionary stage, s' say, we check whether W_e changes below $\gamma_e(k)[s]$ between these two stages.

If there is such a change (we close the gap successfully since we have a W_e change as wanted), then we can travel along the link and put x into E to satisfy $\mathcal{S}_{e,i}$. In this case, the W_e-change lifts $\gamma_e(k)$ to big numbers, and as a consequence, the computation $\Phi_i^{C_e}(x)$ is clear of the γ_e-markers, and can be preserved forever.

If there is no such a W_e change (we close the gap unsuccessfully, because W_e does no change as what we wanted), then at stage s', we put $\gamma_e(k)[s]$ into C_e, to lift $\gamma_e(k)$ to a big number. In this case, we define $g_\beta \restriction \gamma_e(k)[s] = W_e \restriction \gamma_e(k)[s]$. After stage s', we prevent A from changing to preserve $W_e \restriction \gamma_e(k)[s]$, till the next stage at which β finds at another stage s'' at which $\Phi_i^{C_e}(x)$ converges to 0 again. If such a procedure iterates infinitely often, and W_e never provides such changes (between stages s and s'), then g_β is defined infinitely often, and computes W_e correctly.

β has three outcomes, $g <_L w <_L d$, where w denotes the outcome β waits for $\Phi_i^{C_e}(x)$ to converge to 0, g denotes the outcome that β opens gaps infinitely often, and d denotes the outcome that β eventually (during a gap) gets W_e changes and enumerates x into E to make $E(x) = 1 \neq 0 = \Phi_i^{C_e}(x)$.

The construction is a $0'''$ priority argument.

3. Noncuppable degrees and anti-cupping property

In 1963, Shoenfield [72] raised a conjecture that for any given finite partial orderings $P \subseteq Q$, with the least element 0 and the greatest element 1, any embedding of P into c.e. degrees can be extended to an embedding of Q. Shoenfield's Conjecture would imply that the structure $\mathbf{R}$ is homogeneous and has two immediate consequences:

C1. For any c.e. degrees $\mathbf{a}, \mathbf{b}$, if $\mathbf{a}, \mathbf{b}$ are incomparable, then the greatest lower bound of $\mathbf{a}, \mathbf{b}$ does not exist.

C2. For any c.e. degrees $\mathbf{0} < \mathbf{b} < \mathbf{c}$, there is a c.e. degree $\mathbf{a} < \mathbf{c}$ cupping b to $\mathbf{c}$.

However, C1 and C2 are both wrong, and hence, Shoenfield's conjecture does not hold. C1 is refuted by the existence of minimal pairs (Lachlan [39] and independently Yates [87]). As to C2, Yates(unpublished) and Cooper [14] proved the existence of noncuppable degrees.

Definition 3.1. A nonzero c.e. degree $\mathbf{a}$ is noncuppble (in $\mathbf{R}$) if for any c.e. degree $\mathbf{b} < \mathbf{0}'$, $\mathbf{a} \cup \mathbf{b} < \mathbf{0}'$.

Theorem 3.1. *There are noncuppble degrees.*

Proof sketch: We will construct c.e. sets A, E and an auxiliary c.e. set F satisfying the following requirements:

$\mathcal{P}_e$: $E \neq \Phi_e^A$;

$\mathcal{Q}_e$: $A \neq \Phi_e$;

$\mathcal{N}_e$: $\Phi_e^{A,W_e} = K \oplus F \Rightarrow \exists \Gamma_e (K = \Gamma_e^{W_e})$.

Obviously, the $\mathcal{P}$ requirements ensure that A is not complete, the $\mathcal{Q}$ requirements ensure that A is not computable, and the $\mathcal{N}$ requirements ensure that only complete sets can cup A to K.

A $\mathcal{P}$-strategy is a simple Friedberg-Muchinik strategy. Let α be a $\mathcal{P}_e$-strategy. Then α chooses x as a big number, and waits for $\Phi_e^A(x)$ to converge to 0. If $\Phi_e^A(x)$ does not converge to 0, then we have $E(x) = 0 \neq \Phi_e^A(x)$, and $\mathcal{P}_e$ is satisfied. Otherwise, suppose that $\Phi_e^A(x)$ converges to 0 at stage s, then α performs the diagonalization by putting x into E, and preserving the computation $\Phi_e^A(x)$. In this case, $E(x) = 1 \neq 0 = \Phi_e^A(x)$, and again, $\mathcal{P}_e$ is satisfied. A $\mathcal{Q}$-strategy is even simpler.

An $\mathcal{N}_e$-strategy, β say, is devoted to the construction of a (partial) functional Γ_β such that if $K \oplus F = \Phi_e^{A,W_e}$, then $\Gamma_\beta^{W_e} = K$. We now define the length function of agreement as follows:

Definition 3.2. (1) $\ell(\beta, s) = \max\{x : \text{for all } y < x, K \oplus F(y)[s] = \Phi_e^{A,W_e}(y)[s]\}$.

(2) $m(\beta, s) = \max\{\ell(\beta, t) : t < s \text{ and } t \text{ is a } \beta\text{-stage}\}$.

Say that a stage s is β-expansionary if $s = 0$ or s is a β-stage and $\ell(\beta, s) > m(\beta, s)$. We only define Γ_β at β-expansionary stages. That is, if s is β-expansionary, and $\Gamma_\beta^{W_e}(x)[s]$ is not defined, with $2x < \ell(\beta, s)$, then define $\Gamma_\beta^{W_e}(x) = K(x)[s]$ with use s, which is bigger than $\varphi_e(2x)[s]$. After stage s, $\Gamma_\beta^{W_e}(x)$ can be undefined only when W_e changes below $\varphi_e(2x)$. Thus, in the case that x enters K, then we will force W_e to change below $\varphi_e(2x)$ (otherwise, $K \oplus F(2x)$ and $\Phi_e^{A,W_e}(2x)$ will be different, and $\mathcal{N}_e$ is satisfied vacuously), so that we can redefine $\Gamma_\beta^{W_e}(x)$ as $K(x)$ afterwards. β has two outcomes: infinitary outcome 0 and finitary outcome 1.

Problems arise when $\mathcal{Q}$-strategies below the infinitary outcome 0 of α enumerate numbers into A to make A incomputable. Suppose that at stage s, $\Gamma_\beta^{W_e}(x)[s]$ is defined, and a number n less than $\varphi_e(2x)[s]$ is enumerated into A. It may happen that between stage s and the next β-expansionary stage s', K changes at x and W_e has no changes below $\varphi_e(2x)[s]$. As a consequence, $\Gamma_\beta^{W_e}(2x)[s'] \downarrow = 0 \neq 1 = K(x)[s']$, and $\Gamma_\beta^{W_e}$ becomes incorrect.

To avoid this, we will ensure that in the construction, when we put numbers, n say, into A, and n is less than $\varphi_e(2x)[s]$, we need to ensure that $\Gamma_\beta^{W_e}(x)[s]$ is not defined. That is, when $\Gamma_\beta^{W_e}(x)[s]$ is defined and we want to put n into A, then instead of putting n into A immediately, we put a number less than x into F first, to force a W_e change below $\gamma_\beta(x)[s]$, and hence, to make $\Gamma_\beta^{W_e}(x)$ undefined. With this in mind, we delay the definition of Γ_β as follows: when we choose a number n as a fresh number for a Q strategy, ξ say, we also choose a number a_β fresh, and we extend the definition of Γ_β at a stage s only when $\ell(\beta, s)$ is bigger than $2a_\beta + 1$ ($\Phi_e^{A,W_e}(2a_\beta + 1)[s]$ converges). Let $\Gamma_\beta^{W_e}(x)$ be defined at stage s with use $\gamma_\beta(x) = s$.

Now assume that ξ wants to put n into A at stage $s' > s$, then ξ first puts a_β into F, with the intention to undefine all $\Gamma_\beta(x)$, $x \geq a_\beta$. Let $s'' > s'$ be the next β-expansionary stage (if exists), then between stages s' and s'', W_e must have changes below s and this change undefines $\Gamma_\beta^{W_e}(x)$. Thus, at stage s'', β puts x into A, and this enumeration keeps the definition of Γ_β consistent.

Obviously, the noncuppable degrees are downwards closed. Indeed, it is easy to prove that noncuppable degrees form an ideal. We show below that the join of two noncuppable degrees is still noncuppable. Let $\mathbf{b}, \mathbf{c}$ be two noncuppable degrees. For the sake of contradicting, suppose that $\mathbf{b} \cup \mathbf{c}$ is cuppable. That is, there is an incomplete c.e. degree $\mathbf{a}$ such that $\mathbf{a} \cup (\mathbf{b} \cup \mathbf{c}) = \mathbf{0}'$. Then, $(\mathbf{a} \cup \mathbf{b}) \cup \mathbf{c} = \mathbf{0}'$, and as a consequence, $\mathbf{a} \cup \mathbf{b} = \mathbf{0}'$, since $\mathbf{c}$ is noncuppable. However, since $\mathbf{b}$ is also noncuppable, $\mathbf{a} = \mathbf{0}'$. A contradiction. The existence of plus-cupping degrees shows that noncuppable degrees are not downwards dense in the c.e. degrees.

One immediate generalization of noncuppable degrees is the *anti-cupping property*. Given a c.e. degree $\mathbf{a}$, say that $\mathbf{a}$ has anti-cupping property if there is a c.e. degree $\mathbf{b} < \mathbf{a}$ such that $\mathbf{b}$ cups no c.e. degree below $\mathbf{a}$ to $\mathbf{a}$. $\mathbf{b}$ is called a *witness* of the anti-cupping property of $\mathbf{a}$. According to this definition, any noncuppable degree witnesses the anti-cupping property of $\mathbf{0}'$.

Say that $\mathbf{a}$ has *strong anti-cupping property* if there is a c.e. degree $\mathbf{b} < \mathbf{a}$ such that if $\mathbf{a} \leq \mathbf{c} \cup \mathbf{b}$, then $\mathbf{a} \leq \mathbf{c}$. In this case, $\mathbf{b}$ is called a *witness* of the strong anti-cupping property of $\mathbf{a}$.

In [32], Harrington proved that any high c.e. degree has strongly anti-cupping property.

Theorem 3.2. (Harrington, *see* Miller [58]) *Given a high c.e. degree* $\mathbf{a}$, *there is a high c.e. degree* $\mathbf{b} < \mathbf{a}$ *such that for any c.e. degree* $\mathbf{c}$, *if* $\mathbf{a} \leq \mathbf{b} \cup \mathbf{c}$ *then* $\mathbf{c} \geq \mathbf{a}$.

From Theorem 3.2, we know that every high c.e. degree bounds a high noncuppable degree.

Obviously, all degrees with strong anti-cupping property have anti-cupping property. Now we show that there are c.e. degrees with anti-cupping property, but not strong anti-cupping property.

A c.e. degree **c** is *contiguous* if it contains only one c.e. wtt-degree; i.e., all c.e. sets in **c** are all wtt-equivalent. This notion was first proposed by Ladner and Sasso in [43] in 1975. All contiguous degrees are low$_2$, and downwards dense in the c.e. degrees. Recent work of Downey and Lempp [24] and Ambos-Spies and Fejer [3] show that a c.e. degree **a** is contiguous if and only if it is locally distributive if and only if it is not the top of any embedding of lattice N_5 into the c.e. degrees.

Ladner and Sasso in [43] also noticed that any contiguous degree has anti-cupping property. From this observation, we can find degrees with anti-cupping property, but not strongly anti-cupping property, easily. Let **c** be any plus-cupping degree. Since contiguous degrees are downwards dense, there is a contiguous degree **a** below **c**. Thus there is a nonzero c.e. degree **b** below **a** such that **b** is a witness of the anti-cupping property of **a**. However, since **b** is also plus-cupping, there is an incomplete c.e. degree **e** such that $e \cup b = 0'$. Obviously, **a** and **e** are incomparable, and **a** does not have the strong anti-cupping property.

Another extension of anti-cupping property was proposed by Cooper [15] and Slaman and Steel [80]. Different from Harrington's definition, Cooper, Slaman and Steel considered the anti-cupping property in Turing degrees, and we will refer to this as type-2 strong anti-cupping property. Correspondingly, we will call Harrington's strong anti-cupping property type-1.

Say that a c.e. degree **a** has *type-2 strongly anti-cupping property* if there is a c.e. degree **b** below **a** such that **b** cups no degree (not just c.e.) below **a** to **a**. We

Theorem 3.3. (Cooper [15]; Slaman and Steel [80]) *There is a c.e. degree with type-2 strongly anti-cupping property.*

However, unlike the type-1 strongly anti-cupping property, no high degree can have type-2 strongly anti-cupping property, since in [63], Posner proved that if **h** is high, then any nonzero degree below **h** can be cupped to **h**. We note that type-1 and type-2 strongly anti-cupping properties also differ among the low degrees, since Cooper was able to prove that c.e. degrees with type-2 strongly anti-cupping property are downwards dense in the c.e. degrees.

One interesting topic related noncuppable degrees is the so-called "almost-deep" degrees.

Definition 3.3. (Cholak, Groszek and Slaman [11]) A c.e. degree **a** is *almost deep* if **a** cups all low c.e. degrees to low degrees.

In other words, cupping with almost deep degrees preserves lowness. Obviously, any almost deep degree is low.

Theorem 3.4. (Cholak, Groszek and Slaman [11]) *There are nonzero almost deep degrees.*

The motivation of almost deep degrees came from [10], where Bickford and Mills proposed the notion of *deep* degrees as follows: **a** is deep if for any c.e. degree **b**, $(\mathbf{a} \cup \mathbf{b})' = \mathbf{b}'$. In [46], Lempp and Slaman proved that there is only one deep degree, **0**.

Theorem 3.5. (Lempp and Slaman [46]) *There is no nonzero deep degree. That is, for any nonzero c.e. degree **a**, there is a c.e. degree **c** such that* $(\mathbf{c} \cup \mathbf{a})' > \mathbf{c}'$.

In [11], Cholak, Groszek and Slaman observed that **c** in Theorem 3.5 can be low$_2$, and that for any nonzero c.e. degree **a**, there is a non-high c.e. degree **w** such that **a**∪**w** is high. These observations show that joining with a nonzero c.e. degree cannot preserve low$_2$-ness and non-highness. In [36], Jockusch, Li and Yang proved that these two observations can be unified as follows:

Theorem 3.6. (Jockusch, Li and Yang [36]) *For any nonzero c.e. degree **a**, there is a low$_2$ c.e. degree **b** such that **b** ∪ **a** is high.*

4. Slaman triples, cappable degrees and minimal pairs

After seeing the existence of plus-cupping degrees, it is attractive to consider whether there is a nonzero c.e. degree **a** and a single incomplete c.e. degree **c** such that **c** cups every nonzero c.e. degree below **a** to **0**'. If the answer is positive, then $\mathbf{a} \cap \mathbf{c} = \mathbf{0}$ since if **b** were some nonzero c.e. degree below **a** and **c**, then **c** would cup **b** to **0**'. However, since **b** is below **c**, $\mathbf{b} \cup \mathbf{c} = \mathbf{c}$. A contradiction.

Unfortunately, the answer is negative, since Lachlan's Nondiamond Theorem (see [39]) says that that there are no c.e. degrees **a**, **b** such that $\mathbf{a} \cup \mathbf{b} = \mathbf{0}'$ and $\mathbf{a} \cap \mathbf{b} = \mathbf{0}$.

Theorem 4.1. (Lachlan [39]) *For any two nonzero c.e. degrees* **a** *and* **b**, *if* **a** *cups* **b** *to* **0**′, *then there is a nonzero c.e. degree* **c** *below both of them.*

By Lachlan nondiamond theorem, in the plus-cupping theorem, we need infinitely many c.e. degrees to cup nonzero c.e. degrees to **0**′. Slaman introduced a triple **a**, **b**, **c** such that **a** is nonzero, **c** $\not\leq$ **b**, and **b** cups each nonzero c.e. degree below **a** to a degree above **c**, and proved the existence of such triples. We will call such (**a**, **b**, **c**) a Slaman triple.

Theorem 4.2. (Shore and Slaman [76]) *Every high c.e. degree bounds a Slaman triple. That is, if* **h** *is a high c.e. degree, then there are c.e. degrees* **a**, **b**, **c** *below* **h** *such that* **a** *is nonzero,* **c** $\not\leq$ **b** *and* **b** *cups every nonzero c.e. degree below* **a** *to a degree above* **c**.

Obviously, in Theorem 4.2, **a** caps **b** to **0**. We will call **a** the base of this Slaman-triple and **b** the co-base.

Say that two nonzero c.e. degrees **a**, **b** form a *minimal pair* if **0** is the greatest lower bound of **a** and **b**. Theorem 4.2 implies Cooper's minimal pair theorem.

Theorem 4.3. (Cooper [13]) *Every high c.e. degree bounds a minimal pair.*

Historically, the existence of minimal pairs were first proved by Lachlan [39] and independently Yates [87], with the purpose to refute Shoenfield's conjecture (C1). We comment here that even Shoenfield's conjecture is not true, refuting it has led to a significant development of techniques in computability theory, as predicted by Shoenfield himself in [72].

Say a degree **a** is *cappable* if there is a nonzero c.e. degree **b** such that **0** is the greatest lower bound of **a** and **b**, i.e. **a**∩**b** = **0**. That is, a c.e. degree **a** is cappable if and only if **a** is **0** or a half of a minimal pair. As a consequence, the existence of minimal pairs implies the existence of nonzero cappable degrees. Harrington proved that a c.e. degree is either cappable or cuppable.

However, Yates proved in [88] the existence of noncappable degrees, the degrees not as a half of any minimal pair.

Theorem 4.4. (Yates [88]) *There exist incomplete noncappable degrees. That is, there is an incomplete c.e. degree* **a** *such that for any nonzero c.e. degree* **b**, *there is a nonzero c.e. degree* **c** *below both* **a** *and* **b**.

Ambos-Spies [2] extends the notion of noncappable as follows:

Theorem 4.5. (Ambos-Spies [2]) *A nonzero c.e. degree* **a** *is strongly noncappable degree if for any c.e. degree* **b**, **a** ∩ **b** *exists only when* **b** *is comparable with* **a**.

Theorem 4.6. (Ambos-Spies [2]) **0′** *can be split into two low strongly noncappable degrees.*

However, all the strongly noncappable degrees constructed in [2] are low. Lempp [45] proved that strongly noncappable degrees can also be high.

Theorem 4.7. (Lempp [45]) *There is a high strongly noncappable degree.*

Lempp asked in [45] whether in every jump class, there is a strongly non-cappable degree. Ambos-Spies announced a positive answer to this question recently.

Let **M** and **NC** be the set of cappable degrees and noncappable c.e. degrees, respectively. Then **M** is an ideal of **R** and **NC** is a strong filter. Both **M** and **NC** are both nonempty and form a nontrivial partition of the c.e. degrees. The study along this line has led to several important characterizations of cappable and noncappable degrees. In [57], Maass introduces the notion of *the promptly simple sets*, which provides an important dynamic property of those sets with noncappable degrees.

Definition 4.1. (Maass [57])

(1) A coinfinite c.e. set A is *promptly simple* if there is a partial computable function p and a computable enumeration $\{A_s\}_{s \in \omega}$ of A such that for every e, if W_e is infinite, then there are s and x such that $x \in W_{e,\text{at } s} \cap A_{p(s)}$, where $W_{e,\text{at } s}$ is the set of numbers enumerated into W_e at stage s.

(2) A c.e. degree **a** is called *promptly simple*, if it contains a promptly simple set.

Let **PS** be the set of all promptly simple degrees and **LC** be the set of all low-cuppable c.e. degrees. Then:

Theorem 4.8. (Ambos-Spies, Jockusch, Shore and Soare [4]) *A c.e. degree is noncappable if and only if it contains a promptly simple set if and only if it is low-cuppable. That is,* **NC** = **PS** = **LC**.

Thus, a c.e. degree is cappable if and only if it is not low-cuppable. In [52], Li, Wu and Zhang proved that there is a cappable and low$_2$-cuppable degree. By the characterization above, there is a low$_2$-cuppable, but not low-cuppable, degree. This separates the low-cuppable degrees from the low$_2$-cuppable degrees.

It is well-known that every nonzero c.e. degree bounds a cappable degree. The proof is nonuniform, in the sense that if $\mathbf{a} > \mathbf{0}$ is itself cappable, then any c.e. degree below $\mathbf{a}$ is cappable, and if $\mathbf{a}$ is noncappable, then we choose $\mathbf{b}$ as any nonzero cappable degree, and then there is a nonzero c.e. degree $\mathbf{c}$ below both $\mathbf{a}$ and $\mathbf{b}$. Since $\mathbf{c}$ is again cappable, there is a nonzero cappable degree below $\mathbf{c}$ and hence below $\mathbf{a}$.

In [34], Harrington and Soare proved that there are no maximal cappable degrees.

Theorem 4.9. (Harrington and Soare [34]) *If* $\mathbf{a}$ *and* $\mathbf{b}$ *form a minimal pair, then there is a c.e. degree* $\mathbf{c}$ *above* $\mathbf{a}$ *such that* $\mathbf{b}$ *and* $\mathbf{c}$ *also form a minimal pair.*

From Theorem 2.3 and Theorem 4.9, Harrington and Soare pointed our the following general continuity result.

Theorem 4.10. (Harrington and Soare [34]) *Let* $\mathbf{F}(\mathbf{x}, \mathbf{y})$ *be an open formula in the language* $\{<, \cup, \cap, \mathbf{0}, \mathbf{0}'\}$ *and* $\mathbf{a}, \mathbf{b}$ *be two distinct c.e. degrees such that* $\mathbf{F}(\mathbf{a}, \mathbf{b})$ *holds in the upper semilattice* $\mathcal{R}$ *of c.e. degrees. Then there exist c.e. degrees* $\mathbf{a_0}, \mathbf{a_1}, \mathbf{b_0}, \mathbf{b_1}$ *such that*

(1) $\mathbf{a_0} < \mathbf{a} < \mathbf{a_1}$, $\mathbf{b_0} < \mathbf{b} < \mathbf{b_1}$, *and*
(2) *for any c.e. degrees* $\mathbf{u}, \mathbf{v}$ *with* $\mathbf{a_0} < \mathbf{u} < \mathbf{a_1}$ *and* $\mathbf{b_0} < \mathbf{v} < \mathbf{b_1}$, $\mathbf{F}(\mathbf{u}, \mathbf{v})$ *is true.*

In [68], Seetapun improved Theorem 4.9 to a much stronger version (also see Giorgi [31]):

Theorem 4.11. (Seetapun [68]) *For any c.e. degree* $\mathbf{b} \neq \mathbf{0}, \mathbf{0}'$, *there is a c.e. degree* $\mathbf{a}$ *above* $\mathbf{b}$ *such that for any c.e. degree* $\mathbf{c}$, *if* $\mathbf{c}$ *and* $\mathbf{b}$ *form a minimal pair, then* $\mathbf{c}$ *and* $\mathbf{a}$ *form a minimal pair.*

The story for the minimal pairs is totally different. Even though any nonzero c.e. degree bounds cappable degrees, not every c.e. degree bounds a minimal pair.

Theorem 4.12. (Lachlan [42]) *There is a nonzero c.e. degree* $\mathbf{a}$ *such that for any c.e.* $\mathbf{b}$ *and* $\mathbf{c}$ *below* $\mathbf{a}$, $\mathbf{b}$ *and* $\mathbf{c}$ *do not form a minimal pair.* $\mathbf{a}$ *is called a nonbounding degree.*

Obviously, all nonbounding degrees are cappable. Downey, Lempp and Shore [25] proved that nonbounding degrees can be high$_2$ (it cannot be high as we know from Cooper's minimal pair theorem). Following Theorem 4.9, Seetapun argued that there is no maximal nonbounding degree.

Theorem 4.13. (Seetapun) *There are no maximal nonbounding degrees.*

The proof is quite easy. Let **a** be any nonbounding degree, then **a** is cappable. Let **c** be the degree as in Theorem 4.9, then any c.e. degree capping **a** to **0** also caps **c** to **0**. We claim that **c** is also nonbounding. Suppose not. Let **e, f** be a minimal pair below **c**. Then since **a** is nonbounding, either **a**∩**e** = **0** or **a**∩**f** = **0**, which implies that either **c**∩**e** = **0** or **c**∩**f** = **0**. This gives us a contradiction since **e, f** are nonzero degrees below **c**, **c** ∩ **e** = **e** and **c** ∩ **f** = **f**.

From the observation that the base and the co-base of any Slaman triple form a minimal pair, it is easy to see that there is a high$_2$ c.e. degree bounding no Slaman triples. Leonhardi proved in [47] that there is also a high$_2$ c.e. degree bounding no bases of Slaman triples. By almost the same arguments, we can prove that there is no maximal degree bounding no bases of Slaman triples and all such nonbounding degrees are cappable.

We end this section by introducing a variant of Slaman triple, saturated embedding, which is crucial in Slaman and Soare's decidability of the extension of embedding problem in [79]. Say that a triple {**a**, **b**, **a** ∪ **b**} is a *saturated embedding* if for any nonzero **w** below **a**, **w** cups **b** to **a** ∪ **b**. Roughly speaking, Slaman and Soare [79] proved that all obstructions to extensions of embeddings in the language {≤} were either lattice embedding obstructions like the minimal pairs, or saturated embeddings obstructions.

The following question seems to hard to handle:

Question 4.1. Are there nozero c.e. degrees **a** and **b** such that they are incomparable, **a** cups every nonzero c.e. degree below **b** to **a** ∪ **b**, and **b** cups every nonzero c.e. degree below **a** to **a** ∪ **b**?

5. Ideals M and NCup and quotient structures

As mentioned before, **M** and **NCup** are ideals definable in **R**. By the existence of cappable and cuppable degrees, **NCup** is properly contained in **M**.

Now we consider the quotient structures **R**/I, where I is any ideal of **R**. For any c.e. degree **a**, let [**a**] be the equivalence class of **a**. That is,

$[a] = \{b \in \mathbf{R} : \exists x \in I(a \cup x = b \cup x)\}$. We define partial order $\preceq$ on $\mathbf{R}/I$ by: $[a] \preceq [b]$ if and only if $a \leq b \cup x$ for some $x \in I$.

Schwarz [71] started the study of the quotient structure $\mathbf{R}/\mathbf{M}$, the upper semilattice of c.e. degrees modulo cappable degrees. Schwarz proved that the analogues of Friedberg-Muchnik theorem and Sacks splitting theorem are true in $\mathbf{R}/\mathbf{M}$. Schwarz [71] asked whether Shoenfield's conjecture is true in $\mathbf{R}/\mathbf{M}$, and pointed out that there is no minimal pair in $\mathbf{R}/\mathbf{M}$. In [90], Yi provided a negative answer. In [81], Sui and Zhang proved that there is no anti-cupping phenomenon in $\mathbf{R}/\mathbf{M}$.

Research of the quotient structure $\mathbf{R}/\mathbf{NCup}$, the upper semilattice of c.e. degrees modulo noncuppable degrees, was initiated recently by Li, Wu and Yang in [54]. It is easy to prove that there are incomparable elements in $\mathbf{R}/\mathbf{NCup}$. Let a, b be two incomplete c.e. degrees with $a \cup b = 0'$. Then $[a], [b]$ are incomparable, and $[0] \prec [a], [b] \prec [0']$, and $[a] \vee [b] = [0']$, where $\preceq$ is the partial order on $\mathbf{R}/\mathbf{NCup}$ induced from $\leq$.

In [55], Li, Sorbi, Wu and Yang prove that the diamond lattice can be embedded into $\mathbf{R}/\mathbf{NCup}$ preserving 0 and 1. Thus, there are minimal pairs in $\mathbf{R}/\mathbf{NCup}$. As a consequence, $\mathbf{R}/\mathbf{NCup}$ and $\mathbf{R}/\mathbf{M}$ are not elementarily equivalent. Furthermore, Shoenfield's conjecture is also false in $\mathbf{R}/\mathbf{NCup}$.

There are a bunch of questions about $\mathbf{R}/\mathbf{NCup}$. We believe the answer to the following question is positive.

Question 5.1. Is the analogue of Sacks splitting theorem true in $\mathbf{R}/\mathbf{NCup}$.

6. Arslanov's cupping theorem and diamond embeddings

Now we shift our survey to the cupping computably enumerable degrees in the Ershov hierarchy.

In 1981, Posner and Robinson proved in [64] the following cupping theorem in $\mathbf{D}(\leq 0')$.

Theorem 6.1. (Posner and Robinson [64]) *For any nonzero degree* a *below* $0'$, *there is a degree* b *such that*

$$b' = a \cup b = 0'.$$

Slaman and Steel improved this result in [80] by showing that b can be 1-generic.

By the existence of noncuppable degrees, we know that this cupping theorem fails in $\mathbf{R}$. However, Arslanov was able to prove that this cupping theorem is true in $\mathbf{D}_2$, the first level of the Ershov hierarchy beyond $\mathbf{R}$.

Theorem 6.2. (Arslanov's Cupping Theorem) *Every nonzero d.c.e. degree can be joined to* $\mathbf{0}'$ *by a low d.c.e. degree.*

By the existence of noncuppable c.e. degrees, this gives a difference between the elementary theories of $\mathbf{R}$ and $\mathbf{D}_n$ for every $n > 1$, and this difference is a Σ_3-sentence.

Proof sketch of Theorem 6.2:

We will actually prove the following: given any nonzero c.e. degree $\mathbf{a}$, there is a low d.c.e. degree $\mathbf{d}$ cupping $\mathbf{d}$ to $\mathbf{0}'$. To obtain Theorem 6.2, we use Lachlan's observation that any nonzero d.c.e. degree bounds a nonzero c.e. degree. Let $\mathbf{c}$ be any nonzero c.e. degree below $\mathbf{a}$, and let $\mathbf{d}$ be a d.c.e. degree cupping $\mathbf{d}$ to $\mathbf{0}'$. Then $\mathbf{d}$ also cups $\mathbf{a}$ to $\mathbf{0}'$.

Let A be any c.e. set in $\mathbf{a}$, and $\{A_s\}_{s\in\omega}$ be an effective enumeration of A. We will construct a low d.c.e. set D and a partial function Γ satisfying

$\mathcal{G}$: $K = \Gamma^{A,D}$;

$\mathcal{N}_e$: If there are infinitely many stages s such that $\Phi_e^D(e)[s]$ converges, then $\Phi_e^D(e)$ converges.

The $\mathcal{G}$-strategy is devoted to the construction of Γ, and ensures that at any stage s, for any number x, if $\Gamma^{A,D}(x)[s]$ is defined, then $\Gamma^{A,D}(x)[s] = K_s(x)$ and for any $y < x$, $\Gamma^{A,D}(y)[s]$ is also defined, and equals to $K_s(y)$, with use $\gamma(y)[s]$ less than or equal to $\gamma(x)[s]$. If x enters K at stage $s+1$, then we will ensure that A or D must have a change below $\gamma(x)[s]$ at stage $s+1$, making $\Gamma^{A,D}(x)$ undefined. If $\Gamma^{A,D}(x)[s_1]$ is defined, $A_{s_2} \restriction \gamma(x)[s_1] = A_{s_1} \restriction \gamma(x)[s_1]$ and $D_{s_2} \restriction \gamma(x)[s_1] = D_{s_1} \restriction \gamma(x)[s_1]$ for stage $s_2 > s_1$, then $\Gamma^{A,D}(x)[s_2]$ is also defined and equals to $\Gamma^{A,D}(x)[s_1]$, with use $\gamma(x)[s_2] = \gamma(x)[s_1]$.

To satisfy an $\mathcal{N}_e$ requirement, we apply "capricious destruction" method, which is an important technique proposed by Lachlan in his nonsplitting theorem.

First, we define a parameter k, as the threshold. Whenever we find a stage s at which $\Phi_e^D(e)[s]$ converges, we enumerate $\gamma(k)[s]$ into D, to lift the uses $\gamma(z)$ for all $z \geq k$ to big numbers, and at the same time define g on all arguments $y < \gamma(k)[s]$ with $g(y) = A_s(y)$. Wait for A to change below $\gamma(k)[s]$ (i.e. wait for an A-permission below $\gamma(k)[s]$ to lift $\gamma(k)$ to big numbers). Simultaneously, wait for a new stage s' at which $\Phi_e^D(e)[s']$ converges, and iterate this procedure.

Note that enumerating $\gamma(k)[s]$ into D may change the computation $\Phi_e^D(e)$. However, we know that this injury is only caused by $\gamma(k)[s]$, because $\gamma(z)$, $z \geq k$, are all lifted to big numbers. That is, when $\gamma(z)$ is defined later, they are defined as numbers bigger than $\varphi_e(e)[s]$. Now if A changes below $\gamma(k)[s]$ after stage s, we can take $\gamma(k)[s]$ out of D, to recover the computation $\Phi_e^D(e)$ to $\Phi_e^D(e)[s]$. The crucial point here is that the A-change below $\gamma(k)[s]$ makes $\Gamma^{A,D}(k)$ undefined, and when we redefine it later, we define it as numbers bigger than $\varphi_e(e)[s]$. As a consequence, the computation $\Phi_e^D(e)$ is now clear of the γ-uses, and can be preserved forever.

In the argument given above, we assume that K does not change below k, which may turn out to be wrong. When K changes below k, we ignore all the work we have done. That is, we cancel all the parameters, except k. In this case, we say that this $\mathcal{N}_e$ strategy is reset. Since k is fixed, $\mathcal{N}_e$ strategy can be reset at most k many times. Thus we can assume that K does not change below k after a stage large enough.

We argue now that g can be defined at most finitely often. Suppose not, then since A never permits, for a fixed y, we can calculate $A(y)$ as follows: search for a stage s at which $g(y)$ is defined, then $A(y) = A_s(y)$, because otherwise, when $A(y)$ changes later, then as described above, we can take $\gamma(k)[s]$ out of D, satisfy $\mathcal{N}_e$, and hence this $\mathcal{N}_e$ strategy will never define g afterwards. A contradiction. Thus g computes A correctly, and A is computable. This is another contradiction.

Let t be the last stage at which g is defined. There are two cases. One case is that A does not change below $\gamma(k)[t]$ (A does not permit). In this case, $\Phi_e^D(e)$ does not converge at any stage after t, and $\mathcal{N}_e$ is satisfied vacuously. The other case is that a A-change below $\gamma(k)[t]$ allows us to take $\gamma(k)[t]$ out of D, and we can recover $\Phi_e^D(e)$ and preserve it forever. Again, $\mathcal{N}_e$ is satisfied.

The construction is a finitary argument. $\qquad\qquad\qquad\qquad\square$

We comment here that the proof of Theorem 6.2 is nonuniform, due to the nonuniformity of Lachlan's argument. More precisely, if **a** itself is c.e., we can apply Sacks density theorem to show the existence of **c**. If **a** is proper d.c.e., let A be a d.c.e. set in **a** and $\{A_s\}_{s\in\omega}$ be an effective approximation of A, then let **c** be the Turing degree of the Lachlan set $C = \{\langle x, s \rangle : \exists t > s(x \in A_s/A_t)\}$. Then **c** is c.e. and below **a**. Furthermore, **c** is nonzero, as wanted, since **a** is c.e. in **c**, and **a** is a proper d.c.e. degree.

In 1989, Downey provided another difference between **R** and $\mathbf{D}_2$ by showing that the diamond lattice can be embedded into $\mathbf{D}_2$ preserving 0 and 1.

Theorem 6.3. (Downey [23]) *The diamond lattice can be embedded into* $\mathbf{D}_2$ *preserving both* 0 *and* 1.

Note that this difference is at Σ_2 level.

After seeing these two results, and the fact that a c.e. degree is cappable in the c.e. degrees if and only if it is cappable in the d.c.e. degrees, we know that for any nonzero cappable c.e. degree $\mathbf{a}$, there are a nonzero c.e. degree capping $\mathbf{a}$ to $\mathbf{0}$, and a d.c.e. degree $\mathbf{d}$ cupping $\mathbf{a}$ to $\mathbf{0}'$. It is interesting to consider whether there is a d.c.e. degree $\mathbf{d}$ such that $\mathbf{d}$ cups $\mathbf{a}$ to $\mathbf{0}'$, and at the same time, caps $\mathbf{a}$ to $\mathbf{0}$.

Theorem 6.4. (Wu [82]) *There are nonzero c.e. degrees* $\mathbf{a}, \mathbf{c}$, *and a d.c.e. degree* $\mathbf{d} > \mathbf{c}$ *such that* $\mathbf{a} \cap \mathbf{c} = \mathbf{0}$, $\mathbf{a} \cup \mathbf{d} = \mathbf{0}'$, *and* $\mathbf{c}$ *is the greatest c.e. degree* $\mathbf{d}$. *As a consequence,* $\{\mathbf{0}, \mathbf{a}, \mathbf{d}, \mathbf{0}'\}$ *is a diamond.*

In Theorem 6.4, cupping and capping are treated separately, which makes the diamond embedding much easier. The technique employed in Theorem 6.4 can be combined with the gap-cogap method to prove:

Theorem 6.5. (Downey, Li and Wu [26]) *A nonzero c.e. degree is cappable in the c.e. degrees if and only if it is complemented in the d.c.e. degrees.*

This provides a new characterization of cappable c.e. degrees.

7. Isolation and intervals of d.c.e. degrees

$\mathbf{c}$ and $\mathbf{d}$ in Theorem 6.4 are now known as an isolation pair, and we say that $\mathbf{c}$ isolates $\mathbf{d}$ and $\mathbf{d}$ is isolated by $\mathbf{c}$. Say that a d.c.e. degree is isolated if there is a c.e. degree below $\mathbf{d}$ isolating $\mathbf{d}$. Obviously, $\mathbf{c}$ isolates $\mathbf{d}$ if and only if $\mathbf{c}$ is the largest c.e. degree below $\mathbf{d}$.

The existence of the isolated pairs can also be obtained from Kaddah's infima theorem [38] which says that each low c.e. degree is branching in the d.c.e. degrees. Let $\mathbf{c}$ be a low nonbranching c.e. degree, then according to Kaddah's theorem, there are d.c.e. degrees $\mathbf{d}_1, \mathbf{d}_2$ such that $\mathbf{c} = \mathbf{d}_1 \cap \mathbf{d}_2$. It is clear to see that $\mathbf{c}$ isolates at least one of $\mathbf{d}_1$ and $\mathbf{d}_2$.

The notion of isolated degrees was first proposed explicitly by Cooper and Yi in [21], where they commented in the footnote that the existence of isolated degrees can be obtained from Kaddah's result. Both isolated and nonisolated degrees are dense in the c.e. degrees.

Theorem 7.1. (Ding and Qian [22]; LaForte [44]; Arslanov, Lempp and Shore [9]) *The isolated and nonisolated d.c.e. degrees are dense in the c.e. degrees.*

Cooper and Yi [21] and independently Ishmukhametov [35], proved that for any c.e. degree c, and d.c.e. degree d, with $c < d$, there is a d.c.e. degree e such that $c < e < d$. Obviously, if c is a c.e. degree and d is a d.c.e. degree above c, then there is an isolated degree between c and d.

In Wu [83], the notion of bi-isolation was proposed, where the isolated degrees defined by Cooper and Yi are called degrees isolated from below. A d.c.e. degree d is *isolated from above* if there is a c.e. degree c above d, and there is no c.e. degree between c and d.

A d.c.e. degree d is *bi-isolated* if d is isolated from above and from below. It is easy to modify Arslanov's proof to show that there is a bi-isolated degree.

A c.e. degree c is *bi-isolating* if c isolates a d.c.e. degree d_1 from below and isolates a d.c.e. degree d_2 from above. Obviously, c is now the unique c.e. degree between d_1 and d_2. In [83], Wu pointed out that the bi-isolating degrees are dense in the high c.e. degrees. Arslanov asked whether such degrees occur in every jump class. Recently, in [85], we provide a partial solution by showing that such degrees can be low. We end this section by raising Arslanov's question again.

Question 7.1. (Arslanov) Do the bi-isolating degrees occur in every jump class?

8. Maximal degrees and almost universal cupping

After seeing Arslanov's Theorem 6.2, it is natural to ask whether there is a d.c.e. degree cupping every nonzero c.e. degree to $0'$. The answer is negative, since any nonzero d.c.e. degree bounds a nonzero c.e. degree. Now we ask whether there exists an incomplete d.c.e. degree d such that for any c.e. degree c, either $c \leq d$ or $d \cup c = 0'$. We call such d.c.e. degrees *almost universal cupping* degrees. The answer is positive, because of the existence of an incomplete maximal d.c.e. degree.

Theorem 8.1. (Cooper, Harrington, Lachlan, Lempp and Soare [18]) *There is an incomplete maximal d.c.e. degree.*

The proof of Theorem 8.1 is fairly complicated and subtle. Actually, we can prove the existence of almost universal cupping degrees directly.

Theorem 8.2. *There is an incomplete d.c.e. degree* **d** *cupping every c.e. degree not below* **d** *to* **0**$'$.

We construct a d.c.e. set D and an auxiliary c.e. set E satisfying the following requirements:

$$\mathcal{P}_e: E \neq \Phi_e^D.$$
$$\mathcal{R}_e: \Gamma_e^{W_e \oplus D} = K \text{ or } W_e \leq_T D.$$

Obviously, the $\mathcal{P}$ requirements ensure that D is not complete, and the $\mathcal{R}$ requirements ensure that for any c.e. set C not reducible to D, D cups C to K.

A $\mathcal{P}_e$-strategy attempts to find an x such that $E(x) \neq \Phi_e^D(x)$. A single $\mathcal{P}_e$-strategy is the Friedberg-Muchnik strategy.

An $\mathcal{R}_e$ strategy attempts to code K into $W_e \oplus D$ via a functional Γ_e, if W_e is not computable in D. That is, given x, $\mathcal{R}_e$ defines $\Gamma_e^{W_e \oplus D}(x) = K(x)$ first with the use $\gamma_e(x)$ a big number. If later, $K(x)$ changes (from 0 to 1), then we undefine $\Gamma_e^{W_e \oplus D}(x)$ by enumerating $\gamma_e(x)$ (or a number less than $\gamma_e(x)$) into D. If W_e is not computable in D, then $\Gamma_e^{W_e \oplus D}$ will be constructed as a total function and computes K correctly.

However, the $\mathcal{P}$-strategies become much complicated if $\mathcal{P}$ is working below $\mathcal{R}$-strategies.

First, we describe how a $\mathcal{P}$ strategy works below an $\mathcal{R}$ strategy, $\mathcal{R}_i$ say. The construction will proceed on a tree. Let α be an $\mathcal{R}_i$ strategy, and β be a $\mathcal{P}$ strategy below α. The following procedure can happen infinitely often: β finds an x with $\Phi_e^D(x)[s]$ converging at stage s, and puts x into E to make a disagreement between E and Φ^D. This disagreement is *temporary* since it is possible that later, to code K into $W_i \oplus D$, α puts $\gamma_i(y)$ into D, and hence change $\Phi_e^D(x)$.

To circumvent this obstacle, β first defines a parameter k big enough, as in Arslanov's proof, and whenever K changes below k, *reset β*. In the following, we suppose that K does not change below k anymore.

Without loss of generality, suppose that $\Phi_e^D(x)[s_1]$ converges to 0, then instead of enumerating x into E, β puts $\gamma_i(k)[s_1]$ into D, to *undefine* $\gamma_i(z)$ for those $z \geq k_i$, and request that when $\Gamma_i^{W_i \oplus D}(z)$, $z \geq k$, is redefined later, then define $\gamma_i(z)$ big. *Particularly, $\gamma_i(z)$ will be bigger than $\varphi_e(x)[s_1]$.* Define $W_{i,s_1}(x) = \Delta_i^D(x)$ for all those $x \leq \gamma_i(k)[s_1]$ if $\Delta_i(x)$ is not defined. Wait for a W_i-change below $\gamma_i(k)[s_1]$.

We are applying the "capricious destruction" method. That is, the enumeration of $\gamma_i(k)[s_1]$ into D prevents β from being injured by further actions of α, but enumerating $\gamma_i(k)[s_1]$ itself into D may change the computation $\Phi_e^D(x)$.

If W_i changes below $\gamma_i(k)[s_1]$ at stage $s_2 > s_1$, then β can take $\gamma_i(k)[s_1]$ out of D to recover $\Phi_e^D(x)$ to $\Phi_e^D(x)[s_1]$, which equals to 0. The change of W_i below $\gamma_i(k)[s_1]$ undefines all $\Gamma_i(W_i \oplus D; z)$ with $z \geq k$, and when $\Gamma_i^{W_i \oplus D}(z)$ is defined again, $\gamma_i(z)$ will be defined bigger than $\varphi_e(x)[s_1]$, and the enumeration of $\gamma_i(z)$ will not injure $\Phi_e^D(x)$ anymore. Furthermore, β succeeds in preserving a computation $\Phi_e^D(x)$. Now put x into E, and we will have $E(x) = 0 \neq 1 = \Phi_e^D(x)[s_1] = \Phi_e^D(x)$, and $\mathcal{P}_e$ is satisfied permanently.

If W_i has no changes below $\gamma_i(k)[s_1]$, then β's attempt at stage s_1 (to lift $\gamma_i(k)$ to big numbers) fails. However, in this case, we will have

$$W_i \upharpoonright \gamma_i(k)[s_1] = W_{i,s_1} \upharpoonright \gamma_i(k)[s_1],$$

and hence W_i and Δ_i^D equal on each argument below $\gamma_i(k)[s_1]$.

β has outcomes $g_i <_L w <_L d$, where d denotes that β performs the diagonalization eventually, w denotes that β waits forever for Φ_e^D to converge, and g_i denotes that β puts $\gamma_i(k)$ into D infinitely often, and Δ_i^D computes W_i correctly.

The crucial problem arises when we consider two $\mathcal{P}$ strategies, β_1, β_2 working below two $\mathcal{R}$ strategies α_1, α_2. We assume that $\alpha_1 \subset \alpha_2 \subset \beta_1 \hat{\ } g_2 \subset \beta_2$.

Let $k(\beta_i)$ be β_i's threshold. Since β_2 knows that β_1's outcome g_2, β_2 knows that W_{α_2} is computable in D, and believes a computation $\Phi_{\beta_2}^D(x)$ is correct if $\gamma_{\alpha_2}(k_1)$ is bigger use $\varphi(x)$, and β_1's further actions will not injure β_2.

On the other hand, β_2's can actually injure β_1 in the following way. Suppose at stage s_1, β_2 puts $\gamma_{\alpha_1}(k_2)[s_1]$ into D, and at stage $s_2 > s_1$, β_1 puts $\gamma_{\alpha_2}(k_1)[s_2]$ into D (β_1 wants to apply the capricious destruction method to do diagonalization), and β_2 takes $\gamma_{\alpha_1}(k_2)[s_1]$ out of D (β_2 finds a W_{α_1} change and do diagonalization at stage s_2). Suppose that at stage s_3, β_1 finds a W_{α_2} change at x and wants to do diagonalization. Then to recover the computation to the one at stage s_2, β_1 needs to make sure that $\gamma_{\alpha_1}(k_2)[s_1]$ is still in D, which is impossible since we are making D d.c.e.

Note that since the computation β_1 wants to recover depends on the membership $D(\gamma_{\alpha_1}(k_2)[s_1]) = 1$, we know that x is bigger than $\gamma_{\alpha_1}(k_2)[s_1]$. This means that $\Delta^D(x)$ is defined between stages s_1 and s_2. To make the construction work, at stage s_2, when we take $\gamma_{\alpha_1}(k_2)[s_1]$ out of D, we also put s_1 into D. This enumeration is consistent with β_2 because s_1 is bigger than the use of any computation found at stage s_1. On the other hand, enumerating s_1 into D undefines all $\Delta^D(z)$ that are defined between stages s_1 and s_2. Thus, the hard situation described above will not happen.

The construction is a $0'''$ argument.

We now consider a plus-cupping theorem in the d.c.e. degrees. As discussed above, if $\mathbf{a}$ is a plus-cupping degree, there is no single c.e. degree cupping every nonzero c.e. degree below $\mathbf{a}$ to $\mathbf{0}'$. The next theorem says that it can be done in the d.c.e. degrees.

Theorem 8.3. (Wu [86]) *There is a nonzero c.e. degree* $\mathbf{a}$ *and an incomplete d.c.e. degree* $\mathbf{d}$ *such that* $\mathbf{d}$ *cups each nonzero c.e. degree below* $\mathbf{a}$ *to* $\mathbf{0}'$.

The proof of Theorem 8.3 is a gap-cogap argument. Actually, this proof is consistent with the proof of Theorem 8.2, and we can actually prove:

Theorem 8.4. (Wu [86]) *There is a nonzero c.e. degree* $\mathbf{a}$ *and an incomplete d.c.e. degree* $\mathbf{d}$ *such that* $\mathbf{a}$ *cups every c.e. degree not below* $\mathbf{a}$ *to* $\mathbf{0}'$ *and* $\mathbf{d}^-$ *cups each nonzero c.e. degree below* $\mathbf{a}$ *to* $\mathbf{0}'$.

Obviously, Theorem 8.4 implies Li and Yi's cupping theorem immediately.

Theorem 8.5. (Li and Yi [56]) *There are incomplete d.c.e. degrees* $\mathbf{d}_0, \mathbf{d}_1$ *such that for any nonzero c.e. degree* $\mathbf{w}$, *either* $\mathbf{d}_0 \cup \mathbf{w} = \mathbf{0}'$ *or* $\mathbf{d}_1 \cup \mathbf{w} = \mathbf{0}'$.

The following questions are open in the d.c.e. degrees:

Question 8.1. Are c.e. degrees definable in the d.c.e. degrees?

Question 8.2. Are all finite lattices embeddable into the d.c.e. degrees preserving 0 and 1?

Question 8.3. For any $m > n \geq 2$, $\mathbf{D}_m$ is elementarily equivalent to $\mathcal{D}_n$.

Question 8.3 is known as Downey's conjecture. Recently, Arslanov, Kalimullin and Lempp announced that $\mathbf{D}_2$ and $\mathbf{D}_3$ are not elementarily equivalent.

9. Cupping in the ω-c.e. degrees

Say that a degree $\mathbf{a}$ is *universal cupping* if it cups every nonzero c.e. degree to $\mathbf{0}'$. From Lachlan's observation, universal cupping degrees cannot be n-c.e. for any n. Since $\mathbf{0}'$ is splittable above any low degree, universal cupping degrees cannot be low. In [49], Lewis proved that universal cupping degrees can be minimal, and hence low$_2$.

Theorem 9.1. (Lewis [49]) *There is a minimal degree cupping every nonzero c.e. degree to* $\mathbf{0'}$.

Recently, joint with Li and Song, we prove the following cupping theorem:

Theorem 9.2. (Li, Song and Wu [53]) *There is an* ω*-c.e. degree cupping every nonzero c.e. degree to* $\mathbf{0'}$.

The proofs of these two theorems are $\mathbf{0'''}$ arguments. We are interested in whether there two results can be unified:

Question 9.1. Is there an ω-c.e. minimal degree cupping every nonzero c.e. degree to $\mathbf{0'}$.

We end this survey by an application of noncuppable degrees.

In [89], Yates proved that there are degrees **a** such that $\mathbf{0}$ and $\mathbf{0'}$ are the only two c.e. degrees comparable to **a**. Yates' construction was fairly complicated, as commented by Shoenfield, because Yates needed to make the constructed set not computable relative in any c.e. set of incomplete degree. We can actually avoid the hard part of Yates construction by using the existence of noncuppable degrees.

In [80], Slaman and Steel proved that each nonzero degree below $\mathbf{0'}$ has a 1-generic complement. Let **a** be a nonzero noncuppable degree and let **d** be a 1-generic complement, then $\mathbf{0'}$ is the only one c.e. degree above **d**. Since 1-generic degrees bound no nonzero c.e. degrees, $\mathbf{0}$ is the only one c.e. degree below **d**. Thus, $\mathbf{0}$ and $\mathbf{0'}$ are the only two c.e. degrees comparable with **d**, **d** is a Yates degree, and hence Yates degrees can be low. Yates degrees can also be minimal because for any given nonzero incomplete c.e. degree **b**, there is a minimal degree **m** such that $\mathbf{b} \cup \mathbf{m} = \mathbf{0'}$ (see Seetapun and Slaman [70]). Next theorem shows that Yates degrees occur in every jump class.

Theorem 9.3. (Wu [84]) *Let C be an incomputable c.e. set and S be any set c.e. in and above $\emptyset'$, then there is a Δ_2^0 set A such that (1) $A' \equiv_T S$, (2) $K \equiv_T A \oplus C$, (3) any computably enumerable set reducible to A is computable.*

If we let C be a c.e. set with noncuppable degree, then immediately, we have:

Theorem 9.4. (Wu [84]) *Yates degrees occur in every jump class.*

References

1. K. Ambos-Spies. An extension of the nondiamond thoerem in classical and α-recursion theory. *J. Symbolic Logic* **49** (1984), 586–607.

2. K. Ambos-Spies. On pairs of recursively enumerable degrees. *Trans. Amer. Math. Soc.* **283** (1984), 507–531.

3. K. Ambos-Spies and P. A. Fejer. Embeddings of N_5 and the contiguous degrees. *Ann. Pure Appl. Logic* **112** (2001), 151–188.

4. K. Ambos-Spies, C. G. Jockusch, Jr., R. A. Shore, and R. I. Soare. An algebraic decomposition of the recursively enumerable degrees and the coincidence of several degree classes with the promptly simple degrees. *Trans. Amer. Math. Soc.* **281** (1984), 109–128.

5. K. Ambos-Spies, A. H. Lachlan, and R. I. Soare. The continuity of cupping to $\mathbf{0}'$. *Ann. Pure Appl. Logic* **64** (1993), 195–209. .

6. K. Ambos-Spies and R. I. Soare. The recursively enumerable degrees have infinitely many one-types. *Ann. Pure Appl. Logic* **44** (1989), 1–23.

7. M. M. Arslanov. Structural properties of the degrees below $\mathbf{0}'$. *Dokl. Nauk. SSSR* **283** (1985), 270-273.

8. M. M. Arslanov. Open questions about the n-c.e. degrees. In M. Lerman P. Cholak, S. Lempp and R. A. Shore, editors, *Computability Theory and Its Applications: Current Trends and Open Problems*, volume 257 of *Contemporary Mathematics*, 15–22, Providence RI, 2000. AMS.

9. Marat M. Arslanov, Steffen Lempp, and Richard A. Shore. On isolating r.e. and isolated d-r.e. degrees. In *Computability, enumerability, unsolvability*, volume 224 of *London Math. Soc. Lecture Note Ser.*, 61–80. Cambridge Univ. Press, Cambridge, 1996.

10. M. Bickford and C. F. Mills. Lowness properties of r.e. degrees. Unpublished preprint, 1982.

11. P. Cholak, M. Groszek, and T. A. Slaman. An almost deep degree. *J. Symbolic Logic* **66** (2001), 881–901.

12. S. B. Cooper. Jump equivalence of the Δ_2^0 hyperhyperimmune sets. *J. Symbolic Logic* **37** (1972), 598–600.

13. S. B. Cooper. Minimal pairs and high recursively enumerable degrees. *J. Symbolic Logic* **39** (1974), 655–660.

14. S. B. Cooper. On a theorem of C. E. M. Yates. Handwritten notes, 1974.

15. S. B. Cooper. The strong anticupping property for recursively enumerable degrees. J. Symbolic Logic **54** (1989), 527–539.

16. S. B. Cooper. Definability and global degree theory. In J. Oikkonen and J. Väänänen, editors, *Logic Colloquium '90*, volume 2 of *Lecture Notes in Logic*, 25–45, 1991.

17. S. B. Cooper. Local degree theory. In *Handbook of Computability Theory*, 121–153. North-Holland, Amsterdam, 1999.

18. S. B. Cooper, L. Harrington, A. H. Lachlan, S. Lempp, and R. I. Soare. The d.r.e. degrees are not dense. *Ann. Pure Appl. Logic* **55** (1991), 125–151.

19. S. B. Cooper, S. Lempp, and P. Watson. Weak density and cupping in the d-r.e. degrees. *Israel J. Math.* **67** (1989), 137–152.

20. S. B. Cooper and A. Li. On Lachlan's major subdegree problem. to appear.

21. S. B. Cooper and X. Yi. Isolated d.r.e. degrees. University of Leeds, Dept. of Pure Math., 1995. Preprint series, No.17.

22. D. Ding and L. Qian. Isolated d.r.e. degrees are dense in r.e. degree structure. *Arch. Math. Logic* **36** (1996), 1–10.

23. R. Downey. D.r.e. degrees and the nondiamond theorem. *Bull. London Math. Soc.* **21** (1989), 43–50.

24. R. G. Downey and S. Lempp. Contiguity and distributivity in the enumerable Turing degrees. *J. Symbolic Logic* **62** (1997), 1215–1240.

25. R. G. Downey, S. Lempp, and R. A. Shore. Highness and bounding minimal pairs. *Math. Logic Quart.* **39** (1993), 475–491.

26. R. G. Downey, A. Li, and G. Wu. Complementing cappable degrees in the difference hierarchy. *Ann. Pure Appl. Logic* **125** (2004), 101–118.

27. Y. L. Ershov, A hierarchy of sets, Part I. Algebra i Logika **7** (1968), 47-73 (Russian); Algebra and Logic **7** (1968), 24–43 (English translation).

28. Y. L. Ershov, A hierarchy of sets, Part II. Algebra i Logika **7** (1968) 15-47 (Russian), Algebra and Logic **7** (1968), 212–232 (English Translation).

29. P. A. Fejer and R. I. Soare. The plus-cupping theorem for the recursively enumerable degrees. In *Logic Year 1979–80: University of Connecticut*, 49–62, 1981.

30. R. M. Friedberg. Two recursively enumerable sets of incomparable degrees of unsolvability (solution to post's problem, 1944). *Proc. Nat. Acad. Sci. U.S.A.* **43** (1957), 236–238.

31. M. B. Giorgi. *Continuity properties of degree structures*. PhD thesis, University of Leeds, 2001.

32. L. A. Harrington. On Cooper's proof of a theorem of Yates. Notes, 1976.

33. L. A. Harrington. Plus-cupping in the recursively enumerable degrees. Notes, 1978.

34. L. A. Harrington and R. I. Soare. Games in recursion theory and continuity properties of capping degrees. In *Proceedings of the workshop on set theory and the continuum*. Mathematical Sciences Research Institute, Berkeley, 1989. 39–62.

35. Shamil Ishmukhametov. *D.r.e. sets, their degrees and index sets*. PhD thesis, Novosibirsk, Russia, 1986.

36. C. G. Jockusch, Jr., A. Li, and Y. Yang. A join theorem for the computably enumerable degrees. *Trans. Amer. Math. Soc.* **356** (2004), 2557–2568.

37. S. C. Kleene and E. Post, The upper semilattice of degrees of recursive unsolvability, *Ann. Math.* **59** (1954), 1108–1109.

38. D. Kaddah. Infima in the d.r.e. degrees. *Ann. Pure Appl. Logic* **62** (1993), 207–263.

39. A. H. Lachlan. Lower bounds for pairs of recursively enumerable degrees. *Proc. London Math. Soc.* **16** (1966), 537–569.

40. A. H. Lachlan. On the lattice of recursively enumerable sets. *Trans. Amer. Math. Soc.* **130** (1968), 1–37.

41. A. H. Lachlan. A recursively enumerable degree which will not split over all lesser ones. *Ann. Math. Logic* **9** (1975), 307–365.

42. A. H. Lachlan. Bounding minimal pairs. *J. Symbolic Logic* **44** (1979), 626–642.

43. R. E. Landner and L. P. Sasso. The weak truth table degrees of recursively enumerable degree. *Ann. Math. Logic* **8** (1975), 429–448.

44. G. LaForte. The isolated d.r.e. degrees are dense in the r.e. degrees. *Math. Logic Quart.* **42** (1996), 83–103.

45. S. Lempp. A high strongly noncappable degree. *J. Symbolic Logic* **53** (1988), 174–187.

46. S. Lempp and T. A. Slaman. A limit on relative genericity in the recursively enumerable sets. *J. Symbolic Logic* **54** (1989), 376–395.

47. S. D. Leonhardi. Nonbounding and Slaman triples, *Ann. Pure Appl. Logic* **79** (1996), 139–163.

48. M. Lerman. Embeddings into the computably enumerable degrees. In *Computability theory and its applications (Boulder, CO, 1999)*, volume 257 of *Contemp. Math.*, 191–205. Amer. Math. Soc., Providence, RI, 2000.

49. A. Lewis, A single minimal complement for the c.e. degrees. To appear.

50. A. Li. Bounding cappable degrees. *Arch. Math. Logic* **39** (2000), 311–352.

51. A. Li. Definable relations on the computably enumerable degrees. In S. B. Cooper and S. S. Goncharov, editors, *Computability and Models*. Kluwer Academic Publishers, 2002.

52. A. Li, G. Wu, and Z. Zhang. A hierarchy for cuppable degrees. *Illinois J. Math.* **44** (2000), 619–632.

53. A. Li, Y. Song and G. Wu. Universal cupping degrees. *Theory and Applications of Models of Computation 2006*, Lecture Notes in Computer Science **3959** (2006), 721–730.

54. A. Li, G. Wu, and Y. Yang. On the quotient structure of computably enumerable degrees modulo the noncuppable ideal. *Theory and Applications of Models of Computation 2006*, Lecture Notes in Computer Science **3959** (2006), 731–736.

55. A. Li, A. Sorbi, G. Wu, and Y. Yang. Embed the diamond lattice into **R/NCup** preserving 0 and 1. In preparation.

56. A. Li and X. Yi. Cupping the recursively enumerable degrees by d.r.e. degrees. *Proc. London Math. Soc.* **79** (1999), 1–21.

57. W. Maass. Recursively enumerable generic sets. *J. Symbolic Logic* **47** (1982), 809–823.

58. D. Miller. High recursively enumerable degrees and the anticupping property. In *Logic Year 1979–80: University of Connecticut*, 230–245, 1981.

59. A. A. Muchnik. On the unsolvability of the problem of reducibility in the theory of algorithms. *Dokl. Akad. Nauk SSSR*, N. S. **108** (1956), 194–197.

60. A. Nies. Parameter definability is the recursively enumerable degrees. *J. Math. Log.* **3** (2003), 37–65.

61. A. Nies. Definability in the c.e. degrees: questions and results. In *Computability Theory and Its Applications (Boulder, CO, 1999)*, volume 257 of *Contemporary Mathematics*, 207–213. Amer. Math. Soc., Providence, RI, 2000.

62. A. Nies, R. A. Shore, and T. A. Slaman. Interpretability and definability in the recursively enumerable degrees. *Proc. London Math. Soc.* **77** (1998), 241–291.

63. D. B. Posner and R. W. Robinson. High degrees. Ph.D. Thesis, University of California, Berkeley, 1977.

64. D. B. Posner and R. W. Robinson. Degrees joining to $\mathbf{0}'$. *J. Symbolic Logic* **46** (1981), 714–722.

65. E. L. Post. Recursively enumerable sets of positive integers and their decision problems. *Bull. Amer. Math. Soc.* **50** (1944), 284–316.

66. G. E. Sacks. On the degrees less than $\mathbf{0}'$. *Ann. of Math.* **77** (1963), 211–231.

67. G. E. Sacks. The recursively enumerable degrees are dense. *Ann. of Math.* **80** (1964), 300–312.

68. D. Seetapun. *Contributions to Recursion Theory.* PhD thesis, Trinity College, Cambridge, 1991.

69. D. Seetapun. Defeating red. Notes, 1992.

70. D. Seetapun and T. A. Slaman, *Minimal complements*, manuscript.

71. S. Schwarz. The quotient semilattice of the recursively enumerable degrees modulo the cappable degrees, *Trans. Amer. Math. Soc.* **283** (1984), 315–328.

72. J. R. Shoenfield. Applications of model theory to degrees of unsolvability. In J. W. Addison, L. Henkin, and A. Tarski, editors, *The Theory of Models, Proceedings of the 1963 International Symposium at Berkeley*, Studies in Logic and the Foundations of Mathematics, 359–363, Amsterdam, 1965. North–Holland Publishing Co.

73. R. A. Shore. The recursively enumerable degrees. In *Handbook of Computability Theory*, volume 140 of *Stud. Logic Found. Math.*, 169–197. North-Holland, Amsterdam, 1999.

74. R. A. Shore. Natural definability in degree structures. In *Computability Theory and Its Applications (Boulder, CO, 1999)*, volume 257 of *Contemp. Math.*, 255–271. Amer. Math. Soc., Providence, RI, 2000.

75. R. A. Shore and T. A. Slaman. Working below a low$_2$ recursively enumerably degree. *Arch. Math. Logic* **29** (1990), 201–211.

76. R. A. Shore and T. A. Slaman. Working below a high recursively enumerable degree. *J. Symbolic Logic* **58** (1993), 824–859.

77. R. A. Shore and T. A. Slaman. Defining the Turing jump. *Math. Res. Lett.* **6** (1999), 711–722.

78. T. A. Slaman. The global structure of the Turing degrees. In *Handbook of Computability Theory*, volume 140 of *Stud. Logic Found. Math.*, 155–168. North-Holland, Amsterdam, 1999.

79. T. A. Slaman and R. I. Soare. Extension of embeddings in the computably enumerable degrees. *Ann. of Math.* **153** (2001), 1–43.

80. T. A. Slaman and J. R. Steel, *Complementation in the Turing degrees*, Journal of Symbolic Logic **54** (1989), 160–176.

81. Y. Sui and Z. Zhang, The cupping theorem in $\mathbf{R/M}$. *J. Symbolic Logic* **64** (1999), 643–650.

82. G. Wu. Isolation and lattice embeddings. *J. Symbolic Logic* **67** (2002), 1055–1064.

83. G. Wu. Bi-isolation in the d.c.e. degrees. *J. Symbolic Logic* **69** (2004), 409–420.

84. G. Wu. Jump operator and Yates degrees. *J. Symbolic Logic* **71** (2006), 252–264.

85. G. Wu. Intervals containing exactly one c.e. degree, submitted.

86. G. Wu. Cupping computably enumerable degrees in the difference hierarchy. In preparation.

87. C. E. M. Yates. A minimal pair of recursively enumerable degrees. *J. Symbolic Logic* **31** (1966), 159–168.

88. C. E. M. Yates. On the degrees of index sets. *Trans. Amer. Math. Soc.* **121** (1966), 309–328.

89. C. E. M. Yates, Recursively enumerable degrees and the degrees less than $0^{(1)}$. 1967 Sets, Models and Recursion Theory (Proc. Summer School Math. Logic and Tenth Logic Colloq., Leicester, 1965), North-Holland, Amsterdam, 264–271.

90. X. Yi, *Extension of embeddings on the recursively enumerable degrees modulo the cappable degrees*. In: *Computability, enumerability, unsolvability, Directions in recursion theory* (Eds. Cooper, Slaman and Wainer), London Mathematical Society Lecture Note Series **224**, 313–331.